Influenza Virology: Current Topics

EDITED BY

Yoshihiro Kawaoka, DVM, Ph.D.

British Library Cataloguing-in-Publication Data

A catalogue record for this book is available from the British Library

ISBN: 1-904455-06-9

Contents

Contributors v

Preface viii

Chapter 1   Structure and Function of the Influenza Virus RNP 1
Debra Elton, Paul Digard, Laurence Tiley and Juan Ortin

Chapter 2   Entry and Intracellular Transport of Influenza Virus 37
Gary R. Whittaker and Paul Digard

Chapter 3   The Proton Selective Ion Channels of Influenza A
and B Viruses 65
Robert A. Lamb and Lawrence H. Pinto

Chapter 4   Receptor Specificity, Host-Range, and
Pathogenicity of Influenza Viruses 95
*Mikhail N. Matrosovich, Hans-Dieter Klenk
and Yoshihiro Kawaoka*

Chapter 5   Dendritic Cells: Induction and Regulation of the
Adaptive Immune Response to Influenza Virus Infection 139
Kevin L. Legge and Thomas J. Braciale

Chapter 6   Quantitative and Qualitative Characterization of the
CD8+ T cell Response to Influenza Virus Infection 155
Nicole L. La Gruta and Peter C. Doherty

Chapter 7   M2 and Neuraminidase Inhibitors: Anti-Influenza Activity,
Mechanisms of Resistance, and Clinical Effectiveness 169
Larisa V. Gubareva and Frederick G. Hayden

Chapter 8   Influenza Vaccines: Current and Future Strategies 203
Jacqueline M. Katz, Sanjay Garg and Suryaprakash Sambhara

Chapter 9   Epidemiology and Control of Human and Animal Influenza . . . 229
Kanta Subbarao, David E. Swayne and Christopher W. Olsen

Chapter 10   H5 Influenza Viruses. 281
Robert G. Webster

Chapter 11   The Origin and Virulence of the 1918 'Spanish'
Influenza Virus 299
Jeffery K. Taubenberger and Peter Palese

Chapter 12   Signaling and Apoptosis in Influenza Virus-Infected Cells 323
Stephan Ludwig

Chapter 13   Insights into Influenza Virus-Host Interactions through
Global Gene Expression Profiling: Cell Culture
Systems to Animal Models 341
*Marcus J. Korth, John C. Kash, Carole R. Baskin
and Michael G. Katze*

Books of related interest

Papillomavirus Research: From Natural History To Vaccines and Beyond   2006

Antimicrobial Peptides in Human Health and Disease   2005
Biodefense: Principles and Pathogens   2005
Campylobacter: Molecular and Cellular Biology   2005
Cancer Therapy: Molecular Targets in Tumour-Host Interactions   2005
Cytomegaloviruses: Molecular Biology and Immunology   2005
Dictyostelium Genomics   2005
Epstein-Barr Virus   2005
Evolutionary Genetics of Fungi   2005
Foodborne Pathogens: Microbiology and Molecular Biology   2005
HIV Chemotherapy: A Critical Review   2005
Microbe-Host Interface in Respiratory Tract Infections   2005
Mycobacterium: Molecular Biology   2005
Mycoplasmas: Molecular Biology, Pathogenicity and Strategies for Control   2005
Probiotics and Prebiotics: Scientific Aspects   2005
SAGE: Current Technologies and Applications   2005
Microbial Toxins: Molecular and Cellular Biology   2005
Vaccines: Frontiers in Design and Development   2005

Bacterial Spore Formers: Probiotics and Emerging Applications   2004
Brucella: Molecular and Cellular Biology   2004
Computational Genomics: Theory and Application   2004
DNA Amplification: Current Technologies and Applications   2004
Ebola and Marburg Viruses: Molecular and Cellular Biology   2004
Foot and Mouth Disease: Current Perspectives   2004
Internet for Cell and Molecular Biologists (2nd Edition)   2004
Malaria Parasites: Genomes and Molecular Biology   2004
Metabolic Engineering in the Post Genomic Era   2004
Pathogenic Fungi: Structural Biology and Taxonomy   2004
Pathogenic Fungi: Host Interactions and Emerging Strategies for Control   2004
Peptide Nucleic Acids: Protocols and Applications (2nd Edition)   2004

Prions and Prion Diseases: Current Perspectives   2004
Protein Expression Technologies: Current Status and Future Trends   2004
Strict and Facultative Anaerobes: Medical and Environmental Aspects   2004
Real-Time PCR: An Essential Guide   2004
Sumoylation: Molecular Biology and Biochemistry   2004
Tuberculosis: The Microbe Host Interface   2004
Yersinia: Molecular and Cellular Biology   2004

Bioinformatics and Genomes: Current Perspectives   2003
Bioremediation: A Critical Review   2003
Frontiers in Computational Genomics   2003
Genome Mapping and Sequencing   2003
MRSA: Current Perspectives   2003
Multiple Drug Resistant Bacteria: Emerging Strategies   2003
Transgenic Plants: Current Innovations and Future Trends   2003
Regulatory Networks in Prokaryotes   2003

Full details of all these books at: www.horizonbioscience.com

Contributors

Carole R. Baskin
Departments of Microbiology and
Comparative Medicine and Washington
National Primate Research Center,
University of Washington,
Seattle, Washington 98195–8070 USA

Thomas J. Braciale
Carter Immunology Center,
Departments of Pathology and
Microbiology,
University of Virginia,
Charlottesville, VA 22908, USA

Paul Digard
Division of Virology, Department
of Pathology,
University of Cambridge,
Tennis Court Road,
Cambridge CB2 1QP, UK

Peter C. Doherty
Department of Microbiology and
Immunology,
University of Melbourne,
Royal Parade,
Parkville, VIC 3010, Australia.
Department of Immunology,
St. Jude children's Research Hospital,
Memphis, TN 38105, USA

Debra Elton
Division of Virology, Department
of Pathology,
University of Cambridge,
Tennis Court Road,
Cambridge CB2 1QP, UK

Sanjay Garg
Influenza Branch Mailstop G-16
Centers for Disease Control and Prevention
1600 Clifton Road
Atlanta, GA 30333 USA

Larisa V. Gubareva
University of Virginia School of Medicine
P.O.Box 800473, Charlottesville VA
22908, USA

Frederick G. Hayden
University of Virginia School of Medicine
P.O.Box 800473, Charlottesville VA
22908, USA

John C. Kash
Department of Microbiology,
University of Washington, Seattle,
Washington 98195–8070 USA

Jacqueline M. Katz,
Influenza Branch
Mailstop G-16
Centers for Disease Control and
Prevention
1600 Clifton Road
Atlanta, GA 30333 USA

Michael G. Katze
Department of Microbiology and
Washington National Primate
Research Center,
University of Washington, Seattle,
Washington 98195–8070 USA

Yoshihiro Kawaoka
Department of Pathobiological Sciences,
School of Veterinary Medicine,
University of Wisconsin-Madison,
Madison, Wisconsin 53706, USA;
Institute of Medical Science,
University of Tokyo,
4-6-1, Shirokanedai,
Minato-ku, Tokyo 108-8639, Japan

Hans-Dieter Klenk
Institute of Virology,
Philipps University,
Robert-Koch str. 17,
35037 Marburg, Germany

Marcus J. Korth
Department of Microbiology and
Washington National Primate
Research Center,
University of Washington,
Seattle, Washington 98195–8070 USA

Nicole L. La Gruta
Department of Microbiology and
Immunology,
University of Melbourne,
Royal Parade,
Parkville, VIC 3010, Australia

Robert A. Lamb
Howard Hughes Medical Institute,
Dept. Biochemistry, Molecular Biology
and Cell Biology,
Northwestern University,
2205 Tech Drive,
Evanston, IL 60208-3500 USA

Kevin L. Legge
Department of Pathology,
Carver College of Medicine,
University of Iowa,
Iowa City, IA 52242, USA

Stephan Ludwig
Institute of Molecular Virology (IMV)
Westfaelische-Wilhelms-University
Von-Esmarch-Str. 56
48148 Muenster
Germany

Mikhail N. Matrosovich
Institute of Virology, Philipps University,
Robert-Koch str. 17,
35037 Marburg, Germany;
M.P. Chumakov Institute of Poliomyelitis
and Viral Encephalitides,
Russian Academy of Sciences,
142 782, Moscow, Russia

Christopher W. Olsen
Department of Pathobiological Sciences,
School of Veterinary Medicine,
University of Wisconsin-Madison,
2015 Linden Drive,
Madison, WI 53706, USA

Juan Ortin
Centro Nacional de Biotecnologia (CSIC),
Cantoblanco, 28049
Madrid, Spain

Peter Palese
Department of Microbiology,
Mount Sinai School of Medicine,
New York, New York USA

Lawrence H. Pinto
Department of Neurobiology and
Physiology,
Northwestern University
Evanston, Illinois 60208-3500 USA

Suryaprakash Sambhara
Influenza Branch
Mailstop G-16
Centers for Disease Control and
Prevention
1600 Clifton Road
Atlanta, GA 30333 USA

Kanta Subbarao
Laboratory of Infectious Diseases,
National Institute of Allergy and Infectious
Diseases, National Institutes of Health,
Bethesda, Maryland, USA

David Swayne
Southeast Poultry Research Laboratory,
US Department of Agriculture,
Agricultural Research Service,
934 College Station Road,
Athens, Georgia 30605 USA

Jeffery K. Taubenberger
Department of Molecular Pathology,
Armed Forces Institute of Pathology,
1413 Research Blvd., Building 101,
Room 1057, Rockville, MD 20850-3125,
USA

Laurence Tiley
Centre for Veterinary Science,
University of Cambridge,
Madingley Road,
Cambridge CB3 0ES, UK

Robert G. Webster
Rose Marie Thomas Chair,
Division of Virology, Department
of Infectious Diseases,
St. Jude Children's Research Hospital,
USA

Gary Whittaker
Department of Microbiology
& Immunology,
Cornell University,
Ithaca NY 14853, USA

Acknowledgement

The index for this book was created by Gabi Neumann.

Preface

Humans have been afflicted for more than a century with illnesses characteristic of influenza virus infection, and influenza virus has been studied extensively since it was first identified in 1930. Nonetheless, there are still many outstanding questions regarding this virus and the disease it causes. In this book, we have compiled reviews by experts in the field on topics important for understanding influenza virus. This book is neither intended to cover the entire field of influenza research nor serve as a text book. Rather, it is intended to provide information on important influenza-related research themes that have emerged in recent years, and should stimulate research in influenza and other related fields.

Since 1997, H5N1 influenza A viruses have circulated in Asia. Even though a large number of poultry has been destroyed to eradicate the virus and vaccines have been introduced in some countries to minimize the effect of viral infection, these influenza viruses have still transmitted to humans with lethal outcomes. Thus, despite monumental efforts aimed at containment, the H5N1 viruses are expanding their territory, causing a major outbreak in wild waterfowl in China in 2005. Indeed, they have even been transmitted to Europe.

While fewer people succumbed to death in the pandemics of Asian influenza in 1957 and Hong Kong influenza in 1968, 20–50 million people died from Spanish influenza in 1918–19. The human toll related to outbreaks of SARS was less than 1000; yet, the economic burden in affected regions was enormous. Thus, if an influenza pandemic of comparable magnitude to even that of the 1968 Hong Kong influenza outbreak would occur now, panic would erupt, paralyzing society.

Are we prepared for a new influenza pandemic? Despite extensive, coordinated efforts by various agencies and disciplines, both national and international, we are ill-equipped to handle a pandemic. Will we be prepared for such a pandemic in the near future? Perhaps, but it is highly unlikely that adequate supplies of vaccine for the H5N1 viruses will be prepared prior to the occurrence of a pandemic caused by these viruses. Thus, many countries are stockpiling influenza drugs, with the hope that the inevitable emergence of drug-resistant viruses will not nullify those efforts immediately. A better understanding of influenza virus and the disease caused by this virus is clearly indicated and critically important to combating outbreaks that will undoubtedly occur in the near future. It is my hope that the information presented in this book will aid efforts to improve our pandemic preparedness.

Yoshihiro Kawaoka, DVM, Ph.D.
October 2005

CHAPTER 1

Structure and function of the influenza virus RNP

Debra Elton, Paul Digard, Laurence Tiley and Juan Ortin

Abstract

Influenza viruses have negative sense segmented RNA genomes, which are packaged into transcriptionally active ribonucleoproteins (RNPs). These RNPs are transcribed and replicated in the nucleus of host cells. During the replication cycle two types of positive sense RNA are synthesized; capped and polyadenylated messenger RNA and uncapped full length complementary (c)RNA. Complementary RNA acts as the replicative intermediate for synthesis of further negative sense genomic RNA. This cycle is carried out by the viral RNA-dependent RNA polymerase, a heterotrimeric complex which binds RNA through structure and sequence-specific interactions and has multiple functions including capped-RNA-binding activity, RNA endonuclease, polymerase and polyadenylation activities. These activities have specific roles during the viral transcription cycle and are controlled by interactions between the protein components and the RNA promoter structure. The mechanisms involved in the synthesis of viral messenger RNA are fairly well characterised, but less is known about the process of genome replication and the factors that control it. On the other hand, recent advances have been made towards elucidating the structure of the molecular machines responsible for virus RNA synthesis.

INTRODUCTION

Influenza A viruses have single-stranded RNA genomes, consisting of eight segments of negative polarity. Each segment contains conserved sequences at the 3' and 5' termini that share partial sequence complementarity (Skehel and Hay, 1978, Robertson, 1979, Desselberger et al., 1980) and can base pair to form a panhandle structure (Hsu et al., 1987). The transcriptionally active form of the genome is the viral ribonucleoproteins (RNPs), where each vRNA segment is separately encapsidated by the nucleoprotein (NP) and associated with one copy of the viral RNA-dependent RNA polymerase (Figure 1a). NP plays an essential role in maintaining the structure of the RNPs, but is also important for regulation of genome replication. The viral polymerase complex is a heterotrimer composed of two basic proteins, PB1 and PB2, and the more acidic PA (Horisberger, 1980, Detjen et al., 1987, Honda et al., 1990). PB1 carries the polymerase and endonuclease activities, PB2 binds to cap structures on host cell mRNAs but the function of PA remains unclear. The roles of each polypeptide are described in more detail below.

Unlike most negative-sense RNA viruses, transcription of the influenza virus genome takes place in the nucleus of infected cells (Herz et al., 1981). During the infectious cycle, two types of positive sense RNA molecules are transcribed from the RNPs (Figure 1b). Synthesis of capped and polyadenylated messenger RNAs (mRNA) is primed by short capped oligonucleotides of around 10 to 12 nucleotides, which are scavenged from host cell pre-mRNAs by an endonuclease activity contained within the polymerase (Plotch et al., 1981; Doan et al., 1999). Influenza mRNAs therefore contain host cell-derived sequences at their 5' ends. Transcription terminates 15–17 nt before the 5' end of the vRNA segment and the mRNA is polyadenylated by a process of stuttering on a poly (U) track (Figure 1b; Hay et al., 1977a; Robertson et al., 1981, Luo et al., 1991, Poon et al., 1999, Zheng et al., 1999). Once mRNA synthesis and translation have become established, a small proportion of RNA synthesis is dedicated to the production of positive sense copies of the entire viral genome, commonly referred to as cRNA. In contrast to mRNA, synthesis of cRNA involves unprimed initiation (Hay et al., 1982) and readthrough of the polyadenylation signal to produce full-length copies of the vRNA template (Figure 1b). These RNAs are packaged into RNPs in much the same way as vRNA and act as the replicative intermediates for synthesis of new copies of vRNA, required for production of progeny virions.

Much research has been aimed at improving our understanding of the 'switching' mechanism between transcription and replication of the influenza virus genome, to try and define the factors involved and how they regulate viral RNA synthesis. However, the mechanism of synthesis of cRNA, a minor but crucial component of the total influenza RNA synthesized, remains poorly understood. Current theories on the control of transcription and replication will be discussed in this review.

THE INFLUENZA VIRUS RIBONUCLEOPROTEIN
Functions of the viral polypeptides

Four viral proteins are necessary and sufficient for efficient synthesis of influenza virus RNA *in vivo*: the three subunits of the RNA-dependent RNA polymerase and the single-strand RNA binding protein, NP (Huang et al., 1990). Much is now known of the activities of the polymerase complex, particularly during transcription of the vRNA segments, although the functions of PB1 and PB2 are better characterised than those of PA. The individual functions of each RNP component will now be considered.

PB1 is responsible for polymerase activity

PB1, the larger of the two basic P proteins, is the nucleotide polymerase (Ulmanen et al., 1981; Braam et al., 1983). It acts as the backbone of the polymerase complex, containing independent binding sites for both PB2 and PA (Digard et al., 1989), and is also capable of binding NP (Biswas et al., 1998; Medcalf et al., 1999). PB1 shares sequence homology with other nucleic acid polymerases and contains all four of the protein motifs that are conserved amongst RNA-dependent RNA polymerases (Poch et al., 1989, Muller et al., 1994). Mutagenic analysis of these PB1 sequences confirmed that they are required for transcriptional activity (Biswas and Nayak, 1994; Li et al., 2001). UV cross-linking experiments have been widely used to study the binding activities of PB1 and further support the central role of PB1 in polymerase function. The polypeptide has been found in close association with the initial G and subsequent nucleotide residues that are added to the 3' end of the primer fragment during mRNA transcription initiation and elongation (Ulmanen et al., 1981; Braam et al., 1983). When viral RNPs were cross-linked to radiolabelled triphosphate substrates, only PB1 was labelled (Romanos and Hay, 1984, Asano et al., 1995), suggesting that the substrate-binding site is located within this subunit. Together these data provide strong evidence that PB1 is indeed responsible for nucleotide addition during RNA synthesis.

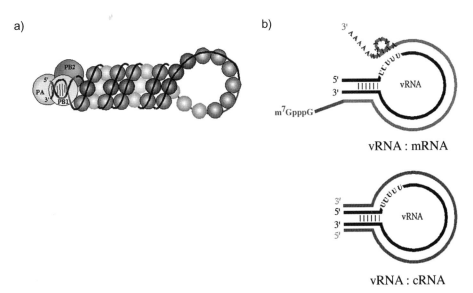

Figure 1.1 Structural features of influenza virus RNPs and RNAs. (a) Cartoon model of RNP organization. The single-stranded vRNA is coiled around the NP monomers to form a hairpin structure. The 5' and 3' ends form a short double-stranded region to which the heterotrimeric RNA-dependent RNA polymerase binds. Adapted from Portela and Digard (2002) by permission of the Society for General Microbiology. (b) Cartoon depicting structural relationships of viral RNA species. Molecules of vRNA (black line) are shown in panhandle configuration. Note juxtaposition of polyadenylation signal (U stretch) to the 5' end of the RNA. Viral mRNAs contain a 5'-extension derived from a host cell mRNA and a long poly(A) tail produced by repetitive transcription of the vRNA poly(U) sequence. In contrast, cRNA molecules are perfect copies of the vRNA template. Adapted with permission from Elton et al., 2002.

The influenza virus polymerase complex binds in a sequence-specific manner to the promoters formed by the conserved termini of each vRNA and cRNA segment (Fodor et al., 1994; Tiley et al., 1994; Li et al., 1998; Gonzalez and Ortin 1999 a, b). Binding sites for the 5' and 3' arms of both the vRNA and cRNA termini are located in the PB1 protein (Li et al., 1998; Gonzalez and Ortin, 1999a, b). Using UV cross-linking experiments, Li et al. (1998) mapped the 5' vRNA binding site to a region centred around two basic residues, R571 and R572, and the binding site for 3' vRNA was mapped to an RNP1 motif containing two phenylalanine residues at positions 251 and 254. However, binding assays using truncated recombinant PB1 proteins defined two regions at the N and C-termini (aa 1–84 and 493–757) that are involved in both 5' and 3' vRNA binding (Gonzalez and Ortin 1999a). The N-terminal region of PB1 has also been implicated in cRNA binding, but an internal region (aa 267–493) substitutes for the C-terminal binding site (Gonzalez and Ortin 1999b). Since the N-terminal region of PB1 binds to both cRNA and vRNA promoters, this site may bind to the proximal stem of the promoter structure (see later), as this region is identical in both RNAs.

In addition to nucleotide polymerase activity, it has been suggested that PB1, rather than PB2, is responsible for the RNA endonuclease activity (Li et al., 2001). Using cross-linking techniques, the authors showed that a capped RNA oligonucleotide that contained thio U close to the point of endonucleolytic cleavage could be cross-linked to PB1 but not PB2 (Li et al., 2001). Amino acids 508 to 522 were proposed to form the active site of the endonuclease. Mutagenesis within this region of PB1 identified three acidic amino acids that were essential for activity and it was suggested that these are analogous to those found in other more characterised endonucleases (Li et al., 2001).

PB2 is the cap-binding protein

The PB2 subunit plays a vital role in transcription of mRNA, having the primary function of binding to the 5' methylated cap of host cell pre-mRNAs before they are cleaved to provide primers for viral mRNA synthesis (Ulmanen et al., 1981, Blaas et al., 1982, Braam et al., 1983; Shi et al., 1996). PB2 is recruited to the polymerase complex through protein-protein interactions with the PB1 subunit (Digard et al., 1989); it also interacts directly with NP but apparently not with PA (Zürcher et al., 1996; Biswas et al., 1998; Medcalf et al., 1999; Poole et al., 2004).

PB2 shares weak sequence homology with cellular cap-binding proteins (de la Luna, 1989; Honda et al., 1999; Li et al., 2001) and UV cross-linking studies using capped RNAs have shown that it closely associates with the cap during mRNA transcription initiation (Ulmanen et al., 1981; Braam et al., 1983). Similarly, photoactivatable 5' cap analogues specifically label PB2 (Blaas et al., 1982; Penn et al., 1982; Li et al., 2001). Furthermore, some PB2 temperature sensitive (ts) mutants show defects in cap recognition *in vitro* (Ulmanen et al., 1983). PB2 is capable of binding cap structures in the absence of the other polymerase proteins (Shi et al., 1996) but, as will be discussed later, this activity of the protein is clearly regulated by other viral components. Efficient binding by PB2 generally requires 5'-cap 1 structures (m7GpppNm), and both the 7-methyl group on the terminal G and the 2'-O-methyl group strongly affect the suitability of the RNA as a substrate for recognition and subsequent cleavage by the virus polymerase (Bouloy et al., 1979, 1980; Kawakami et al., 1985*)*.

Attempts to identify the regions of PB2 responsible for binding to host cell cap structures have given conflicting results. One study used purified viral RNPs and concluded that there were two binding regions within the PB2 sequence, mapping to amino acids 242–282 and 538–577 (Honda et al., 1999). A more recent study identified only one peptide, spanning residues 544–556, that could be cross-linked to a 5' capped oligonucleotide (Li et al., 2001). Here the cross-linking assay was carried out using recombinant PB2 in the presence of 5'vRNA only, possibly avoiding activation of the endonuclease. The peptide mapped forms part of a tryptophan-rich sequence that extends from residues 533 to 564, correlating with the more C-terminal binding region mapped by Honda et al. (1999). More recently still, a systematic mutagenesis approach found that two phenylalanine residues, F363 and F404, were the only conserved aromatic amino acids that were crucial for cap binding activity but not essential for ApG-primed transcription (Fechter et al., 2003). These amino acids were proposed to interact directly with the m^7G of the cap structure and are located between the two regions of PB2 previously associated with cap-binding by UV cross-linking (Honda et al., 1999; Li et al., 2001). The cap-recognition mechanism of PB2 may involve an aromatic sandwich motif, similar to that found in other cap-binding proteins (reviewed by Quiocho et al., 2000).

Following binding of host cap-structures to the viral polymerase, a sequence of 10-12 nucleotides is cleaved from the pre-mRNA by the divalent metal ion-dependent endonuclease activity of the influenza virus polymerase (Plotch et al., 1981, Doan et al., 1999). Initial evidence from antibody inhibition studies suggested that PB2 was the most likely candidate for the viral endonuclease, since sera raised against PB2 effectively inhibited cap-dependent endonuclease activity without affecting cap binding, whereas those against PB1 or PA had no effect (Shi et al., 1995, Blok et al., 1996). However, it was subsequently shown that endonuclease activity could be inhibited by antibodies directed against any of the three P proteins (Masanuga et al., 1999). This highlights the fact that the polymerase complex acts as a unit, and also demonstrates the inherent difficulties associated with use of antibodies to inhibit enzyme activity, since antibodies affecting one component of a multi-protein complex may well have an effect on the activity of one or more of the other components. This may be especially relevant for a complex like the influenza polymerase that shows a very compact structure (see below). More recent work suggests that the endonuclease activity actually resides in PB1 (Li et al., 2001).

PB2 is clearly crucial to viral mRNA transcription, as a result of its cap-recognition function. Several PB2 *ts* mutants show defects at the non-permissive temperature in cap-primed transcription but not ApG primed transcription *in vitro* (Nichol et al., 1981; Ulmanen et al., 1983; Yamanaka et al., 1990b). *In vivo* studies with *ts* mutants have consistently shown that PB2 mutants are defective for synthesis of mRNA, in some cases without concomitant deficiencies in replicative synthesis (Mahy, 1983; Markushkin and Ghendon, 1984, Nichol et al., 1981). Similarly, deliberate mutation of the PB2 gene has been shown to affect mRNA synthesis in a recombinant system (Perales et al., 1996). The role of PB2 in replication of the viral genome is less clear and somewhat controversial. Some studies have concluded that PB2 is not required for genome replication (Nakagawa et al., 1995; Nakagawa et al., 1996, Honda et al., 2002). However, others report that all three polymerase subunits are required to reconstitute an active polymerase complex (Perales and Ortin, 1997; Brownlee and Sharps, 2002; Lee et al., 2002). Consistent with this, some PB2 *ts* mutants are incapable of ApG primed transcription or cannot support genome replication (Nichol et al., 1981; Ghendon et al., 1982; Mahy, 1983), implying a general role for PB2 in the formation of an active polymerase complex. More recently, RNPs reconstituted with mutations within the N-terminal region of PB2 were competent for transcription and cap snatching *in vitro*, but showed diminished RNA synthesis *in vivo* (Gastaminza et al., 2003). Some of these mutations were successfully rescued into recombinant virus and showed normal kinetics for primary transcription but delayed accumulation of cRNA and vRNA, indicating that a fully functional PB2 protein was required for replication as well as transcription (Gastaminza et al., 2003).

PA, responsible for proteolytic activity

The third subunit of the polymerase complex, PA, is the least characterized of the polymerase proteins in terms of function. No specific role in influenza virus transcription or replication has yet been identified, although early work with *ts* mutants suggested involvement in the replication of vRNA (Mahy et al., 1983). More recently, PA has also been implicated in transcription (Fodor et al., 2002).

PA associates with PB1 but not PB2 (Digard et al., 1989), and has been shown to interact with the cellular protein hCLE (Huarte et al., 2001). There is some limited sequence homology between PA and helicases from Escherichia coli and other RNA viruses (de la Luna et al., 1989), although no such activity has yet been demonstrated for the protein. PA is phosphorylated on serine and threonine residues *in vivo* (Sanz-Ezquerro et al., 1998). Since multiple isoforms can be detected in a variety of systems, including virus infection, it is likely that the protein is

5

phosphorylated by a cellular kinase rather than by an influenza virus protein (Sanz-Ezquerro et al., 1998). A number of potential phosphorylation sites exist within the PA sequence and at least one of these, T157, appears to be important as its mutation reduces phosphorylation of the protein (Perales et al., 2000). As this is a potential casein-kinase II site and this enzyme can phosphorylate PA protein *in vitro*, it is reasonable to assume that casein-kinase II is responsible for phosphorylation of PA (Sanz-Ezquerro et al., 1998). Whether this phosphorylation is important for the biological activity of PA remains unclear.

The phenotype of certain *ts* mutants mapped to segment 3 suggested that PA is involved in replication of the viral genome, since the mutants were defective for vRNA synthesis but competent for transcription of mRNA at the non permissive temperature (Krug et al., 1975; Mahy et al., 1981; Mowshowitz, 1981; Thierry and Danos, 1982; Gubareva et al., 1991). However, a recent study by Fodor et al. (2002) indicated that PA is important for both transcription and replication, since point mutations were found that specifically diminished both processes. In particular, one mutation resulted in wild type levels of vRNA but very little mRNA in an *in vitro* transcription/replication assay and was shown to be defective for endonuclease activity (Fodor et al., 2002). A second mutant could transcribe and replicate *in vitro* but a recombinant virus carrying this mutation generated much higher levels of DI RNA than wild type, suggesting a role for PA in RNA elongation (Fodor et al., 2003). This function is consistent with the finding that PA stimulated the binding of PB1 to a model 5'vRNA template (Lee et al., 2002). It is possible that PA mediates this effect by interacting directly with the RNA template as it has previously been shown that PA present in viral RNPs can be UV cross-linked to the 5' end of vRNA (Fodor et al., 1994).

PA is an essential component of most *in vitro* transcription and replication systems (Huang et al., 1990; Kobayashi et al., 1992; Perales and Ortin, 1997; Lee et al., 2002; Brownlee & Sharps, 2002). However, it has also been reported that PA is not necessary for transcriptase activity as a binary complex composed of PB1 and PA was active for unprimed RNA synthesis, whereas PB1 and PB2 together had activities associated with the viral transcriptase (Nakagawa et al., 1995; 1996; Honda et al., 2002). In the absence of evidence for binary polymerase complexes *in vivo*, it was suggested that selective inactivation of PB2 or activation of PA within the complex may enable the polymerase to switch between transcription and replication (Honda et al., 2002).

The most characterised function of PA is its association with protease activity. Recombinant PA decreases its own accumulation levels and causes degradation of co-expressed proteins *in vivo* (Sanz-Ezquerro et al., 1995). Consistent with this, the purified protein is associated with a weak serine protease activity (Hara et al., 2001). Analysis of PA deletion mutants indicated that the amino-terminal third of the polypeptide was sufficient for the induction of generalised proteolysis *in vivo* (Sanz-Ezquerro et al., 1996). However, the *in vitro* chymotrypsin-like activity was found to be centred round serine 624 towards the C-terminus of the protein (Hara et al., 2001).

It was speculated that PA protease activity may be involved in releasing RNPs from the nuclear matrix, thus allowing their transport into the cytoplasm (Hara et al., 2001). Mutagenesis of S624 resulted in the loss of serine protease activity from purified recombinant PA. Consistent with a late defect, the mutation had no effect on transcription or replication of vRNA in a plasmid-based replication assay (Fodor et al., 2002) and had little effect on expression of reporter genes (Fodor et al., 2002; Toyoda et al., 2003). However, it was also possible to recover a recombinant virus by reverse genetics (Fodor et al., 2002), which grew to normal titres in cell culture and was pathogenic for mice, although viral growth was slightly impaired compared to WT (Toyoda et al., 2003). These results argue against a vital role for the serine protease activity in virus growth.

In contrast, Perales et al. (2000) reported that PA-induced proteolysis was important for replication activity of the polymerase complex, but focussed on the proteolytic activity induced by the N-terminal region of PA (Sanz-Ezquerro et al., 1996). In particular, changes at residues T157 and T162 resulted in decreased levels of induced proteolysis and also a deficiency in genome replication, particularly synthesis of cRNA from a vRNA template (Perales et al., 2000). A recombinant virus containing the T157A mutation showed a similar defect in RNA replication (Huarte et al., 2003). However, this may be due to a specific requirement for T157 rather than proteolytic activity, as Naffakh et al. (2001) identified two other point mutations that reduced proteolytic activity but, unlike mutation of T157, had little effect on expression of an influenza-like reporter gene. It is possible that phosphorylation at this site is a factor; since it was previously shown that mutation of T157 resulted in a reduction in the overall phosphorylation level of PA (Perales et al., 2000).

Overall it would seem unlikely that the protease activities centred round S624 and the N-terminus of PA are the same. Serine protease activity apparently plays no part in RNA or indeed virus replication whereas T157, required for the proteolysis induced by the N-terminus of PA, does appear to be important. However, it remains to be seen whether this is linked to protease activity, phosphorylation or some as yet unidentified function of PA.

NP, a structural and functional unit in the RNP

The final viral protein component required for viral RNA synthesis is NP, a highly basic protein that binds to single-stranded RNA in a non-sequence specific manner (Scholtissek and Becht, 1971; Kingsbury et al., 1987, Yamanaka et al., 1990a, Baudin et al., 1994). The RNA-binding activity of NP has been mapped to the terminal third of the protein (Kobayashi et al., 1994, Albo et al., 1995), but this region alone does not bind RNA with high affinity and amino acids throughout the length of the protein are important for optimal binding (Elton et al., 1999b). Analysis of NP-RNA interactions indicates that one NP molecule binds approximately every 25 nucleotides of an RNA strand (Compans et al., 1972, Ortega et al., 2000), though this does not protect the RNA from RNase activity (Duesberg, 1969, Baudin et al., 1994). Chemical probing has shown that NP binds RNA through its sugar-phosphate backbone, leaving the bases exposed to the solvent (Baudin et al., 1994).

In addition to binding single-stranded RNA and self-oligomerisation, NP also associates with a large number of other proteins, both virus-encoded and cellular. Identified interactions with viral proteins include PB1, PB2 and M1, while cellular proteins include importin alpha, CRM1, actin, UAP56 and probably Mx proteins. Many of these interactions are not believed to exert a direct influence on viral RNA synthesis and so will not be considered further here. However, they have been described in another recent review (Whittaker and Digard). NP interacts with the polymerase complex through direct links with both the PB1 and PB2 subunits, but does not appear to make contacts with the PA subunit (Biswas et al., 1998; Medcalf et al., 1999, Poole et al., 2004). In addition to *in vitro* binding studies, there is genetic evidence for a functional link between NP and PB2 (Mandler et al., 1991; Naffakh et al., 2000). The interaction of M1 with NP serves to control both the transcriptional activity of RNPs and their intracellular trafficking. This latter aspect is reviewed elsewhere in this volume (Whittaker and Digard). The inhibitory effect of M1 on viral RNA synthesis was first noted through *in vitro* studies of mRNA transcription carried out by virion-derived RNPs (Zvonarjev and Ghendon, 1980) and later shown to operate in cells (Perez and Donis, 1998). The mechanism by which M1 inhibits transcription is not certain, but may involve M1 polymerization on RNPs affecting an early step in mRNA transcription initiation (Watanabe et al., 1996; Baudin et al., 2001). The one NP-host cell protein interaction identified so far that may have a positive transcriptional function is the association with UAP56, a cellular splicing factor from the DEAD-box helicase family (Momose et al., 2001). Possible roles for this polypeptide in viral transcription will be discussed below. The other cellular polypeptides known to be capable of directly influencing RNA synthesis are the Mx proteins, but as with viral M1, the interaction is inhibitory. Human MxA and murine Mx1 are interferon-inducible proteins that inhibit the multiplication of orthomyxo and other RNA viruses (reviewed by Haller and Kochs, 2002). As well as inhibiting nuclear import of RNPs, nuclear resident Mx proteins also inhibit virus transcription through an interaction with RNPs, probably mediated through NP (Kochs and Haller, 1999; Weber et al., 2000; Turan et al., 2004)

NP plays an essential role in the structural organisation of the RNP and is also required for transcription and replication of the viral genome, as the polymerase cannot effectively use genome length naked viral RNA as template (Honda et al., 1988). In addition, the NP has been implicated as a possible regulator of the switch between transcription and replication of the genome and this is discussed in more detail in the section concerning cRNA synthesis.

Structure of the ribonucleoprotein and the polymerase
The ribonucleoprotein particle
Early electron microscopy studies of virion RNPs showed supercoiled ribbon structures with a terminal loop (Pons et al., 1969; Compans et al., 1972; Heggeness et al., 1982; Jennings et al., 1983). Each ribbon unit is an NP monomer and the polymerase, although not directly detectable, is present at one end of the supercoil, as shown by immunogold labelling (Murti et al., 1988). The presence of the polymerase helps in maintaining the RNA ends linked together (Klumpp et al., 1997) and allows the formation of the closed structure in the RNP. Artificial complexes generated in vitro with NP and RNA show structural and biochemical properties similar to natural RNPs (Yamanaka et al., 1990a). Furthermore, the helical form of the RNP is maintained when the RNA is replaced by negatively charged polymers, suggesting that NP determines RNP organisation rather than the viral RNA (Pons et al., 1969). This is supported by electron microscopy of purified RNA-free NP extracted from RNPs, which showed structures morphologically indistinguishable from intact RNPs (Ruigrok and Baudin, 1995). Each rod-shaped NP monomer has two sites for NP-NP contacts at one end (Ruigrok and Baudin, 1995), and it has subsequently been shown by deletion mutagenesis that two separate regions within the protein are capable of association with full length NP (Elton et al., 1999a).

The virion RNPs are flexible particles with variable length, depending on the RNA segment they contain, and are poor subjects for detailed structural analyses. Recently, however, much smaller recombinant RNPs have been generated by in vivo amplification in cells expressing virus polymerase, NP and a model virus RNA (Ortega et al., 2000). These more uniform populations of RNPs could be studied by electron microscopy and image processing and revealed the presence of circular, elliptic or supercoiled particles, depending on the length of the model RNA included (Ortega et al., 2000). From the length of the viral RNA present in these recombinant RNPs and the number of NP monomers observed, it could be calculated that around 24–25 nt are bound per NP molecule.

A three-dimensional model for one such recombinant RNP has been reported (Martín-Benito et al., 2001), in which nine NP monomers are present in a circular structure and two of the NP monomers are associated with the polymerase complex by means of non-identical

a)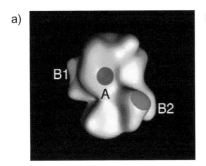
b)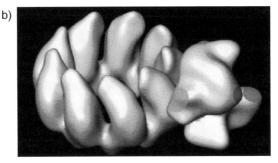

Figure 1.2 Three-dimensional model of the structure of the influenza polymerase complex at a resolution of 23Å (A) and its spatial position within the structure of a model RNP (B). The shaded areas represent specific domains of the PB1 (B1), PB2 (B2) and PA (A) subunits of the complex. Adapted with permission from Area et al., (2004). Copyright (2004) National Academy of Sciences, U.S.A.

contacts (Figure 2). These might represent the NP-PB1 and NP-PB2 interactions identified biochemically (Biswas et al., 1998; Medcalf et al., 1999; Poole et al., 2004). The NP monomers show a banana-like structure and establish a main NP-NP contact at one site, in agreement with previous electron-microscopy data (Ruigrok and Baudin, 1995). In addition, their interaction in the RNP gives rise to a vorticity or handedness in the structure that could form the basis of the superhelical conformation of virus RNPs, provided the length of the enclosed RNA allows. This feature in the model, together with the observation of helical conformations originated from RNA-free NP preparations (Ruigrok and Baudin, 1995), indicate that NP is a major determinant of the overall structure of viral RNPs. The viral RNA included in these recombinant mini-RNPs is mainly negative polarity (Ortega et al., 2000) and its ends are presumably bound by the polymerase complex, but its location within the NP ring is unknown at present.

The polymerase trimer
The enzyme responsible for RNA synthesis in the RNP is the virus polymerase complex, a heterotrimer in which the PB1 subunit constitutes the core to which both PB2 and PA subunits are bound. Various experimental approaches, including two-hybrid and co-immunoprecipitation analyses, pull-down assays and *in vivo* competition experiments have been used to delineate the regions of these subunits involved in complex formation (Pérez and Donis, 1995; Biswas and Nayak, 1996; González et al., 1996; Toyoda et al., 1996; Zürcher et al., 1996; Ohtsu et al., 2002; Poole et al., 2004). In summary, these experiments indicate that the N-terminus of PB1 interacts with the C-terminal region of PA and the C-terminus of PB1 binds the N-terminus of PB2, with additional regions of contact between PB1 and PB2. No interaction has been detected between PA and PB2 proteins. Therefore, this biochemical mapping suggests an N-terminal to C-terminal tandem arrangement of the subunits in the order PA-PB1-PB2, but with a further degree of interlinking between PB1 and PB2.

Our knowledge about the three-dimensional structure of the influenza polymerase complex is very limited, compared to other viral RNA polymerases encoded by positive-stranded or double-stranded RNA viruses. A first three-dimensional model has been recently reported for the polymerase present in recombinant RNPs, i.e. associated to the genomic RNA and the NP (Area et al., 2004) (Figure 2). In contrast to the apparently somewhat linear nature of the inter-subunit interactions determined biochemically, the structure obtained by electron microscopy is very compact and the potential localisation of the subunits is not apparent. The position of specific domains of PB1, PB2 and PA proteins within the polymerase has been determined by three-dimensional reconstruction of RNP-monoclonal antibody complexes or tagged RNPs (Area et al., 2004). The location of both the N-terminal region of PB2 and the C-terminus of PB1 are close to the areas of the polymerase that contact the adjacent NP monomers in the RNP, in agreement with the reports of *in vitro* interactions (Biswas et al., 1998; Medcalf et al., 1999; Poole et al., 2004). On the other hand, the position of the C-terminal region of PA is opposite to the NP-polymerase contacts.

The three-dimensional model reported for the polymerase corresponds to the enzyme present in a mature RNP, which can be activated for transcription *in vitro* and can be rescued into infectious virus *in vivo*. It would represent the enzyme present in virion RNPs, poised for transcription but still not activated. Further work is needed to understand the structural changes in the polymerase and the RNP during the initiation steps in transcription or replication and the structural differences between vRNPs and cRNPs.

STRUCTURE AND FUNCTION OF THE 5' AND 3' ENDS OF THE VIRAL GENOME

Influenza virus non-coding regions (NCRs), i.e. the regions upstream of the start codons and downstream of the stop codons on the plus sense strand, differ in sequence and length between segments and show varying levels of conservation. The terminal 13 nucleotides at the 5' end of vRNA (and therefore 3' cRNA) molecules are identical for all genome segments and are totally conserved between all sub-types of influenza A virus. Immediately adjacent to this region (or within 1–2 bases) is a run of 5–8 consecutive uridine residues that comprise the poly A signal (Robertson 1979; Luo et al. 1991; Li and Palese 1994). The 12 nucleotides at the 3' end of vRNA (and therefore 5' cRNA) molecules are identical between segments (except for an occasional U to C transition in segments 1, 2, 3, 6 and 7; see below) and this conservation is again extended across all sub-types. The conserved terminal sequences, which will be referred to as the core promoter, are partially complementary to each other, and have been shown to be crucial for the promotion and regulation of viral RNA synthesis (Parvin et al., 1989; Li and Palese 1992, Seong and Brownlee 1992a,b, Hagen et al. 1994, Fodor et al. 1994, Neumann and Hobom 1995, Cianci et al. 1995, Lee et al. 2003b). The combination of the two ends of the core promoter plus adjacent 'clamp' sequences (see below) are all that is required for encapsidation, replication, transcription and packaging of a heterologous RNA in the presence of a helper virus or alternative sources of the viral proteins (Luytjes et al., 1989). Mutation of the totally conserved, segment specific, 'clamp' regions adjacent to the core promoter can affect virus replication and packaging (Odagiri and Tashiro, 1997). Sequence conservation extends beyond these regions for some segments, suggesting an important but hitherto unidentified role for these regions (Zheng et al., 1996). Persistence of defective interfering segments requires at least 150 bases of the 5'-end of vRNA (Duhaut and Dimmock, 2002). Relatively little is known about the precise role these sequences play, but there is recent evidence that signals for segment-specific packaging of the genome into infectious virions are located within these regions (Fujii et al., 2003; Watanabe et al., 2003).

Panhandles, forks and corkscrews.

The core promoter and clamp regions of the 5' and 3' NCRs have been studied extensively. The partial complementarity of the termini is not merely a consequence of their role in replicating both positive and negative sense RNA. There is strong evidence for biologically relevant base-pairing between these regions, particularly for the clamp region. Two distinct model conformations for this RNA interaction are currently proposed; the panhandle structure (Figure 3) (Hsu et al., 1987) and the corkscrew structure (Figure 4) (Flick et al., 1996), an embellishment upon the previously defined RNA-fork structure (Fodor et al., 1995). The models are not mutually exclusive. It is possible that the panhandle structure exists in the context of a naked RNA (i.e. before it is bound by the polymerase complex and/or NP), and it is the ability of this RNA to adopt the alternative corkscrew conformation that is crucial for the correct functioning of the polymerase (Flick et al., 1996). Indeed, as discussed below, much of the data supporting the panhandle hypothesis comes from structural analysis of short model RNAs in the absence of protein. The deleted internal region of the segment is replaced by a tetraloop sequence that biases the RNA towards formation of a duplex

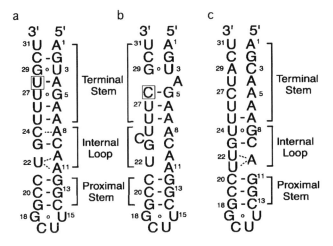

Figure 1.3 Hydrogen bonding interactions between the 5' and 3' ends of influenza viral RNAs as predicted from NMR studies. (a) vRNA (Bae et al., 2001); b) U to C variant at position 4 from the 3' end of vRNA (Lee et al., 2003a); c) cRNA (Park et al., 2003). Standard Watson-Crick base pars are indicated by solid lines, GU base-pairs by open circles and non-canonical base-pairs by dashed lines. Sequence variation (U28C) observed in the 3' end for segments 1–3 and occasionally segments 4 and 6 is indicated by the box.

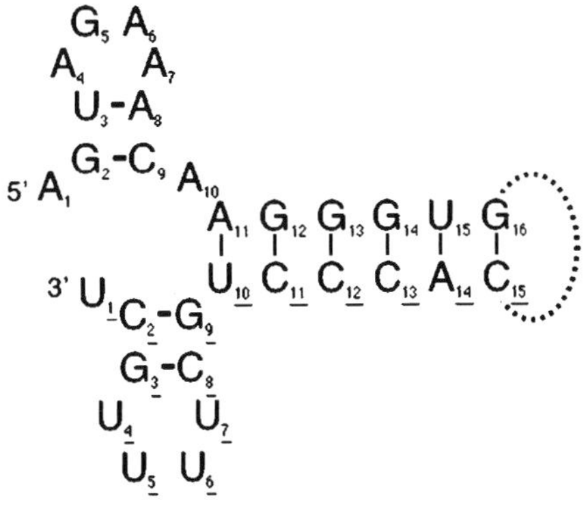

Figure 1.4 The corkscrew model for the termini of vRNA as predicted by analysis of the base-pairing requirements for activity *in vitro* and *in vivo*.

between the 5'- and 3'- termini. This may not accurately reflect what happens *in vivo* where the dynamics of how and when the polymerase interacts with its template is likely to be an important factor in the control of viral RNA synthesis

The panhandle model

The panhandle structure is founded on computer prediction, chemical and enzymatic probing data (Baudin et al., 1994) and NMR analysis (Bae et al., 2001, Park et al., 2003; Lee et al., 2003a). The secondary structures shown in Figure 3 are based upon the solution NMR structures derived for synthetic 31 nucleotide RNAs, incorporating a tetraloop between the 5'- and 3'-ends of vRNA and cRNA sequences (Bae et al., 2001, Park et al., 2003; Lee et al., 2003a). The structures can be considered in three parts: the terminal (or distal) stem, the internal loop and the proximal stem (synonymous with the clamp region of the corkscrew model). The terms proximal and distal have become muddled in the literature. Here we will follow the convention as described in Bae et al. (2001), where proximal is defined as nearer to the centre of the RNA. vRNA exhibits several notable features. The axis of the terminal stem is bent at position 4 towards the major groove, which exposes the minor groove and provides a possible contact point for the polymerase. The A8.C24 mismatch of the internal loop is stabilised by a single non-Watson-Crick hydrogen bond, and stacks with the adjacent bases thus allowing the continuation of the helix into the loop region and the formation of the C9–G23 base pair. The last remaining base of the internal loop (A10) appears to participate in a dynamic (A10-A11).U22 interaction with A10 stacked into the helix, and A11 displaced into the minor groove. This type of RNA configuration has been recognised in the binding sites of at least three other RNA binding proteins: the 16S rRNA binding site for the S8 ribosomal protein; branch-point binding by spliceosomes; and the GA-phage coat protein binding site (Kalurachchi and Nikonowicz, 1998; Smith and Nikonowicz, 1998). The proximal stem adopts a standard Watson-Crick base-pairing conformation. Naturally occurring proximal stem sequences are 4–7 bases long, the first 3 base-pairs of which comprise part of the core promoter (residues 11–13 and 19–21 in Figure 3). Mutations disrupting the base-pairing in the proximal stem adversely effect transcription, but these effects can be largely negated by introducing compensatory mutations on the other RNA strand (Fodor et al., 1995; Flick et al., 1996; Rao et al., 2003) providing strong evidence that this region forms base-pairs at some point during transcription. One exception to this is the G12:C21 base-pair (Figure 3), the substitution of which has been shown to cause a reduction in the level of mRNA synthesis *in vivo* (Fodor et al., 1998), although this effect is not mirrored *in vitro* (Rao et al., 2003).

The single U to C variation in the 3' terminal sequences of some segments 1–3 referred to earlier has been shown to decrease transcription of these segments whilst increasing their replication. The recently determined NMR structure (Lee et al., 2003a) for this sequence variant (Figure 3b) indicates that this minor change results in quite significant alterations in the dynamic behaviour of the stem that markedly reduce its stability. The formation of a new base-pair between G5 and the variant C28 in Figure 3b has several knock on effects. The terminal stem is destabilised; A4 adopts a partially stacked, bulged position; C24 which previously participated in the non-Watson-Crick A8:C24 base-pair, now protrudes towards the major or minor groove. Despite this, the internal loop adopts a similar overall stacked conformation to the U4 variant. The proximal stem is essentially unchanged.

By virtue of the partial complementarity of the 5' and 3' ends of vRNA, cRNA is also able to adopt a related but distinct structure (Figure 3c). The NMR structure (and to a lesser extent that of the U4C variant of vRNA) resembles the RNA fork and corkscrew models (see below), although the putative handles of the corkscrew are absent, as would be predicted for such weak interactions (Park et al., 2003). The key differences between cRNA and vRNA NMR structures are a total lack of base-pairing between the terminal stem residues and an inversion of the internal loop. It is possible that in addition to differential recognition of the primary sequence of v- and cRNA by the polymerase, these structural differences also contribute to the distinct activities of the v- and c-RNA promoters.

The corkscrew model
The RNA fork model (Fodor et al., 1995) proposed that the principal role of the proximal stem or clamp was to facilitate the interaction of the 5'- and 3'-ends of the template whereas the remaining bases functioned as single stranded units containing the polymerase binding sites. Later this was refined into the corkscrew model (Flick et al., 1996, Neumann and Hobom 1995). No physical data supports this model, but it is strongly supported by mutagenesis data from two laboratories (Flick et al., 1996, Flick and Hobom, 1999a, Pritlove et al., 1999, Leahy et al., 2001a,b). Based on the tolerance for compensatory double mutations in the 5'- and 3'-ends, it has been proposed that the first 10 bases of the 5'-end and the last 9 bases of the 3'-end do not participate in the formation of the pan-handle structure described above. Instead, bases G2 and U3 interact with C9 and A8 of the 5'-end to form a 2 base-pair stem with a four base loop (Figure 4; Flick et al., 1996, Pritlove et al., 1999). These short stem-loops constitute the handle of the corkscrew, while the clamp region is the helical screw. The 5' loop corresponds to the novel class of tetraloops (AGNN) that in some yeast RNAs function as binding sites for RNase III (Chanfreau et al., 2000), an enzyme with possible functional similarities to the influenza virus endonuclease. Similarly, the 3'-end can in theory form a 2 base-pair stem with a 4 base (UUUU) loop. Conflicting evidence for the importance of this structure has been reported (see section (d) below) (Flick et al., 1996, Leahy et al., 2001a, Rao et al., 2003).

A significant amount of the data supporting the vRNA corkscrew model was obtained using an assay system based around a so-called promoter 'up-mutant' that has three changes that transform the 9 terminal bases of the 3'-end into the 3'-cRNA sequence (although it is still referred to as 3'-vRNA; Neumann and Hobom 1995; Flick et al., 1996). These mutations overcome a polymerase defect that maps to a single amino acid, E627 in the avian PB2 protein (compared to K627 in human strains). The polymerase from avian strains has particularly low activity on the human wt core promoter sequence. The mutations in the 3' end restore activity to levels similar to those shown on wt templates by human strains of virus (Crescenzo-Chaigne et al., 2002). Some caution must therefore be exercised when interpreting these data especially when comparing them to other systems and structural models.

Functions of the terminal sequences
Role of the 5'-end
The 11 nucleotides at the 5'-end of vRNA are necessary and sufficient for high affinity binding of the influenza polymerase (Lee et al., 2002) and function independently from the 3'-end of the template (Tiley et al., 1994). Therefore neither the panhandle structure nor the complete corkscrew structures are required for polymerase binding. The affinity of the polymerase for 3'-vRNA is low in the absence of the 5' end, but binding is greatly enhanced by the presence of the 5'-vRNA (Tiley et al., 1994, Li et al., 1998, Gonzalez and Ortin, 1999a). This is a result of the complementarity between the bases that form the proximal

stem and the activation of an RNP1-like binding site in PB1 (Li et al., 1998). Binding to the 5' end stimulates the cap-binding activity of the polymerase (Cianci et al., 1995, Li et al., 1998). Recent evidence has revealed that binding to the 5' end also activates the endonuclease (Lee et al., 2003b, Rao et al., 2003), a property that was previously thought to be dependent on both the 5' and 3' vRNA strands. This finding will be discussed in greater detail in the section below describing mRNA transcription.

Detailed mutagenesis and binding studies of the 5'-end (Fodor et al., 1994, Tiley et al., 1994), have defined 6 bases that are required for specific interaction with the polymerase (Figure 5, indicated in capitals). Further studies on this region have shown that a large component of the interaction with the polymerase is structure dependent rather than sequence dependent. Four out of these 6 bases (shaded) form the 2 base-pair stem in the corkscrew model and tolerate compensatory base changes, although different compensatory double mutations give rise to varying levels of activity (Flick et al., 1996; Pritlove et al., 1999) and the polymerase binding site can function at an internal location within an RNA (Flick and Hobom, 1999b).

All activities contingent upon polymerase binding (e.g. enhanced 3'-end binding, cap-binding, endonuclease activity, transcription, polyadenylation and replication) should necessarily be dependent on the 6 bases referred to above. Likewise any activity dependent on 3'-end interactions (endonuclease to some extent, transcription, polyadenylation and replication) should be dependent on the clamp region necessary for enhanced 3'-end binding (minimally 12'-13', Figure 5 in bold capitals). In general where it has been tested this holds true (Fodor et al., 1995; Poon et al., 1998; Leahy et al., 2001b). However, although base A1 is necessary for polymerase binding, a satisfactory explanation for why it does not appear to be necessary for transcription or polyadenylation is lacking (Fodor et al., 1995; Poon et al., 1998). Bases that do not influence polymerase binding, but do affect the downstream activities of the enzyme complex have been identified (Figure 5, shown in bold underlined capitals). Mutation of G5 or A10 reduces endonuclease activity by 80% (Leahy et al., 2001b), while mutation of A7 completely inhibits ApG primed transcription and polyadenylation (Fodor et al., 1995; Poon et al., 1998). It has been suggested that A10 functions as a flexible joint in the corkscrew structure, as it can be mutated (but not deleted) with only partial inhibition of activity (Fodor et al., 1995; Flick et al., 1996; Pritlove et al., 1999; Leahy et al., 2001b). A10 has also been proposed to be the critical structural difference between vRNA and cRNA that specifies selective nuclear export and packaging of vRNA into virions (Tchatalbatchev et al., 2001). G12 and G13 have recently been shown to affect endonuclease activity in a system where endonuclease is only 5'v dependent, suggesting that there is some specific contribution of these bases beyond their role in the clamp region (Rao et al., 2003).

5'-cRNA differs from 5'-vRNA in only four positions: three transitions and one deletion (Figure 5, in bold italics). The polymerase binds strongly to 5'-cRNA utilizing a distinct binding site located at the C-terminus of PB1 (Gonzalez and Ortin, 1999b). Nevertheless, the structural features of 5'-cRNA required for polymerase activity are very similar to 5'-vRNA (Azzeh et al., 2001). It was originally postulated that cap-primed transcription does not normally occur on cRNA templates because they fail to stimulate the endonuclease activity of the polymerase necessary for generating the primer (Cianci et al., 1995). It has been shown that the G to A transition at position 5 and the deletion of A10 (or 11) in combination with the differences in the 3'-end account for the absence of endonuclease stimulation when the polymerase binds cRNA (Leahy et al., 2002). However, it has recently been reported that the cRNA promoter can induce endonuclease activity when supplied with specific substrate mRNAs, but the products are not elongated (Rao et al., 2003). Thus,

1 2 3 4 5 6 7 8 9 10 11 12 13 14 15 16

POL BINDING	AGUagaaACaAgggug
ApG TRANSCRIPTION	aGUagaAACaAGGgug
POLY A	aGUagaAACaAGGgug
ENDONUCLEASE	AGUaGaaACAAGGgug
cRNA (pol binding)	AGCAaaaGC-Agggug

Figure 1.5 The importance of specific bases of the conserved 5' end of vRNA for the template dependent activities of the influenza virus RNA polymerase. Bases indicated in capitals are essential for the indicated polymerase activity; those in lower-case can be mutated without significant effects on activity. Note the discrepancy between the polyadenylation, ApG primed transcription and polymerase binding requirements for base A1 (lowercase, underlined). Shaded bases are implicated in basepairing (G2–C9 and C3–G8) and are tolerant of some compensatory base changes. Bases in bold text are implicated in activities other than binding by the polymerase. G11 and G12 are essential for base-pairing with the 3' end of the template, which is necessary for stimulation of endonuclease activity, transcription and polyadenylation. Bases in bold underlined text are essential for either ApG transcription, polyadenylation or endonuclease activity as indicated. Bases in bold italic type indicate the differences between vRNA and cRNA that are important for the differential activation of the polymerase. Adapted with permission from Elton et al., 2002.

the block to mRNA transcription from cRNA templates still holds but the mechanism is not through lack of activation of endonucleolytic cleavage. One study has shown that cap-primed transcription can occur on cRNA templates (Azzeh et al., 2001). However, these data were generated using the promoter up-mutant that is effectively a 5'-v3'-c hybrid template with a deleted 5'-A10 hinge residue. This may well complicate the interpretation of this result, particularly in an *in vivo* situation where there are polymerase complexes associated with wild-type 5'-ends derived from the helper virus used in the assay system.

A consequence of transcribing the 3'-end of the vRNA template is that all the viral mRNAs contain a copy of the 5'-cRNA sequence immediately downstream of the host derived 5'-capped oligonucleotide primer. The polymerase binds to this sequence through a combined interaction with the cap-structure and the 5'-cRNA sequence (Shih et al., 1995). This has at least two consequences: firstly binding to segment 7 mRNA (that encodes the M1, M2 and putative M3 proteins) prevents the use of the nearby alternative splice donor required for M3 mRNA formation, thus favouring the production of M2. Secondly, it may prevent the undesirable possibility of the polymerase attempting to scavenge capped primers from viral mRNAs (Peng et al., 1996, Shih and Krug 1996). The dependence on the combination of the cap-structure and internal 5'-cRNA sequences suggests that this interaction is somehow different from the usual 5'-cRNA binding mode, which is cap-independent (Tiley et al., 1994).

Role of the 3'-end
The 3'-end of vRNA is required in combination with the 5'-end for efficient ApG primed transcription and, with some mRNA substrates, for endonucleolytic cleavage and formation of the capped primer (Hagen et al., 1994; Honda et al., 2001). Hence the identities of the bases at positions 10–12 (Figure 5) are important because of their complementarity with

the 5'-end required for the formation of the proximal stem or clamp (Fodor et al., 1995; Flick et al., 1996; Leahy et al., 2001a). Position 9 also appears to be important for the interaction with the polymerase (Fodor et al., 1993). Many *in vitro* assays use the dinucleotide ApG, which is complementary to the first 2 bases of the 3'-end and functions as a highly efficient primer to enhance transcription (Plotch and Krug, 1977). For dinucleotide-primed synthesis (which does not involve the endonuclease), none of the first eight bases of the vRNA 3'-end are critically important so long as the dinucleotide primer can base pair with position 2 (Seong and Brownlee, 1992b).

The situation is quite different for cap-primed synthesis. No complementarity between the precursor of the primer and the 3'-end is required for endonucleolytic cleavage (Krug et al., 1980, Hagen et al., 1995) although like ApG, for it to function as a primer it must be able to base-pair with the first 1 to 3 bases of the 3'-end (Hagen et al., 1995). Bases 1, 2, 3, 8 and 9 are involved in the 3' end-dependent activation of the endonuclease that is a prerequisite for cap-primed transcription (Leahy et al., 2001a). This activation is apparently dependent on the ability of the 3'-end to fold into a small stem-loop structure analogous to that defined for the 5'-end (Figure 4). Disruption of this structure does not affect ApG primed transcription or polyadenylation *in vitro* (Fodor et al., 1995, Pritlove et al., 1999) but has been shown to prevent the endonucleolytic cleavage of at least one specific primer. However, with an mRNA primer designed to mimic more closely the sequence used *in vivo*, endonuclease activity was independent of the presence of the 3' end. Nonetheless, bases 2, 3, 8 and 9 were still essential for template activity but not simply due to their ability to form hydrogen bonds, as changes that maintain the small 3' stem loop were also not tolerated (Rao et al., 2003).

It has been widely, but erroneously, reported that a single U to C change 4 bases from the 3' end of vRNA is seen invariantly in segments 1–3 that encode the polymerase. This mutation has been shown to differentially affect mRNA and v/cRNA levels (Lee and Seong, 1998; see below for detail) and may in part explain the different pattern of gene expression/regulation displayed in cell culture by the polymerase proteins from certain laboratory strains (Smith and Hay, 1982). In fact, less than 25% of the current data set for segments 1–3 has this mutation and most of these are old established laboratory strains (Tiley, unpublished observation). Thus, while there is experimental evidence implicating this mutation in the regulation of viral gene expression, the significance for field viruses is far from clear.

The sequence constraints on the function of 3'-end of cRNA are similar to those for 3'-vRNA. The proximal stem is identical between vRNA and cRNA and is required for efficient transcription, with position 12 being particularly critical (Pritlove et al., 1995). The remaining 9 terminal bases are very tolerant of mutations with regard to dinucleotide-primed transcription, with the same caveat of requiring the appropriate dinucleotide. cRNA can also adopt a corkscrew conformation, but the relevance of this structure is unclear in this context, as the polymerase is incapable of mRNA synthesis on such templates (Cianci et al., 1995, Rao et al., 2003.)

The apparent simplicity of the 5'- and 3'-ends of vRNA and cRNA of influenza virus is misleading. The virus has evolved an elegant and highly efficient strategy to control multiple aspects of its replication cycle. The combination of closely related structures with subtle sequence differences and multiple binding sites (both overlapping and non-overlapping) on the polymerase results in a surprising degree of complexity.

MECHANISM OF mRNA TRANSCRIPTION
Initiation

The influenza virus polymerase complex is essentially inactive for any of its associated enzyme functions in the absence of viral RNA. A popular model for the mechanism of mRNA synthesis involves sequential binding of the polymerase to the 5' and 3' termini of vRNA, with each interaction causing allosteric changes to the proteins that result in activation of the cap-binding, endonuclease and nucleotide polymerisation functions in a regulated manner. In this model, the polymerase complex forms by PB1 binding one molecule each of PB2 and PA (Digard et al., 1989; Honda et al., 1990). The polymerase complex then binds to the 5' end of a vRNA segment (Fodor et al., 1994; Tiley et al., 1994), primarily through PB1–RNA interactions (Li et al., 1998; Gonzalezand Ortin, 1999a). This in turn causes a conformational change in the polymerase that activates the cap-binding activity of PB2 (Cianci et al., 1995; Li et al., 1998) and allows the polymerase to bind a host cell mRNA. The 3'-end of the vRNA template then enters the complex through a combination of protein-RNA interactions (again, primarily through PB1; Li et al., 1998; Gonzalez and Ortin, 1999a) and base-pairing between the 5' and 3' sequences. This event stimulates endonuclease activity (Hagen et al., 1994; Cianci et al., 1995; Li et al., 1998, Honda et al., 2002) and the host cap structure, together with 9–15 nucleotides from the 5' end of the mRNA, is cleaved by PB1 (Plotch et al., 1981; Li et al., 2001). The 3'-end of the cleaved mRNA is then used as a primer for transcription initiation by PB1 (Braam et al., 1983; Biswas and Nayak, 1994; Asano et al., 1995), with addition of a guanosine residue directed by the penultimate C residue of the vRNA template (Plotch et al., 1981). Thus the first nucleotide of the template is not transcribed. The nascent mRNA chain is then elongated by sequential addition of ribonucleotides, as directed by the vRNA template. Characterisation of the enzyme's Km for ATP suggests that the polymerase undergoes a transition from an initiation mode to a processive elongation mode between positions 4 and 5 (Klumpp et al., 1998). PB2 dissociates from the cap structure at some point after the first 11–15 nucleotides have been added, when elongation has proceeded past the first templated U residue (Braam et al., 1983). The mechanism responsible for cap release is not known. However, there is evidence that prior to this, binding of the cap-structure to PB2 increases the transcriptional activity of PB1 (Penn and Mahy, 1984; Kawakami et al., 1985).

The key features of the sequential addition model for transcription initiation are that the polymerase assembles on the 5' end of vRNA, cap binding is activated, the polymerase then binds to the 3'end of vRNA and this results in endonuclease activation. Although much support for the sequential model has been generated over the years, more recent studies suggest that this sequence of events is not obligatory. Using an assay where vRNA was added to recombinant polymerase either as a pre-annealed duplex of 5' and 3' vRNA or as sequential components, it was shown that the sequence of assembly had a marked effect on the efficiency of cap-binding but not endonuclease activity (Lee et al., 2003b). Polymerase bound to pre-annealed vRNA template showed high level capped primer binding and endonuclease activity which resulted in enhanced levels of mRNA transcription activity, compared to that from polymerase bound initially to just 5' vRNA. However, the low levels of capped-RNA substrate bound by polymerase associated with only the 5'-end of vRNA were cleaved efficiently, indicating that the 3' end was not required for activation of the endonuclease, and that the original enhancement of endonuclease activity attributed to the presence of the 3'-end actually resulted from increased levels of cap-binding. (Lee et al., 2003b). Another recent study confirmed the 3'-end of vRNA is not required for full endonuclease activity when a capped substrate with a CA sequence upstream of the cleavage site was used as the primer (Rao et al., 2003). Much of the early *in vitro* data on influenza virus transcription was gained using rabbit ß-globin mRNA as a substrate, because of its relatively easy availability, and many other studies have used synthetic cap donor RNAs made from run-off transcripts of the plasmid pGEM using T7 RNA polymerase. However, *in vivo* the polymerase complex shows a particular preference for cleavage of host mRNAs after an A, or in around 20% of the available cloned sequences (for influenza A virus), CA residues (Dhar, 1980; Caton and Robertson, 1980; Beaton and Krug, 1981; Plotch et al., 1981; Shaw and Lamb, 1984). Although ß-globin mRNA and the pGEM transcripts are good substrates for the endonuclease, they are primarily cleaved after a G residue (Plotch et al., 1981) and are not used efficiently as primers (Rao et al., 2003).

Overall therefore, while many of the details of the sequential model of influenza virus mRNA transcription are intact, the details of how the interaction of the polymerase with the conserved terminal sequences of the genome activate its transcriptional activities are in a state of flux. Further work in this area is clearly needed.

Polyadenylation

Processive synthesis of mRNA terminates at a stretch of 5–7 uridine residues about 17 nt from the 5' end of the vRNA template (Robertson et al., 1981), adjacent to the base-paired region of the panhandle structure (Hsu et al., 1987; Figure 1b). The RNA is then polyadenylated by the viral polymerase stuttering at this site, resulting in reiterative copying of the U(5–7) track. Estimates for the length of the polyA tail range from 60-350 residues for mRNA isolated from influenza virus-infected cells (Plotch and Krug, 1977), and up to 120-150 *in vitro* (Perales et al. 1996; Pritlove et al., 1998). Direct experimental evidence that the U track does indeed act as the template for reiterative copying came from a study where the U track was mutated to A_6 and consequently led to synthesis of positive sense capped RNA molecules with poly (U) tails (Poon et al., 1999). Since the U track could be exchanged for an A track without loss of reiterative copying, the sequence itself was not directly responsible for non-processive synthesis by the polymerase complex. Initially it was proposed that the base-paired region of the vRNA panhandle caused a physical block through which the polymerase could not transcribe, resulting in reiterative copying of the adjacent U track (Robertson, 1979; Luo et al., 1991). However, the discovery that the primary polymerase binding site is on the 5'-end of vRNA suggested the attractive hypothesis that the polymerase itself prevents processive transcription through the poly (U) stretch by remaining bound to the 5'-end of its template (Fodor et al., 1994; Tiley et al., 1994). In this model, the continued association of the polymerase with the 5'-end of vRNA creates a loop of untranscribed template that becomes progressively shorter until the polymerase is arrested with its active site over the poly (U) stretch immediately adjacent to its 5' binding site. This steric block forces the polymerase to stutter and polyadenylate the transcript. Experimental evidence in support of this hypothesis comes from the observation that nucleotides required for polymerase binding to the 5' end of vRNA (Fodor et al., 1994; Tiley et al., 1994) are also essential for polyadenylation (Poon et al., 1998; Pritlove et al., 1998).

It is generally believed that there is a mechanism to couple the mode of initiation by the polymerase complex to that of termination. Hay et al., (1982) observed that most full-length transcripts of the vRNA templates were uncapped, while Shaw and Lamb (1984) found that most polyadenylated viral RNAs have host sequences at their 5'-ends. In support of a mechanism to couple initiation and termination, transcripts initiated *in vitro* with a capped primer are also polyadenylated, even in the presence of free NP (Beaton and Krug, 1986). The finding that binding of the polymerase to duplex-form genome termini promotes high levels of cap-primed transcription initiation suggests a mechanism for achieving this coupling, as it is reasonable to suppose that any interaction of the RNA termini in the absence of the polymerase is more likely to happen in *cis* than in *trans* (Lee et al., 2003b). However, this hypothesis is yet to be tested *in vivo* or with cap-donor RNAs containing A or CA at the cleavage site.

MECHANISM OF REPLICATION
Synthesis of cRNA

The process of genome replication is not as well characterised as that of mRNA synthesis. Replication involves the generation of full-length positive sense copies (cRNAs) of the genomic RNA segments, which are then used as templates for amplification of vRNA. Much less cRNA is synthesized during the infectious cycle than mRNA, and it has been estimated that it forms only 5–10% of the total plus-sense RNA present in infected cells (Hay et al., 1977b; Barrett et al., 1979; Herz et al., 1981). Viral mRNAs cannot serve as replicative intermediates for two reasons: they possess host-derived sequences at their 5' ends as a result of cap-snatching (see mechanism of transcription) and secondly, they are truncated at the 3' end where polyadenylation occurs (Figure 1b). The replication of vRNA generates uncapped complete cRNA copies that are not polyadenylated. It therefore requires a change in the mode of initiation by the viral polymerase from one that is dependent upon cap binding and endonuclease activity to one that is cap-independent. The 5'-end of cRNA is triphosphorylated (Hay et al., 1982) and therefore cannot be generated through the endonucleolytic processing of a cap-primed intermediate, as this would leave a monophosphate terminus (Olsen et al., 1996); initiation is therefore presumed to occur using ATP.

It is also necessary for the polymerase to read through the polyadenylation signal towards the 5' end of the vRNA template to generate a complete copy of the genome segment. As well as alterations in the process of transcription, synthesis of cRNA also requires encapsidation of the RNA by NP to form RNP structures, in much the same way as vRNA (Pons, 1971; Hay et al., 1977b). However, the mechanisms that operate to differentiate cRNA from mRNA synthesis remain poorly understood.

RNPs from purified influenza virions are able to transcribe mRNA *in vitro* (Plotch and Krug, 1977) but are unable to support cRNA synthesis, indicating a requirement for other factors besides transcriptionally active RNPs (Skorko et al., 1991). However, the same RNPs introduced into a cell act as templates for cRNA synthesis, but an initial round of mRNA and protein synthesis is required. If protein synthesis is blocked, the RNPs transcribe mRNA but genome replication does not occur (Hay et al., 1977b). In contrast, nuclear extracts prepared from cells infected with influenza virus support the synthesis of both types of positive sense RNA (Beaton and Krug, 1984; 1986; del Rio et al., 1985; Takeuchi et al., 1987; Shapiro and Krug, 1988). RNP complexes recovered from these nuclear extracts by centrifugation make mRNA but are not capable of replication. Replication activity can be restored by addition of the supernatant fraction (Beaton and Krug, 1986; Shapiro and Krug, 1988), and there is evidence that both viral and cellular factors are important.

Role of viral factors
NP
Many different lines of evidence implicate NP as a major factor in replication. Several NP *ts* mutants have been isolated that are defective for replication at the non-permissive temperature (Krug et al., 1975; Scholtissek, 1978; Mahy et al., 1981; Thierry and Danos, 1982) and nuclear

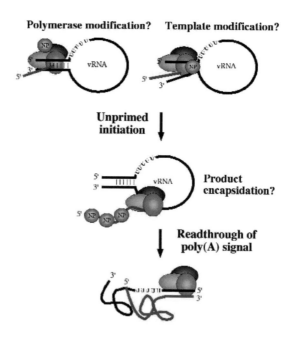

Figure 1.6 Possible roles of NP in cRNA synthesis. In the polymerase modification hypothesis a protein-protein interaction between NP and the polymerase promotes unprimed transcription initiation. In the template modification hypothesis NP disrupts the panhandle structure and so biases the polymerase towards unprimed transcription initiation. The product encapsidation hypothesis predicts that NP is required to co-transcriptionally coat the nascent cRNA molecule. Synthesis of cRNA terminates after processive readthrough of the polyuridine stretch to produce a full-length copy of the template. Reproduced with permission from Elton et al., 2002.

extracts prepared from cells infected with *ts*56 only synthesize mRNA *in vitro* at the non-permissive temperature (Shapiro and Krug, 1988). NP mutations have also been engineered that have differential effects on RNA transcription and replication (Mena et al., 1999). Moreover, antibody-mediated depletion of 'free' NP (not incorporated into RNPs) from infected cell extracts inhibits cRNA synthesis *in vitro* (Beaton and Krug, 1986, Shapiro and Krug, 1988). Thus NP is evidently essential for replication, although its mode of action remains uncertain. However, the addition of exogenous NP to virion RNPs does not induce cRNA synthesis (Skorko et al., 1991), suggesting that NP is not the only factor required.

Several hypotheses have been proposed for the role of NP in the switch between mRNA and cRNA synthesis (Figure 6). The encapsidation hypothesis proposes that NP does not have a regulatory function as such, but is required to co-transcriptionally coat the nascent cRNA segments (Shapiro and Krug, 1988). Other factors alter polymerase activity to change the modes of transcription initiation and termination. This hypothesis has a precedent in that the analogous process of genome replication in the non-segmented negative sense viruses is known to depend on co-transcriptional encapsidation of the replicative intermediate by the N protein (Wagner and Rose, 1995). Alternatively, the template modification hypothesis holds that the interaction of free NP with the template RNA alters its structure and therefore the modes of transcription initiation and termination (Hsu et al., 1987; Fodor et al., 1994; Klumpp et al., 1997). This is plausible, since the terminal sequences of the vRNA template are partially base-paired and recognition of this structure by the polymerase is closely connected with the mechanisms of mRNA transcription initiation and polyadenylation (Cianci et al., 1995, Hagen et al., 1994; Tiley et al., 1994; Pritlove et al., 1998; Poon et al., 1998). A third hypothesis is based on the ability of NP to bind directly to PB1 and PB2: in this model NP modifies the transcriptional function of the polymerase through direct protein-protein contacts (Biswas et al., 1998; Mena et al., 1999, Poole et al., 2004). These hypotheses are not mutually exclusive and require definitive testing, but some recent experiments have given indications as to which mechanisms may operate. The *ts* lesions of two NP mutants (including the *ts*56 mutant mentioned above) that are defective for replication result in *ts* RNA-binding activity without apparent effects on either NP oligomerisation or interactions with the P proteins (Medcalf et al., 1999). This indicates that an NP-RNA interaction is necessary to support genome replication and is consistent with both the encapsidation and template modification hypotheses. In addition, it has recently been reported that NP can interact with PB2 at two different binding regions that also interact with PB1, though not simultaneously (Poole et al., 2004). It is therefore possible that NP may play a part in regulating the interactions between PB1 and PB2 within the polymerase complex (Poole et al., 2004). Replacement of a PB2–PB1 interaction important for mRNA synthesis with a PB2–NP contact could potentially result in a polymerase complex only able to initiate unprimed RNA synthesis, thus favouring replication over transcription. However, it has also been shown that NP is not required for unprimed transcription initiation by the influenza virus RNA polymerase *in vitro* (Lee et al., 2002). This does not rule out the polymerase modification hypothesis, as it is possible NP might promote unprimed initiation above a basal level, but it does not provide any support for it.

PA

In addition to NP, the PA subunit of the polymerase complex has been implicated in replicative synthesis of influenza RNA. Like NP, early evidence for this came from the characterisation of *ts* mutants with lesions mapped to segment 3. Where characterised, these mutants show normal virion transcriptase activities (unlike at least some PB1 and PB2 mutants; Mahy, 1983). However, many PA mutants are defective for synthesis of cRNA and/or vRNA at the non-permissive temperature (Mahy, 1983; Gubareva et al., 1991). In the absence of any other defined role for PA, it thus became a widely accepted hypothesis that the protein was specifically required for genome replication. More recent evidence supporting this suggestion has come from the study of cell lines expressing different combinations of NP and the P proteins. Nakagawa and colleagues found that PB1 and NP alone were able to transcribe cRNA and uncapped polyadenylated RNA in the absence of PB2 or PA (Nakagawa et al., 1995, Nakagawa et al., 1996). However, PA was absolutely required for synthesis of vRNA and also much increased the efficiency of cRNA synthesis. Similar results were subsequently obtained using purified baculovirus-expressed polymerase proteins (Honda et al., 2002). This finding has proved contentious however, as other groups have been unable to detect any form of transcription in the absence of PB2, either *in vivo* or *in vitro* (Perales and Ortin, 1997; Brownlee and Sharps, 2002; Lee et al., 2002). The reasons for this discrepancy remain to be determined.

It has also been suggested that there is a link between PA-induced proteolysis and replication of the viral genome, based on the identification of two PA mutants with reduced levels of induced proteolytic activity that also showed reduced transcriptional activity, especially when supplied with a vRNA sense template (Perales et al., 2000). This prompted the hypothesis that PA is responsible (directly or indirectly) for a proteolytic cleavage event, which converts the influenza virus polymerase from a transcriptase to a replicase (Perales et al., 2000). In this respect, it is interesting that a study has noted the presence of alternative (possibly proteolytic) forms of the PB1 and PA proteins in infected cells but not virions (Akkina et al., 1991). However, a study examining PA proteins from different strains of influenza virus found no correlation between levels of proteolytic activity and the ability to transcribe a model genome segment (Naffakh et al., 2001). Moreover, at lower temperatures the PA from A/PR/8/34 showed reduced levels of induced proteolysis but transcription and replication functioned normally (Naffakh et al., 2001).

Role of host cell factors

Nuclear extracts from uninfected cells are able to stimulate influenza virus RNA synthesis, including that of virion associated RNPs supplied with an exogenous template (Shimizu et al., 1994). It was hypothesised that these cellular factors may be involved in the switch between mRNA transcription and genome replication (Shimizu et al., 1994). Further purification of the activating fraction identified two components that stimulated viral RNA synthesis *in vitro*, termed RNA polymerase activating factor (RAF) 1 and 2 (Momose et al., 1996). RAF-1 was identified as Hsp90, a highly conserved molecular chaperone, and appeared to interact with PB2 (Momose et al., 2002). Recombinant Hsp90 stimulated ApG-primed transcription *in vitro* (Momose et al., 2002) and it was suggested that Hsp90 might facilitate the association of unbound polymerase with template RNA and/or stabilize the complex during its translocation between templates (Momose et al., 2002). It was also suggested that Hsp90 might modulate the activity and structure of the polymerase complex, possibly by altering the association between PB1 and PB2. The second activating factor identified was RAF-2, a fraction containing 36 and 48 kD polypeptides (Momose et al., 1996; 2001). Again this stimulated ApG-primed transcription and p48 was shown to bind NP but not NP-RNA complexes. Mass spectrometry analysis of RAF-2p48 revealed that it is identical to the essential human splicing factor UAP56, also known as BAT1, a putative DEAD-box ATP-dependent RNA helicase (Momose et al., 2001; Linder and Stutz, 2001). Evidence was also presented suggesting that UAP56 acts as a chaperone for NP before it associates with RNA (Momose et al., 2001). Such a function would clearly be relevant to the encapsidation of nascent cRNA molecules by NP. However, recent work has shown that UAP56 is also involved in the nuclear export of mRNA (Luo et al., 2001), which raises intriguing possibilities for its function during virus replication.

In addition, the suggestion has been made that the association of PA with the host-cell protein hCLE plays a role in regulating viral transcription (Huarte et al., 2001). hCLE is ubiquitously expressed in a wide variety of cell lines, and although its function is not known, it shares some sequence similarity with transcriptional regulators (Huarte et al., 2001).

The viral RNPs are therefore capable of interacting with a variety of host cell proteins, some of which have an effect on RNA synthesis, but it requires further investigation to fully establish whether these associations have a specific effect on one form of RNA synthesis over another.

Synthesis of vRNA

The synthesis of vRNA from a cRNA template can be viewed as a simpler process than the transcription or replication of positive sense RNA, since it is the only type of RNA transcribed from a cRNA template. Like cRNA, initiation of vRNA synthesis is unprimed and the products have 5' tri-phosphorylated ends (Young and Content, 1971; Hay et al., 1982; Honda et al., 1998). Until recently, it was generally accepted that the cRNA promoter does not stimulate endonuclease activity in a bound polymerase (Cianci et al., 1995; Honda et al., 2001), providing a mechanism for driving synthesis of only uncapped negative sense RNA. However, as discussed earlier in the context of virus transcription initiation, this may only hold true for certain cap-donor RNAs. Addition of a CA-containing cap donor to a reconstituted polymerase complex bound to 5'cRNA stimulated endonuclease activity to levels approaching those achieved with 5'vRNA (Rao et al., 2003). However, these products are not subsequently extended, again preventing synthesis of non-functional capped vRNA.

The differential activation of the polymerase complex brought about by the vRNA and cRNA promoters may be due to their binding, in part, to different sequences within the PB1 subunit (Gonzalez and Ortin, 1999b). Thus binding to cRNA may prevent the allosteric change in PB1 that is required to activate endonuclease activity in the complex, or in the case of CA-containing primers, the ability to elongate the template. However, the 5'-end of cRNA does stimulate cap-binding activity (Cianci et al., 1995), which may serve to increase overall levels of vRNA transcription through allosteric up-regulation of PB1 activity (Penn and Mahy, 1984; Kawakami et al., 1985). The cRNA template does not contain a polyadenylation signal, so termination of vRNA synthesis would appear to be simpler than that of positive sense RNA. However, as with vRNA, the polymerase binds to the 5' arm of cRNA (Tiley et al., 1994; Cianci et al., 1995; Gonzalez and Ortin, 1999b) and, as discussed above, much evidence indicates that the cRNA promoter also functions as a partial duplex so it is not immediately obvious how synthesis of a full length transcript is achieved. Estimates of the dissociation constants for the interaction of PB1 with the 5'-ends of vRNA and cRNA are similar (2×10^{-8} M and 7×10^{-8} M for v- and cRNA respectively; Gonzalez and Ortin, 1999a, b). However, the interaction of PB1 with the 3'-arm of cRNA is apparently stronger than the corresponding interaction for vRNA (Gonzalez & Ortin, 1999a, b).

The genetics of vRNA synthesis are similar to that of cRNA, with early experiments on *ts* mutants providing evidence that both PA and NP are important (Mahy, 1983). However, the two polarities of genome replication are separable, since mutants have been isolated that can synthesize positive sense RNA but appear to be specifically deficient for synthesis of vRNA (Thierry and Danos, 1982, Markushin and Ghendon, 1984) or vice versa (Mena et al., 1999). In addition, mutations in the NS1 gene show a partial deficiency in the accumulation of vRNA, but not in that of cRNA, suggesting that NS1 acts as a cofactor in the second step of viral RNA replication (Falcón et al., 2004). This may be related to the observed association of NS1 with viral RNPs (Marión et al., 1997).

Analysis of *in vitro* transcription reactions carried out with infected cell extracts has shown that, as with cRNA synthesis, a supply of non-RNP associated NP is required to support vRNA synthesis (Shapiro and Krug, 1988). Similarly, extracts from cells infected with the NP *ts* mutant *ts*56 did not synthesise vRNA at the non-permissive temperature. In the absence of free NP, no discrete vRNA-sized products were synthesized by the infected cell extracts and it was suggested that RNA elongation halted at any point where NP was unavailable because of a failure to encapsidate the nascent vRNA (Shapiro and Krug, 1988). As already noted, the point mutation present in *ts*56 confers temperature-sensitive RNA binding activity on the NP polypeptide (Medcalf et al., 1999), which is consistent with this hypothesis.

Finally, the spatial organisation of vRNA synthesis may differ from that of positive-strand synthesis. One study that examined viral RNA synthesis in a series of cell fractions found that vRNA synthesising activity was only associated with the insoluble 'nuclear matrix', while mRNA and/or cRNA synthesis partitioned into both soluble and insoluble nuclear fractions (Lopez-Turiso et al., 1990). This is consistent with the observation that cRNPs do not undergo nuclear export (Shapiro et al., 1987, Tchatalbachev et al., 2001), although the mechanisms responsible for controlling the intranuclear localisation of RNPs remain to be determined.

TEMPORAL REGULATION OF RNA SYNTHESIS

Influenza virus RNA synthesis is a highly regulated process, both in terms of the three classes of viral RNA and at the level of individual RNA segments. Messenger RNAs can be detected shortly after infection and their rate of synthesis peaks around 2.5h post-infection, depending on the virus strain and cell line used (Hay et al., 1977b; Barrett et al., 1979; Mark et al., 1979; Smith and Hay, 1982; Shapiro et al., 1987). The full length transcripts of the virus genome (cRNAs) can only be detected after mRNA synthesis, consistent with their dependence upon viral protein synthesis (Hay et al., 1977b; Barrett et al., 1979), but their maximal rate of synthesis occurs before that of the viral mRNA (Barrett et al., 1979; Shapiro et al., 1987). Synthesis of vRNA follows cRNA, but continues to increase even after synthesis of the other classes of RNA declines (Hay et al., 1977b; Barrett et al., 1979; Shapiro et al., 1987).

Primary transcription of the virion associated vRNAs into mRNAs after infection of the cell does not seem to be regulated, as approximately equimolar amounts of all eight non-spliced mRNAs are produced (Barrett et al., 1979; Hay et al., 1977b; Inglis and Mahy, 1979). However, within the first hour after infection, mRNAs coding for NP and NS1 are preferentially transcribed (Barrett et al., 1979; Inglis et al., 1979; Smith and Hay, 1982; Shapiro et al., 1987). At later times, after vRNA replication, synthesis of mRNAs coding for the structural proteins HA and M predominates (Hay et al., 1977b; Inglis et al., 1979). The mRNAs coding for the three P proteins are under-represented throughout the replication cycle (Hay et al., 1977b; Smith and Hay, 1982; Enami et al., 1985). The relative abundance of the virus polypeptides appears to be primarily under transcriptional and not translational control, as the amount of each polypeptide is proportional to the amount of each mRNA (Hay et al., 1977b; Inglis and Mahy, 1979). The main exception to this concerns the regulated splicing of mRNA transcripts from segments 7 and 8, which has been reviewed elsewhere (Ortin, 1998).

The cRNA copies of the segments are present in approximately equimolar amounts throughout infection (Hay et al., 1977b; Smith and Hay, 1982; Shapiro et al., 1987). It has been suggested that cRNA replication utilises only the input vRNA segments as templates (Hay et al., 1977b), since maximal rates of cRNA synthesis can be achieved even in the absence of vRNA synthesis (Thierry and Danos, 1982). In contrast, it is generally agreed, with the exception of Enami et al., (1985), that replication of the cRNAs into new vRNA segments is highly regulated. At early times, vRNA segments 5 and 8 are preferentially synthesised (Smith and Hay, 1982; Shapiro et al., 1987), while at later times, the synthesis of segment 7 predominates (Smith and Hay, 1982). From the correlation between the relative synthesis of vRNA and mRNA from each segment, at least at early times, it has been suggested that mRNA synthesis is regulated through vRNA synthesis, and hence from selective replication of the cRNA templates (Smith and Hay, 1982). Therefore, at early times, the relative abundance of vRNA segments 5 and 8 leads to increased production of NP and NS1 mRNAs, and hence the polypeptide products, while later in infection, the production of the M protein is selectively increased by the same mechanism. This may be an over simplification however, and other factors besides the level of vRNA are likely to be involved in regulating the level of individual mRNAs. Transcription of the segments coding for the three polymerase proteins is apparently relatively inefficient because the segment 1, 2, and 3 mRNAs are present in much lower amounts than their respective vRNAs (Hay et al., 1977b; Smith and Hay, 1982).

Influenza virus gene expression is therefore subject to complex control at the transcriptional level, and this presumably arises from modification of the basic virus polymerase activity at different times post infection. Sequence differences between the various segments must be the ultimate regulatory targets, and there is some evidence for this. Position 4 of the 3' end of vRNA is the only nucleotide that varies within the conserved termini of the 8 segments of influenza A. In some strains of virus, segments 1, 2 and 3, that encode the polymerase proteins, and 4 and 6 have a U to C transition at this point (Robertson, 1979). This mutation has been artificially introduced into the neuraminidase gene of influenza virus, and was shown to decrease the level of mRNA synthesis late in the infection cycle, while increasing cRNA and vRNA synthesis (Lee and Seong 1998). This may in part explain the different pattern of gene expression/regulation displayed by the polymerase proteins. Similarly, the non-coding regions of segment 6 outside of the conserved terminal regions have been shown to affect the levels of vRNA and mRNA produced during virus infection (Zheng et al., 1996). However, very little is known about the molecular mechanisms responsible for recognising the relevant RNA sequences. Aberrant regulation of the typical early-late pattern of gene expression can result from several causes, including *ts* mutations in segments 1, 2, 3, 5, or 8 (Gubareva et al., 1991; Herget and Scholtissek, 1993; Mahy et al., 1981; Markushin and Ghendon, 1984; Mukaigawa et al., 1991; Wolstenholme et al., 1980), inhibitors of cellular transcription or translation (Mahy et al., 1977; Minor and Dimmock, 1977; Varich and Kaverin, 1987), or the infection of a non-permissive cell type (Bosch et al., 1978; Smith and Hay, 1982). This can be interpreted as the regulation of influenza virus late gene expression requiring the concerted action of several virus and cellular specific proteins or RNAs, but the data give little information as to the function or precise point of action of any of the components.

CONCLUDING REMARKS

Huge strides have been made in recent years towards understanding the molecular mechanisms that mediate and control influenza virus transcription, particularly since the advent of 'reverse genetic' methods for studying the effect of defined mutations on the function of virus components (Luytjes et al., 1989). However, there are important questions that remain to be answered. The mechanisms that underlie genome replication are still poorly understood, despite a quarter-century of study. Similarly, we know very little about the regulation of viral gene expression. The information regarding structural aspects is scanty. Although three-dimensional models for the RNP and the polymerase have been reported, high-resolution structures for any of the RNP polypeptides would be very welcome. Such issues provide ample scope for exciting research in the field of influenza virus replication, which will surely be aided by the recent development of plasmid-based systems for the creation of entirely recombinant virus (Neumann et al., 1999, Fodor et al., 1999).

Acknowledgements

Work in the authors' laboratories is supported by grants from the Medical Research Council (nos. G0300009 and G9901213 to LT and PD), Biotechnology and Biological Research Council (nos. S18874 to PD and BBS/B/00239 to LT), Wellcome Trust (no. 073126 to PD), Ministerio de Ciencia y Tecnología (no. BMC2001–1223 and PTR1995–1564-OP to JO) and Comunidad de Madrid (no. 08.2/0012.1/2001 to JO).

References

Akkina, R. K., Richardson, J. C., Aguilera, M. C., and Yang, C.-M. (1991). Heterogeneous forms of polymerase proteins exist in influenza A virus-infected cells. Virus Res. *19*,17–30.

Albo, C., Valencia, A., and Portela, A. (1995). Identification of an RNA binding region within the N-terminal third of the influenza A virus nucleoprotein. J. Virol. *69*, 3799–3806.

Area, E., Martín-Benito, J., Gastaminza, P., Torreira, E., Valpuesta, J. M., Carrascosa, J. L., and Ortín, J. (2004). Three-dimensional structure of the influenza virus RNA polymerase, localization of subunit domains. Proc. Natl. Acad. Sci. USA, *101*, 308–313.

Asano, Y., Mizumoto, K., Maruyama, T., and Ishihama, A. (1995). Photoaffinity labeling of influenza virus RNA polymerase PB1 subunit with 8-azido GTP. J. Biochem. *117*, 677–682.

Azzeh, M., Flick, R., and Hobom, G. (2001). Functional analysis of the influenza A virus cRNA promoter and construction of an ambisense transcription system. Virology *289*, 400-410.

Bae, S. H., Cheong, H. K., Lee, J. H., Cheong, C., Kainosho, M., and Choi, B. S. (2001). Structural features of an influenza virus promoter and their implications for viral RNA synthesis. Proc. Natl. Acad. Sci USA *98*, 10602–10607.

Barrett, T., Wolstenholme, A. J., and Mahy, B. W. J. (1979). Transcription and replication of influenza virus RNA. Virology *98*, 211–225.

Baudin, F., Bach, C., Cusack, S., and Ruigrok, R. W. H. (1994). Structure of influenza virus RNP. I. Influenza virus nucleoprotein melts secondary structure in panhandle RNA and exposes the bases to the solvent. EMBO J. *13*, 3158–3165.

Baudin, F., Petit, I., Weissenhorn, W., and Ruigrok, R. W. (2001). In vitro dissection of the membrane and RNP binding activities of influenza virus M1 protein. Virology *281*, 102–108.

Beaton, A. R., and Krug, R. M. (1981). Selected host cell capped RNA fragments prime influenza viral RNA transcription in vivo. Nucl. Acids Res. *9*, 4423–4436.

Beaton, A. R., and Krug, R M. (1984). Synthesis of the templates for influenza virion RNA replication in vitro. Proc Natl Acad Sci USA *81*, 4682–4686.

Beaton, A. R., and Krug, R M. (1986). Transcription antitermination during influenza viral template RNA synthesis requires the nucleocapsid protein and the absence of a 5' capped end. Proc Natl Acad Sci USA *83*, 6282–6286.

Biswas, S. K., and Nayak, D. P (1994). Mutational analysis of the conserved motifs of influenza A virus polymerase basic protein 1. J. Virol. *68*, 1819–1826.

Biswas, S. K., and Nayak, D. P. (1996). Influenza virus polymerase basic protein 1 interacts with influenza virus polymerase basic protein 2 at multiple sites. J. Virol. *70*, 6716–6722.

Biswas, S. K., Boutz, P. L., and Nayak, D. P. (1998). Influenza virus nucleoprotein interacts with influenza virus polymerase proteins. J. Virol. *72*, 5493–5501.

Blaas, D., Patzelt, E., and Keuchler, E. (1982). Identification of the cap binding protein of influenza virus. Nucleic Acid Res. *10*, 4803–4812.

Blok, V., Cianci, C., Tibbles, K., Inglis, S., Kystal, M., and Digard, P. (1996). Inhibition of the influenza virus RNA-dependent RNA polymerase by antisera directed against the carboxy-terminal region of the PB2 subunit. J. Gen. Virol. *77*, 1025–1033.

Bosch, F. X., Hay, A. J., and Skehel, J. J. (1978). RNA and protein synthesis in a permissive and an abortive influenza virus infection. In, Negative Strand Viruses and the Host Cell, B. W. J. Mahy and R. D. Barry, eds. Academic Press, London, pp. 465.

Bouloy, M., Morgan, M. A., Shatkin, A. J., and Krug, R. M. (1979). Cap and internal nucleotides of reovirus mRNA primers are incorporated into influenza viral complementary RNA during transcription in vitro. J. Virol. *32*, 895–904.

Bouloy, M., Plotch, S. J., and Krug, R. M. (1980). Both the 7-methyl and the 2'-O-methyl groups in the cap of mRNA strongly influence its ability to act as primer for influenza virus RNA transcription. Proc. Natl. Acad. Sci. USA *77*, 3952–3956.

Braam, J., Ulmanen, I., and Krug, R. M. (1983). Molecular model of a eucaryotic transcription complex, functions and movements of influenza P proteins during capped RNA-primed transcription. Cell *34*, 609–618.

Brownlee, G. G., and Sharps, J. L. (2002). The RNA polymerase of influenza A virus is stabilized by interaction with its viral RNA promoter. J. Virol. *76*, 7103–7113.

Caton, A. J., and Robertson, J. S. (1980). Structure of the host-derived sequences at the 5' ends of influenza virus mRNA. Nucl. Acids Res. *8*,2591–2603.

Chanfreau, G., Buckle, M., and Jacquier, A. (2000). Recognition of a conserved class of RNA tetraloops by Saccharomyces cerevisiae RNase III. Proc Natl Acad Sci U S A, *97*, 3142–7.

Cianci, C., Tiley, L., and Krystal, M. (1995). Differential activation of the influenza virus polymerase via template RNA binding. J. Virol, *69*, 3995–3999.

Compans, R. W., Content, J., and Duesberg, P 1972. Structure of the ribonucleoprotein of influenza virus. J. Virol. *10*, 795–800.

Crescenzo-Chaigne, B., van der Werf, S., and Naffakh, N. (2002). Differential effect of nucleotide substitution in the 3' arm of the influenza virus vRNA promoter on transcription/replication by avian and human

polymerase complexes is related to the nature of PB2 amino acid 627. Virology *303*, 240-252.

Desselberger, U., Racaniello, V. R., Zazra, J. J., and Palese, P. (1980). The 3' and 5' terminal sequences of inlfuenza virus A, B and C virus RNA segments are highly conserved and show partial inverted complementarity. Gene *8*, 315–328.

Detjen, B. M., St Angelo, C., Katze, M. G., and Krug, R. M. (1987). The three influenza virus polymerase (P) proteins not associated with viral nucleocapsids in the infected cell are in the form of a complex. J. Virol. *61*, 16–22.

Dhar, R., Chanock, R. M. and Lai, C. J. (1980). Nonviral oligonucleotides at the 5' terminus of cytoplasmic influenza viral mRNA deduced from cloned complete genomic sequences. Cell *21*, 495–500.

Digard, P., Blok, V., and Inglis, S. C. (1989). Complex formation between influenza virus polymerase proteins expressed in Xenopus oocytes. Virol. *171*, 162–169.

Doan, L., Handa, B., Roberts, N. A., and Klumpp, K. (1999). Metal ion catalysis of RNA cleavage by the influenza virus endonuclease. Biochem. *38*, 5612–5619.

Duesberg, P. 1969. Distinct subunits of the ribonucleoprotein of influenza virus. J. Mol. Biol. *42*, 485–499.

Duhaut, S. D., and Dimmock, N. J. (2002). Defective segment 1 RNAs that interfere with production of infectious influenza A virus require at least 150 nucleotides of 5' sequence, evidence from a plasmid-driven system. J. Gen. Virol. *83*, 403–411.

Elton, D., Medcalf, E., Bishop, K., and Digard, P. 1999a. Analysis of self-association by the influenza virus nucleoprotein. Virology *260*, 190–200.

Elton, D., Medcalf, E., Bishop, K., Harrison, D., and Digard, P. 1999b. Identification of amino-acid residues of influenza virus nucleoprotein essential for RNA-binding. J. Virol. *73*, 7357–7367.

Elton, D. M., Tiley, L., and Digard, P. (2002). Molecular mechanisms of influenza virus transcription. Recent Res. Devel. Virol. *4*, 1–25.

Enami, M., Fukuda, R., and Ishihama, A. (1985). Transcription and replication of eight segments of influenza virus. Virology *142*, 68–77.

Falcón, A. M., Marión, R. M., Zürcher, T., Gómez, P., Portela, A., Nieto, A., and Ortín, J. (2004). Defective RNA replication and late gene expression in temperature-sensitive (A/Victoria/3/75) influenza viruses expressing deleted forms of NS1 protein. J. Virol. *78*,3880-3888.

Fechter, P., Mingay, L., Sharps, J., Chambers, A., Fodor, E., and Brownlee, G. G. (2003). Two aromatic residues in the PB2 subunit of influenza A RNA polymerase are crucial for cap binding. J. Biol. Chem. *278*, 20381–20388.

Flick, R., and Hobom, G. 1999a. Interaction of influenza virus polymerase with viral RNA in the 'corkscrew' conformation. J. Gen. Virol. *80*, 2565–2572

Flick, R., and Hobom, G. 1999b. Transient bicistronic vRNA segments for indirect selection of recombinant influenza viruses. Virology *262*, 93–103.

Flick R., Neumann, G., Hoffmann, E., Neumeier, E., and Hobolm, G. (1996). Promoter elements in the influenza vRNA terminal structure. RNA *2*, 1046–1057.

Fodor, E., Crow, M., Mingay, L. J., Deng, T., Sharps, J., Fechter, P., and Brownlee, G. G. (2002). A single amino acid mutation in the PA subunit of the influenza virus RNA polymerase inhibits endonucleolytic cleavage of capped RNAs. J. Virol. *76*, 8989–9001.

Fodor, E., Devenish, L., Engelhardt, O. G., Palese, P., Brownlee, G. G., and Garcia-Sastre, A. (1999). Rescue of influenza A virus from recombinant DNA. J. Virol. *73*, 9679–9682.

Fodor, E., Mingay, L. J., Crow, M., Deng, T., and Brownlee, G. G. (2003). A single amino acid mutation in the PA subunit of the influenza virus RNA polymerase promotes the generation of defective interfering RNAs. J. Virol. *77*, 5017–5020.

Fodor, E., Seong, B. L., and Brownlee, G. G. (1993). Photochemical cross-linking of influenza A polymerase to is virion RNA promoter defines a polymerase binding site at residues 9 to 12 of the promoter. J. Gen. Virol. *74*, 1327–1333.

Fodor, E., Pritlove, D. C., and Brownlee, G. G. (1994). The influenza virus panhandle is involved in initiation of transcription. J. Virol. *68*, 4092–4096.

Fodor, E., Pritlove, D. C., and Brownlee, G. G. (1995). Characterization of the RNA-fork model of the virion RNA in the initiation of transcription in influenza virus. J. Virol. *69*, 4012–4019.

Fodor, E., Palese, P., Brownlee, G. G., and Garcia-Sastre, A (1998). Attenuation of influenza A virus mRNA levels by promoter mutations. J. Virol. *72*, 6283–6290.

Fujii, Y., Goto, H., Watanabe, T., Yoshida, T., and Kawaoka, Y. (2003). Selective incorporation of influenza virus RNA segments into virions. Proc. Natl. Acad. Sci. USA *100*, 2002–2007.

Gastaminza, P., Perales, B., Falcon, A. M. and Ortin, J. (2003). Mutations in the N-terminal region of influenza virus PB2 protein affect virus RNA replication but not transcription. J. Virol. *77*, 5098–5108.

Ghendon, Y. Z., Markushin, S. G., Klimov, A. I., and Hay, A. J. (1982). Studies of fowl plague virus temperature-sensitive mutants with defects in transcription. J. Gen. Virol. *63*, 103–111.

Gonzalez, S., and Ortin, J. 1999a. Characterization of influenza virus PB1 protein binding to viral RNA, two separate regions of the protein contribute to the interaction domain. J. Virol. *73*, 631–637.

Gonzalez, S., and Ortin, J. 1999b. Distinct regions of influenza virus PB1 polymerase subunit recognise vRNA and cRNA templates. EMBO J. *18*, 3767–3775.

González, S., Zürcher, T., and Ortín, J. (1996). Identification of two separate domains in the influenza virus PB1 protein responsible for interaction with the PB2 and PA subunits, A model for the viral RNA polymerase structure. Nucleic Acids. Res. *24*, 4456–4463.

Gubareva, L. V., Varich, N. L., Markushin, S. G., and Kaverin, N. V. (1991). Studies on the regulation of influenza-virus RNA replication – a differential inhibition of the synthesis of vRNA segments in shift-up experiments with ts mutants. Arch. Virol. *121*, 9–17.

Hagen, M., Tiley, L., Chung, T. D. Y., and Krystal, M. (1995). The role of template-primer interactions in cleavage and initiation by the influenza virus polymerase. J. Gen. Virol. *76*, 603–611.

Hagen, M., Chung, T. D. Y., Butcher, J. A., and Krystal, M (1994). Recombinant influenza virus, requirement of both 5' and 3' viral ends for endonuclease activity. J. Virol. *68*, 1509–1515.

Haller, O., and Kochs, G. (2002). Interferon-induced mx proteins, dynamin-like GTPases with antiviral activity. Traffic *3*, 710-717.

Hara, K., Shiota, M., Kido, H., Ohtsu, Y., Kashiwagi, T., Iwahashi, J., Hamada, N., Mizoue, K., Tsumura, N., Kato, H., and Toyoda, T. (2001). Influenza virus RNA polymerase PA subunit is a novel serine protease with Ser624 at the active site. Genes Cells *6*, 87–97.

Hay, A. J., Abraham, G., Skehel, J. J., Smith, J. C., and Fellner, P. 1977a. Influenza virus messenger RNAs are incomplete transcripts of the genome RNAs. Nucl. Acids Res. *4*, 4197–209.

Hay, A. J., Lomnizi, B., Bellamy, A., and Skehel, J. J. 1977b. Transcription of the influenza virus genome. Virology *83*, 337–355.

Hay, A. J., Skehel, J. J., and McCauley, J. (1982). Characterization of influenza virus RNA complete transcripts. Virology *116*, 517–522.

Heggeness, M. H., Smith, P. R., Ulmanen, I., Krug, R. M., and Choppin, P. W. (1982). Studies on the helical nucleocapsod of influenza virus. Virology *118*, 466–470.

Herget, M., and Scholtissek, C. (1993). A temperature-sensitive mutation in the acidic polymerase gene of an influenza A virus alters the regulation of viral protein synthesis. J. Gen. Virol *74*, 1789–1794.

Herz, C., Stavnezer, E., Krug, R., and Gurney, T., Jr. (1981). Influenza virus, an RNA virus, synthesizes its messenger RNA in the nucleus of infected cells. Cell *26*, 391–400.

Honda, A.,K. Mizumoto, K., and Ishihama, A. (1998). Identification of the 5' terminal structure of influenza virus genome RNA by a newly developed enzymatic method. Virus Res. *55*,199–206.

Honda, A., Mizumoto, K., and Ishihama A. (1999). Two separate sequences of PB2 subunit constitute the RNA cap-binding site of influenza virus RNA polymerase. Genes Cells *4*, 475–485.

Honda, A., Mizumoto, K., and Ishihama, A. (2002). Minimum molecular architectures for transcription and replication of the influenza virus. Proc. Natl. Acad. Sci. USA *99*, 13166–13177.

Honda, A., Mukaigawa, J., Yokoiyama, A., Kato, A., Ueda, S., Nagata, K., Krystal, M., Nayak, D. P., and Ishihama, A. (1990). Purification and molecular structure of RNA polymerase from influenza virus A/PR8. J. Biochem. *107*, 624–628.

Honda, A., Endo, A., Mizumoto, K., and Ishihama, A. (2001). Differential roles of viral RNA and cRNA in functional modulation of the influenza virus RNA polymerase. J. Biol. Chem. *276*, 31179–31185.

Honda, A., Ueda, K., Nagata, K. and Ishima, A. (1988). RNA polymerase of influenza virus: role of NP in RNA chain elongation. J. Biochem. *104*, 1021–1026.

Horisberger, M. A. (1980). The large P proteins of influenza A viruses are composed of one acidic and two basic polypeptides. Virology 107, 302–5.

Hsu, M. T., Parvin, J. D., Gupta, S., Krystal, M., and Palese, P. (1987). Genomic RNAs of influenza viruses are held in circular conformation in virions and in infected cells by a terminal panhandle. Proc. Natl. Acad. Sci. USA. *84*, 8140-8144.

Huang, T. S., Palese, P., and Krystal, M. (1990). Determination of influenza virus proteins required for genome replication. J. Virol. *64*, 5669–5673.

Huarte, M., Falcon, A., Nakaya, Y., Ortin, J., Garcia-Sastre, A., and Nieto, A. (2003). Threonine 157 of influenza virus PA polymerase subunit modulates RNA replication in infectious viruses. J. Virol. *77*, 6007–6013.

Huarte, M., Sanz-Ezquerro, J. J., Roncal, F., Ortin, J., and Nieto, A. (2001). PA subunit from influenza virus polymerase complex interacts with a cellular protein with homology to a family of transcriptional activators. J. Virol. 75, 8597–8604.

Inglis, S. C., and Mahy, B. W. J. (1979). Polypeptides specified by the influenza virus genome. 3. Control of synthesis in infected cells. Virology 95, 154–164.

Inglis, S. C., Barrett, T., Brown, C. M., and Almond, J. W. (1979). The smallest genome RNA segment of influenza virus contains two genes that may overlap. Proc. Natl. Acad. Sci. USA 76, 3790-3794.

Jennings, P. A., Finch, J. T., Winter, G., and Robertson, J. S. (1983). Does the higher order structure of the influenza virus ribonucleoprotein guide the sequence rearrangements in influenza viral RNA. Cell 34, 619–627.

Kalurachchi K., and Nikonowicz E. P. (1998). NMR structure determination of the binding site for ribosomal protein S8 from Escherichia coli 16 S rRNA. J. Mol. Biol. 280, 639–654.

Kawakami, K., Mizumoto, K., Ishihama, A., Shinozaki-Yamaguchi, K., and Miura, K. (1985). Activation of influenza virus associated RNA polymerase by cap structure (m7GpppNm). J. Biochem. 97, 655–661.

Kingsbury, D. W., Jones, I. M., and Murti, K. G. (1987). Assembly of influenza ribonucleoprotein in vitro using recombinant nucleoprotein. Virology 156, 396–403.

Klumpp, K., Ruigrok, R. W. H., and Baudin, F. (1997). Roles of the influenza virus polymerase and nucleoprotein in forming a functional RNP structure. EMBO J. 16, 1248–1257.

Klumpp, K., Ford, M. J., and Ruigrok, R. W. H. (1998). Variation in ATP requirement during influenza virus transcription. J. Gen. Virol. 79, 1033–1045.

Kobayashi, M., Toyoda, T., Adyshev, D. M., Azuma, Y., and Ishihama, A. (1994). Molecular dissection of influenza virus nucleoprotein, deletion mapping of the RNA binding domain. J. Virol. 68, 8433–8436.

Kobayashi, M., Tuchiya, K., Nagata, K., and Ishihama, A. (1992). Reconstitution of influenza virus RNA polymerase from three subunits expressed using recombinant baculovirus system. Virus Res. 22, 235–245.

Kochs, G., and Haller, O. (1999). GTP-bound human MxA protein interacts with the nucleocapsids of Thogoto virus (Orthomyxoviridae). J. Biol. Chem. 274; 4370-4376.

Krug, R. M., Broni, B. A., LaFiandra, A. J., Morgan, M. A., and Shatkin, A. J. (1980). Priming and inhibitory activities of RNAs for the influenza viral transcriptase do not require base pairing with the virion template RNA. Proc Natl Acad Sci U S A 77, 5874–5878.

Krug, R. M., Ueda, M., and Palese, P. (1975). Tempereature sensitive mutants of influenza WSN virus defective in virus-specific RNA synthesis. J. Virol. 16, 790-796.

Leahy, M. B., Dobbyn, H. C., and Brownlee, G. G. 2001a. Hairpin loop structure in the 3' arm of the influenza A virus virion RNA promoter is required for endonuclease activity. J. Virol. 75, 7042–7049.

Leahy, M. B., Pritlove, D. C., Poon, L. L., and Brownlee, G. G. 2001b. Mutagenic analysis of the 5' arm of the influenza A virus virion RNA promoter defines the sequence requirements for endonuclease activity. J. Virol. 75, 134–142.

Leahy, M. B., Zecchin, G., and Brownlee, G. G. (2002). Differential activation of influenza A virus endonuclease activity is dependent on multiple sequence differences between the virion RNA and cRNA promoters. J. Virol. 76, 2019–2023.

Lee, M. K., Bae, S. H., Park, C. J., Cheong, H. K., Cheong, C., and Choi, B. S. 2003a. A single-nucleotide natural variation (U4 to C4) in an influenza A virus promoter exhibits a large structural change, implications for differential viral RNA synthesis by RNA-dependent RNA polymerase. Nucl. Acids Res. 31, 1216–1223.

Lee, M. T., Bishop, K., Medcalf, L., Elton, D., Digard, P., and Tiley,L. (2002). Definition of the minimal viral components required for the initiation of unprimed RNA synthesis by influenza virus RNA polymerase. Nucl. Acids Res. 30, 429–438.

Lee, M. T., Klumpp, K., Digard, P., and Tiley, L. 2003b. Activation of influenza virus RNA polymerase by the 5' and 3' terminal duplex of genomic RNA. Nucl. Acids Res. *31*, 1624–1632.

Lee, K. H., and Seong, B. L. (1998). The position 4 nucleotide at the 3' end of the influenza virus neuraminidase vRNA is involved in temporal regulation of transcription and replication of neuraminidase RNAs and affects the repertoire of influenza virus surface antigens. J. Gen. Virol. *79*, 1923–1934.

Li, X., and Palese, P. (1992). Mutational analysis of the promoter required for influenza virus virion RNA synthesis. J. Virol. *66*, 4331–4338.

Li, X., and Palese, P. (1994). Characterization of the polyadenylation signal of influenza virus RNA. J. Virol. *68*, 1245–1249.

Li, M-L., Ramirez, B. C., and Krug, R. M. (1998). RNA-dependent activation of primer RNA production by influenza virus polymerase, different regions of the same protein subunit constitute the two required RNA binding sites. EMBO J. *17*, 5844–5852.

Li, M-L., Rao, P., and Krug, R. M. (2001). The active sites of the influenza cap-dependent endonuclease are on different polymerase subunits. EMBO J. *20*, 2078–2086.

Linder, P., and Stutz, F. (2001). mRNA export, travelling with DEAD box proteins. Curr. Biol. *11*, R961–R963.

Lopez-Turiso, J. A., Martinez, C., Tanaka, T., and Ortin, J. (1990). The synthesis of influenza virus negative-strand RNA takes place in insoluble complexes present in the nuclear matrix fraction. Virus Res. *16*, 325–338.

de la Luna, S., Martinez, C., and Ortin, J. (1989). Molecular cloning and sequencing of influenza virus A/Victoria/3/75 polymerase genes, sequence evolution and prediction of possible functional domains. Virus Res. *13*, 143–155.

Luo, G., Luytjes, W., Enami, M., and Palese, P. (1991). The polyadenylation signal of influenza virus RNA involves a stretch of uridines followed by the RNA duples of the panhandle structure. J. Virol. *65*, 2861–2867.

Luo, M.-J., Zhou, Z., Magni, K., Christoforides, C., Rappsilber, J., Mann, M., and Reed, R. (2001). Pre-mRNA splicing and mRNA export linked by direct interactions

between UAP56 and Aly. Nature *413*, 644–647.

Luytjes, W., Krystal, M., Enami, M., Parvin, J. D., and Palese, P. (1989). Amplification, expression and packaging of a foreign gene by influenza virus. Cell *59*, 1107–1113.

Mahy, B. W. J., Carroll, A. R., Brownson, J. M. T., and McGeoch, D. J. 1977. Block to influenza virus replication in cells preirradiated with ultraviolet light. Virology *83*, 150-162.

Mahy, B. W. J. (1983). Mutants of Influenza Virus. In, Genetics of Influenza Virus. P. Palese and D. W. Kingsbury eds. Springer-Verlag, New York. p 192–254.

Mahy, B. W. J., Barret, T., Nichol, S. T., Penn, C. R., and Wolstenholme (1981). Analysis of the functions of influenza virus genome RNA segments by use of temperature-sensitive mutants of fowl plague virus. In, The Replication of Negative Strand Viruses. D. H. L Bishop and R. W. Compans eds. Elsevier North Holland Inc.

Mandler, J., Müller, K., and Scholtissek, C. (1991). Mutants and revertants of an avian influenza A virus with temperature-sensitive defects in the nucleoprotein and PB2. Virology *181*, 512–519.

Marión, R. M., Zürcher, T., de la Luna, S., and Ortín, J. (1997). Influenza virus NS1 protein interacts with viral transcription-replication complexes in vivo. J. Gen. Virol. *78*, 2447–2451.

Mark, G. E., Taylor, J. M., Broni, B., and Krug, R. M. (1979). Nuclear accumulation of influenza viral RNA transcripts and the effects of cycloheximide, actinomycin D, and α-amanitin. J. Virol. *29*, 744–752.

Markushin, S. G., and Ghendon, Y. Z. (1984). Studies of fowl plague virus temperature-sensitive mutants with defects in synthesis of virion RNA. J Gen Virol. *65*, 559–575.

Martin-Benito, J., Area, E., Ortega, J., Llorca, O., Valpuesta, J. M., Carrascosa, J. L., and Ortín, J. (2001). Three-dimensional reconstruction of a recombinant influenza virus ribonucleoprotein particle. EMBO Reports *2*, 313–317.

Masunaga, K., Mizumoto, K., Kato, H., Ishihama, A., and Toyoda, T. (1999). Molecular mapping of influenza virus RNA polymerase by site-specific antibodies. Virology *256*, 130-141.

Medcalf, L., Poole, E., Elton, D., and Digard, P. (1999). Temperature-sensitive lesions in two influenza A viruses defective for replicative transcription disrupt RNA binding by the nucleoprotein. J. Virol. *73*, 7349–7356.

Mena, I., Jambrina, E., Albo, C., Perales, B., Ortin, J., Arrese, M., Vallejo, D., and Portela, A. (1999). Mutational analysis of influenza A virus nucleoprotein, identification of mutations that affect RNA replication. J. Virol. *73*, 1186–1194.

Minor, P. D., and Dimmock, N. J. 1977. Selective inhibition of influenza virus protein synthesis by inhibitors of DNA function. Virology *78*, 393–406.

Momose, F., Handa, H., and Nagata, K. (1996). Identification of host factors that regulate the influenza virus RNA polymerase activity. Biochimie *78*, 1103–1108.

Momose, F., Basler, C. F., O'Neill, R. E., Iwamatsu, A., Palese, P., and Nagata, K. (2001). Cellular splicing factor RAF-2p48/NPI-5/BAT-1/UAP56 interacts with the influenza virus nucleoprotein and enhances viral RNA synthesis. J. Virol. *75*, 1899–1908.

Momose, F., Naito, T., Yano, K., Sugimoto, S., Morikawa, Y., and Nagata, K. (2002). Identification of Hsp90 as a stimulatory host factor involved in influenza virus RNA synthesis. J. Biol. Chem. *277*, 45306–45314.

Mowshowitz, S. L. (1981). RNA synthesis of temperature-sensitive mutants of WSN influenza virus. In, Replication of negative strand viruses (Bishop, D. W. I., Compans, R. W., eds), 317–323. Elsevier/North Holland.

Mukaigawa, J., Hatada, E., Fukuda, R., and Shimizu, K. (1991). Involvement of the influenza A virus PB2 protein in the regulation of viral gene expression. J. Gen. Virol. *72*, 2661–2670.

Muller, R., Poch, O., Delarue, M., Bishop, D. H., and Bouloy, M. (1994). Rift valley fever L segment, correction of the sequence and possible functional role of newly identified regions conserved in RNA-dependent polymerases. J. Gen. Virol. *75*, 1345–1352.

Murti, K. G., Webster, R. G., and Jones, I. M. (1988). Localization of RNA polymerases on influenza virus ribonucleoproteins by immunogold labeling. Virology *164*, 562–566.

Naffakh, N., Massin, P., Escriou, N., Crescenzo-Chaigne, B., and van der Werf, S. (2000). Genetic analysis of the compatability between polymerase proteins from human and avian strains of influenza A viruses. J. Gen. Virol. *81*, 1283–1291.

Naffakh, N., Massin, P., and van der Werf, S. (2001). The transcription/replication activity of the polymerase of influenza A viruses is not correlated with the level of proteolysis induced by the PA subunit. Virology *285*, 244–252.

Nakagawa, Y., Kimura, N., Toyoda, T., Mizumoto, K., Ishihama, A., Oda, K., and Nakagawa, S. (1995). The RNA polymerase PB2 subunit is not required for replication of the influenza virus genome but is involved in capped mRNA synthesis. J. Virol *69*, 728–733.

Nakagawa Y., Oda, K., and Nakada, S. (1996). The PB1 subunit alone can catalyze cRNA synthesis, and the PA subunit in addition to the PB1 subunit is required for viral RNA synthesis in replication of the influenza virus genome. J. Virol *70*, 6390-6394.

Neumann, G., and Hobom, G. (1995). Mutational analysis of influenza virus promoter elements in vivo. J. Gen. Virol. *76*, 1709–1717

Neumann, G., Watanabe, T., Ito, H., Watanabe, S., Goto, H., Gao, P., Hughes, M., Perez, D. R., Donis, R., Hoffmann, E., Hobom, G. and Kawaoka, Y. (1999). Generation of influenza A viruses entirely from cloned cDNAs. Proc. Natl. Acad. Sci. USA *96*, 9345–9350.

Nichol, S. T., Penn, C. R., and Mahy, B. W. J. (1981). Evidence for the involvement of influenza A (fowl plague Rostock) virus protein P2 in ApG and mRNA primed in vitro RNA synthesis. J. Gen. Virol. *57*, 407–413.

Odagiri, T., and Tashiro, M. (1997). Segment-specific noncoding sequences of the influenza virus genome RNA are involved in the specific competition between defective interfering RNA and its progenitor RNA segment at the virion assembly step. J. Virol. *71*, 2138–2145.

Olsen, D. B., Benseler, F., Cole, J. L., Stahlhut, M. W., Dempski, R. E., Darke, P. L., and Kuo, L. C. (1996). Elucidation of basic mechanistic and kinetic properties of influenza endonuclease using chemically synthesized RNAs. J. Biol. Chem. *271*,7435–7439.

Ohtsu, Y., Honda, Y., Sakata, Y., Kato, H and Toyoda, T. (2002). Fine mapping of the subunit binding sites of influenza virus RNA polymerase. Microbiol. Immunol. *46*, 167–175.

Ortega, J., Martin-Benito, J., Zurcher, T., Valpuesta, J. M., Carrascosa, J. L., and Ortin, J. (2000). Ultrastructural and functional analyses of recombinant influenza virus ribonucleoproteins suggest dimerization of nucleoprotein during virus amplification. J. Virol. *74*, 156–163.

Ortin, J. (1998). Multiple levels of post-transcriptional regulation of influenza virus gene expression. Seminars Virol. *8*, 335–342.

Park, C-J., Bae, S-H., Lee, M-K., Varani, G., and Choi, B-S. (2003). Solution structure of the influenza A virus cRNA promoter, implications for differential recognition of viral promoter structures by RNA-dependent RNA polymerase. Nucl. Acids Res. *31*, 2824–2832.

Parvin, J. D., Palese, P., Honda, A., Ishihama, A., and Krystal, M. (1989). Promoter analysis of the influenza virus polymerase. J. Virol. *63*, 5142–5152.

Peng, Q. H, Galarza, J. M, Shi, L. C., and Summers, D. F. (1996). Influenza A virus RNA-dependent RNA polymerase cleaves influenza mRNA in vitro. Virus Res. *42*, 149–158.

Penn, C. R., Blaas, D. Kuechler, E., and Mahy, B. W. (1982). Identification of the cap-binding protein of two strains of influenza A/FPV. J. Gen. Virol. *62*, 177–180.

Penn, C. R., and Mahy, B. W. J. (1984). Capped mRNAs may stimulate the influenza virus polymerase by allosteric modulation. Virus Res. *1*, 1–13.

Perales, B., Sanz-Ezquerro, J.-J., Gastaminza, P., Ortega, J., Santarén, J. F., Ortín, J., and Nieto, A. (2000). The replication activity of influenza virus polymerase is linked to the capacity of the PA subunit to induce proteolysis. J. Virol. *74*, 1307–1312.

Perales, B., de la Luna, S., Palacios, I., and Ortin, J. (1996). Mutational analysis identifies functional domains in the influenza A virus PB2 polymerase subunit. J. Virol. *70*, 1678–1686.

Perales, B., and Ortin, J. (1997). The influenza A virus PB2 polymerase subunit is required for the replication of viral RNA. J. Virol. *71*, 1381–1385.

Pérez, D. R., and Donis, R. O. (1995). A 48-amino-acid region of influenza A virus PB1 protein is sufficient for complex formation with PA. J. Virol. *69*, 6932–6939.

Perez, D. R., and Donis, R. O. (1998). The matrix 1 protein of influenza A virus inhibits the transcriptase activity of a model influenza reporter genome in vivo. Virology *249*, 52–61.

Plotch, S. J., Bouloy, M., Ulmanen, I., and Krug, R. M. (1981). A unique cap (m7GpppXm)-dependent influenza virion endonuclease cleaves capped RNAs to generate the primers that initiate viral RNA transcription. Cell *23*, 847–858.

Plotch, S. J., and Krug, R. M. 1977. Influenza Virion Transcriptase, Synthesis in vitro of large, polyadenylic acid containing complementary RNA. J. Virol. *21*, 24–34.

Poch, O., Sauvaget, I., Delarue, M., and Tordo, N (1989). Identification of four conserved motifs among the RNA-dependent polymerase encoding elements. EMBO J. *8*, 3867–3874.

Pons, M. W. 1971. Isolation of influenza virus ribonucleoprotein from infected cells. Demonstration of the presence of negative-stranded RNA in viral RNP. Virology *46*, 149–160.

Pons, M. W., Schultze, I. T., and Hirst, G. K. 1969. Isolation and characterization of the ribonucleoprotein of influenza virus. Virology *39*, 250-259.

Poole, E., Elton, D., Medcalf, L., and Digard, P. (2004). Functional domains of the influenza A virus PB2 protein, identification of NP- and PB1-binding sites. Virology *321*, 120-133.

Poon, L. M., Pritlove, D., Sharps, J., and Brownlee, G. G. (1998). The RNA polymerase of influenza virus, bound to the 5' end of virion RNA, acts in cis to polyadenylate mRNA. J. Virol. *72*, 8214–8219.

Poon, L. M., Pritlove, D., Fodor, E., and Brownlee, G. G. (1999). Direct evidence that the poly(A) tail of influenza A virus mRNA is synthesized by reiterative copying of a U track in the virion RNA template. J. Virol. *73*, 3473–3476.

Portela, A., and Digard, P. (2002). The influenza virus nucleoprotein, a multifunctional RNA-binding protein pivotal to virus replication. J. Gen. Virol. *83*, 723–734.

Pritlove, D. C., Fodor, E., Seong, B. L. and Brownlee, G. G. (1995). In vitro transcription and polymerase binding studies of the termini of influenza A virus cRNA, evidence for a cRNA panhandle. J. Gen. Virol. *76*, 2205–2213.

Pritlove, D. C., Poon, L. L. M., Fodor, E., Sharps, J., and Brownlee, G. G. (1998). Polyadenylation of influenza virus mRNA transcribed in vitro from model virion RNA templates, requirement for 5' conserved sequences. J. Virol. *72*, 1280-1286.

Pritlove D. C., Poon L. L., Devenish L. J., Leahy, M. B., and Brownlee, G. G. (1999). A hairpin loop at the 5' end of influenza a virus virion RNA is required for synthesis of poly(A)(+) mRNA in vitro. J Virol. *73*, 2109–2114.

Quiocho, F. A., Hu, G., and Gershon, P. D. (2000). Structural basis of mRNA cap recognition by proteins. Curr. Opin. Struct. Biol. *10*, 78–86.

Rao, P., Uan, W., and Krug, R. M. (2003). Crucial role of CA cleavage sites in the cap-snatching mechanism for initiating viral mRNA synthesis. EMBO J. *22*, 1188–1198.

del Rio, L., Martinez, C., Domingo, E., and Ortin, J. (1985). In vitro synthesis of full-length influenza virus complementary RNA. EMBO J. *4*, 243–247.

Robertson, J. S. (1979). 5' and 3' terminal nucleotide sequences of the RNA genome segments of influenza virus. Nucleic Acids Res. *6*, 157–163.

Robertson, J. S., Schubert, M., and Lazzarini, R. A. (1981). Polyadenylation site for influenza virus mRNA. J. Virol. *38*, 157–163.

Romanos, M. A., and Hay, A. J. (1984). Identification of the influenza virus transcriptase by affinity- labeling with pyridoxal 5'-phosphate. Virology *132*, 110-117.

Ruigrok, R. W. H., and Baudin, F. (1995). Structure of influenza virus ribonucleoprotein particles. II. Purified RNA-free influenza virus ribonucleoprotein forms structures that are indistinguishable from the intact influenza virus ribonucleoprotein particles. J. Gen. Virol. *76*, 1009–1014.

Sanz-Ezquerro J. J., Fernandez Santaren, J., Sierra, T., Aragon, T., Ortega, J., Ortin, J., Smith, G. L., and Nieto, A. (1998). The PA influenza virus polymerase subunit is a phosphorylated protein. J. Gen. Virol. *79*, 471–478.

Sanz-Ezquerro J. J., de la Luna, S., Ortin, J., and Nieto, A. (1995). Individual expression of influenza virus PA protein induces degradation of coexpressed proteins. J. Virol. *69*, 2420-2426.

Sanz-Ezquerro J. J., Zurcher, T., de la Luna, S., Ortin, J., and Nieto, A. (1996). The amino-terminal one-third of the influenza virus PA protein is responsible for the induction of proteolysis. J. Virol. *70*, 1905–1911.

Scholtissek, C., and Becht, H. 1971. Binding of ribonucleic acids to the RNP-antigen protein of influenza viruses. J. Gen. Virol. *10*, 11–16.

Scholtissek, C. (1978). The genome of influenza virus. Curr. Topics Microbiol. Immunol. *80*, 139–169.

Seong, B. L., and Brownlee, G. G. 1992a. A new method for reconstituting influenza polymerase and RNA in vitro, a study of the promoter elements for cRNA and vRNA synthesis in vitro and viral rescue in vivo. Virology *186*, 247–260

Seong, B. L., and Brownlee, G. G. 1992b. Nucleotides 9 to 11 of the influenza A virion RNA promoter are crucial for activity in vitro. J. Gen. Virol. *73*, 3115–3124.

Shapiro, G. I., Gurney, T., and Krug, R. M. (1987). Influenza virus gene expression, control mechanisms at early and late times of infection and nuclear- cytoplasmic transport of virus-specific RNAs. J. Virol. *61*, 764–773.

Shapiro, G. I. and Krug, R. M. (1988). Influenza virus RNA replication in vitro, synthesis of viral template RNAs and virion RNAs in the absence of an added primer. J. Virol. *62*, 2285–2290.

Shaw, M. W., and Lamb, R. A. (1984). A specific sub-set of host-cell mRNAs prime influenza virus mRNA synthesis. Virus Res. *1*, 455–467.

Shi, L., Galarza, J. M., and Summers, D. F. (1996). Recombinant-baculovirus-expressed PB2 subunit of the influenza A virus RNA polymerase binds cap groups as an isolated subunit. Virus Res. *42*, 1–9.

Shi, L., Summers, D. F., Peng, Q., and Galarza, J. M. (1995). Influenza A virus RNA polymerase subunit PB2 is the endonuclease which cleaves host cell mRNA and functions only as the trimeric enzyme. Virology *208*, 38–47.

Shih, S. R., and Krug, R. M. (1996). Surprising function of the three influenza viral polymerase proteins, selective protection of viral mRNAs against the cap-snatching reaction catalysed by the same polymerase proteins. Virology *226*, 430-435.

Shih, S. R., Nemeroff, M. E., and Krug, R. M. (1995). The choice of alternative 5' splice sites in influenza virus M1 mRNA is regulated by the viral polymerase complex. Proc. Natl. Acad. Sci. U. S. A. *92*, 6324–6328.

Shimizu, K., Handa, H., Nakada, S., and Nagata, K. (1994). Regulation of influenza virus RNA polymerase activity by cellular and viral factors. Nucleic Acids Res. *22*, 5047–5053.

Skehel, J. J., and Hay, A. J. (1978). Nucleotide sequences at the 5' termini of influenza virus RNAs and their Transcripts. Nucleic Acids Res. *4*, 1207–1218.

Skorko, R., Summers, D. F., and Galarza, J. M. (1991). Influenza A virus in vitro transcription, roles of NS1 and NP proteins in regulating RNA synthesis. Virology *180*, 668–677.

Smith, G. L., and Hay, A. J. (1982). Replication of the influenza virus genome. Virology *118*, 96–108.

Smith, J. S., and Nikonowicz, E. P. (1998). NMR structure and dynamics of an RNA motif common to the spliceosome branch-point helix and the RNA-binding site for phage GA coat protein. Biochemistry *37*, 13486–13498.

Takeuchi, K., Nagata, K., and Ishihama, A. (1987). In vitro synthesis of influenza viral RNA, characterization of an isolated nuclear system that supports transcription of influenza viral RNA. J. Biochem. *101*, 837–845.

Tchatalbachev, S., Flick, R., and Hobom, G. (2001). The packaging signal of influenza viral RNA molecules. RNA *7*, 979–989.

Thierry, F., and Danos, O. (1982). Use of specific single stranded DNA probes cloned in M13 to study the RNA synthesis of four temperature-sensitive mutants of HK/68 influenza virus. Nucl. Acids Res. *10*, 2925–2938.

Tiley, L. S., Hagen, M., Matthews, J. T., and Krystal, M (1994). Sequence-specific binding of the influenza virus RNA polymerase to sequences located at the 5' ends of the viral RNAs. J. Virol *68*, 5108–5116.

Toyoda, T., Hara, K., and Imamura, Y. (2003). Ser624 of the PA subunit of influenza A virus is not essential for viral growth in cells and mice, but required for the maximal viral growth. Arch. Virol. *148*, 1687–1696.

Toyoda, T., Adyshev, D. M., Kobayashi, M., Iwata, A., and Ishihama, A. (1996). Molecular assembly of the influenza virus RNA polymerase, determination of the subunit-subunit contact sites. J. Gen. Virol., *77*, 2149–2157.

Turan, K., Mibayashi, M., Sugiyama, K., Saito, S., Numajiri, A., and Nagata, K. (2004). Nuclear MxA proteins form a complex with influenza virus NP and inhibit the transcription of the engineered influenza virus genome. Nucleic Acids Res. *32*, 643–652.

Ulmanen,I., Broni, B. A., and Krug, R. M. (1981). The role of two of the influenza virus core P proteins in recognizing cap 1 structures (m7GpppNm) on RNAs and in initiating viral RNA transcription. Proc. Natl. Acad. Sci. USA *78*, 7355–7359.

Ulmanen, I., Broni, B. A., and Krug, R. M. (1983). Influenza virus temperature-sensitive cap (m7GpppNm)-dependent endonuclease. J. Virol. *45*, 489–503.

Varich, N. L., and Kaverin, N. V. (1987). Regulation of the replication of influenza virus RNA segments, partial suppression of protein synthesis restores the 'early' replication pattern. J. Gen. Virol. *68*, 2879–2887.

Wagner, R. R., and Rose, J. K. (1995). Rhabdoviridae, the viruses and their replication. In, B. N. Fields, D. M. Knipe, P. M. Howey et al. (ed.), Fields Virology. Lippincott-Raven, Philadelphia, Pa.

Watanabe, K., Handa, H., Mizumoto, K., and Nagata, K. (1996). Mechanism for inhibition of influenza virus RNA polymerase activity by matrix protein. J. Virol. *70*, 241–247.

Watanabe, T., Watanabe, S., Noda, T., Jujii, Y., and Kawaoka, Y. (2003). Exploitation of nucleic acid packaging signals to generate a novel influenza-based vector stably expressing two foreign genes. J. Virol. *77*, 10575–10583.

Weber, F., Haller, O., and Kochs, G. (2000). MxA GTPase blocks reporter gene expression of reconstituted Thogoto virus ribonucleoprotein complexes. J. Virol. *74*, 560-563.

Wolstenholme, A. J., Barret, T., Nichol, S. T., and Mahy, B. W. J. (1980). Influenza virus-specific RNA and protein syntheses in cells infected with temperature-sensitive mutants defective in the genome segment encoding nonstructural proteins. J. Virol. *35*, 1–7.

Yamanaka, K., Ishihama, A., and Nagata, K. 1990a. Reconstitution of influenza virus RNA-nucleoprotein complexes structurally resembling native viral ribonucleoprotein cores. J. Biol. Chem. *265*, 11151–11155.

Yamanaka, K., Ogasawara, N., Ueda, M., Yoshikawa, H., Ishihama, A., and Nagata, K. 1990b. Characterization of a temperature-sensitive mutant in the RNA polymerase PB2 subunit gene of influenza A/WSN/33 virus. Arch Virol. *114*, 65–73.

Young, R. J., and Content, J. 1971. 5'-terminus of influenza virus RNA. Nature, New Biology *230*, 140-142.

Zheng, H., Lee, H., Palese, P., and Garcia-Sastre, A. (1999). Influenza A virus RNA polymerase has the ability to stutter at the polyadenylation site of a viral RNA template during RNA replication. J. Virol. *73*, 5240-5243.

Zheng, H. Y., Palese, P., and Garcia-Sastre, A. (1996). Nonconserved nucleotides at the 3' and 5' ends of an influenza A virus RNA play an important role in viral RNA replication. Virology *217*, 242–251.

Zürcher, T., de la Luna, S., Sanz-Ezquerro, J. J., Nieto, A., and Ortín, J. (1996). Mutational analysis of the influenza virus A/Victoria/3/75 PA protein, Studies of interaction with PB1 protein and identification of a dominant negative mutant. J. Gen. Virol., *77*, 1745–1749.

Zvonarjev, A. Y., and Ghendon, Y. Z. (1980). Influence of membrane (M) protein on influenza virus virion transcriptase activity in vitro and its susceptibility to rimantidine. J Virol. *33*, 583–586.

CHAPTER 2

Entry and intracellular transport of influenza virus

Gary R. Whittaker and Paul Digard

Abstract

All viruses need to recognize and enter target cells in order to cause infection. For the influenza viruses, an initial interaction with cell surface carbohydrate is followed by receptor-mediated endocytosis that traffics the virion into the endosomal pathway. Exposure to low pH in maturing endosomes triggers fusion of viral and cellular membranes leading to cytoplasmic uncoating of the virion. The released viral genomic ribonucleoproteins (RNPs) are imported into the nucleus where they are transcribed and replicated. Progeny RNPs are later exported to the cytoplasm and eventually arrive at the apical plasma membrane, where final virus assembly and budding take place. As with all viruses, infection is cyclical, and several important events occur during virus assembly and release that have profound effects on the entry process. In this review we survey both virus entry and intracellular transport of the viral components, highlighting both recent discoveries and the interdependence of virus assembly and entry. Where appropriate, we also highlight differences between members of the Orthomyxovirus family.

INTRODUCTION

The orthomyxovirus family comprises the influenza A, B and C viruses, which are major pathogens of humans and other species, as well as the thogotovirus genus of tick-borne viruses and the aquatic infectious salmon anaemia virus (ISAV). These viruses share the common properties of possessing a segmented single-strand negative sense RNA genome encapisdated into ribonucleoproteins (RNPs) that are further packaged into enveloped virions (Cox et al., 2000). They also all share the strategy of entering the cell via pH-dependent endocytosis, exiting the cell via the plasma membrane, and transcribing and replicating their genomes in the nuclei of infected cells. This necessitates a great deal of temporally regulated intracellular trafficking events to support efficient virus replication. Although many aspects of these events have been understood at least superficially for some time, our knowledge of virus entry and trafficking has increased dramatically in recent years, especially with regard to the cellular mechanisms co-opted by the virus. This review aims to summarise and discuss present views of these topics.

VIRUS ENTRY
Influenza virus receptors

Approximately fifty years ago, the receptor requirements for binding of influenza virus were identified as sialic acid residues on the surface of cells (Gottschalk, 1959). It has since become established that the virus can bind via its hemagglutinin spike glycoprotein to cell surface sialic acid residues that are present on either glycoprotein or glycolipid (Skehel and

Wiley, 2000). The specific conformation of the sialic acid linkage (α-2, 3 vs. α-2, 6) has also been established to control species tropism of the virus, with avian strains preferentially recognizing α-(2, 3)-linked sialic acid, and human strains α-(2, 6)-linked sialic acid (Steinhauer and Wharton, 1998; Suzuki et al., 2000); see chapter 4 this volume. Based on its crystal structure, the viral hemagglutinin (HA) is known to bind to sialic acid substrates via surface pockets at the membrane-distal region of HA (Skehel and Wiley, 2000).

Influenza A virus is known to enter cells via the apical surface of polarized epithelial cells (Gottlieb et al., 1993). One of the major factors determining the route of entry is the availability of α-(2, 6)-linked sialic acid on the apical plasma membrane of human respiratory epithelial cells (Baum and Paulson, 1990). Recent work with primary human airway epithelial cells has shown distinct tropism for human vs. avian strains of virus, with the human viruses initially entering non-ciliated cells with high levels of α-(2, 6)-linked sialic acid, whereas avian viruses show a preference for ciliated cells which have α-(2, 3)-linked sialic acid at sufficient density to allow sub-optimal entry and infection (Matrosovich et al., 2004).

The majority of assays to analyze sialic acid binding have relied on systems that are non-permissive for virus infection (e.g. erythrocytes, liposomes, isolated molecules etc.). Because of this, it remains formally possible that the receptors used by the virus to infect host cells are different to those mediating agglutination and binding to red blood cells. Indeed some mutations in the sialic acid binding pocket of HA (e.g. Y98F) clearly abrogate binding to red cells, but do not affect virus infectivity (Martin et al., 1998). In addition, it has recently been shown that influenza can infect de-sialylated MDCK cells, albeit at a relatively low efficiency (Stray et al., 2000). These data confirm earlier reports using desialyated cells suggesting that, whereas hemagglutination is prevented, such cells retain the ability to bind virus and become infected (de Lima et al., 1995). Kinetic modeling of virus binding and endocytosis with MDCK cells has revealed two kinds of binding sites for influenza virus (Nunes-Correia et al., 1999), with the possible presence of low-affinity binding sites that correspond to non-sialic acid mediated virus-cell interactions. In other viruses families, the receptor-host cell interactions are much more complex than originally predicted (Mettenleiter, 2002; Wimmer, 1994), and so it remains an open question whether influenza virus requires additional receptors in addition to sialic acid. In addition to recognition by HA, the viral neuraminidase (NA) also recognizes sialic acid by virtue of the receptor destroying activity of NA. For successful virus infection, there is a requirement for a functional balance in sialic acid recognition between HA and NA (Wagner et al., 2002).

Influenza C virus has unique receptor requirements and binds to 9-O-acetylneuraminic acid on host cells (Rogers et al., 1986). Correspondingly, it has a different virion glycoprotein composition, with a single envelope protein (hemagglutinin-esterase-fusion, HEF) functionally replacing both the HA and NA of influenza A and B (Rosenthal et al., 1998). One possible outcome of this differential receptor interaction is that whereas influenza A virus is known to preferentially infect the apical surface of polarized epithelial cells (Gottlieb et al., 1993), influenza C virus can infect from both the apical and basolateral surface, presumably due to wider expression of the virus receptor (Schultze et al., 1996).

ISAV also possesses a dual-function haemagglutinin-esterase glycoprotein (Falk et al., 2004). The receptor recognition and destruction functions are directed against 4-O-acetylated sialic acids (Hellebo et al., 2004). Unlike influenza C virus, ISAV encodes an additional glycoprotein of unknown function (Falk et al., 2004). Although trypsin treatment has been shown to increase the infectivity of ISAV (Falk et al., 1997), there is no evidence for proteolytic cleavage of the haemagglutinin-esterase protein (Falk et al., 2004; Krossoy et al., 2001), making it possible that in ISAV, receptor recognition and fusion functions are on separate molecules.

The tick-borne orthomyxoviruses Thogoto and Dhori virus contain a single glycoprotein that mediates both receptor binding and fusion. However in this case, the envelope protein (GP) is related to the gp64 glycoprotein of baculoviruses (Morse et al., 1992; Portela et al., 1992). It is possible that reassortment or recombination occurred between an orthomyxovirus and a baculovirus in an insect host during evolution of the *Thogotoviridae*.

Internalization and endocytic sorting

Following initial interaction with its receptor, studies in MDCK and other tissue culture cells show that the virus enters cells by receptor-mediated endocytosis. Classic morphological studies were performed by electron microscopy in the early 1980s to show uptake of viruses into intracellular vesicles (Matlin et al., 1981; Patterson et al., 1979; Yoshimura et al., 1982). These vesicles were later shown to be a pre-lysosmal compartment with the low pH necessary for fusion of the virus (Yoshimura and Ohnishi, 1984).

More recently, molecular approaches have been applied to decipher some of the specific interactions occurring during the endocytosis of influenza virus (Sieczkarski and Whittaker, 2002a). As with most endocytic events, initial internalization of the virus from the cell surface requires dynamin, a GTPase important in pinching off the neck of forming endocytic vesicles (Roy et al., 2000). Although originally thought to enter solely through the clathrin-mediated pathway of endocytosis, recent studies have shown that influenza virus can also use non-clathrin-dependent pathways for productive entry and infection (Sieczkarski and Whittaker, 2002b); (Figure 1). These experiments are in accordance with early morphological studies that showed viruses entering smooth-surfaced vesicles (Matlin et al., 1981) as well as coated vesicles. Subsequently, the virus is trafficked through both early and late endosomes (Sieczkarski and Whittaker, 2003). Recent work using live cell imaging of influenza virus entry in CHO cells has revealed a multi-stage process, including actin-dependent movement in the cell periphery and rapid, dynein-directed translocation of endosome-containing viruses to the perinuclear region (Lakadamyali et al., 2003).

Based on experiments designed to inhibit ubiquitin-dependent endocytic sorting (Khor et al., 2003), there appears to be a distinct role for influenza virus sorting to late endosomes/multi-vesicular body (MVB). This is in contrast most other pH-dependent enveloped virus with less stringent pH requirements, which appear to enter and fuse from early endosomes. For influenza virus, cellular protein kinase-mediated signaling events, e.g. mediated by protein kinase C (PKC), are likely to be important during entry (Kunzelmann et al., 2000; Sieczkarski et al., 2003). In addition, inhibition of the lipid kinase phosphatidylinositol-3-OH kinase (PI3K) with wortmannin, results in a block to influenza virus entry (X. Sun and G.R. Whittaker, unpublished). Both of these kinases seem to regulate transit through the endosomal network, with inhibition resulting in accumulation of virions in late endocytic compartments (Figure 1).

Membrane fusion

In addition to receptor binding, a second major function of the influenza hemagglutinin (HA) is to mediate fusion of virus and cell membranes (Skehel and Wiley, 2000). Such a fusion process is essential to deliver the genome of any enveloped virus into the cell, and influenza HA has proven to be a paradigm of virus-cell fusion, as well as providing a mechanistic framework for other fusion events; e.g. those mediated by cellular SNARE proteins. The role of HA in influenza virus entry has been reviewed extensively elsewhere (Cross et al., 2001) and the reader is referred here for more detail.

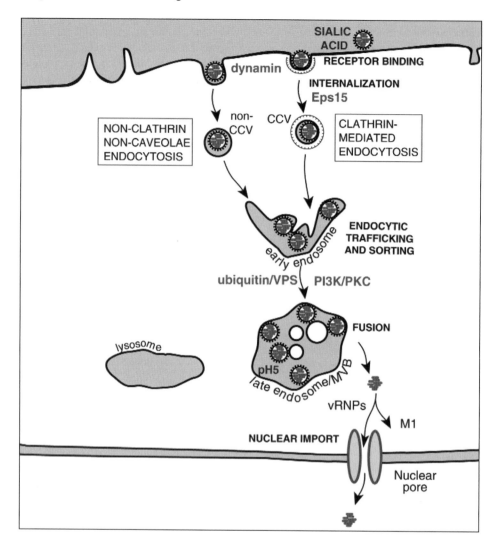

Figure 2.1 Routes of influenza virus entry into host cells. Influenza virus binds to cell surface sialic acid and is internalized by clathrin-dependent or independent endocytosis. The virus is trafficked to late endosomes, where fusion occurs in a pH-dependent manner. The virus uncoats and the genomic RNPs enter the nuclear via nuclear pores.

In the influenza virion, HA exists in a metastable state, with the hydrophobic fusion peptide hidden towards the base of the molecule. Fusion is triggered *in vivo* by exposure to the low pH environment of the endosome. The HA molecule has a high degree of alpha-helical secondary structure, and in the process of fusion it undergoes a major conformational change, accompanied by the formation of a 'coiled coil' of alpha helices (Carr and Kim, 1993), which re-orientates the fusion peptide to the outermost part of the HA molecule and initiates the fusion event. HA shares many features with fusion proteins of other viruses (e.g. retroviruses and paramyxovirus) and is the founding member of the Class I family of fusion proteins (Colman and Lawrence, 2003), which have common structural features but differing activation requirements. Whereas human immunodeficiency virus (HIV) Env and simian virus 5 (SV5) F proteins have extensive 'six-helix bundles' in their active fusogenic state, influenza HA has a much smaller six-helix bundle and has been proposed to employ a 'leash in the groove' mechanism to bring the virus envelope and endosomal membrane into close enough proximity to initiate fusion (Park et al., 2003).

Influenza virus fusion has been extensively studied by biophysical techniques. The most common method involves labeling of the virus envelope with a fluorescent probe and binding of the labelled virus to the surface of erythrocytes (or alternatively liposomes or tissue culture cells), followed by induction of fusion by artificially dropping the external pH. For probes such as octadecyl rhodamine (R18), fusion is monitored by dequenching of R18 as the probe dilutes into the target membrane. Under these conditions, fusion occurs with rapid kinetics ($T^{1/2}$ of approximately 50 seconds or less: see Figure 2a). In contrast, to this essentially *in vitro* approach, few detailed studies have been made with *in vivo* fusion assays, i.e. with fusion occurring from within endosomes. Despite limitations with R18-based assays due to non-specific probe transfer, fusion can be observed when cells are maintained at neutral pH (Stegmann et al., 1987; Stegmann et al., 1993). Under such conditions, fusion occurs with a defined lag phase of 5–15 min (during which the virus is

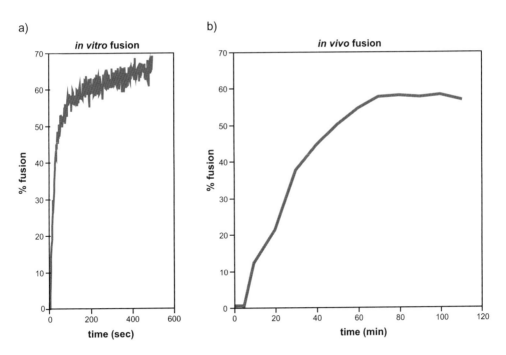

Figure 2.2 pH-dependence of influenza virus fusion. a) Fluorescence dequenching of R18-labelled influenza virus (strain WSN), with Chinese hamster ovary (CHO) cells '*in vitro*'. Virus was bound at pH 7.0 and fusion induced by dropping the pH to 5.0. b) Fluorescence dequenching of R18-labelled influenza virus (strain NIB26), with Madin-Darby canine kidney (MDCK) cells '*in vivo*'. Virus was bound at pH 7.0 and 4°C, and endocytosis induced by raising the temperature to 37°C. Data has been modified from (Stegmann et al., 1993). In both cases fusion efficiency was calculated following treatment with Triton X-100 to obtain 100% dequenching.

presumably being trafficked through the endosomal network) accompanied by a slow fusion kinetics ($T^{1/2}$ of approximately 35 min: see Figure 2b). Based on *in vitro* studies, influenza virus fusion is generally considered to have an optimum pH of approximately 5.0, with fusion effectively occurring only between pH 4.5 and pH 5.5 (Hoekstra and Klappe, 1993). Such a pH requirement fits well with *in vivo* data showing selective entry of influenza virus through late endosomes (Sieczkarski and Whittaker, 2003). This is in contrast to most other pH-dependent enveloped viruses, which can fuse in the range of pH 6.2–6.5, i.e. within early or recycling endosomes (Russell and Marsh, 2001).

Virus fusion relies on several features of HA that are initiated during virus assembly in the previous replication cycle. First, HA is cleaved immediately N-terminal to the fusion peptide. This cleavage of HA0 to HA1 and HA2 is essential for the subsequent conformational change to occur in the endosome and allow fusion peptide exposure (Steinhauer, 1999). Second, the HA trimerizes, with HA trimers functioning in a cooperative manner during fusion (Danieli et al., 1996). Third, the partitioning of HA into lipid microdomains during budding is essential for subsequent fusogenic activity (Takeda et al., 2003). In this scenario, the HA (present in a trimeric form in the virus envelope) must be organized cooperatively, possibly in a specific lipid environment, for fusion to occur. The influenza virus envelope contains high levels of cholesterol, depletion of which severely affects the ability of virions to infect cells (Sun and Whittaker, 2003; Takeda et al., 2003). Such depletion may affect virus fusion directly (Sun and Whittaker, 2003) and/or other early events in virus replication.

The fusion activity of influenza C virus resides within the single HEF envelope protein. The crystal structure of HEF shows many similarities to HA, despite limited sequence similarity (Rosenthal et al., 1998). In general, the fusion activity of influenza C virus is thought to follow a similar pattern to that of influenza A. However, the kinetics and pH threshold of fusion may be significantly different, with half-maximal fusion occurring between pH 5.6 and 6.1 for influenza C viruses, accompanied by a relatively prolonged lag phase relative to influenza A and B viruses (Formanowski et al., 1990).

For ISAV, there is also a requirement for low pH during virus entry. The virus has been shown to fuse efficiently at pH 5.6 or below, to localize to endosomes and is sensitive to lysosomotropic agents such as chloroquine and bafilomycin A (Eliassen et al., 2000).

Virus uncoating

Once fusion has occurred from the endosomal compartment, the uncoated virus is released into the cytoplasm (Bui et al., 1996; Martin and Helenius, 1991a; Martin and Helenius, 1991b). In addition to being a trigger for fusion, endosome acidification is essential for a second event during virus entry – virus uncoating. Within the endosome, H+ ions are transferred into the interior of the virion via the M2 ion channel present in the virus envelope (Pinto et al., 1992; Sugrue and Hay, 1991); see chapters 6 and 8, this volume. Acidification of the virus interior is the target for the anti-influenza drug amantadine, which blocks the M2 ion channel, thus preventing the interior of the virus from encountering low pH (Bukrinskaya et al., 1982; Hay et al., 1985; Martin and Helenius, 1991a). This prevents the release of the influenza matrix protein (M1) from the genomic ribonucleoproteins (vRNPs) during virus uncoating (Bui et al., 1996), so that the vRNPs do not enter the nucleus and replication cannot occur.

For influenza B, it has recently been shown that the virus expresses a unique ion channel (BM2), which acts similarly to the influenza A M2 protein to promote virus uncoating (Mould et al., 2003). The most striking functional difference between the A M2 and BM2 ion channel proteins lies in the resistance of BM2 to amantadine. One likely reason for this is that the pore-lining residues facing the ectodomain differ between the two proteins. Similarly, influenza C encodes its own ion channel (CM2), the activity of which is insensitive to amantadine (Hongo et al., 2004). The influenza B virus NB protein has also been suggested to act as an ion channel during virus uncoating (Fischer and Sansom, 2002). Although bacterially expressed NB protein reconstituted into planar membranes has ion channel activity (Fischer et al., 2001), such experiments are subject to potential artifacts (Lamb and Pinto, 1997). Furthermore, the ion channel activity of NB was reported to be blocked by amantadine (Fischer et al., 2001), yet amantadine does not inhibit the replication of the influenza B virus (Davies et al., 1964). It is possible that NB expression in oocytes upregulates an endogenous Cl^- conductance (Shimbo et al., 1996), so accounting for the observed experimental data for NB ion channel activity.

Cytosolic transport

Following genome uncoating, many viruses take advantage of cytoskeletal components, especially microtubules, for transport to their replication sites (Sodeik, 2000). In the case of influenza A virus, there is no evidence for association of released vRNPs with the cytoskeleton (Martin and Helenius, 1991b). However, while still in an uncoated form in the endosome, the virus is subject to cytoskeleton-mediated endosome trafficking. Therefore, inhibitors of both microtubules and the actin network inhibit virus entry (Gottlieb et al., 1993; Lakadamyali et al., 2003). In particular, the relative role of the actin cytoskeleton is highly dependent on the cell type studied, with virus entry in polarized epithelial cells being actin-dependent and non-polarized cells actin-independent (X. Sun, S.B. Sieczarski and G.R. Whittaker, unpublished). The lack of direct movement of vRNPs on microtubules may in part reflect the requirement for a relatively low pH for fusion and the residence of influenza virus in both early and late endosomes – the net result being the delivery of the uncoated virus to the vicinity of the nuclear pore solely in an endosome-mediated manner.

Influenza A virus shows one other aspect of specific cytoplasmic trafficking. Well over a decade after complete genomic sequences were available it was fortuitously discovered that in some strains of virus a small open reading frame in the +1 frame of the PB1 gene was expressed to produce an 87 residue polypeptide with pro-apoptotic functions (Chen et al., 2001). In some cells, this PB1-F2 protein is targeted to mitochondria by a positively charged amphipathic α-helix similar to signals identified in other known mitochondrial proteins (Gibbs et al., 2003).

INFLUENZA VIRUS AND THE NUCLEUS

Although early experiments showed the accumulation of substantial amounts of influenza virus antigens in the nuclei of infected cells and the pharmacology of virus infection hinted at the involvement of nuclear DNA (Barry et al., 1962; Breitenfeld and Schafer, 1957), the actual site of viral RNA synthesis remained controversial for many years. For over a decade, numerous investigations produced conflicting results (including sequential studies from the same laboratories) and it was not until the early 1980s that the question was regarded as finally answered. A carefully controlled study, utilising non-aqueous cell-fractionation (Herz et al., 1981) and *in situ* analysis of the nuclear localisation of the RNA products produced by short period pulse labelling (Jackson et al., 1982), provided independent convincing evidence that influenza virus transcription occured in the nucleus.

Nuclear import of viral components

RNPs from the infecting virus particle must enter the nucleus before viral RNA synthesis can occur (Figure 5). Similarly, to support genome replication, the protein components of the RNP must undergo nuclear import after synthesis in the cytoplasm. However NP and the three subunits of the viral polymerase are all too large to freely enter the nucleus by diffusion (Paine, 1975) and, once assembled with even the smallest RNA segment, the RNP far exceeds this limit. Therefore an active nuclear transport mechanism is necessary. Nuclear localisation signals (NLSs) have been identified in all three P proteins (Mukaigawa and Nayak, 1991; Nath and Nayak, 1990; Nieto et al., 1994) as well as in NP. The localisation signals in influenza A virus NP have been by far the most intensively studied and the interactions of this protein with the importin α-mediated host cell nuclear import machinery (Gorlich and Kutay, 1999) have proved to be unexpectedly complicated. One of the first reports of a discrete amino-acid sequence in a protein that functioned to direct its nuclear import concerned NP; amino-acids 327–345 of the protein (Figure 3a; CAS) were proposed to contain a nuclear accumulation signal, based on the behaviour of deletion mutants expressed in Xenopus oocytes (Davey et al., 1985). A decade later, yeast two-hybrid screens found that NP interacted with two members of a family of cellular polypeptides that at the time of publication were of uncertain function (O'Neill and Palese, 1995). Almost simultaneously however, these proteins were identified as the importin α components of the nuclear import machinery (Görlich et al., 1994; Moroianu et al., 1995). More recent work has shown that all non-tissue specific isoforms of human importin α can import NP into the nucleus without marked preference (M. Kohler, D. Elton; unpublished experiments cited in (Kohler et al., 2001), despite the fact that *in vitro* binding experiments suggest a range of binding affinities (Melen et al., 2003). Subsequent mutational analysis of NP did not implicate the 'oocyte NLS' as being responsible for binding importin α, but instead identified a short sequence at the N-terminus of the protein (Figure 3a: NLS I) that also functioned as a transferable NLS in mammalian cells (Wang et al., 1997). Mutation of this NLS in the context of full length NP did not prevent nuclear import, indicating the presence of other signal(s) in the polypeptide (Neumann et al., 1997; Wang et al., 1997). Consistent with this, a sequence matching a canonical bi-partite cellular NLS has been identified in influenza A and Thogoto virus NPs (Figure 3a; NLS II) and shown to be active in the absence of NLS I (Weber et al., 1998). Systematic deletion analysis of NP also suggests the presence of another potential NLS located between amino acids 320-400 (Bullido et al., 2000). No evidence has been found to suggest that the originally identified 'oocyte' NLS functions as such in mammalian cells, consistent with its failure to bind importin α in yeast-2-hybrid assays (Neumann et al., 1997; Wang et al., 1997). In fact, the sequence appears to act in opposition to the NLSs to cause cytoplasmic accumulation of NP (Digard et al., 1999; Weber et al., 1998) (Figure 3a; cytoplasmic accumulation signal [CAS]). Although NLS I in NP appears to be the strongest of the NLSs present in the polypeptide, it does not show a clear resemblance to the basic amino-acid sequences usually associated with NLS function (Dingwall and Laskey, 1991; Wang et al., 1997). Perhaps consistent with the atypical nature of this signal, NP has been shown to bind to the minor rather than major NLS-binding pockets on importin α 3 and 5 (Melen et al., 2003). NP is phosphorylated during infection and one of the sites of phosphorylation has been mapped to a serine residue in NLS I (Arrese and Portela, 1996). This has an obvious potential regulatory role for the activity of the NLS I. However, mutation of the potential phosphorylation site has a modest effect on the function of NLS I in isolation, but not in the context of full length NP (Bullido et al., 2000).

Reconstitution experiments have shown that NP is necessary and sufficient to direct nuclear import of a viral RNA segment (O'Neill et al., 1995). However, each of the three

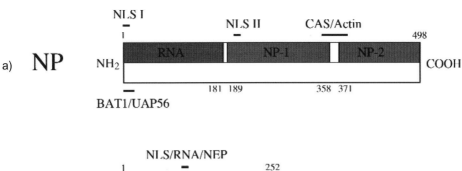

a) NP

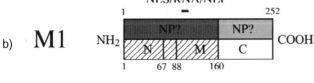

b) M1

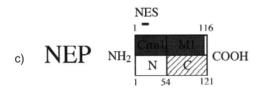

c) NEP

Figure 2.3 Functional maps and domain structures of influenza A virus NP (a) , M1 (b) and NEP (c) polypeptides. The polypeptides are represented as linear molecules with functional (top row) and structural (bottom row) domains (as labelled) indicated by shaded boxes. Numerals refer to amino-acid coordinates. Hatched boxes indicate regions for which high-resolution structural information is available. Black bars indicate the location of trafficking and other signals.

polymerase proteins also contain NLSs capable of directing their independent entry into the nucleus (Mukaigawa and Nayak, 1991; Nath and Nayak, 1990; Nieto et al., 1994). As well as being necessary for the nuclear import of newly synthesised polymerase molecules, this may provide further redundancy for the nuclear import of the infecting genomic RNAs. In addition, the import function of the multiple NLS signals present on an RNP are clearly regulated by the viral M1 protein, as an amantadine-induced failure of M1 to dissociate from RNPs during virus uncoating inhibits RNP nuclear import (Kemler et al., 1994; Martin and Helenius, 1991a; Martin and Helenius, 1991b); a block that can be reversed by acidification of the cytoplasm to cause M1-RNP dissociation (Bui et al., 1996). Nuclear import of RNPs at the start of infection is also targetted by cellular innate antiviral mechanisms. The human MxA protein has been shown to inhibit nuclear entry of Thogoto virus RNPs (Kochs and Haller, 1999b). MxA also inhibits multiplication of influenza A virus, but probably at a later, transcriptional, stage (Pavlovic et al., 1992). Nevertheless, both inhibitory mechanisms result from an interaction of MxA with the RNP, most likely with NP itself (Kochs and Haller, 1999a; Turan et al., 2004; Weber et al., 2000).

Other influenza virus proteins are also found in the nucleus during infection, including M1 itself, NS1 and NEP (Bucher et al., 1989; Greenspan et al., 1985; Krug and Etkind,

1973; Patterson et al., 1988; Young et al., 1983). Although one study concluded that M1 enters the nucleus by diffusion after virus entry (Martin and Helenius, 1991b), GST fusions of M1 and NS2 (which have molecular masses well above the NPC diffusion threshold) efficiently enter the nucleus when expressed in eukarytotic cells (G. Whittaker unpublished data). Despite the relatively small sizes of M1 and NS2 (between 28 and 14 kDa), their nuclear import is still apparently a signal-mediated process. Two NLSs have been mapped in NS1 (Greenspan et al., 1988) and one in M1 (Ye et al., 1995) (Figure 3b). The NLS in M1 has been intensively studied and several other functions attributed to this short stretch of basic amino acids. Firstly, M1 shows a sequence-independent RNA-binding activity, which has been mapped to these amino-acids (Eister et al., 1997; Wakefield and Brownlee, 1989). A subsequent study concluded that this NLS signal-mediated RNA-binding activity was in part responsible for the interaction of M1 with RNPs (Ye et al., 1999). Recently, the same region of M1 has been identified as the sequence recognised by the viral NEP polypeptide (Akarsu et al., 2003). Thus the 5 amino-acids 101-RKLKR-105 of the influenza A M1 protein are likely to be crucial for the trafficking functions of the protein (Figure 3b). The same region of the protein has also been proposed to be important for the interaction of M1 with membranes, on the basis of recent evidence suggesting that this interaction is electrostatic in nature rather than hydrophobic (Ruigrok et al., 2000). Perhaps unsurprisingly, many mutations introduced into this sequence are incompatible with virus viability (Hui et al., 2003; Liu and Ye, 2002); this subject will be returned to when the nuclear export of RNPs is discussed.

Intranuclear distribution of viral polypeptides.

In many cell types the nucleus is the largest single organelle, often with a greater volume than the remainder of the cell. Although this large compartment is not further subdivided by membranes it does contain multiple organisational domains (Dundr and Misteli, 2001). In the herpesvirus field in particular, there is much data showing that viruses take advantage of specific features of the nuclear architecture (Everett, 2001). The study of influenza viruses is less advanced in this regard, but there is evidence that they also do not use the nucleus as a large undifferentiated space. Unlike many other viruses, the orthomyxoviruses do not cause gross structural changes to the nucleus. However, there is evidence that they do interact with structural as well as enzymatic elements of the compartment. Biochemical fractionation and *in situ* autoradiography of pulse-labelled RNA indicate that a substantial fraction of RNPs are bound tightly to insoluble 'nuclear matrix'or chromatin components of the nucleus (Bui et al., 2000; Bukrinskaya et al., 1979; Jackson et al., 1982; López-Turiso et al., 1990). Immunofluorescent analysis of the intracellular distribution of NP in infected cells during the first half or so of the infectious cycle generally shows predominantly nuclear staining (Breitenfeld and Schafer, 1957; Martin and Helenius, 1991a), consistent with the known location of viral RNA synthesis. In recent years, the advent of techniques for higher resolution optical microscopy are permitting this general picture to be refined. Single optical sections taken through the nuclei of infected fibroblast cells usually show NP distributed in a speckled pattern throughout the nucleus with the exception of nucleoli (Elton et al., 2001; Ma et al., 2001; Martin and Helenius, 1991b) (Figure 4a). However, if nuclear export of RNPs is blocked by treatment of the cells with the toxin leptomycin B (LMB; see below) then apparent redistribution of NP/RNPs to the nuclear periphery is seen (Elton et al., 2001; Ma et al., 2001) (Figure 4b). More recent analysis has shown that this change in intranuclear distribution of NP is not limited to drug-treated cells but is a normal feature of infection

3.5h   9h + LMB   9h − LMB

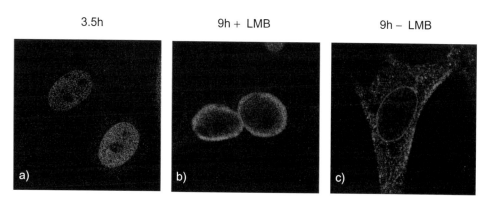

Figure 2.4 Timecourse of NP intracellular localisation. Baby hamster kidney fibroblasts were infected with influenza virus A/PR/8/34 virus and fixed and stained for NP (green) and nucleoporin 62 (red) at the indicated times post infection. Cells in (b) were treated with 11 nM leptomycin B (LMB) from 1 h post infection. This figure is also reproduced in colour in the colour section at the end of the book

best seen in cells of epithelial origin. In this type of cell, NP is distributed throughout the nucleus for the first two hours or so of infection (using reasonably high multiplicities of infection), but then redistributes to the nuclear periphery over the next few hours before the onset of nuclear export (Elton et al., 2005). The molecular mechanisms that control and mediate this specific intranuclear distribution of the protein have yet to be elucidated but it is worth noting that the viral M1 and NEP polypeptides do not show a similar redistribution in response to LMB treatment (Elton et al., 2001; Ma et al., 2001) or during normal infection of epithelial cells (Elton et al., 2005). Furthermore, NP expressed in the absence of other viral proteins shows a similar ability to localise to the periphery of the nucleus, suggesting that it is interacting with a cellular structure (Elton et al., 2005). Although the identity of the cellular structure(s) RNPs bind to is not known, the interaction must be regulated, as late in infection the RNPs dissociate and exit the nucleus. Nuclear export of RNPs is discussed in detail below, but it is worth noting that there is *in vitro* evidence that the M1 protein binds to histones (Zhirnov and Klenk, 1997) and the authors proposed the interesting hypothesis that this activity of M1 serves to displace RNPs from chromatin prior to their nuclear export. It has also been recently proposed that M1, NS1 and NEP associate with ND10 domains during infection (Sato et al., 2003). The functional consequences of the interaction of RNPs or other viral components with specific nuclear structures also remain poorly understood, but the dependence of viral mRNA synthesis on host RNA pol II transcription coupled with the recent advances being made in the understanding of how the nucleus is divided into functional domains makes this an attractive area for research. One study has suggested a correlation between RNP attachment to the insoluble nuclear matrix and the ability to synthesise negative-sense genomic RNA, with the less tightly bound RNPs biased towards mRNA synthesis (López-Turiso et al., 1990). The factors that control the balance between transcription and replication of the influenza virus genome are poorly understood (see chapter 5 this volume) and it is a fascinating possibility that the virus might use the spatial organisation of the nucleus to differentiate the two processes.

Nuclear export of RNPs

The NEP hypothesis

Progeny influenza virions are assembled at the apical plasma membrane (see below) and therefore RNPs must undergo nuclear export during the final stages of infection so they can be incorporated into the budding particles. The virus faces the same problem as at the start of infection; how to get very large RNP particles across a nuclear envelope with relatively small gated channels to the cytoplasm. There is also the additional problem that nuclear export must be carefully regulated, as the negative-sense RNPs have two opposing spatially separated functions: to serve as templates for transcription in the nucleus and to be packaged into new virus particles in the cytoplasm. Inappropriately early export of newly assembled RNPs would therefore adversely affect viral gene expression. The dramatic switch in NP localisation in infected cells during the second half of the infectious cycle (Figure 4c) makes it clear that RNP export is both a regulated and active process, and this process has been the subject of intense study in recent years. Unlike certain nuclear-replicating viruses that exit this compartment by budding through the inner nuclear membrane (*e.g.* the herpesviruses and some plant rhabdoviruses) influenza viruses appear to use a conventional cellular nuclear export pathway to deliver their RNPs to the cytoplasm. Nuclear export of influenza A virus RNPs is inhibited by the toxin leptomycin B (Elton et al., 2001; Ma et al., 2001; Watanabe et al., 2001) (Figure 4b), indicating the involvement of the cellular CRM1-dependent nuclear export pathway. The current, widely accepted hypothesis for how RNPs are selectively targetted to this particular export pathway late in infection involves the viral M1 and NEP polypeptides. In this model, the late pattern of viral gene expression leads to increased levels of M1 and NEP in the nucleus. M1 binds to RNPs and silences their transcriptional activity (Perez and Donis, 1998; Zvonarjev and Ghendon, 1980) and following this, NEP binds to the M1-RNP complexes (Yasuda et al., 1993). NEP contains a functional nuclear export signal that interacts with cellular CRM1 and this directs the transport of the entire complex to the cytoplasm (Figure 5) (Neumann et al., 2000; O'Neill et al., 1998). The M1/NEP export hypothesis therefore requires the interaction of three viral (NP, M1, NEP) and one cellular (CRM1) polypeptides. No high resolution structural information is available for NP and nor is it known where M1 binds on the protein. However, a variety of other functional regions of the protein have been mapped, including (but not limited to) the minimal RNA-binding and oligomerization domains (Figure 3a; recently reviewed by (Portela and Digard, 2002). Partial crystal structures are available for both M1 and NEP which reveal details of their domain organisation and provides a framework within which the results of genetic analyses can be interpreted (Akarsu et al., 2003; Arzt et al., 2001; Arzt et al., 2004; Harris et al., 2001; Sha and Luo, 1997) (Figure 3b, c). To a certain extent, these structural and functional analyses support a 'daisy chain' model for how the four proteins interact (Akarsu et al., 2003) (Figure 5b). The following section will consider the biochemical evidence for and against this model; subsequent sections will discuss functional data pertaining to the NEP export hypothesis.

Protein-protein interactions involved in the NEP hypothesis.

Biochemical characterisation of M1 and NEP provides ample evidence for the proposed protein-protein interactions necessary for the NEP export hypothesis. Three partial crystal structures are available for M1; one solved at acidic and two at neutral pH (Akarsu et al., 2003; Arzt et al., 2001; Arzt et al., 2004; Harris et al., 2001; Sha and Luo, 1997). All three studies show a very similar structure of two predominantly α-helical subdomains (designated

Import   Export

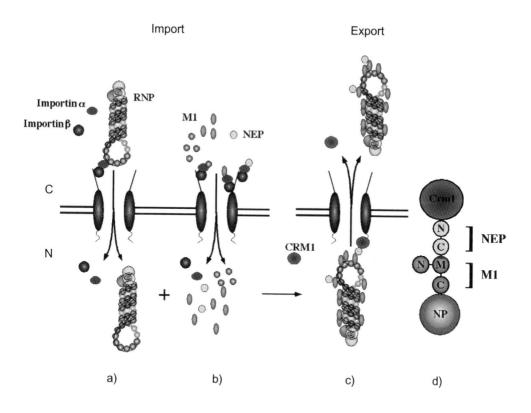

a)   b)   c)   d)

Figure 2.5 Influenza virus nucleocytoplasmic trafficking and the NEP hypothesis for RNP export. (a) RNPs and (b) monomeric NP, M1 and NEP are imported from the cytoplasm (C) to the nucleus (N) through NLS-mediated interactions with host cell importin α followed by importin ß docking with the nuclear pore complex (grey ovals). (c) An NEP-M1-RNP complex forms in the nucleus and interacts with host cell CRM1 vias the NES in NEP. (d) The daisy chain model for the NP-M1-NEP-Crm1 complex. The C-terminal domain of M1 interacts with NP, while the M domain interacts with the C-terminal domain of NEP through the basic NLS/RNA-binding sequence (indicated in red). The N-terminal domain of NEP contains the NES sequence and interacts with Crm1. This figure is also reproduced in colour in the colour section at the end of the book

N and M in Figure 3b) separated by a linker sequence. No high resolution structural information is available for the C-terminal third of the protein, but circular dichroism studies indicate that it too contains a substantial amount of α-helix (Arzt et al., 2001); it is therefore supposed to form a separate (C)-terminal domain (Figure 3b). These partial crystal structures provide no evidence for a major pH-dependent conformational shift in M1, but biophysical analysis of the putative linker peptide between the M and C domains suggests that the packing arrangement of these domains may be sensitive to pH (Okada et al., 2003).

The interaction between M1 and RNPs has been extensively characterized (Melnikov et al., 1985; Watanabe et al., 1996; Ye et al., 1989; Ye et al., 1999; Ye et al., 1987; Zvonarjev and Ghendon, 1980), although the region of M1 responsible for binding to complexes of NP and RNA is disputed. As mentioned above, the study of Ye et al (1999) concluded that M1 binds to RNPs through a combination of M1-NP and M1-RNA interactions, of which the latter is mediated by the basic NLS sequence. The N-terminal domain of M1 was identified as important for the protein-protein contact with NP (Figure 3b; NP-N?). Consistent with these data, two studies examining the inhibitory effect of M1 on virus transcription (which is thought to result from M1 binding to the RNP) concluded that the N and M domains were sufficient (Perez and Donis, 1998; Watanabe et al., 1996). However, while a further study agreed that the M1 NLS/RNA-binding sequence was important for transcription inhibition, the authors concluded that only the C-terminal domain of the M1 bound RNPs (Baudin et al., 2001) (Figure 3b; NP-C?). This discrepancy is at present unresolved. Three studies have examined the region of M1 to which NEP binds, culminating in the identification of the M1 NLS sequence as being responsible (Akarsu et al., 2003; Ward et al., 1995; Yasuda et al., 1993) (Figure 3c). Although this result suggested the 'daisy chain' model for RNP-M1-NEP interactions in which separate domains of M1 bind its partner polypeptides (Akarsu et al., 2003) (Figure 5b), it is difficult to reconcile with the finding that the same M1 sequence is responsible for binding to RNPs (Ye et al., 1999).

Since the original demonstration that NEP interacted with M1 (Yasuda et al., 1993), biophysical and structural analysis of the protein has indicated that it consists of two domains, of which the N-terminal one is relatively unstructured (Akarsu et al., 2003; Lommer and Luo, 2002). The C-terminal domain of NEP has been crystallized and has been shown to be responsible for binding M1 (Akarsu et al., 2003; Ward et al., 1995) (Figure 3c). The NEP protein of influenza B virus has similarly been shown to bind to RNPs and be incorporated into virions, but it is not clear if M1 is required for the interaction (Imai et al., 2003).

The M1/NEP export hypothesis holds that NEP acts as the adaptor molecule between the viral RNP 'cargo' and the cellular nuclear export apparatus. Consistent with this, the NEP proteins from influenza A, B and C viruses have shown to possess similar nuclear export activity to the human immunodeficiency virus type 1 rev protein (O'Neill et al., 1998; Paragas et al., 2001). Initial examination of the cellular binding partners for influenza A NEP using a yeast 2-hybrid screen identified a variety of nucleoporins (O'Neill et al., 1998). Subsequently the NEP molecules from influenza A, B and C have been shown to bind to Crm1 by a variety of techniques (Neumann et al., 2000; Paragas et al., 2001; Akarsu et al., 2003). The N-terminal domain of influenza A NEP has been shown to bind to Crm1 (Akarsu et al., 2003; Neumann et al., 2000; Paragas et al., 2001), consistent with the presence of a leucine-rich stretch of amino-acids resembling the type of nuclear export signal (NES) known to be recognised by Crm1 (Fornerod et al., 1997; O'Neill et al., 1998) (Figure 3c). However, despite NEP being able to form a ternary complex with Crm1 and cellular Ran as expected for an authentic export substrate (Akarsu et al., 2003), the leucine rich NES sequence in NEP does not appear to be required for Crm1-binding (Neumann et al., 2000). Nevertheless, it is required for the export function of the protein and for virus viability (Neumann et al., 2000; O'Neill et al., 1998).

Overall therefore, there is ample evidence supporting the existence of the necessary protein-protein interactions required for the M1/NEP export hypothesis, although in a number of places the molecular details need clarifying. There is also substantial functional data to support the hypothesis. The original definition of M1 as an essential factor for RNP export was made on the basis that export did not occur in the absence of M1 expression, either because of treatment of the cells with a drug (H7) that inhibits late viral gene expression or because of infection with defective viruses apparently lacking segment 7 (Martin and Helenius, 1991a). RNP export also failed when M1 function was inhibited by the microinjection of a specific antiserum (Martin and Helenius, 1991a). A subsequent study found that the H7-induced block to RNP export could be reversed by exogenously supplied M1, as long as the inhibitor was also removed (Bui et al., 2000). H7 is a broad spectrum kinase inhibitor (Hidaka et al., 1984), which implies that a phosphorylation reaction is also required, but it remains to be determined whether the substrate is a viral and/or a cellular protein (Bui et al., 2000). The kinetics of M1 accumulation also closely parallel those of RNP nuclear export (Martin and Helenius, 1991a).

Analyses of viruses containing mutant M1 genes are mostly consistent with the M1/NEP export hypothesis. Two independent studies of the temperature-sensitive (ts) virus A/WSN/33 ts51 identified the same single amino-acid mutation in the N-terminal domain of M1 which caused nuclear retention of M1 at the non-permissive temperature (Enami et al., 1993; Rey and Nayak, 1992). However, one study concluded that RNP nuclear export was essentially unaffected by this (Rey and Nayak, 1992), while the other showed a defect in nuclear export of vRNA (Enami et al., 1993). Subsequent examination of the same virus by a third laboratory confirmed nuclear retention of the mutant M1 protein at the non-permissive temperature and identified its aberrantly high levels of phosphorylation as the cause (Whittaker et al., 1995). In the hands of these authors, the defect in M1 trafficking caused a delay but not a complete block in RNP export and also resulted in the re-import of RNPs back into the nucleus (Whittaker et al., 1996; Whittaker et al., 1995). The advent of plasmid-based 'reverse genetics' systems (Fodor et al., 1999; Neumann et al., 1999) has greatly facilitated the creation of recombinant influenza viruses from cDNA and this technique has recently been applied to the NLS sequence in M1 (Hui et al., 2003; Liu and Ye, 2002). As referred to above, several mutations introduced into this sequence were apparently lethal to virus viability. One study noted that unrescuable mutations were associated with an apparent failure of the mutant M1 molecules to enter the nucleus (though their cytoplasmic distribution also appeared abnormal) and failure of RNPs to exit the nucleus (Liu and Ye, 2002). Although these data are consistent with the M1/NEP export hypothesis, which requires M1 to enter the nucleus before binding to the RNPs, the other study found no correlation between loss of nuclear localisation and the ability of the mutant M1 polypeptide to support virus replication (Hui et al., 2003). Furthermore, the NLS sequence could largely be replaced by an unrelated, neutral sequence (the so called 'late domain' sequence first identified as important for retroviral budding). The authors concluded that the M1 NLS function is not critical for virus replication; a finding that does not fit well with the M1/NEP export hypothesis. Also, irrespective of how much M1 is required to enter the nucleus to support RNP export, the loss of the sequence identified as the NEP-binding site on M1 (Akarsu et al., 2003) with no major reduction in virus growth (Hui et al., 2003) is utterly discrepant with the 'daisy chain' model of how an export-competent complex is formed (Figure 5b).

Further evidence for the importance of M1 to RNP trafficking comes from a series of experiments examining the effect of the cellular heat-shock response on virus infection. RNP export is inhibited in cells cultured at 41°C and this correlates with the apparent failure of M1 to bind to RNPs (Sakaguchi et al., 2003). A similar outcome was observed when cells were treated with prostaglandin A1, a hormone known to induce the synthesis of various heat-shock proteins, indicating that RNP export does not fail because of temperature lability of a virus or cellular component (Hirayama et al., 2004). Some evidence was also shown to suggest the association of cellular HSP70 with RNPs under these conditions and it was proposed that HSP70 blocks M1 binding to RNPs (Hirayama et al., 2004).

Direct functional evidence for influenza A NEP involvement in RNP nuclear export comes from microinjection of anti-NEP IgG (O'Neill et al., 1998) and, most compellingly, from the creation of a virus lacking NEP. This virus grows normally in the presence of complementing NEP and can initiate infection of normal cells, but RNP export then fails and only a single round of infection is achieved (Neumann et al., 2000). Further supporting evidence for the postulated role of NEP in RNP trafficking comes from treatment of influenza A or B infected cells with the mitogen activated protein kinase inhibitor U0126. This MAPK/ERK kinase inhibitor reduces the titre of released virus by around 90% and this correlates with much reduced nuclear export of RNPs (Ludwig et al., 2004; Pleschka et al., 2001). In contrast to the effect of the PKC inhibitor H7 however, nuclear retention of RNPs after UO126 treatment was not associated with a reduction in M1 and NEP synthesis. Instead, the drug reduced the Rev-like nuclear export activity in a non-influenza virus based assay system of influenza A NEP. The similar Rev-like functionalities of the influenza B and C NEP proteins suggests the presence of a similar RNP export mechanism in influenza A, B and C (Paragas et al., 2001). However, it should be noted that Thogoto virus does not obviously code for an NEP analogue (Wagner et al., 2001).

Overall, despite the large amount of evidence in favour of the NEP hypothesis, it is less clear that it explains all facets of RNP nuclear export. A recent paper provides evidence that caspase activity is required for efficient nuclear export of RNPs (Wurzer et al., 2003). Since caspase activity may increase the apparent size of nuclear pores (Faleiro and Lazebnik, 2000), this may represent an alternative or an additional mechanism to NEP-mediated nuclear export of RNPs (Wurzer et al., 2003). Since the various drug treatments that inhibit RNP export by targetting the Crm1, MEK and caspase pathways have relatively modest individual effects on titres of replicated virus, between 50–95% knockdown; (Elton et al., 2001; Pleschka et al., 2001; Wurzer et al., 2003), the possibility of redundancy in the export mechanism is not implausible. Consistent with the existence of alternative pathways, several studies show RNP export occuring in the presence of much reduced amounts or even the complete absence of NEP (Bui et al., 2000; Elton et al., 2001; Huang et al., 2001; Wolstenholme et al., 1980; Wurzer et al., 2003). In one study, RNP export even occurred despite a *u.v.* radiation-induced block to late viral gene expression that reduced M1 expression to undetectable levels (Mahy et al., 1977). Similarly, as discussed above, mutations affecting M1 trafficking do not always affect RNP export as the NEP hypothesis might predict (Hui et al., 2003; Rey and Nayak, 1992; Whittaker et al., 1995). Many of the apparent discrepancies between these data and the straightforward NEP-mediated hypothesis for RNP export can perhaps be resolved by postulating that only small, non-stoichiometric quantities of M1 and/or NEP are required for their trafficking functions. This is consistent with the small quantities of NEP that become packaged into virions (Richardson and Akkina, 1991). However, NP itself contains a nuclear export signal, interacts with CRM1 and shows Crm1 and caspase-dependent nucleo-cytoplasmic trafficking in the absence of other viral proteins (Akarsu et al., 2003; Elton et al., 2001; Neumann et al., 1997; Whittaker et al., 1996; Wurzer et al., 2003). No direct evidence exists to show that these intrinsic trafficking activities of NP are important in the context of virus infection and it remains to be determined if they reflect functional or perhaps evolutionary redundancy. Whether RNP export proceeds exclusively or only partially through a NEP-mediated pathway, it is not clear how to explain the observation that only RNPs containing vRNA molecules are exported from the nucleus (Shapiro et al., 1988; Tchatalbachev et al., 2001). The study of hybrid v- and cRNA molecules indicated that this export specificity resides in the 5'-arm of the conserved promoter sequence, suggesting that the viral polymerase plays a role in controlling nuclear export (Tchatalbachev et al., 2001). No evidence has been presented yet to suggest the influenza virus polymerase interacts directly with viral or cellular export mediators, but it is also possible that a polymerase bound to cRNA might negatively regulate trafficking of its RNP.

THE SECOND CYTOSOLIC PHASE OF INFECTION; VIRUS ASSEMBLY

Once in the cytoplasm, the internal virion components must be assembled into progeny virions by budding through membranes containing HA, NA and M2. All the viral envelope proteins are synthesised, matured and trafficked to the plasma membrane by the standard exocytic pathway and only an outline of this process will be considered here. In infected epithelial cells, the process of virus assembly is polarised and only occurs at the apical plasma membrane (Bachi et al., 1969; Rodriguez-Boulan and Sabatini, 1978). This polarity is thought to be determined in major part by specific sorting of the viral membrane proteins to the apical plasma membrane (PM) as each of the HA, neuraminidase (NA) and M2 ion channel are independently targetted there (Gottlieb et al., 1986; Hughey et al., 1992; Kundu et al., 1996; Rodriguez-Boulan and Pendergast, 1980; Roth et al., 1983). The HA and NA glycoproteins are also targetted to lipid rafts (Barman and Nayak, 2000; Scheiffele et al., 1997) and several studies indicate that virus particles are assembled at and incorporate lipid rafts (Scheiffele et al., 1999; Simpson-Holley et al., 2002; Takeda et al., 2003; Zhang et al., 2000). As mentioned earlier, incorporation of lipid rafts into virus particles has functional consequences for the entry process of these new virions into cells (Sun and Whittaker, 2003; Takeda et al., 2003). The sequence elements in HA and NA that target them to the apical plasma membrane and to lipid rafts lie in their trans-membrane domains. Mutational analyses show subtle differences in the sequences required for the two activities (reviewed by (Barman et al., 2001). The M2 protein, on the other hand, is not targeted to lipid rafts, possibly explaining the low abundance of this protein in the virus envelope (Zhang et al., 2000). Although the viral membrane proteins are apically targeted, it is not clear if the internal virion components show similar specific trafficking within the cytosol. Evidence for this is discussed below.

Cytosolic trafficking; a role for the cytoskeleton?

Influenza virus morphogenesis is imperfectly understood, but a widely accepted hypothesis involves the M1 protein acting as an adapter molecule between the membrane and the RNPs, through its ability to interact simultaneously with lipid membranes, the cytoplasmic tails of the viral glycoproteins and RNPs (reviewed by (Barman et al., 2001). With this model for virion assembly, directed transport of the cytoplasmic virion components is not required for polarised virus budding because the cytoplasmic tails of the viral glycoproteins that M1 binds to are only present at the apical surface of the plasma membrane. Consistent with this, an immuno-electron microscopic study found NP and M1 throughout the cytoplasm of MDCK cells with very little evidence of their concentration at the underside of the membrane (Patterson et al., 1988). In contrast, a more recent immunofluorescent analysis found that a proportion of M1 colocalised with maturing HA and therefore trafficked along the cytosolic face of exocytic pathway vesicles to the apical PM (Ali et al., 2000). Consistent with directed trafficking of M1, a further study showed its specific accumulation at the apical PM, but in this case, independent of HA targetting (Mora et al., 2002). Also consistent with directed trafficking of the internal virion components, studies employing immunofluorescence show apparent specific accumulation of NP at the PM of both infected fibroblasts and epithelial cells (Avalos et al., 1997; Mora et al., 2002; Simpson-Holley et al., 2002; Whittaker et al., 1996). Furthermore, destruction of the apical targetting signal in HA does not lead to a loss of polarity in viral budding, or a loss of NP accumulation at the apical PM (Barman et al., 2003; Mora et al., 2002). Similarly, polarised virus budding still occurs in viruses lacking an intact NA gene (Liu et al., 1995). Possibly this reflects functional redundancy between HA, NA and M2 for targetting virus assembly to the apical surface, as has been demonstrated for the process of particle formation itself (Jin et al., 1997).

Alternatively, it is possible that RNPs and/or M1 independently localise to the apical surface. There is no definitive evidence for this but there is suggestive evidence for cytoskeletal involvement during the assembly phase of virus infection. Cell fractionation and immunofluorescence experiments suggest that M1 and NP are associated with the cytoskeleton at late times in infection and both antigens redistribute in cells treated with drugs that disrupt microfilament turnover (Avalos et al., 1997; Bucher et al., 1989; Husain and Gupta, 1997; Simpson-Holley et al., 2002). NP shows the same behaviour when expressed in the absence of other viral proteins, suggesting a direct interaction with the cytoskeleton (Avalos et al., 1997). Consistent with this, purified NP has been shown to bind polymerised (F)-actin *in vitro* either as the free protein or when in complex with RNA (Digard et al., 1999). In contrast, M1 apparently requires other viral components for its association with the cytoskeleton (Avalos et al., 1997). *In vitro*, NP is an actin-bundling protein that changes the mechanical properties of actin (Digard et al., 1999; Digard et al., 2001). However, the biological significance of these virus-cytoskeleton interactions remains enigmatic. The microfilament network remains remarkably unaffected in virus infected cells (Arcangeletti et al., 1997). With the exception of cytochalasin B, treatment of cells with drugs that disrupt actin treadmilling do not inhibit virus multiplication as assessed by plaque titration of released virus (Roberts and Compans, 1998; Simpson-Holley et al., 2002). Even in the case of cytochalasin B, virus budding as such was not affected, rather that release of assembled particles from cells was inhibited because of a defect in NA maturation (Griffin and Compans, 1979). The late stages of virus multiplication are similarly insensitive to microtubule inhibitors, but there is some evidence suggesting that intermediate filaments are necessary (Arcangeletti et al., 1997). All actin poisons so far tested can however affect the morphology of the virus particles.

Polarised epithelial cells infected with some strains of influenza virus (generally those without long histories of laboratory passage) produce filamentous virions of lengths exceeding 10 μm (Mosley and Wycoff, 1946). Interference with actin turnover by the drugs cytochalasin D, latrunculin A (which inhibit actin polymerization by binding to F- and monomeric actin respectively) and jasplakinolide (an inhibitor of actin depolymerization) inhibits filament production without affecting the production of spherical particles (Roberts and Compans, 1998; Simpson-Holley et al., 2002). The reason for this specific effect is not known but has been postulated to reflect the dependency of filament budding on proper apical membrane polarity or correct organisation of lipid raft structures (Roberts and Compans, 1998; Simpson-Holley et al., 2002). Reorganisation of surface HA and sub-membranous M1 and NP around aggregates of cortical actin was seen after drug treatment of cells infected with filamentous and non-filamentous strains of virus, indicating that in either case the sites of virus assembly are linked to the actin cytoskeleton (Simpson-Holley et al., 2002). Whether this linkage occurs in all cell types and is mediated via NP-actin interactions and/or HA/NA association with lipid rafts remains unknown. Mutations have been identified in influenza A NP that lessen its affinity for F-actin (Digard et al., 1999). These mutations also affect trafficking of NP when expressed in the absence of other viral polypeptides, biasing the protein towards nuclear accumulation (Digard et al., 1999; Weber et al., 1998), leading to the suggestion that actin-binding serves to retain exported RNPs in the cytoplasm. Preliminary reverse-genetics analysis of the NP-actin binding mutants indicates that at least one of the changes is incompatible with virus viability (D. Elton, P. Digard unpublished experiments) but further work is necessary in this area. However, a defect in budding observed with a persistent strain of influenza C virus has been linked with a failure of NP to associate with the actin cytoskeleton (Hechtfischer et al., 1999).

CONCLUDING REMARKS

Our understanding of the cellular structures and processes that the influenza viruses interact with as their lifecycle tours them through the cell has made substantial advances in recent years. One of the most significant has been the elucidation of the function of the NS2/NEP protein; with hindsight it now seems strange that for a virus encoding less than a dozen gene products, and with such profound effects on global health, such basic functions remain as a '?' in virology text books for such a long time. Another, even more recent, advance (that falls out of the scope of this chapter) promises to answer a question of even longer standing; the mechanism by which the influenza viruses ensure packaging of a full complement of their segmented genomes. Despite these advances many question still remain unresolved. One area in which we still remain in a similar state of ignorance to 20 or more years ago concerns the role of phosphorylation in the virus lifecycle. Several influenza A proteins are known to be phosphoproteins, including NP, M1 and NEP, and in the case of NP, at least one modification site has been identified (Portela and Digard, 2002). Pharmacological intervention suggests that cellular kinases play roles at multiple points during virus multiplication, including as discussed above, the entry and RNP nuclear export processes. Recent work also suggests the involvement of signaling kinases in virus budding (Hui and Nayak, 2002). Nevertheless, in not one instance can we say with any degree of certainty that the virus has evolved a strategy in which phosphorylation of one of its gene products acts as a control element. One would hope that the advent of ever more precise kinase inhibitory drugs, mass-spectrometry as a means of characterizing protein modification and the increasing availability of dominant negative forms of cellular kinases will remedy this state of affairs.

Acknowledgements

Work in the authors' laboratories is supported by grants from the Medical Research Council (nos. G0300009 and G9901213 to PD), Biotechnology and Biological Research Council (no. S18874 to PD), Wellcome Trust (no. 073126 to PD), the American Lung Association and the National Institutes of Health (no. R01AI48678 to GW). We thank Dr Debra Elton for helpful comments on the manuscript.

References

Akarsu, H., Burmeister, W. P., Petosa, C., Petit, I., Muller, C. W., Ruigrok, R. W., and Baudin, F. (2003). Crystal structure of the M1 protein-binding domain of the influenza A virus nuclear export protein (NEP/NS2). EMBO J. *22*, 4646–4655.

Ali, A., Avalos, R. T., Ponimaskin, E., and Nayak, D. P. (2000). Influenza virus assembly, effect of influenza virus glycoproteins on the membrane association of M1 protein. J. Virol. *74*, 8709–8719.

Arcangeletti, M. C., Pinardi, F., Missorini, S., De Conto, F., Conti, G., Portincasa, P., Scherrer, K., and Chezzi, C. (1997). Modification of cytoskeleton and prosome networks in relation to protein synthesis in influenza A virus-infected LLC-MK2 cells. Virus Res. *51*, 19–34.

Arrese, M., and Portela, A. (1996). Serine 3 is critical for phosphorylation at the N-terminal end of the nucleoprotein of influenza virus A/Victoria/3/75. J. Virol. *70*, 3385–3391.

Arzt, S., Baudin, F., Barge, A., Timmins, P., Burmeister, W. P., and Ruigrok, R. W. (2001). Combined results from solution studies on intact influenza virus M1 protein and from a new crystal form of its N-terminal domain show that M1 is an elongated monomer. Virology. *279*, 439–446.

Arzt, S., Petit, I., Burmeister, W. P., Ruigrok, R. W., and Baudin, F. (2004). Structure of a knockout mutant of influenza virus M1 protein that has altered activities in membrane binding, oligomerisation and binding to NEP (NS2). Virus Res. *99*, 115–119.

Avalos, R. T., Yu, Z., and Nayak, D. P. (1997). Association of influenza virus NP and M1 proteins with cellular cytoskeletal elements in influenza virus-infected cells. J. Virol. *71*, 2947–2958.

Bachi, T., Gerhard, W., Lindenmann, J., and Muhlethaler, K. 1969. Morphogenesis of influenza A virus in Ehrlich ascites tumor cells as revealed by thin-sectioning and freeze-etching. J Virol. *4*, 769–776.

Barman, S., Adhikary, L., Kawaoka, Y., and Nayak, D. P. (2003). Influenza A virus hemagglutinin containing basolateral localization signal does not alter the apical budding of a recombinant influenza A virus in polarized MDCK cells. Virology. *305*, 138–152.

Barman, S., Ali, A., Hui, E. K., Adhikary, L., and Nayak, D. P. (2001). Transport of viral proteins to the apical membranes and interaction of matrix protein with glycoproteins in the assembly of influenza viruses. Virus Res. *77*, 61–69.

Barman, S., and Nayak, D. P. (2000). Analysis of the transmembrane domain of influenza virus neuraminidase, a type II transmembrane glycoprotein, for apical sorting and raft association. J Virol. *74*, 6538–6545.

Barry, R. D., Ives, D. R., and Cruickshank, J. G. 1962. Participation of deoxyribonucleic acid in the multiplication of influenza virus. Nature. *194*, 1139–1140.

Baudin, F., Petit, I., Weissenhorn, W., and Ruigrok, R. W. H. (2001). *In vitro* dissection of the membrane and RNP binding activities of influenza virus M1 protein. Virology. *281*, 102–108.

Baum, L. G., and Paulson, J. C. (1990). Sialyloligosaccharides of the respiratory epithelium in the selection of human influenza virus receptor specificity. Acta Histochem Suppl. *40*, 35–38.

Breitenfeld, P. M., and Schafer, W. 1957. The formation of fowl plague virus antigens in infected cells, as studied with fluorescent antibodies. Virology. *4*, 328–345.

Bucher, D., Popple, S., Baer, M., Mikhail, A., Gong, Y.-F., Whitaker, C., Paoletti, E., and Judd, A. (1989). M protein (M1) of influenza virus, antigenic analysis and intracellular localization with monoclonal antibodies. J. Virol. *63*, 3622–3633.

Bui, M., Whittaker, G., and Helenius, A. (1996). Effect of M1 protein and low pH on nuclear transport of influenza virus ribonucleoproteins. J. Virol. *70*, 8391–8401.

Bui, M., Wills, E., Helenius, A., and Whittaker, G. R. (2000). The role of influenza virus M1 protein in nuclear export of viral ribonucleoproteins. J. Virol. *74*, 1781–1786.

Bukrinskaya, A. G., Vorkounova, N. K., Kornilayeva, G. V., Narmanbetova, R. A., and Vorkunova, G. K. (1982). Influenza virus uncoating in infected cells and effect of rimantadine. J. Gen. Virol. *60*, 49–59.

Bukrinskaya, A. G., Vorkunova, G. K., and Vorkunova, N. K. (1979). Cytoplamsic and nuclear input virus RNPs in influenza virus-infected cells. J. Gen. Virol. *45*, 557–567.

Bullido, R., Gómez-Puertas, P., Albo, C., and Portela, A. (2000). Several protein regions contribute to determine the nuclear and cytoplasmic localization of the influenza A virus nucleoprotein. J. Gen. Virol. *81*, 135–142.

Carr, C. M., and Kim, P. S. (1993). A spring-loaded mechanism for the conformational change of influenza hemagglutinin. Cell. *73*, 823–832.

Chen, W., Calvo, P. A., Malide, D., Gibbs, J., Schubert, U., Bacik, I., Basta, S., O'Neill, R.,

Schickli, J., Palese, P., et al. (2001). A novel influenza A virus mitochondrial protein that induces cell death. Nat Med. *7*, 1306–1312.

Colman, P. M., and Lawrence, M. C. (2003). The structural biology of type I viral membrane fusion. Nat Rev Mol Cell Biol. *4*, 309–319.

Cox, N. J., Fuller, F., Kaverin, N., Klenk, H.-D., Lamb, R. A., Mahy, B. W. J., McCauley, J., Nakamura, K., Palese, P., and Webster, R. (2000). Family Orthomyxoviridae. In Virus Taxonomy, Classification and Nomenclature of Viruses. VIIth Report of the International Committee on Taxonomy of Viruses, M. H. V. v. Regenmortel, C. M. Fauquet, D. H. L. Bishop, E. B. Carstens, M. K. Estes, S. M. Lemon, J. Maniloff, M. A. Mayo, D. J. McGeoch, C. R. Pringle, and R. B. Wickner, eds. Academic Press San Diego, pp. 585–597.

Cross, K. J., Burleigh, L. M., and Steinhauer, D. A. (2001). Mechanisms of cell entry by influenza virus. Expert Rev Mol Med. *3*, 1–18.

Danieli, T., Pelletier, S. L., Henis, Y. I., and White, J. M. (1996). Membrane fusion mediated by the influenza virus hemagglutinin requires the concerted action of at least three hemagglutinin trimers. J Cell Biol. *133*, 559–569.

Davey, J., Dimmock, N. J., and Colman, A. (1985). Identification of the sequence responsible for the nuclear accumulation of the influenza virus nucleoprotein in Xenopus oocytes. Cell. *40*, 667–675.

Davies, W. L., Grunert, R. R., Haff, R. F., McGahen, J. W., Neumayer, E. M., Paulshock, M., Watts, J. C., Wood, T. R., Hermann, E. C., and Hoffmann, C. E. 1964. Antiviral Activity of 1–Adamantanamine (Amantadine). Science. *144*, 862–863.

de Lima, M. C., Ramalho-Santos, J., Flasher, D., Slepushkin, V. A., Nir, S., and Duzgunes, N. (1995). Target cell membrane sialic acid modulates both binding and fusion activity of influenza virus. Biochim Biophys Acta. *1236*, 323–330.

Digard, P., Elton, D., Bishop, K., Medcalf, E., and Pope, A. (1999). Modulation of nuclear localization of the influenza virus

nucleoprotein through interaction with actin flilaments. J. Virol. *73*, 2222–2231.

Digard, P., Elton, D., Simpson-Holley, M., and Medcalf, E. (2001). Interaction of the influenza virus nucleoprotein with F-actin. In Options for the control of influenza IV, A. Osterhhaus, N. J. Cox, and A. W. Hampson, eds. Excerpta Medica International Congress Series, Elsevier Science B. V Amsterdam, pp. 503–512.

Dingwall, C., and Laskey, R. A. (1991). Nuclear targeting sequences – a consensus? Trends Biochem Sci. *16*, 478–481.

Dundr, M., and Misteli, T. (2001). Functional architecture in the cell nucleus. Biochem J. *356*, 297–310.

Eister, C., Larsen, K., Gagnon, J., Ruigrok, R. W., and Baudin, F. (1997). Influenza virus M1 protein binds to RNA through its nuclear localization signal. J Gen Virol. 78 (Pt 7), 1589–1596.

Eliassen, T. M., Froystad, M. K., Dannevig, B. H., Jankowska, M., Brech, A., Falk, K., Romoren, K., and Gjoen, T. (2000). Initial events in infectious salmon anemia virus infection, evidence for the requirement of a low-pH step. J Virol. *74*, 218–227.

Elton, D., Amorim, M. J., Medcalf, L. and Digard P. (2005). Genome gating, polarised intranuclear trafficking of influenza virus ribonucleoproteins. Biology Letters *1*, 113–117.

Elton, D., Simpson-Holley, M., Archer, K., Medcalf, L., Hallam, R., McCauley, J., and Digard, P. (2001). Interaction of the influenza virus nucleoprotein with the cellular CRM1-mediated nuclear export pathway. J. Virol. *75*, 408–419.

Enami, K., Qiao, Y., Fukuda, R., and Enami, M. (1993). An influenza virus temperature-sensitive mutant defective in the nuclear-cytoplasmic transport of the negative-sense viral RNAs. Virology. *194*, 822–827.

Everett, R. D. (2001). DNA viruses and viral proteins that interact with PML nuclear bodies. Oncogene. *20*, 7266–7273.

Faleiro, L., and Lazebnik, Y. (2000). Caspases disrupt the nuclear-cytoplasmic barrier. J Cell Biol. *151*, 951–959.

Falk, K., Aspehaug, V., Vlasak, R., and Endresen, C. (2004). Identification and characterization of viral structural proteins of infectious salmon anemia virus. J Virol. *78*, 3063–3071.

Falk, K., Namork, E., Rimstad, E., Mjaaland, S., and Dannevig, B. H. (1997). Characterization of infectious salmon anemia virus, an orthomyxo-like virus isolated from Atlantic salmon (Salmo salar L.). J Virol. *71*, 9016–9023.

Fischer, W. B., Pitkeathly, M., and Sansom, M. S. (2001). Amantadine blocks channel activity of the transmembrane segment of the NB protein from influenza B. Eur Biophys J. *30*, 416–420.

Fischer, W. B., and Sansom, M. S. (2002). Viral ion channels, structure and function. Biochim Biophys Acta. *1561*, 27–45.

Fodor, E., Devenish, L., Engelhardt, O. G., Palese, P., Brownlee, G. G., and García-Sastre, A. (1999). Rescue of influenza A virus from recombinant DNA. J. Virol. *73*, 9679–9682.

Formanowski, F., Wharton, S. A., Calder, L. J., Hofbauer, C., and Meier-Ewert, H. (1990). Fusion characteristics of influenza C viruses. J Gen Virol. 71 (Pt 5), 1181–1188.

Fornerod, M., Ohno, M., Yoshida, M., and Mattaj, I. W. (1997). CRM1 is an export receptor for leucine-rich nuclear export signals. Cell. *90*, 1051–1060.

Gibbs, J. S., Malide, D., Hornung, F., Bennink, J. R., and Yewdell, J. W. (2003). The influenza A virus PB1-F2 protein targets the inner mitochondrial membrane via a predicted basic amphipathic helix that disrupts mitochondrial function. J Virol. *77*, 7214–7224.

Gorlich, D., and Kutay, U. (1999). Transport between the cell nucleus and the cytoplasm. Annu Rev Cell Dev Biol. *15*, 607–660.

Görlich, D., Prehn, S., Laskey, R. A., and Hartmann, E. (1994). Isolation of a protein that is essential for the first step of nuclear protein import. Cell. *79*, 767–778.

Gottlieb, T. A., Gonzalez, A., Rizzolo, L., Rindler, M. J., Adesnik, M., and Sabatini, D. D. (1986). Sorting and endocytosis of viral glycoproteins in transfected polarized epithelial cells. J Cell Biol. *102*, 1242–1255.

Gottlieb, T. A., Ivanov, I. E., Adesnik, M., and Sabatini, D. D. (1993). Actin microfilaments

play a critical role in endocytosis at the apical but not the basolateral surface of polarized epithelial cells. J Cell Biol. *120*, 695–710.

Gottschalk, A. 1959. Chemistry of Virus Receptors. In The Viruses, Biochemical Biological and Biophysical properties, F. M. Burnet, and W. M. Stanley, eds. Academic Press New York, pp. 51–61.

Greenspan, D., Krystal, M., Nakada, S., Arnheiter, H., Lyles, D. S., and Palese, P. (1985). Expression of influenza virus NS2 nonstructural protein in bacteria and localization of NS2 in infected eucaryotic cells. J. Virol. *54*, 833–843.

Greenspan, D., Palese, P., and Krystal, M. (1988). Two nuclear location signals in the influenza virus NS1 nonstructural protein. J. Virol. *62*, 3020–3026.

Griffin, J. A., and Compans, R. W. (1979). Effect of cytochalasin B on the maturation of enveloped viruses. J Exp Med. *150*, 379–391.

Harris, A., Forouhar, F., Qiu, S., Sha, B., and Luo, M. (2001). The crystal structure of the influenza matrix protein M1 at neutral pH, M1-M1 protein interfaces can rotate in the oligomeric structures of M1. Virology. *289*, 34–44.

Hay, A. J., Wolstenholme, A. J., Skehel, J. J., and Smith, M. H. (1985). The molecular basis of the specific anti-influenza action of amantadine. EMBO J. *4*, 3021–3024.

Hechtfischer, A., Meier-Ewert, H., and Marschall, M. (1999). A persistent variant of influenza C virus fails to interact with actin filaments during viral assembly. Virus Res. *61*, 113–124.

Hellebo, A., Vilas, U., Falk, K., and Vlasak, R. (2004). Infectious salmon anemia virus specifically binds to and hydrolyzes 4–O-acetylated sialic acids. J Virol. *78*, 3055–3062.

Herz, C., Stavnezer, E., Krug, R. M., and Gurney, T. (1981). Influenza virus, an RNA virus, synthesizes its messenger RNA in the nucleus of infected cells. Cell. *263*, 391–400.

Hidaka, H., Inagaki, M., Kawamoto, S., and Sasaki, Y. (1984). Isoquinolonesulfonamides, novel and potent inhibitors of cyclic nucelotide dependent protein kinase C. Biochemistry. *23*, 5036–5041.

Hirayama, E., Atagi, H., Hiraki, A., and Kim, J. (2004). Heat shock protein 70 is related to thermal inhibition of nuclear export of the influenza virus ribonucleoprotein complex. J Virol. *78*, 1263–1270.

Hoekstra, D., and Klappe, K. (1993). Fluorescencence assays to monitor fusion of enveloped viruses. In Methods in Enzymology, Vol. *220*, Membrane Fusion Techniques Part A, N. Duzgunes, ed. Academic Press San Diego, pp. 261–276.

Hongo, S., Ishii, K., Mori, K., Takashita, E., Muraki, Y., Matsuzaki, Y., and Sugawara, K. (2004). Detection of ion channel activity in Xenopus laevis oocytes expressing Influenza C virus CM2 protein. Arch Virol. *149*, 35–50.

Huang, X., Liu, T., Muller, J., Levandowski, R. A., and Ye, Z. (2001). Effect of influenza virus matrix protein and viral RNA on ribonucleoprotein formation and nuclear export. Virology. *287*, 405–416.

Hughey, P. G., Compans, R. W., Zebedee, S. L., and Lamb, R. A. (1992). Expression of the influenza A virus M2 protein is restricted to apical surfaces of polarized epithelial cells. J Virol. *66*, 5542–5552.

Hui, E. K., Barman, S., Yang, T. Y., and Nayak, D. P. (2003). Basic residues of the helix six domain of influenza virus M1 involved in nuclear translocation of M1 can be replaced by PTAP and YPDL late assembly domain motifs. J Virol. *77*, 7078–7092.

Hui, E. K., and Nayak, D. P. (2002). Role of G protein and protein kinase signalling in influenza virus budding in MDCK cells. J Gen Virol. *83*, 3055–3066.

Husain, M., and Gupta, C. M. (1997). Interactions of viral matrix protein and nucleoprotein with host cell cytoskeletal actin in influenza viral infection. Curr Sci. *73*, 40–47.

Imai, M., Watanabe, S., and Odagiri, T. (2003). Influenza B virus NS2, a nuclear export protein, directly associates with the viral ribonucleoprotein complex. Arch Virol. *148*, 1873–1884.

Jackson, D. A., Caton, A. J., McCready, S. J., and Cook, P. R. (1982). Influenza virus RNA is synthesized at fixed sites in the nucleus. Nature. 296, 366–368.

Jin, H., Leser, G. P., Zhang, J., and Lamb, R. A. (1997). Influenza virus hemagglutinin and neuraminidase cytoplasmic tails control particle shape. EMBO J. *16*, 1236–1247.

Kemler, I., Whittaker, G., and Helenius, A. (1994). Nuclear import of microinjected influenza virus ribonucleoproteins. Virology. *202*, 1028–1033.

Khor, R., McElroy, L., and Whittaker, G. R. (2003). The ubiquitin-vacuolar protein sorting pathway is selectively required for endocytosis and infection of pH-dependent enveloped viruses. Traffic. *4*, 857–868.

Kochs, G., and Haller, O. 1999a. GTP-bound human MxA protein interacts with the nucleocapsids of Thogoto virus (Orthomyxoviridae). J. Biol. Chem. *274*, 4370-4376.

Kochs, G., and Haller, O. 1999b. Interferon-induced human MxA GTPase blocks nuclear import of Thogoto virus nucleocapsids. Proc Natl Acad Sci U S A. *96*, 2082–2086.

Kohler, M., Gorlich, D., Hartmann, E., and Franke, J. (2001). Adenoviral E1A protein nuclear import is preferentially mediated by importin alpha3 in vitro. Virology. *289*, 186–191.

Krossoy, B., Devold, M., Sanders, L., Knappskog, P. M., Aspehaug, V., Falk, K., Nylund, A., Koumans, S., Endresen, C., and Biering, E. (2001). Cloning and identification of the infectious salmon anaemia virus haemagglutinin. J Gen Virol. *82*, 1757–1765.

Krug, R. M., and Etkind, R. M. 1973. Cytoplasmic and nuclear virus-specific proteins in influenza virus-infected MDCK cells. Virology. *56*, 334–348.

Kundu, A., Avalos, R. T., Sanderson, C. M., and Nayak, D. P. (1996). Transmembrane domain of influenza virus neuraminidase, a type II protein, possesses an apical sorting signal in polarized MDCK cells. J Virol. *70*, 6508–6515.

Kunzelmann, K., Beesley, A. H., King, N. J., Karupiah, G., Young, J. A., and Cook, D. I. (2000). Influenza virus inhibits amiloride-sensitive Na$^+$ channels in respiratory epithelia. Proc. Natl. Acad. Sci. USA. *97*, 10282–10287.

Lakadamyali, M., Rust, M. J., Babcock, H. P., and Zhuang, X. (2003). Visualizing infection of individual influenza viruses. Proc Natl Acad Sci U S A. *100*, 9280-9285.

Lamb, R. A., and Pinto, L. H. (1997). Do Vpu and Vpr of human immunodeficiency virus type 1 and NB of influenza B virus have ion channel activities in the virus life cycles? Virology. *229*, 1–11.

Liu, C., Eichelberger, M. C., Compans, R. W., and Air, G. M. (1995). Influenza type A virus neuraminidase does not play a role in viral entry, replication, assembly, or budding. J Virol. *69*, 1099–1106.

Liu, T., and Ye, Z. (2002). Restriction of viral replication by mutation of the influenza virus matrix protein. J Virol. *76*, 13055–13061.

Lommer, B. S., and Luo, M. (2002). Structural plasticity in influenza virus protein NS2 (NEP). J Biol Chem. *277*, 7108–7117.

López-Turiso, J. A., Martínez, C., Tanaka, T., and Ortín, J. (1990). The synthesis of influenza virus negative-strand RNA takes place in insoluble complexes present in the nuclear matrix fraction. Virus Res. *16*, 325–336.

Ludwig, S., Wolff, T., Ehrhardt, C., Wurzer, W. J., Reinhardt, J., Planz, O., and Pleschka, S. (2004). MEK inhibition impairs influenza B virus propagation without emergence of resistant variants. FEBS Lett. *561*, 37–43.

Ma, K., Roy, A. M., and Whittaker, G. R. (2001). Nuclear export of influenza virus ribonucleoproteins, identification of an export intermediate at the nuclear periphery. Virology. *282*, 215–220.

Mahy, B. W. J., Carroll, A. R., Brownson, J. M. T., and McGeoch, D. J. 1977. Block to influenza virus replication in cells preirradiated with ultraviolet light. Virology. *83*, 150–162.

Martin, J., Wharton, S. A., Lin, Y. P., Takemoto, D. K., Skehel, J. J., Wiley, D. C., and Steinhauer, D. A. (1998). Studies of the binding properties of influenza hemagglutinin receptor-site mutants. Virology. *241*, 101–111.

Martin, K., and Helenius, A. 1991a. Nuclear transport of influenza virus

ribonucleoproteins, the viral matrix protein (M1) promotes export and inhibits import. Cell. *67*, 117–130.

Martin, K., and Helenius, A. 1991b. Transport of incoming influenza virus nucleocapsids into the nucleus. J. Virol. *65*, 232–244.

Matlin, K. S., Reggio, H., Helenius, A., and Simons, K. (1981). Infectious entry pathway of influenza virus in a canine kidney cell line. J. Cell. Biol. *91*, 601–613.

Matrosovich, M. N., Matrosovich, T. Y., Gray, T., Roberts, N. A., and Klenk, H. D. (2004). Human and avian influenza viruses target different cell types in cultures of human airway epithelium. Proc Natl Acad Sci USA.

Melen, K., Fagerlund, R., Franke, J., Kohler, M., Kinnunen, L., and Julkunen, I. (2003). Importin alpha nuclear localization signal binding sites for STAT*1*, STAT*2*, and influenza A virus nucleoprotein. J Biol Chem. *278*, 28193–28200.

Melnikov, S., Mikheeva, A. V., Leneva, I. A., and Ghendon, Y. Z. (1985). Interaction of M protein and RNP of fowl plague virus in vitro. Virus Res. *3*, 353–365.

Mettenleiter, T. C. (2002). Brief overview on cellular virus receptors. Virus Res. *82*, 3–8.

Mora, R., Rodriguez-Boulan, E., Palese, P., and Garcia-Sastre, A. (2002). Apical budding of a recombinant influenza A virus expressing a hemagglutinin protein with a basolateral localization signal. J Virol. *76*, 3544–3553.

Moroianu, J., Blobel, G., and Radu, A. (1995). Previously identified protein of uncertain function is karyopherin α and together with karyopherin β docks import substrate at nuclear pore complexes. Proc. Natl. Acad. Sci. USA. *92*, 2008–2011.

Morse, M. A., Marriott, A. C., and Nuttall, P. A. (1992). The glycoprotein of Thogoto virus (a tick-borne orthomyxo-like virus) is related to the baculovirus glycoprotein gp64. Virology. *186*, 640-646.

Mosley, V. M., and Wycoff, R. W. G. 1946. Electron micrography of the virus of influenza. Nature. *157*, 263.

Mould, J. A., Paterson, R. G., Takeda, M., Ohigashi, Y., Venkataraman, P., Lamb, R. A., and Pinto, L. H. (2003). Influenza B virus BM2 protein has ion channel activity that conducts protons across membranes. Dev Cell. *5*, 175–184.

Mukaigawa, J., and Nayak, D. P. (1991). Two signals mediate nuclear localization of influenza virus (A/WSN/33) polymerase basic protein 2. J Virol. *65*, 245–253.

Nath, S. T., and Nayak, D. P. (1990). Function of two discrete regions is required for nuclear localization of polymerase basic protein 1 of A/WSN/33 influenza virus (H1 N1). Mol Cell Biol. *10*, 4139–4145.

Neumann, G., Castrucci, M. R., and Kawaoka, Y. (1997). Nuclear import and export of influenza virus nucleoprotein. J. Virol. *71*, 9690–9700.

Neumann, G., Hughes, M. T., and Kawaoka, Y. (2000). Influenza A virus NS2 protein mediates vRNP nuclear export through NES-independent interaction with hCRM1. EMBO J. *19*, 6751–6758.

Neumann, G., Watanabe, T., Ito, H., Watanabe, S., Goto, H., Gao, P., Hughes, M., Perez, D. R., Donis, R., Hoffmann, E., et al. (1999). Generation of influenza A viruses entirely from cloned cDNAs. Proc Natl Acad Sci U S A. *96*, 9345–9350.

Nieto, A., de la Luna, S., Barcena, J., Portela, A., and Ortin, J. (1994). Complex structure of the nuclear translocation signal of influenza virus polymerase PA subunit. J. Gen. Virol. *75*, 29–36.

Nunes-Correia, I., Ramalho-Santos, J., Nir, S., and Pedroso de Lima, M. C. (1999). Interactions of influenza virus with cultured cells, detailed kinetic modeling of binding and endocytosis. Biochemistry. *38*, 1095–1101.

O'Neill, R. E., Jaskunas, R., Blobel, G., Palese, P., and Moroianu, J. (1995). Nuclear import of influenza virus RNA can be mediated by viral nucleoprotein and transport factors required for protein import. J. Biol. Chem. *270*, 22701–22704.

O'Neill, R. E., and Palese, P. (1995). NPI-*1*, the human homolog of SRP-*1*, interacts with influenza virus nucleoprotein. Virology. *206*, 116–125.

O'Neill, R. E., Talon, J., and Palese, P. (1998). The influenza virus NEP (NS2 protein) mediates the nuclear export of viral ribonucleoproteins. EMBO J. *17*, 288–296.

Okada, A., Miura, T., and Takeuchi, H. (2003). Zinc- and pH-dependent conformational transition in a putative interdomain linker region of the influenza virus matrix protein M1. Biochemistry. *42*, 1978–(1984).

Paine, P. L. 1975. Nucleocytoplasmic movement of fluorescent tracers microinjected into living salivary gland cells. J Cell Biol. *66*, 652–657.

Paragas, J., Talon, J., O'Neill, R. E., Anderson, D. K., Garcia-Sastre, A., and Palese, P. (2001). Influenza B and C virus NEP (NS2) proteins possess nuclear export activities. J Virol. *75*, 7375–7383.

Park, H. E., Gruenke, J. A., and White, J. M. (2003). Leash in the groove mechanism of membrane fusion. Nat Struct Biol. *10*, 1048–1053.

Patterson, S., Gross, J., and Oxford, J. S. (1988). The intracellular distribution of influenza virus matrix protein and nucleoprotein in infected cells and their relationship to haemagglutinin in the plasma membrane. J. Gen. Virol. *69*, 1859–1872.

Patterson, S., Oxford, J. S., and Dourmashkin, R. R. (1979). Studies on the mechanism of influenza virus entry into cells. J. Gen. Virol. *43*, 223–229.

Pavlovic, J., Haller, O., and Staeheli, P. (1992). Human and mouse Mx proteins inhibit different steps of the influenza virus multiplication cycle. J Virol. *66*, 2564–2569.

Perez, D. R., and Donis, R. O. (1998). The matrix 1 protein of influenza A virus inhibits the transcriptase activity of a model influenza reporter genome in vivo. Virology. *249*, 52–61.

Pinto, L. H., Holsinger, L. J., and Lamb, R. A. (1992). Influenza virus M2 protein has ion channel activity. Cell. *69*, 517–528.

Pleschka, S., Wolff, T., Ehrhardt, C., Hobom, G., Planz, O., Rapp, U. R., and Ludwig, S. (2001). Influenza virus propagation is impaired by inhibition of the Raf/MEK/ERK signalling cascade. Nature Cell Biol.

Portela, A., and Digard, P. (2002). The influenza virus nucleoprotein, a multifunctional RNA-binding protein pivotal to virus replication. J Gen Virol. *83*, 723–734.

Portela, A., Jones, L. D., and Nuttall, P. (1992). Identification of viral structural polypeptides of Thogoto virus (a tick-borne orthomyxo-like virus) and functions associated with the glycoprotein. J. Gen. Virol. *73*, 2823–2830.

Rey, O., and Nayak, D. (1992). Nuclear retention of M1 protein in a temperature-sensitive mutant of influenza A/WSN/33 virus does not affect nuclear export of viral ribonucleoproteins. J. Virol. *66*, 5815–5824.

Richardson, J. C., and Akkina, R. K. (1991). NS$_2$ protein of influenza virus is found in purified virus and phosphorylated in infected cells. Arch. Virol. *116*, 69–80.

Roberts, P. C., and Compans, R. W. (1998). Host cell dependence of viral morphology. Proc. Natl. Acad. Sci. USA. *95*, 5746–5751.

Rodriguez-Boulan, E., and Sabatini, D. D. (1978). Asymmetric budding of viruses in epithelial monlayers, a model system for study of epithelial polarity. Proc Natl Acad Sci U S A. *75*, 5071–5075.

Rodriguez-Boulan, E., and Pendergast, M. (1980). Polarized distribution of viral envelope proteins in the plasma membrane of infected epithelial cells. Cell. *20*, 45–54.

Rogers, G. N., Herrler, G., Paulson, J. C., and Klenk, H. D. (1986). Influenza C virus uses 9–O-acetyl-N-acetylneuraminic acid as a high affinity receptor determinant for attachment to cells. J Biol Chem. *261*, 5947–5951.

Rosenthal, P. B., Zhang, X., Formanowski, F., Fitz, W., Wong, C. H., Meier-Ewert, H., Skehel, J. J., and Wiley, D. C. (1998). Structure of the haemagglutinin-esterase-fusion glycoprotein of influenza C virus. Nature. *396*, 92–96.

Roth, M. G., Compans, R. W., Giusti, L., Davis, A. R., Nayak, D. P., Gething, M. J., and Sambrook, J. (1983). Influenza virus hemagglutinin expression is polarized in cells infected with recombinant SV40 viruses carrying cloned hemagglutinin DNA. Cell. *33*, 435–443.

Roy, A.-M. M., Parker, J. S., Parrish, C. R., and Whittaker, G. R. (2000). Early stages of influenza virus entry into Mv-1 lung cells, involvement of dynamin. Virology. *267*, 17–28.

Ruigrok, R. W. H., Barge, A., Durrer, A., Brunner, J., Ma, K., and Whittaker, G. R. (2000). Membrane interaction of influenza M1 protein. Virology. *267*, 1781–1786.

Russell, D. G., and Marsh, M. (2001). Endocytosis in pathogen entry and replication. In Endocytosis, M. Marsh, ed. Oxford University Press Oxford, pp. 247–280.

Sakaguchi, A., Hirayama, E., Hiraki, A., Ishida, Y., and Kim, J. (2003). Nuclear export of influenza viral ribonucleoprotein is temperature-dependently inhibited by dissociation of viral matrix protein. Virology. *306*, 244–253.

Sato, Y., Yoshioka, K., Suzuki, C., Awashima, S., Hosaka, Y., Yewdell, J., and Kuroda, K. (2003). Localization of influenza virus proteins to nuclear dot 10 structures in influenza virus-infected cells. Virology. *310*, 29–40.

Scheiffele, P., Rietveld, A., Wilk, T., and Simons, K. (1999). Influenza viruses select ordered lipid domains during budding from the plasma membrane. J. Biol. Chem. *274*, 2038–2044.

Scheiffele, P., Roth, M. G., and Simons, K. (1997). Interaction of influenza virus haemagglutinin with sphingolipid-cholesterol membrane domains via its transmembrane domain. EMBO J. *16*, 5501–5508.

Schultze, B., Zimmer, G., and Herrler, G. (1996). Virus entry into a polarized epithelial cell line (MDCK), similarities and dissimilarities between influenza C virus and bovine coronavirus. J Gen Virol. 77 (Pt 10), 2507–2514.

Sha, B., and Luo, M. (1997). Structure of a bifunctional membrane-RNA binding protein, influenza virus matrix protein M1. Nature Struct. Biol. *4*, 239–244.

Shapiro, G. I., Gurney, T., and Krug, R. M. (1988). Influenza virus gene expression, control mechanisms at early and late times of infection and nuclear-cytoplasmic transport of virus-specific RNAs. J. Virol. *61*, 764–773.

Shimbo, K., Brassard, D. L., Lamb, R. A., and Pinto, L. H. (1996). Ion selectivity and activation of the M$_2$ ion channel of influenza virus. Biophys. J. *70*, 1335–1346.

Sieczkarski, S. B., Brown, H. A., and Whittaker, G. R. (2003). The role of protein kinase C βII in influenza virus entry via late endosomes. J. Virol. *77*, 460–469.

Sieczkarski, S. B., and Whittaker, G. R. 2002a. Dissecting virus entry via endocytosis. J. Gen. Virol. *83*, 1535–1545.

Sieczkarski, S. B., and Whittaker, G. R. 2002b. Influenza virus can enter and infect cells in the absence of clathrin-mediated endocytosis. J. Virol. *76*, 10455–10464.

Sieczkarski, S. B., and Whittaker, G. R. (2003). Differential requirements of Rab5 and Rab7 for endocytosis of influenza and other enveloped viruses. Traffic. *4*, 333–343.

Simpson-Holley, M., Ellis, D., Fisher, D., Elton, D., McCauley, J., and Digard, P. (2002). A functional link between the actin cytoskeleton and lipid rafts during budding of filamentous influenza virions. Virology. *301*, 212–225.

Skehel, J. J., and Wiley, D. C. (2000). Receptor binding and membrane fusion in virus entry, the influenza hemagglutinin. Annu Rev Biochem. *69*, 531–569.

Sodeik, B. (2000). Mechanisms of viral transport in the cytoplasm. Trends Microbiol. *8*, 465–472.

Stegmann, T., Morselt, H. W. M., Scholma, J., and Wilschut, J. (1987). Fusion of influenza virus in an intracellular acidic compartment measured by fluorescence dequenching. Biochem. Biophys. Acta. *904*, 165–170.

Stegmann, T., Schoen, P., Bron, R., Wey, J., Bartoldus, I., Ortiz, A., Nieva, J. L., and Wilschut, J. (1993). Evaluation of viral membrane fusion assays. Comparison of the octadecylrhodamine dequenching assay with the pyrene excimer assay. Biochemistry. *32*, 11330–11337.

Steinhauer, D. A. (1999). Role of hemagglutinin cleavage for the pathogenicity of influenza virus. Virology. *258*, 1–20.

Steinhauer, D. A., and Wharton, S. A. (1998). Structure and function of the haemagglutinin. In Textbook of Influenza, K. G. Nicholson, R. G. Webster, and A. J. Hay, eds. Blackwell Science Oxford.

Stray, S., Cummings, R. D., and Air, G. M. (2000). Influenza virus infection of desialylated cells. Glycobiology. *10*, 649–658.

Sugrue, R. J., and Hay, A. J. (1991). Structural characteristics of the M2 protein of influenza A viruses, evidence that it forms a tetrameric channel. Virology. *180*, 617–624.

Sun, X., and Whittaker, G. R. (2003). Role for influenza virus envelope cholesterol in virus entry and infection. J Virol. *77*, 12543–12551.

Suzuki, Y., Ito, T., Suzuki, T., Holland, R. E., Jr., Chambers, T. M., Kiso, M., Ishida, H., and Kawaoka, Y. (2000). Sialic acid species as a determinant of the host range of influenza A viruses. J Virol. *74*, 11825–11831.

Takeda, M., Leser, G. P., Russell, C. J., and Lamb, R. A. (2003). Influenza virus hemagglutinin concentrates in lipid raft microdomains for efficient viral fusion. Proc Natl Acad Sci U S A. *100*, 14610-14617.

Tchatalbachev, S., Flick, R., and Hobom, G. (2001). The packaging signal of influenza viral RNA molecules. RNA. *7*, 979–989.

Turan, K., Mibayashi, M., Sugiyama, K., Saito, S., Numajiri, A., and Nagata, K. (2004). Nuclear MxA proteins form a complex with influenza virus NP and inhibit the transcription of the engineered influenza virus genome. Nucleic Acids Res. *32*, 643–652.

Wagner, E., Engelhardt, O. G., Gruber, S., Haller, O., and Kochs, G. (2001). Rescue of recombinant Thogoto virus from cloned cDNA. J Virol. *75*, 9282–9286.

Wagner, R., Matrosovich, M., and Klenk, H. D. (2002). Functional balance between haemagglutinin and neuraminidase in influenza virus infections. Rev Med Virol *12*, 159–166.

Wakefield, L., and Brownlee, G. G. (1989). RNA-binding properties of influenza A virus matrix protein M1. Nucleic Acids Res. *17*, 8569–8580.

Wang, P., Palese, P., and O'Neill, R. E. (1997). The NPI-1/NPI-3 (karyopherin α) binding site on the influenza A virus nucleoprotein is a nonconventional nuclear localization signal. J. Virol. *71*, 1850–1856.

Ward, A. C., Castelli, L. A., Lucantoni, A. C., White, J. F., Azad, A. A., and Macreadie, I.

G. (1995). Expression and analysis of the NS$_2$ protein of influenza A virus. Arch. Virol. *140*, 2067–2073.

Watanabe, K., Handa, H., Mizumoto, K., and Nagata, K. (1996). Mechanism for inhibition of influenza virus RNA polymerase activity by matrix protein. J. Virol. *70*, 241–247.

Watanabe, K., Takizawa, N., Katoh, M., Hoshida, K., Kobayashi, N., and Nagata, K. (2001). Inhibition of nuclear export of ribonucleoprotein complexes of influenza virus by leptomycin B. Virus Res. *77*, 31–42.

Weber, F., Haller, O., and Kochs, G. (2000). MxA GTPase blocks reporter gene expression of reconstituted Thogoto virus ribonucleoprotein complexes. J Virol. *74*, 560–563.

Weber, F., Kochs, G., Gruber, S., and Haller, O. (1998). A classical bipartite nuclear localization signal on Thogoto and influenza A virus nucleoprotein. Virology. *250*, 8–18.

Whittaker, G., Bui, M., and Helenius, A. (1996). Nuclear trafficking of influenza virus ribonucleoproteins in heterokaryons. J. Virol. *70*, 2743–2756.

Whittaker, G., Kemler, I., and Helenius, A. (1995). Hyperphosphorylation of mutant influenza virus matrix (M1) protein causes its retention in the nucleus. J. Virol. *69*, 439–445.

Wimmer, E. (1994). Cellular Receptors for Animal Viruses Cold Spring Harbor Laboratory Press Cold Spring Harbor.

Wolstenholme, A. J., Barrett, T., Nichol, S. T., and Mahy, B. W. J. (1980). Influenza virus-specific RNA and protein syntheses in cells infected with temperature-sensitive mutants defective in the genome segment encoding nonstructural proteins. J. Virol. *35*, 1–7.

Wurzer, W. J., Planz, O., Ehrhardt, C., Giner, M., Silberzahn, T., Pleschka, S., and Ludwig, S. (2003). Caspase 3 activation is essential for efficient influenza virus propagation. EMBO J. *22*, 2717–2728.

Yasuda, J., Nakada, S., Kato, A., Toyoda, T., and Ishihama, A. (1993). Molecular assembly of influenza, association of the NS2 protein with virion matrix. Virology. *196*, 249–255.

Ye, Z., Baylor, N. W., and Wagner, R. R. (1989). Transcription-inhibition and RNA-binding domains of influenza A virus matrix protein mapped with anti-idiotypic antibodies and synthetic peptides. J. Virol. *63*, 3586–3594.

Ye, Z., Liu, T., Offringa, D. P., McInnes, J., and Levandowski, R. A. (1999). Association of influenza virus matrix protein with ribonucleoproteins. J. Virol. *73*, 7467–7473.

Ye, Z., Pal, R., Fox, J. W., and Wagner, R. R. (1987). Functional and antigenic domains of the matrix (M1) protein of influenza A virus. J. Virol. *61*, 239–246.

Ye, Z., Robinson, D., and Wagner, R. R. (1995). Nucleus-targeting domain of the matrix protein (M1) of influenza virus. J. Virol. *69*, 1964–1970.

Yoshimura, A., Kuroda, K., Kawasaki, K., Yamashina, S., Maeda, T., and Ohnishi, S.-I. (1982). Infectious cell entry mechanism of influenza virus. J. Virol. *43*, 284–293.

Yoshimura, A., and Ohnishi, S. (1984). Uncoating of influenza virus in endosomes. J. Virol. *51*, 497–504.

Young, J. F., Desselberger, U., Palese, P., Ferguson, B., Shatzman, A. R., and Rosenberg, M. (1983). Efficient expression of influenza virus NS1 nonstructural proteins in Escherichia coli. Proc. Natl. Acad, Sci, USA. *80*, 6105–6109.

Zhang, J., Pekosz, A., and Lamb, R. A. (2000). Influenza virus assembly and lipid raft microdomains, a role for the cytoplasmic tails of the spike gycoproteins. J. Virol. *74*, 4634–4644.

Zhirnov, O. P., and Klenk, H.-D. (1997). Histones as a target for influenza virus matrix protein M1. Virology. *235*, 302–310.

Zvonarjev, A. Y., and Ghendon, Y. Z. (1980). Influence of membrane (M) protein on influenza A virus virion transcriptase activity in vitro and its susceptibility to rimantadine. J. Virol. *33*, 583–586.

CHAPTER 3

The proton selective ion channels of influenza A and B viruses

Robert A. Lamb and Lawrence H. Pinto

Abstract

Influenza A and B viruses each encode via very different coding strategies a small oligomeric integral membrane protein, M2 of influenza A virus and BM2 of influenza B virus, and each protein is a proton selective ion channel. M2 and BM2 proteins have very different amino acid sequences but they share two key amino acid residues in the channel pore, a histidine and a tryptophan. These two residues provide a model of elegant simplicity for ionic selectivity and gating of these minimalistic ion channels. The activity of the ion channels are required during virus uncoating in the acidic environment of the endosome, to permit acidification of the interior of the virion particle which brings about protein-protein dissociation. The ion channels also equilibrate the acidic pH of the lumen of the *trans* Golgi network with the cytoplasm, during their own transport through the exocytic pathway. The influenza A virus M2 ion channel protein is the target of the antiviral drug amantadine and the drug blocks directly ion channel activity. Thus, once the atomic structures of the M2 and BM2 ion channel proteins are known, it makes the channels attractive targets for rational drug design. The M2 and BM2 ion channel proteins may be multifunctional as the available data suggests the M2 cytoplasmic tail is involved in influenza virus assembly.

INTRODUCTION

Influenza viruses are enveloped negative strand RNA viruses classified as *Orthomyxoviridae*. The genera include *Influenzavirus A* and *Influenzavirus B*. Human influenza A virus was first isolated in 1933 (Smith et al., 1933) and in 1940 Francis (Francis, 1940) detected a new type of influenza virus that had no antigenic relationship to influenza A virus which was called type B (Horsfall et al., 1940). Influenza B virus, unlike influenza A virus, does not undergo periodic major antigenic shifts in its major surface glycoproteins, hemagglutinin (HA) and neuraminidase (NA). In part this may be due to the inability of influenza B virus to undergo genetic reassortment with influenza A virus (Compans et al., 1977) coupled with the absence of an animal reservoir for influenza B virus (Kilbourne, 1987).

Influenza A and B viruses both have negative stranded RNA genomes consisting of eight RNA segments. For influenza A virus (A/PR/8/34) the genome chain length totals *13*,588 nucleotides whereas for influenza B virus the genome totals *14*,639 nucleotides. The difference in genome size is largely due to the fact that the 5' and 3' untranslated regions of influenza B virus are longer than those of influenza A virus. For both influenza A and B viruses RNA segments 1–3 encode the three polymerase proteins, PB*1*, PB2 and PA, RNA segment 4 encodes HA, RNA segment 5 encodes the nucleocapsid protein (NP), RNA segment 6 encodes NA, RNA segment 7 encodes the matrix protein (M1) and an ion channel protein (M2 for influenza A virus and BM2 for influenza B virus) and RNA segment 8 encodes two proteins NS1 and NEP that are translated from unspliced and spliced mRNAs using overlapping reading frames (Lamb and Krug, 2001).

THE INFLUENZA A VIRUS M2 ION CHANNEL PROTEIN
The influenza A virus M2 protein, identification of the M2 mRNA and demonstration that the M2 protein is an integral membrane protein

The first evidence that RNA segment 7 of influenza A virus encodes two proteins came from nucleotide sequencing studies. The gene is 1027 nucleotides in chain length and has one large open reading frame (ORF) of 237 residues encoding the M1 protein. In addition, a smaller ORF of 88 residues was identified in the +1 reading frame (Allen et al., 1980; Lamb and Lai, 1981; Winter and Fields, 1980). By using appropriate radioactive amino acid precursors and influenza virus recombinants of known genotype, the M2 protein was identified in influenza virus-infected cells and it was shown to be encoded by the second ORF (Lamb and Choppin, 1981) with the M1 protein encoded by a collinear transcript mRNA with alternative splicing of the M1 mRNA producing two mRNAs, one encoding the M2 protein while the other, designated M mRNA3, is not known to encode any protein product (Inglis and Brown, 1981; Lamb et al., 1981). The extent of splicing is controlled through the action of a cellular splicing factor SF2/ASF (Shih and Krug, 1996).

The predicted amino acid sequence of the M2 protein contains 97 residues with a single internal hydrophobic domain of 19 residues that have the potential to act as a membrane-spanning domain. In 1985, it was shown that the M2 protein is an integral membrane protein that is abundantly expressed at the plasma membrane of virus-infected cells (Lamb et al., 1985). The M2 protein spans the membrane once, and by using domain specific antibodies and specific proteolysis, it was shown that the M2 protein is orientated such that it has 23 N-terminal extracellular residues (the N-terminal methionine residues is removed (Tobler et al., 1999)) a 19 residue transmembrane (TM) domain and a 54 residue cytoplasmic tail (Lamb et al., 1985). All strains of influenza A virus encode the M2 protein and the TM domain is the most conserved region of the M2 protein sequence (Ito et al., 1991).

The presence of an N-terminal extracellular domain in the absence of a cleavable signal sequence indicates that the M2 protein is a model type III integral membrane protein. The M2 protein is inserted into the membrane of the endoplasmic reticulum co-translationally and its insertion into membranes is dependent on the signal recognition particle (Hull et al., 1988). However, assuming as expected the M2 protein is presented to the ER membrane as a loop, it is still not known how the N-terminus is 'flipped' across the membrane.

The effect of amantadine on influenza virus replication

Influenza virus particles are internalized into cells by receptor-mediated endocytosis. After exposure of virions to the low-pH-environment found in endosomal compartments, the HA undergoes a low-pH-induced conformational change which results in a protein refolding event that causes insertion of the hydrophobic fusion peptide into a target membrane and subsequent membrane fusion (Bullough et al., 1994; Chen et al., 1999; Skehel et al., 1982; reviewed in Skehel and Wiley, 2000). The consequence of this membrane fusion event is that the viral RNPs are released into the cytoplasm and then the RNPs are transported to the nucleus to begin mRNA transcription (reviewed in Lamb and Krug, 2001).

Early effects of amantadine on influenza virus replication

Amantadine (1-aminoadamantane hydrochloride) at micromolar concentrations specifically inhibits influenza A virus replication (Davies et al., 1964). Virtually all isolates of influenza A virus are predicted to be sensitive to amantadine inhibition including 1918 Spanish influenza virus (Reid et al., 2002). Notable exceptions that are insensitive to amantadine are A/WSN/*33*, A/PR8/34 and some of the recent (1997 and 2003) highly pathogenic avian H5N1 isolates. For all influenza A virus strains the amantadine block to virus replication occurs at an early stage between the steps of virus penetration and uncoating (Bukrinskaya et al., 1982; Skehel et al., 1978). It has been found that in the presence of amantadine, the M1 protein fails to dissociate from the RNPs (Bukrinskaya et al., 1982; Martin and Helenius, 1991) and the transport of the RNP complex to the nucleus does not occur (Martin and Helenius, 1991). Viral mutants resistant to amantadine contain amino acid changes that map to the M2 TM domain, suggesting that the M2 protein is the target of the drug (Hay et al., 1985).

Late effects of amantadine on influenza virus replication

In addition to the 'early' effect of amantadine, the drug has a second 'late' effect on the replication of some subtypes of avian influenza viruses which have an HA that is cleaved intracellularly and have a high pH optimum (pH 5.8–6.0) of fusion (e.g. fowl plague virus [FPV]). A large body of data indicates that addition of amantadine to virus-infected cells causes a premature conformational change in HA as the HA molecules are transported through the TGN (Ciampor et al., 1992a; Ciampor et al., 1992b; Grambas et al., 1992; Grambas and Hay, 1992; Sugrue et al., 1990a). By immunological and biochemical criteria this form of HA is indistinguishable from the low-pH-induced form of HA. The low-pH-induced conformational transition in HA is thought to occur because the intralumenal pH of the TGN compartment has been lowered below the threshold needed to induce the acid pH transition of HA (Skehel et al., 1982; Skehel and Wiley, 2000). The consequence of the irreversible conformational change in HA, which brings about the extrusion of the hydrophobic fusion peptide at the wrong time in the infectious cycle and in the wrong subcellular compartment, is that the HA oligomers aggregate and viral budding is greatly restricted (Ruigrok et al., 1991; reviewed in Wiley and Skehel, 1987).

Rationale for the influenza virus M2 protein having ion channel activity

Taken together, the data summarized above led to the hypothesis that the function of the influenza virus M2 protein is to act as an ion channel that modulates the pH of intracellular compartments (Hay, 1992; Sugrue and Hay, 1991; reviewed in Lamb et al., 1994). The M2 protein was speculated to keep the pH of the TGN lumen above the threshold for the low pH conformational change. When the M2 ion channel is blocked by amantadine, the TGN lumenal pH is predicted to be lowered, and this causes HA to undergo its low pH transition. As the same mutations in the M2 protein TM domain abolish susceptibility to both the 'early' and 'late' effects of amantadine, a rational explanation is that the M2 protein was a component of virions and would have the same function as M2 synthesized in virus-infected cells. It is generally believed that once the virion particle has been endocytosed, the ion channel activity of the virion-associated M2 protein permits the flow of ions from the endosome into the virion interior to disrupt protein-protein interactions and free the RNPs from the M1 protein.

An attractive feature of this hypothesis for virus uncoating is that the low pH of endosomes where uncoating occurs, in contrast to the neutral pH at the plasma membrane

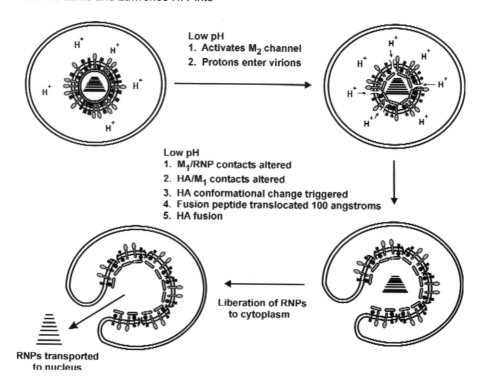

Figure 3.1 Schematic diagram of the proposed role of the M2 ion channel activity in virus entry. The M2 ion channel activity is thought to facilitate the flow of protons from the lumen of the endosome into the virion interior, bringing about dissociation of protein-protein interactions between the HA cytoplasmic tail and M1, M1 and lipid and/or RNPs and M1 from the RNPs. (Modified from Lamb and Krug, 2001

where assembly occurs, would push the equilibrium of the RNP-M1 disassembly-assembly process in favor of uncoating (reviewed in Lamb et al., 1994). In addition to the data discussed above which indicates that amantadine blocks the separation of the M1 protein from the RNAs, some data from an *in vitro* analysis of virion disruption is probably related to natural uncoating. To solubilize the M1 protein from purified virions it has become customary to use detergent and high salt concentrations. Interestingly, Zhirnov, 1990 reported that if the pH of the detergent buffer is lowered to pH 5.5, there was no need to add nonphysiological salt concentrations to achieve M1 protein solubilization; these data were confirmed (Takeda et al., 2002).

The M2 protein is a component of the influenza virion and is the target of amantadine

The finding by Hay and collaborators that influenza viruses resistant to amantadine contained mutations in the M2 protein TM domain, indicated that the M2 protein is the target of amantadine (Hay et al., 1985). As treatment of virions with amantadine prior to infection blocks influenza virus replication, it was strongly suspected that the M2 protein would be identified as a component of virions. Development of high titer monoclonal antibodies to the M2 protein (e.g. MAb 14C2) greatly facilitated the detection of M2 protein in virions and it was shown that on average 20-60 molecules of M2 are incorporated into virus particles (Zebedee and Lamb, 1988). Thus, M2 protein is greatly under-represented in virions in comparison to HA, as on the infected cell surface there are about 1×10^6 molecules of M2 and 4×10^6 molecules of HA and there are ~500 trimers per spherical virion. The final test that proved that M2 is a component of virions came when its presence in virion preparations was demonstrated by immuno-gold electromicroscopy (Hughey et al., 1992; Jackson et al., 1991).

Is M2 essential for the virus life cycle?

The amantadine-sensitive ion channel activity of influenza A virus M2 protein was discovered through understanding the two steps in the virus life cycle that are inhibited by the antiviral drug amantadine, virus uncoating in endosomes and M2 protein mediated equilibration of the intralumenal pH of the trans Golgi network. However, recently it was reported that influenza virus can undergo multiple cycles of replication without M2 ion channel activity (Watanabe et al., 2001). An M2 protein containing a deletion in the TM domain (M2-del29-31) has no detectable ion channel activity (Holsinger et al., 1994) yet a mutant virus was obtained containing this deletion. Watanabe and coworkers (2001) reported that the M2-del29-31 virus replicated as efficiently as wild type (wt) virus. To investigate this unexpected finding further, the effect of amantadine on the growth of four influenza viruses was tested, A/WSN/33; N31S M2WSN, a mutant in which asparagine residue at position 31 in the M2 TM domain was substituted with a serine residue; MUd/WSN which possesses seven RNA segments from WSN plus the RNA segment 7 derived from A/Udorn/72; and A/Udorn/72. N31S M2WSN was amantadine sensitive whereas A/WSN/33 was amantadine resistant indicating M2 residue N31 is the sole determinant of resistance of A/WSN/33 to amantadine. The growth of influenza viruses inhibited by amantadine was compared to the growth of a M2-del29-31 virus. It was found that the M2-del29-31 virus was debilitated in growth to an extent similar to that of influenza virus grown in the presence of amantadine. Furthermore, in a test of biological fitness, it was found that wt virus outgrew almost completely M2-del29-31 virus in four days after co-cultivating a *100*:1 ratio of M2-del29-31 virus to wt virus, respectively (Takeda et al., 2002). Thus, it seems reasonable to conclude that the M2 ion channel protein, which is conserved in all known strains of influenza virus, evolved its function because it contributes to the efficient replication of the virus in a single cycle.

The M2 protein is a homotetramer modified post-translationally by palmitoylation and phosphorylation

Analysis of the M2 protein on non-reducing gels and by chemical cross-linking showed that the native form of the M2 protein is a homotetramer consisting of either a pair of disulfide-linked dimers or disulfide-linked tetramers (Holsinger and Lamb, 1991; Panayotov and Schlesinger, 1992; Sugrue and Hay, 1991). Site-specific mutagenesis studies showed that the M2 protein forms intermolecular disulfide-bonds at cysteine residues 17 and 19 (Holsinger and Lamb, 1991), is post-translationally modified by palmitoylation (Sugrue et al., 1990b; Veit et al., 1991) at Cys50 (Holsinger et al., 1995), and is also post-translationally modified by phosphorylation (Sugrue and Hay, 1991) at serine residue 64 (Holsinger et al., 1995).

Apical transport of the M2 protein but exclusion from membrane rafts

Influenza virus buds from the apical surface of polarized epithelial cells and thus it might be anticipated that M2 protein would be targeted to the apical surface of virus-infected polarized cells. The signal that specifies apical sorting has been the subject of much discussion and experimentation. The very small size of the M2 protein ectodomain (23 residues) made it of great interest to determine if the M2 protein would be properly sorted in the Golgi apparatus when it was expressed in cells from cDNA in the absence of the other influenza virus proteins. It was shown both in virus-infected cells and when the M2 protein is expressed from cDNA that the M2 protein is targeted to the apical cell surface (Hughey et al., 1992) in spite of the small size of the ectodomain.

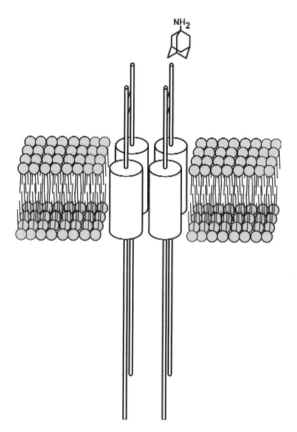

Figure 3.2 Schematic diagram of the influenza virus M2 protein ion channel in a membrane. The influenza virus M2 protein is a disulfide linked homotetramer with each chain consisting of 97 amino acid residues with 23 residues exposed extracellularly, a 19 residue TM domain and a 54 residue cytoplasmic tail. The disufide bonds can form either between the same subunit partner as shown, or once the first bond is made the second disulfide can link to another partner to form the fully disulfide-linked tetramer. The M2 protein has a pH-activated ion channel activity that conducts protons. The channel is specifically blocked by the anti-viral drug amantadine hydrochloride. The pore of the channel has been shown to be the TM domain of the M2 protein. The ion channel activity is essential for the uncoating of influenza virus in endosomal compartments in the infected cell. (Modified from Lamb and Krug, 2001).

Many cellular membranes contain dynamic liquid-ordered phase microdomains, often called lipid rafts, that are enriched in sphingolipids and cholesterol and that can preferentially incorporate or exclude certain proteins (Brown and London, 1998; Simons and Ikonen, 1997). These lipid microdomains are thought to act as platforms to allow concentration of selected proteins together on membranes, and in many cases rafts appear to be used by enveloped viruses as nucleation points to facilitate the efficient concentration and assembly of viral proteins in preparation for budding (for recent reviews, see Chazal and Gerlier, 2003; Nayak and Barman, 2002; Suomalainen, 2002).

The precise size of the lipid raft is a topic of much discussion (Munro, 2003; Pralle et al., 2000; Varma and Mayor, 1998). However, from the point of view of influenza virus budding, the direct relationship of the sphingomyelin/cholesterol enriched budding patch (the viral budozone) to a raft is a semantic argument. It is quite clear that influenza virus assembles at discrete patches at the plasma membrane which are visible by electron microscopy (Takeda et al., 2003). Viruses containing mutated HA proteins that fail to concentrate at these sites of budding, assemble and bud inefficiently (Takeda et al., 2003).

Budding from membrane rafts may provide a mechanism for controlling which viral proteins are incorporated efficiently into virions. Influenza virions normally are densely packed with glycoproteins HA (approximately 500 trimers per virion) and NA, but contain a relatively small amount of the M2 membrane protein (5 to 15 tetramers per virion) despite abundant cell surface expression of M2 protein in infected cells (Zebedee and Lamb, 1988). Whereas HA and NA proteins associate with membrane rafts, M2 protein is largely excluded from raft microdomains, hence providing an explanation for its considerable under-representation in virus as compared to HA and NA proteins (Zhang et al., 2000)

Demonstration by electrophysiological methods of the M2 protein ion channel activity

Direct evidence that the M2 protein has ion channel activity was obtained by expression of the M2 protein in oocytes of *Xenopus laevis* and measuring membrane currents (Pinto et al., 1992). The M2 ion channel activity was found to be blocked by amantadine, thereby providing direct evidence for the mechanism of action of the drug (Holsinger et al., 1994; Pinto et al., 1992; Wang et al., 1993; Wang et al., 1994b). Altered M2 proteins containing changes in the TM domain, which when found in virus lead to resistance to amantadine, exhibited ion channel activities that were not affected by the drug (Pinto et al., 1992). Specific changes in the M2 protein TM domain were found to alter the kinetics and ion selectivity of the channel, providing strong evidence that the M2 TM domain constitutes the pore of the channel (Pinto et al., 1992). This notion is supported by the observation that when a peptide corresponding to the M2 protein TM domain was incorporated into planar membranes, a proton translocation susceptible to inhibition by amantadine could be detected (Duff and Ashley, 1992).

It was important to show that the measurements that were made in oocytes of the influenza virus ion channel activity refected the activity found in mammalian cells. Thus, the M2 protein was expressed in CV-1 cells by using an SV40-M2 recombinant virus and the whole cell membrane currents were recorded. It was found that the whole cell current was activated by low pH and inhibited by the M2 ion channel-specific blocker, amantadine hydrochloride (Wang et al., 1994a). The M2 ion channel activity was also measured in mouse erythroleukemia (MEL) cells that stably expressed the M2 protein (Chizhmakov et al., 1996; Chizhmakov et al., 2003).

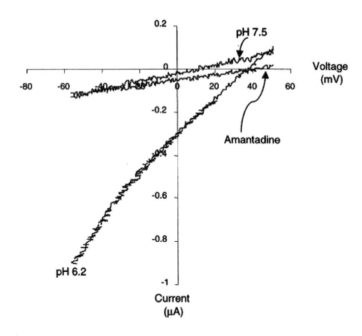

Figure 3.3 Current-voltage relationship of the M2 ion channel expressed in oocytes of *Xenop*:*s laevis* measured using a two-electrode voltage clamp apparatus. Note that the current is increased at lowered pH (pH 6.2) but is blocked by amantadine

Further evidence that the ion channel activity is intrinsic to the M2 protein came from the reconstitution of baculovirus expressed and purified M2 protein into planar bilayers (Tosteson et al., 1994) and into lipid vesicles (Lin et al., 1997; Lin and Schroeder, 2001; Schroeder et al., 1994).

Nature of the functional tetramer

As discussed above the native form of the M2 protein is minimally a homotetramer consisting of a pair of either disulfide-linked dimers or disulfide linked tetramers. In studies with chemical cross-linking reagents (Holsinger and Lamb, 1991) and when large amounts of M2 protein were purified on sucrose gradients (Schroeder et al., 1994) a small amount of a larger complex (150-180 kD) was identified that appeared to contain only M2 molecules and could represent a higher-order structure of M2 oligomers. The biologically active oligomer of the vast majority of cellular ion channel proteins spans the membrane many times (e.g. Ca^{2+} and Na^+ channels have 24 TM domains, K^+ channels have 4 subunits each of 6 TM domains). Thus, it was exceedingly important to determine, unambiguously, the subunit stoichiometry of the M2 ion channel. This was done by studying the currents of oocytes that expressed mixtures of the wt M2 protein (epitope tagged), and the mutant protein M2-V27S that is resistant to the inhibitor amantadine. The composition of mixed oligomers of the two proteins expressed at the plasma membrane of individual oocytes was quantified after antibody-capture of the cell surface expressed molecules and it was found that the subunits mixed freely. When the ratio of wt to mutant protein subunits was 0.85: 0.15, the amantadine sensitivity was reduced to 50% and for a ratio of 0.71:0.29 to 20%.

These results are consistent with the amantadine-resistant mutant subunits being dominant and the oligomeric state being a tetramer (Sakaguchi et al., 1997).

Analysis of the post-translational modifications of the influenza virus M2 protein

The M2 protein forms disulfide bonds through ectodomain cysteine residues 17 and 19 (Holsinger and Lamb, 1991), is palmitoylated through a thioether linkage on cytoplasmic tail cysteine residue 50 (Holsinger et al., 1995) and is phosphorylated on cytoplasmic tail serine residue 64 (Holsinger et al., 1995). When the membrane currents of oocytes of *Xenopus laevis* expressing wt and site-specifically altered forms of the M2 protein, that are incapable of being post-translationally modified was measured, the ion channel activity was unaffected. Thus, the available data do not indicate a functional role for post-translational modifications of the M2 protein in its ion channel activity (Holsinger et al., 1995). The use of reverse genetics systems permitted investigation of disulfide bond formation of M2 and phosphorylation of M2 in the life cycle of the virus. Ablation of the disulfide-forming potential still yielded a virus that was fully infectious in mice or ferrets, and even though the disulfide bond between M2 cysteine residue 17 is completely conserved in all strains of influenza virus it does not appear to be a functional requirement for virus infectivity (Castrucci et al., 1997). Furthermore, ablation of M2 phosphorylation is not required for the growth of the virus in mice (Thomas et al., 1998).

M2 ion channel structure

To develop a functional model of the M2 protein structure, cysteine-scanning was used to generate a series of mutants with successive substitutions in the TM segment of the protein, and the mutants were expressed in *Xenopus laevis* oocytes. The effect of the mutations on reversal potential, ion currents and amantadine resistance were measured. Fourier analysis revealed a periodicity consistent with a 4 stranded coiled-coil or helical bundle. A three-dimensional model of this structure suggests a possible mechanism for the proton selectivity of the M2 channel of influenza virus in which a proton entering the channel may protonate a neutral H37 at the imidazole δ–nitrogen facing the exterior of the virus. Deprotonation of the ε-nitrogen would allow translocation of the proton into the viral interior. Finally, tautomerization or a ring flip would complete the cycle (Pinto et al., 1997).

To understand further the structural arrangement of the M2 TM domains the cysteine substitution mutants were subjected to oxidative disulfide cross-linking to identify residues in close proximity. Oxidative treatment of M2 protein in membranes using iodine resulted in maximum cross-linking at TM domain residues *27*, 34 and 41. Oxidation of M2 protein in membranes using the catalyst Cu(II)(*1*,10-phenanthroline)*3*, resulted in cross-linking of many TM domain residues when the reaction was allowed to proceed at 37°C, suggesting that rotational movements of the TM domains in the membrane can occur. However, analysis of the kinetics of disulfide-linked dimer formation showed that TM domain residues *27*, *30*, *34*, 37 and 41 formed the most rapidly. Furthermore, when oxidation was performed at 4°C, maximum cross-linking occurred at TM domain residues *27*, *30*, *34*, 37 and 41 (Bauer et al., 1999). These positions correspond to the *a* and *d* positions of a heptad repeat that had been identified in the functional model (Pinto et al., 1997). Thus, these biochemical data are consistent with the TM domain region of the M2 tetramer forming a four-helix bundle. Analysis of the disulfide bonds which formed when oxidation of M2 protein in membranes was performed at pH 5.*2*, showed greatly reduced cross-linking at TM domain residues *40*,

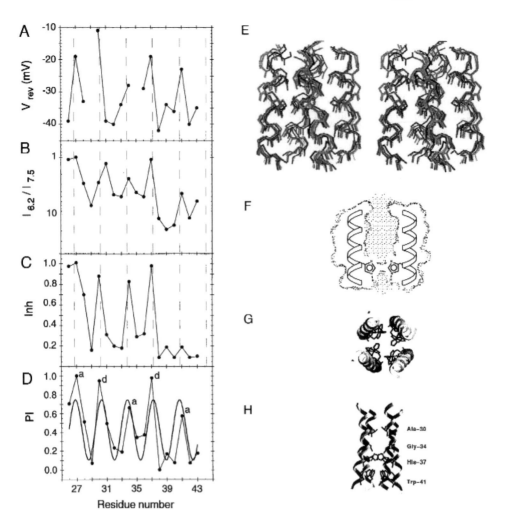

Figure 3.4 A–D. The variation in Reversal Voltage (V_{rev}) (A), Activation by low pH ($I_{6.2}/I_{7.5}$) (B), Inhibition by amantadine (Inh) (C) and Perturbation Index, a sum of the deviations of A–C from the wt values (D), as a function of position in the TM helix sequence. The smooth curve through the PI(n) data (D) represents the best fit to a cosine function. The dashed lines in (A) through (C) show peak positions expected for a 3.5-residue periodicity. Labels a and d in (D) represent heptad positions in the sequence. The perturbational index (PI(n)) is calculated from the three individual parameters. To combine these three parameters into PI(n), the individual values were normalized so their average values would be zero, then weighted according to their standard deviations. The values of the three parameters for each position were then averaged, and the combined PI(n) scale was normalized to range from 0 to 1. Because the periodic variation of Inh was limited to approximately the N-terminal half of the TM region, only residues 26–38 were included in calculating PI(n). (From Pinto et al., 1997).

E. Stereo view of the seven representative models of the M_2 proton channel using Molscript. The backbone atoms of residues 26 to 41 are shown. (Modified from Pinto et al., 1997).

F. Side view of the M_2 *proton* channel showing a slice through the structure (the other two helices are not shown for clarity). A possible mechanism for the movement of protons through the His layer of the M_2 TM channel lumen is indicated (Modified from Pinto et al., 1997).

G to H. Model of the proposed TM domain of the M2 protein showing top view as seen from the extra-cellular side (G) and a cross-section in the plane of the lipid bilayer (H). Residues identified as facing the ion-conducting aqueous pore are indicated (30, 34, 37, 41). The model of the M2 channel is taken from that calculated in Pinto et al., 1997. The figure was generated using the program GRASP. (Modified from Shuck et al., 2000).

42 and 43 than that found at pH 7.4 (Bauer et al., 1999). This pH-dependent change in cross-linking of residues towards the cytoplasmic side of the TM domain parallels the activation of the M2 ion channel at low pH and is consistent with a conformational change expected from a molecular 'gate' found in this region (see below).

To identify residues of the M2 TM domain that line the presumed central ion-conducting pore, the TM domain cysteine substitution mutants were utilized again. The accessibility of the cysteine mutants to modification by the water-soluble sulfhydryl-specific reagents methane-thiosulfonate ethylammonium (MTSEA) and MTS tetraethylammonium (MTSET) was tested. Extracellular application of MTSEA evoked decreases in the conductances measured from two mutants, M2-A30C and M2-G34C. The changes observed were not reversible on washout, indicative of a covalent modification. Inhibition by MTSEA, or by the larger reagent MTSET, was not detected for residues closer to the extracellular end of the channel than Ala-*30*, indicating the pore may be wider near the extracellular opening. To investigate the accessibility of the cysteine mutants to reagents applied intracellularly, oocytes were microinjected directly with reagents during recordings. The conductance of the M2-W41C mutant was decreased by intracellular injection of a concentrated MTSET solution. However, intracellular application of MTSET caused no change in the conductance of the M2-G34C mutant, a result in contrast to that obtained when the reagent was applied extracellularly. These data suggest that a constriction in the pore exists between residues 34 and 41 that prevents passage of the MTS reagent. These findings are consistent with the proposed role for His-37 as the selectivity filter (see below). Taken together these data confirm our model (above) that Ala-*30*, Gly-*34*, His-*37*, and Trp-41 line the channel pore (Shuck et al., 2000).

To obtain structural information about the M2 ion channel at atomic resolution, solid-state NMR studies have been performed. A peptide representing the TM domain of M2 was incorporated into hydrated dioleoylphosphatidylcholine (DOPC) and dimyristoylphosphatidyl (MMPC) lipid bilayers orientated between thin glass plates. NMR data suggested that the helices of a four helix bundle are tilted at 33–37° with respect to the bilayer (Kovacs and Cross, 1997; Kovacs et al., 2000; Wang et al., 2001). Other NMR studies suggest the M2 channel is closed by the proximity of the four indole side chains of His37 (Nishimura et al., 2002) a finding consistent with the electrophysiology and mutagenesis studies described below. When the intact M2 protein was expressed in bacteria, purified and reconstituted into DMPC/DMPG liposomes and NMR studies performed, the TM helix was found to have a tilt angle of 25° and form a symmetrical or at least pseudosymmetrical tetrameric bundle (Tian et al., 2002). Further NMR studies suggest that a region of the cytoplasmic tail (Arg45-Gly62), adjacent to the M2 TM domain forms an amphipathic α-helix that lies flat on the cytoplasmic face of the lipid bilayer (Tian et al., 2003).

Biological evidence for a role for the TM domain histidine residue

In mammalian cells infected with influenza virus ~10^6 molecules of M2 protein accumulate at the cell surface (Zebedee et al., 1985), and if these were all active channels it seemed likely that the continuous ion flux through the channel would be extremely deleterious to the cell.

Indeed, expression of the M2 protein is deleterious to several host cell systems. Infection of Sf9 cells with a recombinant baculovirus expressing the M2 protein leads to inhibition of the replication of the recombinant baculovirus (Black et al., 1993a; Black et al., 1993b; Tosteson et al., 1994), and expression of M2 protein in oocytes of *Xenopus laevis* leads to premature cell death (Giffin et al., 1995; Lamb et al., 1994) and high level expression of the M2 protein in yeast results in growth impairment (Kurtz et al., 1995). In addition, expression of M2 protein in *E. coli* causes membrane permeability changes (Guinea and Carrasco, 1994).

The conductance of the ion channel in oocytes expressing the wt M2 protein is not activated by changes in membrane voltage (Pinto et al., 1992) and there is no known ligand for which M2 protein acts as a receptor. Thus, these considerations led to an investigation of other means of activation of the wt M2 protein-associated conductance. Because the intracellular sites of action of the M2 protein are the endosome and the TGN compartments, both of which are acidic compartments, it seemed possible that M2 protein ion conductance would be regulated by changes in pH. When this notion was tested it was found that the M2 protein-associated ion channel currents increased monotonically with decreasing pH, and a change from pH 7.4 to pH 5.4 increased the inward current 7–10 fold (Pinto et al., 1992). The only amino acid in the TM domain of the M2 protein capable of being protonated within the range of pH that results in modulation of the ion channel activity is histidine residue 37 ($pK^a \sim 5.7$). When His37 was replaced with alanine an ion channel activity was observed that is not activated by low pH but is partially blocked by amantadine (Pinto et al., 1992), suggesting that protonation of the imidazole of this amino acid may be responsible for the activation by low pH. This hypothesis was tested by replacing His37 with either glutamate or glycine, (residues that throughout the pH range are negatively charged or uncharged, respectively). According to the hypothesis, mutant ion channels with these substitutions should not be activated by lowered pH within the range pH 5–7. It was found that these mutant proteins did indeed form ion channels that were not activated by lowered pH, but possessed limited amantadine sensitivity (Wang et al., 1995). These data are consistent with the model of protonization of His*37*, deprotonization and tautomerization discussed above.

Proton ion selectivity and the mechanism for proton conduction of the M2 ion channel

Under normal conditions the only ion conducted by the M2 channel is the proton (Chizhmakov et al., 1996; Lin and Schroeder, 2001; Mould et al., 2000b; Shimbo et al., 1996). To test the possibility that protons interact directly with the M2 ion channel protein when traversing its pore region, advantage was taken of some differences in the physical properties of H_2 and D_2 (Mould et al., 2000b). D_2O is 1.25 times more viscous than H_2O. If protons pass through the M2 channel as hydronium ions then only a modest decrease in current on changing from H_2O to D_2O solvent would be expected according to the ratio of their viscosities. On the other hand, if protons traverse the M2 channel via a proton wire, involving the exchange of protons between H_2O molecules occupying the channel pore, or if protons interact directly with the channel, then a larger conductance decrease upon changing from H_2O to D_2O should occur since there exists a large difference in both the mobility and the zero point energies of protons and deuterons respectively. The magnitude of the conductance ratio between protons and D^+ was found to be 1.8–2.3, values that are much higher than the ratio of the viscosities of these liquids (1.3). These results, together with the finding that His37 is essential for channel activity, are most simply explained by a conduction mechanism in which protons interact with the pore-lining His37 residue (Mould et al., 2000b) or form short-lived proton 'wires' across the His37 barrier (Smondyrev and Voth, 2002). The single channel conductance of the M2 channel is very low (Lin and

Schroeder, 2001; Mould et al., 2000b) (10^{-14} to 10^{-15} Siemens) and it is not likely that long lived 'wires' form, as this mechanism would result in a high proton conductance

Activation of the M2 ion channel is controlled by the Trp41 residue

Cells expressing the M2 protein develop large, inward H^+ currents when they are bathed in media of low pH, but, surprisingly, do not display outward H^+ currents when the pH of the medium is returned to a high value, even if the cells have been acidified considerably (Mould et al., 2000a; Pinto et al., 1992). The most likely explanation for this finding is that the pore of the M2 channel is closed by high extracellular (extravirion) pH (pH_{out}). The property of opening an ion channel in response to a chemical stimulus such as low pH is activation, and the part of the molecule that opens and closes the pore is the gate. Evidence that the indole side chain of Trp41 is a 'gate' that can retain H^+ within the cell has been obtained (Tang et al., 2002).

The cysteine scanning mutagenesis experiments described above showed that the currents of one of the mutant proteins, M2-W41C, were much larger than those of wt M2 protein. This was in contrast to cysteine substitutions at other pore-lining residues, for which the currents were smaller than those of wt. Therefore the presence of outward H^+ currents from acidified oocytes that expressed the M2-W41C, A, and F mutant proteins was examined. It was found that outward currents flowed into solutions of high pH from acidified oocytes,

Figure 3.5 Model for activation of the M2 ion channel showing only the TM domain from residues 24 to 44; the selectivity filter His37 and the gate Trp41 are shown as side chains. For clarity these residues are shown only for two of the four subunits; the residues facing the viewer and the subunit closest viewer are omitted. Upper panel. Shows scheme for the closed M2 channel. The channel is closed when pH_{out} is high because His37 is not charged and Trp41 obstructs the pore near its cytoplasmic end. Bottom panel shows the M2 channel in the open state. With low pH_{out} His37 is charged, allowing roation of Trp41 to a conformation parallel to the pore's axis, permitting H^+ to flow. (Modified from Tang et al., 2002).

in contrast to the case for oocytes expressing the wt M2 protein that failed to conduct current into high pH medium. These results are similar to those obtained when acidified cells were treated with the protonophore FCCP (which transports H^+ across the membrane by diffusion), which also showed outward H^+ currents. The ability to conduct outward current, however, is very specific, as cells expressing the M2-W41Y mutant protein are not capable of conducting outward currents into solution of high pH. These findings suggest that the wt protein is held in the closed state when pH_{out} is high by the action of the indole side-chain obstructing the pore, and is opened when pH_{out} is lowered by the removal of the side chain from the pore (Tang et al., 2002).

A simple model for activation of the M2 protein channel was proposed (Tang et al., 2002). When pH_{out} is lowered the His37 selectivity filter becomes protonated and as a result the indole of Trp41 rotates to permit H^+ to flow. This movement may be accomplished by cation-pi interactions (Okada et al., 2001; Zhong et al., 1998a; Zhong et al., 1998b) as suggested from Raman spectroscopy measurements. After returning to high pH_{out} outward current will not flow because the deprotonation of the His_{37} selectivity filter in high pH medium causes the indole of Trp41 to return to its pore-blocking position. If Trp41 is mutated to have a smaller sized side chain, pore blockage cannot occur. Thus, His37 acts as the detector of low pH_{out} and Trp41 acts as the gate. These data are entirely consistent with the previous observations in which cysteine scanning mutagenesis showed a pH-dependent propensity of residues 40-43 to form inter-subunit disulfide bonds on oxidation (Bauer et al., 1999). This proposed mechanism for influenza virus M2 ion channel gating has several advantages for the virus, (1) the functions of selectivity and activation are built into only two residues of this small viral protein and (2) transient exposure to low pH_{out} will result in lasting acidification of the virion because protons are retained by the tryptophan gate, increasing the effectiveness of the small number of M2 molecules found in the virion.

Characterization of the amantadine block

The amino acid sequence of the M2 protein TM domain is fairly well conserved in human, swine, equine and avian strains. Nonetheless, there are some amino acid differences and it seemed likely that these could lead to altered properties of the channel. For example, in tissue culture cells different strains of influenza virus show a difference in susceptibility to inhibition of virus growth by amantadine, influenza A/chicken/Germany/34 (FPV Rostock) is less sensitive to amantadine than A/chicken/Germany/27 (Weybridge). Studies on the nature of the amantadine block of the ion channel activity of the M2 protein of subtypes FPV Rostock, Weybridge and A/Udorn/72, when expressed in oocytes of *Xenopus laevis*, indicated that amantadine is effective as a closed state blocker and less effective once the channel has been activated by lowered pH (Wang et al., 1993). The forward rate constant for the three M2 proteins was Udorn > Weybridge > Rostock (Udorn = 600-900 M^{-1} sec^{-1}). Because the amantadine block is nearly irreversible (the calculated reverse reaction rate for amantadine is 3×10^{-4} sec^{-1}) it is not practical to measure an equilibrium constant. However, by making measurements of the ion channel activity of oocytes expressing the M2 protein with amantadine after a fixed time (2 min.) of exposure to amantadine, it was possible to calculate the isochronic apparent inhibitory constant, $appK_i$ and it was found that Udorn < Weybridge < Rostock (Wang et al., 1993). These data are consistent with the observed difference in susceptibility of the FPV subtypes to inhibition of virus growth by amantadine, with Weybridge being more sensitive than Rostock (Hay et al., 1985). Perhaps the amantadine ammonium nitrogen shares a hydrogen bond with an imidazole of His*37*, making a 'surrogate proton' available to His37. The Hill coefficient for amantadine block is *1*, indicating that one molecule of amantadine blocks one active channel oligomer (Wang et al., 1993).

Cu(II) inhibition of the M$_2$ ion channel

The above studies with cysteine substitution mutants showed that a constriction exists in the pore of the channel between residues G34 and W*41*, but the inactivity of the M2-H37C mutant protein made it necessary to adopt another means to study the possible role of His37 in the channel. As His-containing proteins are often found to coordinate transition elements, transition elements were screened for their ability to affect the function of the M2 protein and it was found that one element, Cu(II), inhibited the channel at low concentrations (Gandhi et al., 1999). This inhibition was slowly reversible and was quite specific to Cu(II), Cu(I), Zn(II), Ni(II), Pt(II), Mg(II) and Mn(II) were ineffective in inhibiting the channel. Two results indicate that the inhibition was due to coordination by His37. First, the inhibition of either the M2-H37A or M2-H37G mutant protein by Cu(II) was incomplete and short-lived. Second, extracellular application of a specific inhibitor of M2 channel function that interacts with residues external to His*37*, BL-1743 (Tu et al., 1996), prevented the subsequent inhibition by extracellularly applied Cu^{2+}. The effect of application of BL-1743 was reversible, and after the compound was removed, Cu^{2+} was once again able to inhibit the channel. These results demonstrate that the His37 residue lines the pore of the channel and is capable of coordinating with ions present in the pore.

M2 ion channels from different influenza virus sub-types have different levels of activity

The activities of the FPV Rostock, FPV Weybridge and Udorn M2 proteins have been evaluated in mammalian cells using as an assay the M2-mediated alteration in the conformational form of HA (either low-pH or native form) (Grambas et al., 1992; Grambas and Hay, 1992; Takeuchi and Lamb, 1994). The data obtained indicate that the ion channel activity of FPV Rostock M2 is more active than that of either the FPV Weybridge or Udorn M2 proteins. In addition, when FPV Rostock and FPV Weybridge viruses containing M2 TM domain point mutations that confer amantadine resistance were investigated, using as an assay their effect on the percentage native FPV HA that could be identified, it was found that the primary structure of the whole M2 TM domain influenced the consequence of specific residue changes. For example, the change G34E in FPV Rostock M2 protein was found to diminish M2 protein activity whereas the G34E change in FPV Weybridge M2 protein was found to increase the M2 protein activity (Grambas et al., 1992). Estimates of ion channel specific activity, by measuring the whole cell currents of oocytes expressing M2 proteins with changes in the TM domain as a function of M2 protein expression levels (Holsinger et al., 1994), lends further support to the notion that the relationship of the role of a specific residue at a particular position in the M2 ion channel to the overall molecular architecture of the M2 ion channel is complex. For example, in both FPV Rostock and FPV Weybridge, the change S31N lowers M2 protein activity but with Udorn M2 protein it leads to amantadine resistance with little change in ion channel specific activity. In contrast, the change A30P in FPV Weybridge, leads to a viable virus (Hay et al., 1985), whereas the change A30P in Udorn M2 protein effectively abolishes ion channel activity (Holsinger et al., 1994). In addition, when M2 residue 27 in FPV Rostock is threonine it leads to amantadine resistance (Hay et al., 1985) whereas the presence of threonine in the equivalent position in Udorn M2 protein leads to channel activity that is amantadine sensitive (Holsinger et al., 1994). Direct electrophysiological measurements of the proton channels formed by influenza A virus strains 'Weybridge' and 'Rostock' showed various proton conductances and these differences were determined by three amino acids in the TM domain, V27S, F38L and D44N (Chizhmakov et al., 2003).

M2 expression alters the pH in the TGN and affects transport through the Golgi apparatus

To obtain evidence that the M2 protein could alter the pH of intracellular compartments in the absence of an influenza virus infection, the effect of expressing M2 on the biogenesis of HA was examined (Takeuchi and Lamb, 1994). When FPV HA was expressed from cDNA in cells, it was found that the majority of the HA molecules were in the low-pH form of HA (Takeuchi and Lamb, 1994). The lysosomaltropic agent, ammonium chloride stabilized the accumulation of HA in its pH neutral form (Takeuchi and Lamb, 1994) and increased its fusion ability (Ohuchi et al., 1994). As anticipated co-expression of HA and the FPV M2 protein stabilized the accumulation of HA in its pH neutral form, thus confirming that expression of M2 affects intracellular pH (Takeuchi and Lamb, 1994). It was observed that transfection of increasing amounts of M2 cDNA caused the inhibition of cleavage of the HA0 to HA1 and HA2 (Takeuchi and Lamb, 1994). This observation raised the possibility that over-expression of M2 has an effect on intracellular processing of FPV HA. It was found that high level expression of the M2 protein slowed the rate of intracellular transport of HA and other integral membrane proteins and this is due to the M2 ion channel activity (Henkel et al., 1999; Henkel and Weisz, 1998; Sakaguchi et al., 1996). The deleterious effect of M2 ion channel activity on the rate of intracellular protein transport is similar to the effect of monensin, which at micromolar concentrations delays intracellular protein transport through the Golgi apparatus. However, whereas monensin causes dilation of all cisternae of the Golgi apparatus, M2 protein expression causes a preferential dilation of the TGN (Henkel et al., 2000; Sakaguchi et al., 1996) an observation which is consistent with the M2 ion channel activity being activated by the low pH found in the TGN (Anderson et al., 1984; Anderson and Orci, 1988). Thus, the data provide direct evidence for a way in which a virus infection damages cells.

Roles of the M2 cytoplasmic tail in ion channel activity and in virus assembly

The role of the M2 cytoplasmic tail on ion channel activity has been studied using cytoplasmic tail truncation mutants expressed in oocytes of *Xenopus laevis*. Translational stop codons were introduced into the M2 cDNA at residues *46, 52, 62, 72, 77, 82, 87,* and *92*. The deletion mutants were designated, trunc#, according to the amino acid position that was changed to a stop codon. When the conductance of the truncation mutants was measured over time, trunc72, trunc77 and trunc92 behaved comparably to wt M2 protein (a decrease of only 4% over 30 min). In contrast, the conductance of trunc82 decreased by 28%, 27% for trunc62 and 81% for trunc52 channels. Complete closure of the channel could be observed in some cells for trunc62 and trunc52 within 30 min. These data suggest that a role of the cytoplasmic tail region of the M2 ion channel is to stabilize the pore against premature closure while the ectodomain is exposed to low pH (Tobler et al., 1999).

As discussed above solid-state NMR spectroscopy experiments of intact M2 incorporated into DMPC/DMPG lipid bilayers have suggested that the cytoplasmic tail contains an amphipathic helix (Tian et al., 2003) and computer modeling has suggested there are two amphipathic helices (Saldanha et al., 2002). It has been suggested that the amphipathic helix may stabilize the tetrameric TM helical bundle by interacting with the cytoplasmic side of the lipid bilayer (Tian et al., 2003).

An indication of another function of the M2 protein in the influenza virus life cycle came from studies with a monoclonal antibody (14C2) specific for the N-terminal domain of M2. When this antibody was included in an agarose overlay of a standard plaque assay titration, it was found to restrict the size of plaque growth of a variety of influenza A virus strains (Zebedee and Lamb, 1988). Variant viruses resistant to the antibody were isolated, and they were found to have compensating changes in the cytoplasmic tail of M2 as well as in the N-terminal domain of the M1 protein (Zebedee and Lamb, 1989), suggesting that the antibody may be interfering with critical M1/M2 interactions during virus assembly and budding. Addition of the 14C2 MAb to virus-infected cells reduced the surface expression level of the M2 protein, possible because the MAb causes internalization of surface expressed M2 protein (Hughey et al., 1995). It seems possible that the M2 protein is multi-functional and its cytoplasmic tail forms protein-protein interactions, most likely with the M1 protein, necessary for virus assembly. Consistent with this idea it was found, using a reverse genetics system that was state of the art at the time of the experiments, that one residue could be deleted from the C-terminus of M2 protein and viable virus rescued, however, it was not possible to rescue viruses lacking 5 or 10 C-terminal residues (Castrucci and Kawaoka, 1995). This observation suggests that the M2 protein C-terminus is essential for virus-replication but not for ion channel activity. Given that the reverse genetics systems currently in use are much more efficient (Fodor et al., 1999; Neumann et al., 1999) than those used in (1995). it will be interesting to re-examine this topic.

The M2 ectodomain, suggested roles in virus assembly and in generating cross-reactive immune responses

The role of the M2 ectodomain in ion channel activity has not been examined in detail. Introduction of a site for addition of N-linked carbohydrate into the ectodomain, such that the ectodomain is glycosylated, causes alterations in ion channel activity that are not fully characterized (unpublished observations).

It has been suggested that there is an important role for the M2 ectodomain in incorporation of the M2 protein into virions, based on analysis of chimeric molecules between M2 and a paramyxovirus F protein (Park et al., 1998). However, interpretation of these data is somewhat complex given that M2 is a tetramer and F proteins are trimeric.

M2 ectodomain residues 1–10 are conserved among virtually all influenza A virus subtypes and it was suggested in 1985 that the M2 ectodomain might be a cross-reactive protective antigen (Lamb et al., 1985). Antibodies to expressed M2 protein have been detected in serum samples from humans and ferrets infected with influenza A viruses and passive transfer of the M2 ectodomain-specific 14C2 MAb between mice reduced the level of replication of influenza A virus in mice (Treanor et al., 1990). Several studies suggest that antibodies raised to a complete or to a sub-set of M2 ectodomain residues confer protective immunity in mice (Fiers et al., 2004; Liu et al., 2003; Neirynck et al., 1999). Further studies are needed to investigate the usefulness of M2 ectodomain antibodies as a vaccine candidate.

THE INFLUENZA B VIRUS BM2 ION CHANNEL PROTEIN
Influenza B virus encodes two small integral membrane proteins NB and BM2

The influenza B virus genome encodes two extra proteins. Influenza B virus RNA segment 6, in addition to coding for the B/NA protein, also encodes via a bicistronic mRNA and using an overlapping reading frame, the 100 residue NB glycoprotein (Shaw et al., 1982). Initiation of translation of NB occurs using an AUG codon positioned four nucleotides before the AUG codon used to initiate the NA protein (Shaw et al., 1983; Williams and Lamb, 1989). Influenza B virus RNA segment 7 in addition to encoding the M1 protein encodes the BM2 protein (Horvath et al., 1990). The influenza B virus BM2 protein is translated from an open reading frame that is +2 nucleotides with respect to the reading frame of the M1 protein and it is conserved in all isolates of influenza B virus. The BM2 protein contains 109 residues and in influenza B virus-infected cells the BM2 protein has an apparent M_r of 12–15 kDa (Horvath et al., 1990). A further study of the BM2 protein reported by Odagiri and coworkers indicates that it is post-translationally modified by phosphorylation, that BM2 is incorporated into purified virions and that the BM2 protein is localized predominantly to the cytoplasm (Odagiri et al., 1999).

The influenza B virus matrix protein can be removed from the RNP core by low pH (pH 5.5) treatment (Mould et al., 2003; Zhirnov, 1990). Thus, there has been considerable speculation as to whether influenza B virus would also require an ion channel activity and the question arose as to which viral protein would have this activity.

It had been speculated by analogy to the known function of the influenza A virus M2 protein, that NB protein has ion channel activity. Indeed, when the NB protein was expressed in bacteria and reconstituted into planar membranes it was reported to have

Figure 3.6 (*opposite*)

(A) Schematic diagram to indicate the conservation of the BM2 TM domain His and Trp residues as compared to residues in the influenza A virus M2 proton-selective ion channel protein. Histidine side chains and tryptophan are shown. In BM2 the serine side chains are shown. (Modified from Paterson et al., 2003)

(B) Specific immunofluorescent staining of BM2, M2gBM2 and BM2–Flag indicates BM2 adopts an NoutCin orientation in membranes. Eukaryotic expression vectors expressing M2gBM2 (BM2 containing an altered ectodomain), BM2-Flag (BM2-with a C-terminal epitope tag). BM2 cDNAs and its derivative cDNAs were transfected into HeLa-CD4-LTR-β-gal cells and at 18 h post transfection cells were fixed with 2% formaldehyde. For permeabilization (Perm.) cells were treated with 0.1% saponin. Cells were stained with primary antibodies to the antibody tags (M2 14C2 or Flag antibody). (Modified from Paterson et al., 2003).

(C and D). Time course of acidification and recovery from acidification of BM2 and BM2-H19C in oocytes and mammalian cells. (C). Oocytes expressing wt BM2, BM2-H19C and A/M2 were bathed in solutions of pH 8.5 or pH 5.8 (at the time shown by the dark bar) and the pH_{in} of these oocytes was measured with a pH micro-electrode while the membrane voltage was held at –20 mV with a 2-electrode voltage-clamp apparatus. Traces are averages from 5 oocytes. (D). Time course of acidification of mammalian cells expressing the BM2 protein. HeLa T4 cells were co-transfected to express transiently the BM2 and BM2-H19C proteins together with EGFP indicator protein. Each of the plots represents the average fluorescence emission at 530 nm of 7 cells excited at 485 nm which has been normalized to the intensity measured initially in pH 7.4 (Modified from Mould et al., 2003).

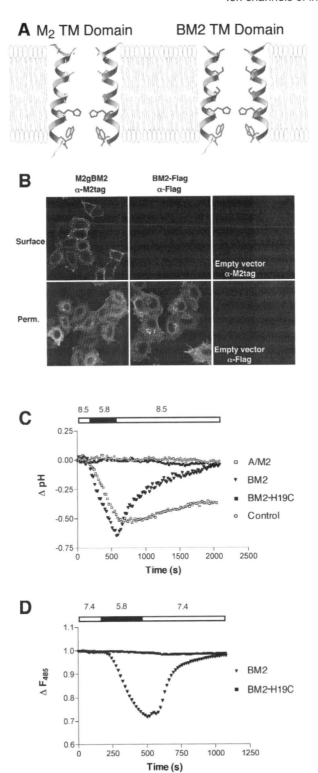

an ion channel activity (Fischer et al., 2001; Sunstrom et al., 1996). However, these planar bilayer experiments are subject to artifacts (reviewed in Lamb and Pinto, 1997). Furthermore, the ion channel activity reported by Fischer and coworkers (2001) was reported to be blocked by amantadine (Fischer et al., 2001) yet amantadine does not inhibit the replication of influenza B virus. In addition, expression of NB does not cause acidification of oocytes incubated in low pH media (Mould et al., 2003). Lastly, by using a reverse genetics system it has been found possible to delete the NB coding region from the influenza B virus genome and yet the virus was recovered and is viable in tissue culture but shows some growth defects in mice (Hatta and Kawaoka, 2003). Nonetheless, NB is not essential for influenza virus survival.

In 2003 it was reported that in the amino acid sequence of BM2 there is a hydrophobic region (residues 7–25) that could act as a TM anchor (Paterson et al., 2003). Analysis of properties of the BM2 protein, including detergent solubility, insolubility in alkali pH *11*, flotation in membrane fractions, and epitope tagging immuno-cytochemistry indicated BM2 protein is the fourth integral membrane protein encoded by influenza B virus in addition to HA, NA and the NB glycoprotein (Paterson et al., 2003). Biochemical analysis indicated that the BM2 protein adopts an $N_{out}C_{in}$ orientation in membranes and fluorescence microscopy indicated BM2 is expressed at the cell surface. As the BM2 protein possesses only a single hydrophobic domain and lacks a cleavable signal sequence it is another example of a Type III integral membrane protein, in addition to M2, NB and CM2 proteins of influenza A, B and C viruses, respectively (Paterson et al., 2003). Chemical cross-linking studies indicate that the BM2 protein is oligomeric, most likely a tetramer (Paterson et al., 2003). Very recently, Odagiri's laboratory (Watanabe et al., 2003) confirmed that BM2 is an integral membrane protein.

If the amino acid sequence of the BM2 TM domain region (7)ILSICSFILSALHFMA WTI(25) is modeled as an α-helix then S*9*, S*12*, S*16*, H*19*, and W23 could all form the same face of an α-helix and thus these residues could form the aqueous pore of an oligomeric ion channel. Remarkably, the two key residues for activation and gating (H37 and W41) in the influenza A virus M2 ion channel are found at the same spacing H-X-X-X-W (H19 and W23) in the BM2 TM domain. Thus, it seemed possible that the BM2 protein would have an ion channel activity.

BM2 protein is capable of acidifying oocytes and mammalian cells

To investigated whether the BM2 protein can cause cytoplasmic acidification BM2 and BM2-H19C proteins, and as a control a known proton-selective ion channel, the M2 protein of influenza A virus, were expressed in *Xenopus laevis* oocytes and also in mammalian cells. For oocytes, pH micro-electrodes were used to measure the response of intracellular pH to lowering external bathing solution pH (Mould et al., 2000b; Shimbo et al., 1996). Oocytes were bathed in Barth's solution at pH 8.*5*, pH 5.8 and returned to pH 8.5 while cells were voltage clamped at –20 mV (Figure 5A). Oocytes expressing BM2 protein acidified when exposed to low extracellular pH with a time course that was faster than that observed for oocytes expressing the M2 protein of influenza A virus. In addition, itracellular pH of BM2-expressing oocytes fully recovered upon return to pH 8.5 within 1000 sec, unlike the slow and incomplete recovery for the M2 proton channel of influenza A virus observed during this interval, suggesting a difference in the efflux of protons through the M2 channel of influenza A virus and the BM2 channels into solutions of alkaline pH. Oocytes expressing the BM2-H19C protein failed to acidify upon lowering the external solution pH, a finding which is consistent with H19 being important for the function of BM2 ion channel activity, and suggesting a possible similarity in ion conduction mechanism between the BM2 protein and the M2 protein of influenza A virus (Mould et al., 2003).

To show that the BM2 protein exhibited a similar acidification behavior in mammalian cells (and to rule out the unlikely possibility that in oocytes the BM2 protein causes upregulation of an endogenous proton channel), the acidification experiment was repeated using HeLa cells. Here, pH_{in} was monitored by co-expression of EGFP, which exhibits a pH-sensitive fluorescence (Llopis et al., 1998). Consistent with the results obtained in Xenopus oocytes, HeLa cells expressing the BM2 protein acidified when the extracellular pH was lowered from pH 7.4 to pH 5.8 and recovered after returning the bathing solution to pH 7.4. HeLa cells expressing the BM2-H19C mutant protein also failed to acidify under the same conditions (Mould et al., 2003).

BM2 protein has ion channel activity that is activated by low pH

Similarities in the acidification profile of cells expressing the BM2 protein and the M2 protein of influenza A virus in solutions of low pH suggest that the BM2 protein has an ion channel activity. This possibility was tested by using voltage clamp techniques to measure whole-cell membrane currents in Xenopus oocytes expressing wt BM2 or BM2 mutant proteins (Flag epitope tagged) in response to lowering bathing solution pH. Oocytes were maintained at pH 8.5 to minimize intracellular acidification and at pH 8.5, oocytes expressing the wt BM2 protein had almost no membrane currents at –20 mV. Voltage ramps (–40 mV to + 60 mV) revealed that at more positive potentials, BM2 expressing oocytes have a significant outward current which by comparison was larger than that observed for cells expressing the M2 protein of influenza A virus (Mould et al., 2000b; Shimbo et al., 1996). The shift in reversal voltage observed of 83.7 mV $\pm$ 9.7 mV (n=5) is consistent with a shift in the proton equilibrium potential. As only pH was changed and no other ions, it suggests protons are conducted; however, a formal determination of the BM2 ionic selectivity awaits full ion substitution experiments. As an inhibitor of the BM2 ion channel activity is not available, the small endogenous oocyte currents could not be subtracted, and this is one of the reasons a change in reversal voltage of +58 mV per pH unit for a pure proton conductance was not observed. A second reason is that oocytes expressing proton channels acidify near the plasma membrane while being bathed in low pH solution (Mould et al., 2000a). Oocytes expressing the BM2-H19C protein did not show pH-activated currents, suggesting that this mutation renders the BM2 ion channel inactive. The ion channel activity of BM2 and BM2-H19C that were not Flag epitope tagged was also tested and the data obtained were indistinguishable from those obtained for the tagged proteins. A trivial explanation for these results would be failure of the mutant protein to be expressed at the cell surface, but the expression of the protein was monitored by immunofluorescence microscopy using an antibody to the C-terminal FLAG epitope and it was found that the mutant protein indeed was expressed at the cell surface. Thus, replacement of TM domain residue His_{19} with cysteine eliminates the membrane currents and acidification induced by bathing in solutions of low pH (Mould et al., 2003).

As discussed above recent data for the M2 protein of influenza A virus indicate a model of elegant simplicity in which the bulky indole side chain of residue W41 regulates channel opening by functioning as a minimalistic gate that opens and closes the pore (Tang et al., 2002). Thus, given the conservation of spacing of H37 and W41 in the M2 channel influenza A virus and H19 and W23 in the BM2 channel, mutant BM2-W23C was constructed. When whole-cell membrane currents were measured for oocytes expressing the BM2-W23C protein, large predominantly outward currents were observed. The smaller change in reversal voltage of oocytes expressing the BM2-W23C protein on lowering pH of the bathing medium from pH 8.5 to pH 5.8 treatment (10 mV ± 3.8 mV, n = 5) than for wt BM2 suggests that the BM2-W23C is less proton selective than wt BM2 protein. The difference between the I-V relationship of oocytes expressing wt BM2 and BM2-W23C measured in pH 8.5 solution after prior bathing in pH 5.8 suggests that BM2 residue W23, like residue W41 of the M2 protein of influenza A virus plays an important, yet subtly different, role in channel gating.

The BM2 protein is essential for influenza B virus viability

Recently, further evidence for the importance of the BM2 protein has been obtained when it was found by using influenza B virus reverse genetics systems that the BM2 protein is an essential protein for virus replication as it cannot be deleted from the influenza B virus genome and rescued virus obtained (Hatta et al., 2004; Jackson et al., 2004). These data suggest a critical and essential function of BM2 in the life cycle of the virus.

Does influenza C virus have an ion channel activity?

It is not know if there is a rational basis for influenza C virus to require an ion channel during its uncoating process. Influenza C virus does encode a small intergral membrane protein, CM2.

The influenza C virus CM2 protein is a small glycosylated integral membrane protein (115 residues) that spans the membrane once. The CM2 protein forms disulfide linked dimers and tetramers and is oriented in membranes in an $N_{out}C_{in}$ orientation. CM2 contains a single site for N-linked carbohydrate addition which is frequently further modified with lactosaminoglycans. The CM2 protein is abundantly expressed at the cell surface of virus-infected and cDNA-transfected cells and is incorporated into influenza C virions (Hongo et al., 1997; Pekosz and Lamb, 1997). No functional activity of the CM2 protein has been reported.

Acknowledgements

We thank our many students, post-doctoral fellows and collaborators for their contribution to this work. We especially thank Reay G. Paterson who stepped in to fill a void and perform important experiments at the right time on the BM2 protein. Research in the authors' laboratories is supported by Research Grants from the National Institute of Allergy and Infectious Diseases R37-AI-20201 (RAL), R01 AI-23173 (RGP and RAL) and R01 AI-31882 (LHP). RAL is an Investigator of the Howard Hughes Medical Institute.

References

Allen, H., McCauley, J., Waterfield, M., and Gething, M. J. (1980). Influenza virus RNA segment 7 has the coding capacity for two polypeptides. Virology. 107, 548–551.

Anderson, R. G., Falck, J. R., Goldstein, J. L., and Brown, M. S. (1984). Visualization of acidic organelles in intact cells by electron microscopy. Proc. Natl. Acad. Sci. USA. 81, 4838–4842.

Anderson, R. G., and Orci, L. (1988). A review of acidic intracellular compartments. J. Cell Biol. 106, 539–543.

Bauer, C. M., Pinto, L. H., Cross, T. A., and Lamb, R. A. (1999). The influenza virus M_2

ion channel protein, probing the structure of the transmembrane domain in intact cells by using engineered disulfide cross-linking. Virology. *254*, 196–209.

Black, R. A., Rota, P. A., Gorodkova, N., Cramer, A., Klenk, H.-D., and Kendal, A. P. 1993a. Production of the M2 protein of influenza A virus in insect cells is enhanced in the presence of amantadine. J. Gen. Virol. *74*, 1673–1677.

Black, R. A., Rota, P. A., Gorodkova, N., Klenk, H.-D., and Kendal, A. P. 1993b. Antibody response to the M2 protein of influenza A virus expressed in insect cells. J. Virol. *67*, 1203–1210.

Brown, D. A., and London, E. (1998). Functions of lipid rafts in biological membranes. Annu. Rev. Cell Dev. Biol. *14*, 111–136.

Bukrinskaya, A. G., Vorkunova, N. K., Kornilayeva, G. V., Narmanbetova, R. A., and Vorkunova, G. K. (1982). Influenza virus uncoating in infected cells and effects of rimantadine. J. Gen. Virol. *60*, 49–59.

Bullough, P. A., Hughson, F. M., Skehel, J. J., and Wiley, D. C. (1994). Structure of influenza haemagglutinin at the pH of membrane fusion. Nature. *371*, 37–43.

Castrucci, M. R., Hughes, M., Calzoletti, L., Donatelli, I., Wells, K., Takada, A., and Kawaoka, Y. (1997). The cysteine residues of the M2 protein are not required for influenza A virus replication. Virology. *238*, 128–134.

Castrucci, M. R., and Kawaoka, Y. (1995). Reverse genetics system for generation of an influenza A virus mutant containing a deletion of the carboxyl-terminal residue of M2 protein. J. Virol. *69*, 2725–2728.

Chazal, N., and Gerlier, D. (2003). Virus entry, assembly, budding, and membrane rafts. Microbiol. Mol. Biol. Rev. *67*, 226–237.

Chen, J., Skehel, J. J., and Wiley, D. C. (1999). N- and C-terminal residues combine in the fusion-pH influenza hemagglutinin HA2 subunit to form an N cap that terminates the triple-stranded coiled coil. Proc. Natl. Acad. Sci. USA. *96*, 8967–8972.

Chizhmakov, I. V., Geraghty, F. M., Ogden, D. C., Hayhurst, A., Antoniou, M., and Hay, A. J. (1996). Selective proton permeability and pH regulation of the influenza virus M2

channel expressed in mouse erythroleukaemia cells. J. Physiol. *494*, 329–336.

Chizhmakov, I. V., Ogden, D. C., Geraghty, F. M., Hayhurst, A., Skinner, A., Betakova, T., and Hay, A. J. (2003). Differences in conductance of M2 proton channels of two influenza viruses at low and high pH. J. Physiol. *546*, 427–38.

Ciampor, F., Bayley, P. M., Nermut, M. V., Hirst, E. M., Sugrue, R. J., and Hay, A. J. 1992a. Evidence that the amantadine-induced, M2-mediated conversion of influenza A virus hemagglutinin to the low pH conformation occurs in an acidic *trans* Golgi compartment. Virology. *188*, 14–24.

Ciampor, F., Thompson, C. A., Grambas, S., and Hay, A. J. 1992b. Regulation of pH by the M2 protein of influenza A viruses. Virus Res. *22*, 247–258.

Compans, R. W., Bishop, D. H. L., and Meier-Ewert, H. 1977. Structural components of influenza C virions. J. Virol. *21*, 658–665.

Davies, W. L., Grunert, R. R., Haff, R. F., McGahen, J. W., Neumayer, E. M., Paulshock, M., Watts, J. C., Wood, T. R., Herman, E. C., and Hoffman, C. E. 1964. Antiviral activity of 1-adamantanamine (amantadine). Science. *144*, 862–863.

Duff, K. C., and Ashley, R. H. (1992). The transmembrane domain of influenza A M2 protein forms amantadine-sensitive proton channels in planar lipid bilayers. Virology. *190*, 485–489.

Fiers, W., De Filette, M., Birkett, A., Neirynck, S., and Min Jou, W. (2004). A 'universal' human influenza A vaccine. Virus Res. *103*, 173–176.

Fischer, W. B., Pitkeathly, M., and Sansom, M S. (2001). Amantadine blocks channel activity of the transmembrane segment of the NB protein from influenza B. Eur. Biophys. J. *30*, 416–420.

Fodor, E., Devenish, L., Engelhardt, O. G., Palese, P., Brownlees, G. G., and Garcia-Sastre, A. (1999). Rescue of influenza A virus from recombinant DNA. J. Virol. *73*, 9679–9682.

Francis, T. J. 1940. A new type of virus from epidemic influenza. Science. *92*, 405–406.

Gandhi, C. S., Shuck, K., Lear, J. D., Dieckmann, G. R., DeGrado, W. F., Lamb, R. A., and Pinto, L. H. (1999). Cu(II) inhibition of the proton translocation machinery of the influenza A virus M2 protein. J. Biol. Chem. *274*, 5474–5482.

Giffin, K., Rader, R. K., Marino, M. H., and Forgey, R. W. (1995). Novel assay for the influenza virus M2 channel activity. FEBS Lett. *357*, 269–274.

Grambas, S., Bennett, M. S., and Hay, A. J. (1992). Influence of amantadine resistance mutations on the pH regulatory function of the M2 protein of influenza A viruses. Virology. *191*, 541–549.

Grambas, S., and Hay, A. J. (1992). Maturation of influenza A virus hemagglutinin – estimates of the pH encountered during transport and its regulation by the M2 protein. Virology. *190*, 11–18.

Guinea, R., and Carrasco, L. (1994). Influenza virus M2 protein modifies membrane permeability in *E. coli* cells. FEBS Lett. *343*, 242–246.

Hatta, M., Goto, H., and Kawaoka, Y. (2004). Influenza B virus requires BM2 protein for replication. J. Virol. *78*, 5576–5583.

Hatta, M., and Kawaoka, Y. (2003). The NB protein of influenza B virus is not necessary for virus replication in vitro. J. Virol. *77*, 6050–6054.

Hay, A. J. (1992). The action of adamantanamines against influenza A viruses, Inhibition of the M2 ion channel protein. Semin. Virol. *3*, 21–30.

Hay, A. J., Wolstenholme, A. J., Skehel, J. J., and Smith, M. H. (1985). The molecular basis of the specific anti-influenza action of amantadine. EMBO J. *4*, 3021–3024.

Henkel, J. R., Gibson, G. A., Poland, P. A., Ellis, M. A., Hughey, R. P., and Weisz, O. A. (2000). Influenza M2 proton channel activity selectively inhibits trans-Golgi network release of apical membrane and secreted proteins in polarized Madin-Darby canine kidney cells. J. Cell. Biol. *148*, 495–504.

Henkel, J. R., Popovich, J. L., Gibson, G. A., Watkins, S. C., and Weisz, O. A. (1999). Selective perturbation of early endosome and/or trans-Golgi network pH but not lysosome pH by dose-dependent expression of influenza M2 protein. J. Biol. Chem. *274*, 9854–9860.

Henkel, J. R., and Weisz, O. A. (1998). Influenza virus M2 protein slows traffic along the secretory pathway. pH perturbation of acidified compartments affects early Golgi transport steps. J. Biol. Chem. *273*, 6518–6524.

Holsinger, L. J., and Lamb, R. A. (1991). Influenza virus M2 integral membrane protein is a homotetramer stabilized by formation of disulfide bonds. Virology. *183*, 32–43.

Holsinger, L. J., Nichani, D., Pinto, L. H., and Lamb, R. A. (1994). Influenza A virus M2 ion channel protein, A structure-function analysis. J. Virol. *68*, 1551–1563.

Holsinger, L. J., Shaughnessy, M. A., Micko, A., Pinto, L. H., and Lamb, R. A. (1995). Analysis of the posttranslational modifications of the influenza virus M2 protein. J. Virol. *69*, 1219–1225.

Hongo, S., Sugawara, K., Muraki, Y., Kitame, F., and Nakamura, K. (1997). Characterization of a second protein (CM2) encoded by RNA segment 6 of influenza C virus. J. Virol. *71*, 2786–2792.

Horsfall, F. L. J., Lennette, E. H., Rickard, E. R., Andrewes, C. H., Smith, W., and Stuart-Harris, C. H. 1940. The nomenclature of influenza. Lancet. ii, 413.

Horvath, C. M., Williams, M. A., and Lamb, R. A. (1990). Eukaryotic coupled translation of tandem cistrons, identification of the influenza B virus BM2 polypeptide. EMBO J. *9*, 2639–2647.

Hughey, P. G., Compans, R. W., Zebedee, S. L., and Lamb, R. A. (1992). Expression of the influenza A virus M2 protein is restricted to apical surfaces of polarized epithelial cells. J. Virol. *66*, 5542–5552.

Hughey, P. G., Roberts, P. C., Holsinger, L. J., Zebedee, S. L., Lamb, R. A., and Compans, R. W. (1995). Effects of antibody to the influenza A virus M2 protein on M2 surface expression and virus assembly. Virology. *212*, 411–421.

Hull, J. D., Gilmore, R., and Lamb, R. A. (1988). Integration of a small integral membrane protein, M2, of influenza virus into the endoplasmic reticulum, analysis of the internal signal-anchor domain of a protein with an ectoplasmic NH_2 terminus. J. Cell Biol. *106*, 1489-1498.

Inglis, S. C., and Brown, C. M. (1981). Spliced and unspliced RNAs encoded by virion RNA segment 7 of influenza virus. Nucleic Acids Res. *9*, 2727–2740.

Ito, T., Gorman, O. T., Kawaoka, Y., Bean, W. J., Jr., and Webster, R. G. (1991). Evolutionary analysis of the influenza A virus M gene with comparison of the M1 and M2 proteins. J. Virol. *65*, 5491–5498.

Jackson, D., Zurcher, T., and Barclay, W. (2004). Reduced incorporation of the influenza B virus BM2 protein in virus particles decreases infectivity. Virology. *322*, 276–285.

Jackson, D. C., Tang, X. L., Murti, K. G., Webster, R. G., Tregear, G. W., and Bean, W. J., Jr. (1991). Electron microscopic evidence for the association of M2 protein with the influenza virion. Arch. Virol. *118*, 199-207.

Kilbourne, E. D. (1987). 'Influenza.' Plenum Medical Book Company, New York.

Kovacs, F. A., and Cross, T. A. (1997). Transmembrane four-helix bundle of influenza A M2 protein channel, Structural implications from helix tilt and orientation. Biophys. J. *73*, 2511–2517.

Kovacs, F. A., Denny, J. K., Song, Z., Quine, J. R., and Cross, T. A. (2000). Helix tilt of the M2 transmembrane peptide from influenza A virus, an intrinsic property. J. Mol. Biol. *295*, 117–125.

Kurtz, S., Luo, G., Hahnenberger, K. M., Brooks, C., Gecha, O., Ingalls, K., Numata, K.-I., and Krystal, M. (1995). Growth impairment resulting from expression of influenza virus M2 protein in *Saccharomyces cerevisiae*, Identification of a novel inhibitor of influenza virus. Anti. Agents Chem. *39*, 2204–2209.

Lamb, R. A., and Choppin, P. W. (1981). Identification of a second protein (M_2) encoded by RNA segment 7 of influenza virus. Virology. *112*, 729–737.

Lamb, R. A., Holsinger, L. J., and Pinto, L. H. (1994). The influenza A virus M_2 ion channel protein and its role in the influenza virus life cycle. *In* 'Receptor-Mediated Virus Entry into Cells' (E. Wimmer, Ed.), pp. 303–321.

Cold Spring Harbor Laboratory Press, Cold Spring Harbor, N. Y.

Lamb, R. A., and Krug, R. M. (2001). *Orthomyxoviridae*, the viruses and their replication. *In* 'Fields Virology (Fourth Edition)' (D. M. Knipe, and P. M. Howley, Eds.), pp. 1487–1531. Lippincott, Williams and Wilkins, Philadelphia.

Lamb, R. A., and Lai, C.-J. (1981). Conservation of the influenza virus membrane protein (M_1) amino acid sequence and an open reading frame of RNA segment 7 encoding a second protein (M_2) in H1N1 and H3N2 strains. Virology. *112*, 746–751.

Lamb, R. A., Lai, C.-J., and Choppin, P. W. (1981). Sequences of mRNAs derived from genome RNA segment 7 of influenza virus, colinear and interrupted mRNAs code for overlapping proteins. Proc. Natl. Acad. Sci. USA. *78*, 4170–4174.

Lamb, R. A., and Pinto, L. H. (1997). Do Vpu and Vpr of human immunodeficiency virus type 1 and NB of influenza B virus have ion channel activities in the viral life cycles. Virology. *229*, 1–11.

Lamb, R. A., Zebedee, S. L., and Richardson, C. D. (1985). Influenza virus M_2 protein is an integral membrane protein expressed on the infected-cell surface. Cell. *40*, 627–633.

Lin, T.-I., Heider, H., and Schroeder, C. (1997). Different modes of inhibition by adamantane amine derivatives and natural polyamines of the functionally reconstituted influenza virus M_2 proton channel protein. J. Gen. Virol. *78*, 767–774.

Lin, T. I., and Schroeder, C. (2001). Definitive assignment of proton selectivity and attoampere unitary current to the M2 ion channel protein of influenza A virus. J. Virol. *75*, 3647–3656.

Liu, W., Li, H., and Chen, Y. H. (2003). N-terminus of M2 protein could induce antibodies with inhibitory activity against influenza virus replication. FEMS Immunol. Med. Microbiol. *35*, 141–146.

Llopis, J., McCaffery, J. M., Miyawaki, A., Farquhar, M. G., and Tsien, R. Y. (1998). Measurement of cytosolic, mitochondrial, and Golgi pH in single living cells with green fluorescent proteins. Proc. Natl. Acad. Sci. USA. *95*, 6803–6808.

Martin, K., and Helenius, A. (1991). Transport of incoming influenza virus nucleocapsids into the nucleus. J. Virol. *65*, 232–244.

Mould, J. A., Drury, J. E., Frings, S. M., Kaupp, U. B., Pekosz, A., Lamb, R. A., and Pinto, L. H. 2000a. Permeation and activation of the M2 ion channel of influenza A virus. J. Biol. Chem. *275*, 31038–31050.

Mould, J. A., Li, H.-C., Dudlak, C. S., Lear, J. D., Pekosz, A., Lamb, R. A., and Pinto, L. H. 2000b. Mechanism for proton conduction of the M2 ion channel of influenza A virus. J. Biol. Chem. *275*, 8592–8599.

Mould, J. A., Paterson, R. G., Takeda, M., Ohigashi, Y., Venkataraman, P., Lamb, R. A., and Pinto, L. H. (2003). Influenza B virus BM2 protein has ion channel activity that conducts protons across membranes. Dev. Cell. *5*, 175–184.

Munro, S. (2003). Lipid rafts, elusive or illusive? Cell. *115*, 377–388.

Nayak, D. P., and Barman, S. (2002). Role of lipid rafts in virus assembly and budding. Adv. Virus Res. *58*, 1–28.

Neirynck, S., Deroo, T., Saelens, X., Vanlandschoot, P., Jou, W. M., and Fiers, W. (1999). A universal influenza A vaccine based on the extracellular domain of the M2 protein. Nat. Med. *5*, 1157–1163.

Neumann, G., Watanabe, T., Ito, H., Watanabe, S., Goto, H., Gao, P., Hughes, M., Perez, D. R., Donis, R., Hoffmann, E., Hobom, G., and Kawaoka, Y. (1999). Generation of influenza A viruses entirely from cloned cDNAs. Proc. Natl. Acad. Sci. USA. *96*, 9345–9350.

Nishimura, K., Kim, S., Zhang, L., and Cross, T. A. (2002). The closed state of a H+ channel helical bundle combining precise orientational and distance restraints from solid state NMR. Biochemistry. *41*, 13170–13177.

Odagiri, T., Hong, J., and Ohara, Y. (1999). The BM2 protein of influenza B virus is synthesized in the late phase of infection and incorporated into virions as a subviral component. J. Gen. Virol. *80*, 2573-2581.

Ohuchi, M., Cramer, A., Vey, M., Ohuchi, R., Garten, W., and Klenk, H.-D. (1994). Rescue of vector-expressed fowl plague virus hemagglutinin in biologically active form by acidotropic agents and coexpressed M2 protein. J. Virol. *68*, 920-926.

Okada, A., Miura, T., and Takeuchi, H. (2001). Protonation of histidine and histidine-tryptophan interaction in the activation of the M2 ion channel from influenza a virus. Biochemistry. *40*, 6053–6060.

Panayotov, P. P., and Schlesinger, R. W. (1992). Oligomeric organization and strain-specific proteolytic modification of the virion M2 protein of influenza A H1N1 viruses. Virology. *186*, 352–355.

Park, E. K., Castrucci, M. R., Portner, A., and Kawaoka, Y. (1998). The M2 ectodomain is important for its incorporation into influenza A virions. J. Virol. *72*, 2449–2455.

Paterson, R. G., Takeda, M., Ohigashi, Y., Pinto, L. H., and Lamb, R. A. (2003). Influenza B virus BM2 protein is an oligomeric integral membrane protein expressed at the cell surface. Virology. *306*, 7–17.

Pekosz, A., and Lamb, R. A. (1997). The CM2 protein of influenza C virus is an oligomeric integral membrane glycoprotein structurally analogous to influenza A virus M2 and influenza B virus NB proteins. Virology. *237*, 439–451.

Pinto, L. H., Dieckmann, G. R., Gandhi, C. S., Papworth, C. G., Braman, J., Shaughnessy, M. A., Lear, J. D., Lamb, R. A., and DeGrado, W. F. (1997). A functionally defined model for the M_2 proton channel of influenza A virus suggests a mechanism for its ion selectivity. Proc. Natl. Acad. Sci. USA. *94*, 11301–11306.

Pinto, L. H., Holsinger, L. J., and Lamb, R. A. (1992). Influenza virus M_2 protein has ion channel activity. Cell. *69*, 517–528.

Pralle, A., Keller, P., Florin, E. L., Simons, K., and Horber, J. K. (2000). Sphingolipid-cholesterol rafts diffuse as small entities in the plasma membrane of mammalian cells. J Cell Biol. *148*, 997–1008.

Reid, A. H., Fanning, T. G., Janczewski, T. A., McCall, S., and Taubenberger, J. K. (2002). Characterization of the 1918 'Spanish' influenza virus matrix gene segment. J. Virol. *76*, 10717–10723.

Ruigrok, R. W. H., Hirst, E. M. A., and Hay, A. J. (1991). The specific inhibition of influenza

A virus maturation by amantadine, an electron mircoscopic examination. J. Gen. Virol. *72*, 191–194.

Sakaguchi, T., Leser, G. P., and Lamb, R. A. (1996). The ion channel activity of the influenza virus M_2 protein affects transport through the Golgi apparatus. J. Cell Biol. *133*, 733-747.

Sakaguchi, T., Tu, Q., Pinto, L. H., and Lamb, R. A. (1997). The active oligomeric state of the minimalistic influenza virus M_2 ion channel is a tetramer. Proc. Natl. Acad. Sci. USA. *94*, 5000-5004.

Saldanha, J. W., Czabotar, P. E., Hay, A. J., and Taylor, W. R. (2002). A model for the cytoplasmic domain of the influenza A virus M2 channel by analogy to the HIV-1 Vpu protein. Protein Pept. Lett. *9*, 495–502.

Schroeder, C., Ford, C. M., Wharton, S. A., and Hay, A. J. (1994). Functional reconstitution in lipid vesicles of influenza virus M2 protein expressed by baculovirus, Evidence for proton transfer activity. J. Gen. Virol. *75*, 3477–3484.

Shaw, M. W., Choppin, P. W., and Lamb, R. A. (1983). A previously unrecognized influenza B virus glycoprotein from a bicistronic mRNA that also encodes the viral neuraminidase. Proc. Natl. Acad. Sci. USA. *80*, 4879–4883.

Shaw, M. W., Lamb, R. A., Erickson, B. W., Briedis, D. J., and Choppin, P. W. (1982). Complete nucleotide sequence of the neuraminidase gene of influenza B virus. Proc. Natl. Acad. Sci. USA. *79*, 6817–6821.

Shih, S. R., and Krug, R. M. (1996). Novel exploitation of a nuclear function by influenza virus, The cellular SF2/ASF splicing factor controls the amount of the essential viral M2 ion channel protein in infected cells. EMBO J. *15*, 5415–5427.

Shimbo, K., Brassard, D. L., Lamb, R. A., and Pinto, L. H. (1996). Ion selectivity and activation of the M_2 ion channel of influenza virus. Biophys. J. *70*, 1335–1346.

Shuck, K., Lamb, R. A., and Pinto, L. H. (2000). Analysis of the pore structure of the influenza A virus M(2) ion channel by the substituted-cysteine accessibility method. J Virol. *74*, 7755–61.

Simons, K., and Ikonen, E. (1997). Functional rafts in cell membranes. Nature. *387*, 569–572.

Skehel, J. J., Bayley, P. M., Brown, E. B., Martin, S. R., Waterfield, M. D., White, J. M., Wilson, I. A., and Wiley, D. C. (1982). Changes in the conformation of influenza virus hemagglutinin at the pH optimum of virus-mediated membrane fusion. Proc. Natl. Acad. Sci. USA. *79*, 968–972.

Skehel, J. J., Hay, A. J., and Armstrong, J. A. (1978). On the mechanism of inhibition of influenza virus replication by amantadine hydrochloride. J. Gen. Virol. *38*, 97–110.

Skehel, J. J., and Wiley, D. C. (2000). Receptor binding and membrane fusion in virus entry, the influenza hemagglutinin. Annu. Rev. Biochem. *69*, 531–569.

Smith, W., Andrewes, C. H., and Laidlaw, P. P. 1933. A virus obtained from influenza patients. Lancet. ii, 66–68.

Smondyrev, A. M., and Voth, G. A. (2002). Molecular dynamics simulation of proton transport through the influenza A virus M2 channel. Biophys. J. *83*, 1987–(1996).

Sugrue, R. J., Bahadur, G., Zambon, M. C., Hall-Smith, M., Douglas, A. R., and Hay, A. J. 1990a. Specific structural alteration of the influenza haemagglutinin by amantadine. EMBO J. *9*, 3469–3476.

Sugrue, R. J., Belshe, R. B., and Hay, A. J. 1990b. Palmitoylation of the influenza A virus M2 protein. Virology. *179*, 51–56.

Sugrue, R. J., and Hay, A. J. (1991). Structural characteristics of the M2 protein of the influenza A viruses, Evidence that it forms a tetrameric channel. Virology. *180*, 617–624.

Sunstrom, N. A., Premkumar, L. S., Premkumar, A., Ewart, G., Cox, G. B., and Gage, P. W. (1996). Ion channels formed by NB, an influenza B virus protein. J. Membr. Biol. *150*, 127–132.

Suomalainen, M. (2002). Lipid rafts and assembly of enveloped viruses. Traffic. *3*, 705–709.

Takeda, M., Leser, G. P., Russell, C. J., and Lamb, R. A. (2003). Influenza virus hemagglutinin concentrates in lipid raft microdomains for efficient viral fusion. Proc. Natl. Acad. Sci. USA. *100*, 14610–14617.

Takeda, M., Pekosz, A., Shuck, K., Pinto, L. H., and Lamb, R. A. (2002). Influenza a virus M2 ion channel activity is essential for efficient replication in tissue culture. J. Virol. 76, 1391–1399.

Takeuchi, K., and Lamb, R. A. (1994). Influenza virus M_2 protein ion channel activity stabilizes the native form of fowl plague virus hemagglutinin during intracellular transport. J. Virol. 68, 911–919.

Tang, Y., Zaitseva, F., Lamb, R. A., and Pinto, L. H. (2002). The gate of the influenza virus M2·proton channel is formed by a single tryptophan residue. J. Biol. Chem. 277, 39880–39886.

Thomas, J. M., Stevens, M. P., Percy, N., and Barclay, W. S. (1998). Phosphorylation of the M2 protein of influenza A virus is not essential for virus viability. Virology. 252, 54–64.

Tian, C., Gao, P. F., Pinto, L. H., Lamb, R. A., and Cross, T. A. (2003). Initial structural and dynamic characterization of the M2 protein transmembrane and amphipathic helices in lipid bilayers. Protein Sci. 12, 2597–2605.

Tian, C., Tobler, K., Lamb, R. A., Pinto, L. H., and Cross, T. A. (2002). Expression and initial structural insights from solid-state NMR of the M2 proton channel from influenza A virus. Biochemistry. 41, 11294–11300.

Tobler, K., Kelly, M. L., Pinto, L. H., and Lamb, R. A. (1999). Effect of cytoplasmic tail truncations on the activity of the M_2 ion channel of influenza A virus. J. Virol. 73, 9695–9701.

Tosteson, M. T., Pinto, L. H., Holsinger, L. J., and Lamb, R. A. (1994). Reconstitution of the influenza virus M_2 ion channel in lipid bilayers. J. Membr. Biol. 142, 117–126.

Treanor, J. J., Tierney, E. L., Zebedee, S. L., Lamb, R. A., and Murphy, B. R. (1990). Passively transferred monoclonal antibody to the M_2 protein inhibits influenza A virus replication in mice. J. Virol. 64, 1375–1377.

Tu, Q., Pinto, L. H., Luo, G., Shaughnessy, M. A., Mullaney, D., Kurtz, S., Krystal, M., and Lamb, R. A. (1996). Characterization of inhibition of M2 ion channel activity by BL-1743, an inhibitor of influenza A virus. J. Virol. 70, 4246–4252.

Varma, R., and Mayor, S. (1998). GPI-anchored proteins are organized in submicron domains at the cell surface. Nature. 394, 798–801.

Veit, M., Klenk, H.-D., Kendal, A., and Rott, R. (1991). The M2 protein of influenza A virus is acylated. Virology. 184, 227–234.

Wang, C., Lamb, R. A., and Pinto, L. H. 1994a. Direct measurement of the influenza A virus M_2 protein ion channel activity in mammalian cells. Virology. 205, 133-140.

Wang, C., Lamb, R. A., Pinto, L. H. (1995). Activation of the M_2 ion channel of influenza virus, A role for the transmembrane domain histidine residue. Biophys. J. 69, 1363–1371.

Wang, C., Takeuchi, K., Pinto, L. H., and Lamb, R. A. (1993). Ion channel activity of influenza A virus M_2 protein, Characterization of the amantadine block. J. Virol. 67, 5585–5594.

Wang, J., Kim, S., Kovacs, F., and Cross, T. A. (2001). Structure of the transmembrane region of the M2 protein H(+) channel. Protein Sci. 10, 2241–2250.

Wang, J. J., Lu, Y.-L., and Ratner, L. 1994b. Particle assembly and Vpr expression in human immunodeficiency virus type 1-infected cells demonstrated by immunoelectron microscopy. J. Gen. Virol. 75, 2607–2614.

Watanabe, S., Imai, M., Ohara, Y., and Odagiri, T. (2003). Influenza B virus BM2 protein is transported through the trans-Golgi network as an integral membrane protein. J. Virol. 77, 10630–10637.

Watanabe, T., Watanabe, S., Ito, H., Kida, H., and Kawaoka, Y. (2001). Influenza A virus can undergo multiple cycles of replication without M2 ion channel activity. J. Virol. 75, 5656–5662.

Wiley, D. C., and Skehel, J. J. (1987). The structure and function of the hemagglutinin membrane glycoprotein of influenza virus. Annu. Rev. Biochem. 56, 365–394.

Williams, M. A., and Lamb, R. A. (1989). Effect of mutations and deletions in a bicistronic mRNA on the synthesis of influenza B virus NB and NA glycoproteins. J. Virol. 63, 28–35.

Winter, G., and Fields, S. (1980). Cloning of influenza cDNA into M13, The sequence of the RNA segment encoding the A/PR/8/34 matrix protein. Nucleic Acids Res. *8*, 1965–1974.

Zebedee, S. L., and Lamb, R. A. (1988). Influenza A virus M_2 protein, monoclonal antibody restriction of virus growth and detection of M_2 in virions. J. Virol. *62*, 2762–2772.

Zebedee, S. L. and Lamb, R. A. (1989). Growth restriction of influenza A virus by M_2 protein antibody is genetically linked to the M_1 protein. Proc. Natl. Acad. Sci. USA. *86*, 1061–1065.

Zebedee, S. L., Richardson, C. D., and Lamb, R. A. (1985). Characterization of the influenza virus M_2 integral membrane protein and expression at the infected-cell surface from cloned cDNA. J. Virol. *56*, 502–511.

Zhang, J., Pekosz, A., and Lamb, R. A. (2000). Influenza virus assembly and lipid raft microdomains, a role for the cytoplasmic tails of the spike glycoproteins. J. Virol. *74*, 4634–4644.

Zhirnov, O. P. (1990). Solubilization of matrix protein M1/M from virions occurs at different pH for orthomyxo- and paramyxoviruses. Virology. *176*, 274–279.

Zhong, Q., Husslein, T., Moore, P. B., Newns, D. M., Pattnaik, P., and Klein, M. L. 1998a. The M2 channel of influenza A virus, a molecular dynamics study. FEBS Lett. *434*, 265–271.

Zhong, W., Gallivan, J. P., Zhang, Y., Li, L., Lester, H. A., and Dougherty, D. A. 1998b. From ab initio quantum mechanics to molecular neurobiology, a cation-pi binding site in the nicotinic receptor. Proc. Natl. Acad. Sci. USA. *95*, 12088–12093.

CHAPTER 4

Receptor specificity, host-range, and pathogenicity of influenza viruses

Mikhail N. Matrosovich, Hans-Dieter Klenk and Yoshihiro Kawaoka

Abstract

Influenza viruses attach to target cells via multivalent interactions of the viral hemagglutinin protein with sialyloligosaccharide moieties of cellular glycoconjugates. The interactions between the virus and cellular receptors and extracellular inhibitors determine virus host-range and tissue tropism. Sialic acids are ubiquitous on the surface of most avian and mammalian cells. Therefore, in addition to mediating infection of susceptible cells, influenza viruses can bind to a variety of other cell types leading to significant biological responses, such as polyclonal activation of B-lymphocytes, deactivation of neutrophils, and stimulation of inflammatory responses. Here, we discuss current knowledge of the influenza virus interactions with cellular receptors at the molecular level, outline methods used to characterize receptor specificity of influenza viruses, and give an overview of available data on the role of virus receptor specificity in host range restriction, interspecies transmission, and pathogenicity.

RECEPTORS AND RECEPTOR DETERMINANTS OF INFLUENZA VIRUSES

Virus receptors are defined as cell-surface molecules used by viruses to attach to target cells and initiate infection. The parts of these cellular molecules that directly interact with the virus are known as receptor determinants (Paulson, 1985). Specific receptors for influenza viruses are still not known, but sialic acids were identified as the receptor determinants long ago (Klenk et al., 1955; Rosenberg et al., 1956).

Sialic acids (Sias) are a family of 9-carbon acid sugars that are abundant at the outer cell membranes and in biological fluids of animals. They occupy terminal positions on oligosaccharide chains of glycoproteins and glycolipids. For reviews on the chemistry of sialic acids, their occurrence in nature, and their biological functions, see Schauer, 1982; Varki, 1992; Schauer et al., 1995; Reuter and Gabius, 1996; Kelm and Schauer, 1997; Schauer and Kamerling, 1997; Varki, 1997.

Glycoproteins and gangliosides as virus receptors

Glycoproteins and gangliosides carry similar sialyloligosaccharide sequences and can bind influenza viruses in vitro (reviewed by Paulson, 1985; Suzuki, 1994). However, the relative roles played by glycoproteins and gangliosides in virus infection *in vivo* are not yet defined (for a discussion, see Bukrinskaya, 1982; Paulson, 1985; Herrler et al., 1995). Incubation of sialidase-treated cells with exogenous gangliosides restores influenza A and C virus binding and penetration into such cells (Bergelson et al., 1982; Bukrinskaya et al., 1982; Herrler and Klenk, 1987), suggesting that gangliosides alone can serve as functional receptors of influenza viruses.

Glycoproteins alone can also mediate virus infection, since mutant cells that lack the ability to synthesize gangliosides are susceptible to influenza A and Sendai virus infection (Matrosovich et al., 2000a; Ablan et al., 2001).

Sialyloligosaccharides as receptor determinants
Sialic acid species
The common distinct features of sialic acids are a carboxylic group, an amido group and a glycerol tail (C_7-C_9) attached to a pyranose ring at positions 2, 5, and 6, respectively (Figure 1). More than 40 members of this family are currently recognized, and they differ by substituents at N_5 and O_4, O_7, O_8, and O_9 (Varki, 1992, 1997; Schauer and Kamerling, 1997).

Two acyl substituents at N_5 are found in nature. N-acetylneuraminic acid (Neu5Ac) is the most common derivative and a biosynthetic precursor of other sialic acids. Enzymatic hydroxylation of the 5-N-acetyl group leads to N-glycolylneuraminic acid (Neu5Gc), which is found in all mammals except humans (Schauer, 1982; Varki, 1992).

Other natural sialic acids arise from substitution of one or more of the hydroxyl groups of Neu5Ac and Neu5Gc with acetyl groups and, less often, with methyl, lactyl, or sulfate groups. Among these O-acylated analogs, only those with 4-O-acetyl, 7-O-acetyl and 9-O-acetyl substitutions have been tested for their ability to bind to influenza viruses (Levinson et al., 1969; Higa et al., 1985; Pritchett and Paulson, 1989; Sauter et al., 1989, Matrosovich et al., 1992; Sauter et al., 1992; Matrosovich et al., 1997). The former two species bind to some influenza A virus strains; 9-O-acetylated sialic acid species do not bind to influenza A viruses (Higa et al., 1985) but do serve as receptor determinants of influenza C viruses (Rogers et al., 1986; Herrler and Klenk, 1987; Herrlerr et al., 1995). The role of other natural sialic acid derivatives has not been defined. Most of them, probably, either cannot support binding of influenza viruses (see below), or do not play a significant role in binding because they are present in low quantities on target cells. Thus, despite a significant diversity of sialic acid species, the non-O-acetylated Neu5Ac and Neu5Gc (the latter, in non-human hosts) appear to be the main species used by influenza A viruses to infect their natural hosts.

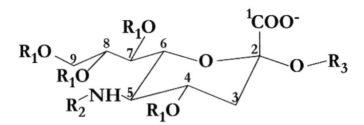

Figure 4.1 Structure of sialic acids. The 9-carbon backbone common to all sialic acids is shown. Natural substituents of sialic acids: R_1, acetyl (Sia positions 4,7,8,9), lactyl (9), methyl (8), sulfate (8,9), phosphate, sialic acid (8,9), fucose (4), glucose (8), or galactose (4). R_2, acetyl or glycolyl. R_3 Gal (3'/4'/6'), GalNAc (6'), GlcNAc (4'/6'), Sia (8'/9'). Modified from Varki (1992, 1997).

Types of α-Glycosidic linkage between sialic acids and penultimate sugars and diversity of asialic parts of sialyloligosaccharides

Sialic acids exist in nature predominantly as components of sialylglycoconjugates and form exclusively α-glycosidic linkages. D-galactose (Gal), N-acetyl-D-galactosamine (GalNAc) and, less often, N-acetyl-D-glucosamine (GlcNAc), and sialic acid (Sia) function as the penultimate sugars of oligosaccharide acceptors to which the sialic acid is attached. In general, sialic acids are α2-3- or α2-6-linked to Gal and GalNAc, α2-6-linked to GlcNAc, or α2-8-linked to the second sialic acid residue. Few other types of linkage have been identified, for example, Sia(α2-4)Gal, Sia(α2-4)GlcNAc, and Sia(α2-9)Sia, but these occur less often, and their possible recognition by influenza viruses has not been studied. Influenza viruses cannot bind to gangliosides that carry Neu5Ac(α2-8)Neu5Ac-moieties (Suzuki et al., 1986;1992), or to tetrasaccharide Neu5Ac(α2-8)Neu5Ac(α2-3)Gal(β1-4)Glc in solution (Matrosovich et al., 1993). Therefore, variants of the sialic acid glycosidic linkage to the penultimate sugar that could play a major role in influenza virus binding in nature are likely limited to Sia(α2-3/6)Gal, Sia(α2-3/6)GalNAc, and Sia(α2-6)GlcNAc. Sialic acids are usually attached at the terminal position of oligosaccharide chains; in fact, internal location of sialic acids prevents their binding by influenza viruses (Rogers and Paulson, 1983; Suzuki et al., 1986; Sauter et al., 1989; Suzuki, 1994; Gambaryan et al., 1995; Suzuki et al., 1992).

Although the number of structurally distinct sialyldisaccharide residues is limited, they can be present in highly variable microenvironments (see review by Schauer and Kamerling, 1997), and their binding can be substantially affected by the structure of more distant parts of the oligosaccharide chain (Paulson, 1985; Suzuki, 1994; Gambaryan et al., 1995; Müthing, 1996; Miller-Podraza et al., 2000; Gambaryan et al., 2003). The diversity of the 'primary' structures of sialic acid-containing natural molecules is increased by some degree of conformational flexibility of oligosaccharides, variations in number and positions of the saccharide chains in the glycoproteins, proximity and orientation of the sugar chains with respect to the cell surface, and their different steric accessibility in the context of other cellular components. A universal term 'presentation' is often used to accommodate all of these variables that affect recognition of cell-surface carbohydrate receptors by animal and viral lectins (Weis, 1997). The presentation of influenza virus receptors is not well understood with current knowledge being mostly limited to the virus recognition of the terminal sialic acid moiety and the penultimate galactose residue as discussed below.

STRUCTURAL FEATURES OF THE HA RECEPTOR-BINDING SITE AND MOLECULAR INTERACTIONS INVOLVED IN RECEPTOR RECOGNITION

Hemagglutinin (HA) is the major surface glycoprotein of influenza A and B viruses responsible for virus attachment to susceptible cells and for mediating fusion between viral and endosomal membranes. HA is also the primary target for the humoral immune response in the infected host. HA is a type I integral membrane protein with an N-terminal signal sequence, a membrane-anchor domain, and a short cytoplasmic tail at the C-terminus. The protein is posttranslationally modified by the addition of N-linked glycans (Keil et al, 1984, 1985), covalent attachment of palmitic acid to cysteine residues located in the cytoplasmic tail (Veit et al., 1990, 1991), and cleavage of the signal peptide. Monomeric HA molecules associate non-covalently in homotrimers during maturation. The final processing step is cleavage of HA monomers into two subunits, HA1 and HA2, connected by a single disulfide bond. For reviews, see Rott and

Klenk (1987); Wiley and Skehel (1987); Wilson and Cox (1990); Lamb and Krug (1996); Garten and Klenk (1999); Steinhauer (1999); Skehel and Wiley (2000).

Polyvalency of virus binding to receptors

The influenza virus particle harbors about 400 to 500 HA trimeric spikes (Ruigrock, 1998), and hence the virion is capable of simultaneous interactions via several receptor-binding sites with multiple copies of cell-surface receptors. The characteristic feature of such cooperative polyvalent binding is that the high avidity of binding is mediated by multiple low-affinity individual contacts. This type of attachment is typical for many carbohydrate-binding viruses, bacteria, and lectins (Lonberg-Holm and Philippson, 1974; Karlsson, 1989; Matrosovich, 1989; Kiessling and Pohl, 1996; Weis, 1997). The interactions of individual HAs with monovalent sialyloligo-saccharide determinants are very weak, with dissociation constants in the range of 0.1–5 mM (Pritchett et al., 1987; Sauter et al., 1989; Gambaryan et al., 1995; Gambaryan et al., 1999). However, the binding to the virus can be increased 10^2-fold by bridging two sialosides with a synthetic spacer (Glick et al, 1991) or greater than 10^3-fold by coupling multiple copies of sialo-sides to a macromolecular carrier (Matrosovich et al, 1990; Spaltestein and Whitesides, 1991).

Because influenza viruses are polyvalent, their attachment to the cell depends on receptor density; viruses with lower affinity for the receptors require higher receptor density for efficient binding (Herrler and Klenk, 1987; Martin et al., 1998). Because several HA trimers likely participate in virus attachment, membrane-distal portions of HA far outside the sialic acid binding site can markedly affect the binding. For example, N-linked glycans at the top of the HA molecule usually have profound effects on receptor-binding activity as they sterically interfere with the virus fit to multiple copies of the receptor (see page 104).

The sialic acid-binding pocket and recognition of the sialic acid moiety

The crystal structure of the HA ectodomain (containing intact HA1 and the major portion of HA2) and of its complexes with a number of natural and synthetic sialic acid compounds has been determined by X-ray analysis. Most studies were done using the high-growth reassortant virus, X-31, that bears the HA and NA of the human virus A/Aichi/2/68 (H3N2) (reviewed by Wiley and Skehel, 1987; Skehel and Wiley, 2000). More recently, structures of human H1, avian H3 and H5, and swine H1 and H9 HAs have also been determined (Ha et al., 2002; 2003; Gamblin et al., 2004; Stevens et al., 2004). The structures of the other 11 antigenic HA subtypes of influenza A viruses and of type B viruses remain unresolved; however, a 40% to 70 % amino acid homology between these HAs and H3 HA suggests that the main structural features of all HAs are the same. Throughout this review, the H3 amino numbering system is used, which is based on the alignment presented in Figure 6. For the location of amino acids on the 3-D HA model, see Figures 2–4.

HA consists of a fibrous stem, formed by amino acids that belong to both the HA1 and HA2 subunits, and of a globular head that is composed entirely of amino acid residues from HA1. The receptor-binding pocket that accommodates the sialic acid residue is a shallow depression on the protein surface located on the membrane distal end of the globular head. This pocket is formed by amino acid residues that are conserved among different virus strains (Figures 2 and 6) and is surrounded by amino acids that vary in response to the immune pressure exerted by the host (Wiley and Skehel, 1987; Wilson and Cox, 1990). Some residues in the region of the receptor-binding site (RBS) are likely conserved because they are essential for virus interactions with the receptor, others may be important to maintain the structural integrity of the protein and/or of the RBS.

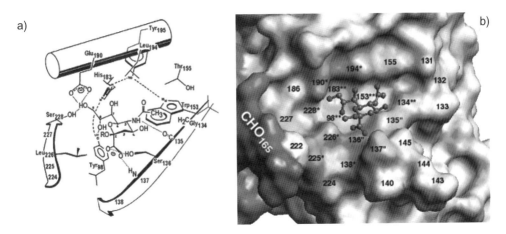

Figure 4.2 Scheme (**a**) and molecular model (**b**) of the HA sialic acid-binding pocket based on the crystal structure of X-31 virus HA complexes with sialosides (Weis et al., 1988; Sauter et al., 1992). Figure **a** is modified from Kelm et al., 1992. Dotted lines indicate possible hydrogen bonds between sialic acid and the RBS, dashed lines show potential hydrogen bond within the protein. Figure **b** shows position of sialic acid moiety (ball and stick model) in the binding pocket (solvent accessible surface). Different colors of amino acids are solely for the identification purposes. Two stars next to the amino acid number indicate that amino acid is conserved among all influenza A viruses. One star indicates that amino acid is conserved among avian viruses and changed in human influenza A viruses. " indicates that atoms interacting with sialic acid moiety are conserved. CHO_{165} – a portion of N-linked glycan attached at Asn_{165} of the HA that was resolved by X-ray analysis. Figures 2b, 3, and 4 were generated using DS ViewerPro 5.0 (Accelrys Inc.) This figure is also reproduced in colour in the colour section at the end of the book

    Various natural and synthetic sialosides modified at distinct positions of the Neu5Ac moiety have been tested for their binding to the X-31 strain (Hanson et al., 1992; Sauter et al., 1992), to the closely related human H3N2 virus, Memphis/102/72, (Pritchett et al., 1987; Pritchett and Paulson, 1989; Kelm et al., 1992), and to a variety of other influenza A and B viruses (Rogers et al., 1986; Matrosovich et al., 1991, 1992, 1993; Suzuki, 1994; Tuzikov et al., 1997). A combination of structural and binding data has identified the essential interactions between the Neu5Ac moiety and the RBS and the contributions of different substituents in the sialic acid molecule to binding.

    The bottom of the X31 HA RBS is formed by Tyr_{98}, Trp_{153}, His_{183}, and Tyr_{195}, which create a network of hydrogen bonds (Figure 2). Side chains of Glu_{190} and Leu_{194} and a side chain of Thr_{155} define the rear of the site. Residues 134 to 138 form the 'right' side of the pocket, and residues 224 to 228 form the 'left' side. The sialic acid is bound with one side of the pyranose ring in tight contact to the protein and the other side facing the solvent. Each of the ring substituents unique for sialic acid (carboxylic group, 5-acetamido group, and C_7-C_9 polyhydroxylated moiety) interacts with the HA. The carboxylic group is hydrogen bonded to the hydroxylic group of Ser_{136} and to the main chain amide group of Asn_{137}. The 5-N-acetyl group is located over the indole ring of Trp_{153} and participates in van der Waals and hydrophobic interactions with this moiety. The 5-acetamido nitrogen forms a hydrogen bond with the main chain carbonyl at position 135. The C-8-hydroxyl is hydrogen bonded to Tyr_{98}, the amino acid conserved among all influenza A virus strains and substituted by

Phe$_{98}$ in type B HAs. The C-9-hydroxyl is within hydrogen bonding distance from Tyr$_{98}$, Glu$_{190}$, Ser$_{228}$, and His$_{183}$ in the crystal structure of X-31 HA, however, neither of the possible hydrogen bonds appears to be strong, since substitution of the 9-OH of the sialic acid moiety by 9-H does not change the binding of the modified analog to the virus (Kelm et al., 1992). This feature is likely a unique characteristic of the X-31 virus and of the other human H3 strains isolated in the first years after the 1968 pandemic. Most other influenza A and B viruses tested, including the H3N2 human strains isolated after 1973, require the C-9- hydroxylic group of sialic acid for the efficient binding (Matrosovich et al., 1991).

The carboxylic group and the acyl group at N$_5$ are indispensable for the binding of sialic acid to all influenza A and B viruses (Kelm et al., 1992; Sauter et al., 1992; Matrosovich et al., 1993). The atoms contacting these groups in the receptor-binding site (α-CH$_2$-group of Gly$_{134}$, side chain hydroxylic group of Ser/Thr in position 136, and side chain atoms of Trp$_{153}$) are strictly preserved during the evolution of the viruses in different hosts (see Figure 6). In contrast, the atomic counterparts of the sialic acid in the 'left' side of the RBS (positions 190, 225, 226, and 228) are more variable. In particular, avian viruses always carry Glu$_{190}$, Gly$_{225}$, Gln$_{226}$, and Gly$_{228}$, whereas human viruses typically have Asp$_{190}$ (H1N1 strains and H3N2 viruses isolated after 1992), Gln$_{190}$ (type B), Asp$_{225}$ (H1N1 viruses), Leu/Ile/Val in position 226 and Ser$_{228}$ (H3N2 and H2N2 viruses). A laboratory mutant of X-31 virus with a deletion of residues 224–230 was still able to bind receptors and grew in embryonated eggs (Daniels et al., 1987), and H7N2 viruses with the deletion of HA residues 221–228 were isolated from domestic chickens (Suarez et al., 1999). These variations indicate that the left side of the RBS is less critical for interactions with the sialic acid residue.

The tight packing of hydroxylic groups at C$_9$, C$_8$, and C$_7$ of Neu5Ac to the amino acids in the RBS suggests that acylation of these hydroxyls would interfere with binding. Indeed, none of the various influenza A viruses tested to date bind to 9-O-acetylated sialic acid (Higa et al., 1985). Data on binding to 7-O-acetylated N-acetylneuraminic acid are limited to a study on a single strain (X-31) that showed the virus did not bind to this sialic acid species (Sauter et al., 1989, 1992). No data are available on the recognition of 8-O-acylated analogs, however, the crystal structure of the HA-sialic acid complex (Figure 2) clearly indicates that these analogs could not fit into the binding pocket.

In contrast to the C$_7$-C$_9$ glycerol tail, the hydroxyl at C$_4$ does not interact with the HA of X-31 and faces toward the solvent. Consistent with this, 4-O-acetylation of Neu5Ac does not appreciably influence receptor binding by this strain and by most other H3-subtype human viruses (Pritchett and Paulson, 1989; Sauter et al., 1989, 1992; Matrosovich et al., 1992). However, some human H3 strains, all H1 human viruses, and type B viruses do not bind 4-O-acetylated sialic acids (Levinson et al., 1969; Matrosovich et al., 1992, 1998). The crystal structure of the X-31 HA in complex with the methyl glycoside of Neu4, 5Ac$_2$ (Sauter et al., 1992) shows that the carbonyl oxygen of the 4-O-acetyl group occupies the space between the main chain oxygen of Gly$_{135}$ and the hydroxylic group of the Ser$_{145}$. Mutations in positions 145, 190 and 226 of H3 subtype viruses abolish the binding of Neu4, 5Ac$_2$ (Matrosovich et al., 1992, 1998). All three mutations likely cause steric interference of the 4-O-acetyl substituent with its atomic counterparts in the RBS, either directly (145), or allosterically (190 and 226). With the exception of horses, no natural host of influenza virus is known to synthesize Neu4, 5Ac$_2$ (Kammerling and Schauer, 1997; Schauer et al., 1982). Therefore, there appears to be neither positive nor negative selective pressures on virus recognition of this sialic acid modification in birds, pigs, and humans. This could explain significant type, subtype, and strain variations with respect to virus recognition of Neu4, 5Ac$_2$.

N-glycolyl neuraminic acid differs from Neu5Ac by the substitution of the methyl group (CH_3) of the 5N-acetamido moiety for a bulkier and more hydrophylic hydroxymethyl group (CH_2-OH). There is a correlation between the ability of the H3-subtype human viruses to bind Neu5Gc and the nature of the amino acid at position 155 of HA (Higa et al, 1985; Anders et al., 1986; Tuzikov et al., 1997; Matsuda et al., 1999). This is consistent with the close proximity of the N-acyl group to the side chain of amino acid 155 (Figure 2).

Recognition of the type of the Sia-Gal glycosidic linkage

Sialyloligosaccharides bearing terminal Sia(α2-3)Gal- and Sia(α2-6)Gal-moieties have different molecular shapes, and influenza viruses can discriminate between them. Human and avian influenza viruses are especially selective in this respect, with human viruses preferentially binding to Sia(α2-6)Gal-terminated receptors and avian viruses preferring Sia(α2-3)Gal-containing ones (see below).

HA-receptor interactions in the region of the glycosidic linkage

Comparison of the binding of different influenza viruses to free Neu5Ac and to monovalent synthetic and natural sialosides suggests that in addition to binding to terminal sialic acid, influenza viruses also interact with penultimate galactose (Matrosovich et al., 1993, 1997, 1999, 2000; Gambaryan et al., 1995, 1999). Depending on the type of Sia-Gal linkage, these interactions can strengthen the HA-receptor complex. For example, avian viruses display at least a tenfold higher affinity for 3'-sialyllactose (Neu5Ac(α2-3)Gal(β1-4)Glc, 3'SL) than for free Neu5Ac. This property clearly differentiates avian influenza viruses from human and swine viruses, which bind Neu5Ac and 3'SL with comparable affinity, that is, they do not appreciably interact with 3-linked Gal. By contrast with 3'SL, avian viruses typically bind to 6'-sialyllactose (Neu5Ac(α2-6)Gal(β1-4)Glc, 6'SL) and to 6'-sialyl(N-acetyllactosamine) (Neu5Ac(α2-3)Gal(β1-4)GlcNAc, 6'SLN) with less affinity than they bind to Neu5Ac, which suggest the interactions between the avian HA and the 6-linked Gal moiety are energetically unfavorable.

Insights into the molecular mechanisms of these interactions come from the crystal structures of the HA complexes with the sialylpentassacharides LSTa and LSTc (Eisen, 1997; Ha et al., 2002, 2003). For H3 and H5 avian HAs, the α(2-3)-linked sialyloligo-saccharide (LSTa) is bound with the C3 methylene group of the linkage and the whole 3-linked Gal being projected upward (Figure 3). This is the so-called *trans* (or *syn*) conformation of the α(2-3)-glycosidic linkage, which corresponds to the lowest energy conformation of non-complexed 2-3-oligosaccharides in solution (Breg et al., 1989; Poppe et al., 1989). In this conformation, the glycosidic oxygen and the axial 4-OH group of the Gal hydrogen bond to the side chain amide and carbonyl groups of the Gln_{226}. The α(2-6)-sialyloligosaccharide (LSTc) binds to avian HAs in its own lowest energy *cis* (or *anti*) conformation (Poppe et al., 1992). In this case, the hydrophobic C6 methylene group of the α(2-6)-linkage projects downward toward the polar atoms of the side chain of Gln_{226}. These unfavorable polar-nonpolar interactions between 6-linked Gal and Gln_{226} appear to account, in part, for the poor binding of avian viruses to 2-6-linked receptors. Given the unfavorable contacts between the 6-linked Gal and Gln_{226}, the 2-6-linked oligosaccharide binds less deeply in the RBS than does the 2-3-linked analog (Ha et al., 2003). As a result, potential hydrogen bonds between the analog and RBS are generally longer (weaker) in the 2-6-complex than in 2-3-complex. In other words, the receptor preference of avian viruses is based on the energetically optimal fit of the most populated *trans* conformer of

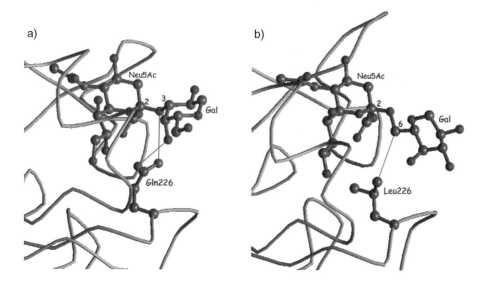

Figure 4.3 HA interactions with (α2-3)-linked and (α2-6)-linked galactose residues.
a. Neu5Ac(α2-3)Gal moiety of pentasaccharide LSTa [Neu5Ac(α2-3)Gal(β1-3)GlcNAc(β1-3)Gal(β1–4)Glc] (ball and stick model) in the receptor-binding site of A/Duck/Ukraine/63 (H3N8) (Ha et al., 2003). The galactose residue is bound in the minimum-energy *syn* conformation of the glycosidic linkage that allows hydrogen bonding (*red* lines) of the glycosidic oxygen and the 4-OH group of Gal to the side chain atoms of Gln_{226}.
b. Neu5Ac(α2-6)Gal moiety of pentasaccharide LSTc [Neu5Ac(α2-6)Gal(β1-4)GlcNAc(β1-3)Gal(β1-4)Glc] in the RBS of X31 HA (Eisen et al., 1997). The Gal residue binds in the minimum-energy *anti* conformation, in which the C6-methylene group of Gal participates in van-der-Waals and hydrophobic interactions (*grey* line) with the nonpolar side chain of Leu_{226}. The HA protein backbone is depicted by cyan tubes on both figures. This figure is also reproduced in colour in the colour section at the end of the book

the Neu5Acα(2-3)Gal moiety and on a poor fit of the minimum energy *cis* conformer of the Neu5Acα(2-6)Gal to the avian HA RBS.

The RBS of human X31 virus (H3N2) differs from the avian HA RBS by a few amino acids, in particular, X31 HA carries Leu_{226} rather than Gln_{226}. Both LSTa and LSTc bind to the X31 HA in a *cis* conformation of the Sia-Gal linkage (Eisen et al., 1997). The C3/C6 methylene group of the glycosidically linked Gal faces towards Leu_{226} and could participate in van-der-Waals and hydrophobic interactions with the nonpolar side chain of Leu_{226} (Figure 3b). Indeed, human H3N2 viruses typically bind α2-6-linked sialosides including the methyl-α-sialoside with greater affinity than they bind free sialic acid (Matrosovich et al., 1993; 1997). These data confirm that the RBS of human H3 viruses binds to the first methylene group of the asialic part of sialosides. Thus, human viruses bind the Neu5Acα(2-6)Gal-moiety with greater affinity than do avian viruses due to the energetically favourable interactions of the 6-linked Gal with Leu_{226}. At the same time, a lack of Gln_{226} prevents the human virus HA from hydrogen bonding to the glycosidic oxygen and to the 4-OH group of 3-linked Gal, thus destroying the 'α2-3-specific recognition motif' (Ha et al., 2002) of the avian RBS.

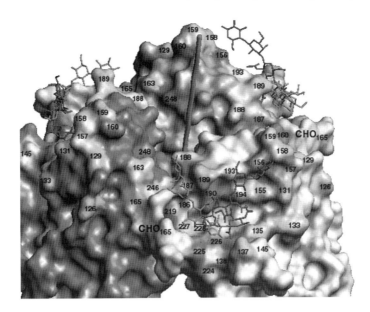

Figure 4.4 X31 HA complexes with pentasaccharides LSTc (*red*) and LSTa (*green*) (stick models) superimposed on the same HA model (solvent accessible surface). The sialic acid moiety is shown in magenta. Individual HA monomers are tinted in shades of gray. Positions of some amino acid residues that were shown to affect the receptor-binding activity and are discussed in this review are colored and numbered. The figure is based on crystallographic data of Eisen et al. (1997). This figure is also reproduced in colour in the colour section at the end of the book

Different orientation of the asialic parts of oligosaccharides

Conformational analysis of free sialyloligosaccharides in solution and the crystal structures of their complexes with HA show that the asialic parts of the 2-6-linked and 2-3-linked oligosaccharides have different spatial orientation. For example, in the X31 HA complex with the 2-6-analog LSTc (Eisen et al., 1997), the third saccharide, GlcNAc, is positioned directly over the terminal Neu5Ac with the 4th and 5th saccharide residues exiting the RBS to the right in a relatively close proximity to the rim of the RBS formed by amino acid residues Thr_{131}, Lys_{156}, Ser_{157}, Gly_{158} and Ser_{193} (Figure 4, LSTc). By contrast, the 2-3-linked asialic portion of the LSTa is situated to the left of the sialic acid residue in a groove formed by the side chain atoms of Thr_{187}, Gln_{189}, Ser_{219}, Trp_{222}, Ser_{186}, Ser_{227}, the backbone atoms of Gly_{218}, Arg_{220}, Ser_{227}, and Ser_{228}, and the carbohydrate attached to the Asn_{165} of the neighboring HA monomer (Figure 4, LSTa). These differences suggest that, depending on the type of Sia-Gal linkage, and also on the structure of the oligosaccharide core, different regions of the HA surface come in contact with the oligosaccharide chains and with the protein or lipid part of the receptor. Although virus interactions with distant parts of oligosaccharide and protein/lipid moieties of receptors have not been well studied to date, it seems evident that they can contribute substantially to virus recognition of the 2-3 versus 2-6 linked receptor determinants by either stabilizing the binding in the presence of energetically favourable bonds, or by diminishing it in the presence of sterical conflicts. Some examples of these phenomena will be given throughout this review.

Effects of N-linked glycans

The receptor-binding characteristics of HA can be affected by the number, position, and host-determined structure of N-linked carbohydrates located in the vicinity of the RBS and/or on the tip of the HA globular head (reviewed by Schulze, 1997; Klenk et al., 2002). Schulze and colleagues studied human virus A/WSN/33 (H1N1) and its variant that lacks one N-linked glycan from its HA (Crecelius et al., 1984; Deom et al., 1986) and found that oligosaccharide attached to Asn_{129} (CHO_{129}) decreased the affinity of the virus for cellular receptors and receptor analogs. The effect of CHO_{129} was dependent on the host cell, a carbohydrate attached to the virus grown in Madin-Darby bovine kidney (MDBK) cells had a much stronger unfavorable effect on binding than CHO_{129} attached to virus grown in chicken embryo fibroblasts (CEF). The HA protein in MDBK-cells was shown to contain about 4000 Daltons of carbohydrate more than that found on the HA from CEF. Moreover, the MDBK-produced carbohydrates were more branched than those attached in CEF (Deom and Schulze, 1985). These data suggest that bulky oligosaccharides on the HA of MDBK-grown virus sterically interfere with the interaction between the virus and receptors. Similarly, the HA of influenza viruses grown in Madin-Darby canine kidney cells (MDCK) contains more bulky N-glycans than do HAs from egg-grown viruses (Inkster et al., 1993; Romanova et al., 2003), and the receptor-binding activity of MDCK-grown viruses is generally lower than that of their egg-grown counterparts (Gambaryan et al., 1998a, 1999).

Examples of how HA glycans modulate the receptor-binding activity of various influenza A and B viruses have been reported (Günther et al, 1993; Matrosovich et al., 1997; Gubareva et al., 1998; Gambaryan et al., 1998a, 1999; Matrosovich et al., 1998, 1999; Govorkova et al., 1999; Ohuchi et al., 1999; Wagner et al., 2000). At least two mechanisms for such modulation can be suggested: i) steric hindrance of the receptor-binding site by an adjacent carbohydrate that interferes with the recognition of the receptor determinants; and ii) steric conflicts between the carbohydrate chains and the remote domains of the receptor molecules or the cell surface, thereby interfering with polyvalent virus binding. Most of the experimental data indicate that the presence of carbohydrates on the HA head typically decreases the binding affinity. An unusual mechanism of modulation of HA activity by N-glycans was reported by Ohuchi et al. (1995, 1997), who found that the HA of A/FPV/Rostock/34 (H7N1) expressed in the absence of viral neuraminidase could not agglutinate erythrocytes because of the sialylation of glycans attached at Asn_{133} and Asn_{158}. This finding indicates that sialylation of the N-glycans may provide an additional obstacle for HA interaction with receptors and that viral NA may play a part in overcoming this problem.

The effects of carbohydrates on virus binding affinity markedly depend on the nature of the receptor to which the virus binds (Crecelius et al., 1984; Aytay and Schulze, 1991), in particular, on the type of Sia-Gal linkage (Günther et al., 1993). This concept was confirmed by Gambaryan et al. (1998a), who evaluated the effects of CHO_{131} on the affinity of H1N1 human virus for a panel of receptor analogs and found that this carbohydrate interfered with the binding of the virus to Neu5Ac(α2-6)Gal-containing receptors, but had little or no effect on the binding of the virus to Neu5Ac(α2-3)Gal-containing receptors. Based on the model of Eisen et al. (1997), this specificity could be explained by the close proximity of CHO_{131} to the asialic portion of 2-6-linked receptors and its relatively long distance from the asialic part of 2-3-specific receptors (see Figure 4). Some other specific examples of the effects of N-linked glycans will be discussed in later sections.

Effects of charged amino acids on the HA globular head

Studies on three distinct groups of virus receptor-binding variants revealed noticeable effects of charged amino acid substitutions on the top of the HA globular head upon virus binding to macromolecular receptors and cells (Gambaryan et al., 1998b, 1999; Kaverin et al., 2000). Namely, mutations that decreased the negative charge of the amino acid side chain (for example, Glu→Gly), created a new positive charge (Asn→Lys), or that reversed the charge from negative to positive (Glu→Lys) always correlated with an increased binding affinity. At least some of these charged mutations apparently did not affect the virus interactions with the specific receptor determinants, as the affinity for free sialic acid and low molecular weight sialyloligosaccharides did not change. At the same time, all charged mutations enhanced the virus interaction with the negatively charged macromolecule dextran sulphate, which does not carry specific sialic acid determinants. Hence, charged amino acid substitutions appear to affect virus receptor-binding affinity by modulating the electrostatic interactions between the virus particle and the negatively charged sialylglycoconjugates and cells (Gambaryan et al., 1998b, 1999).

ASSAYS FOR CHARACTERISATION OF THE RECEPTOR-BINDING ACTIVITY OF INFLUENZA VIRUSES

Because the biological receptors of influenza viruses on their target cells have not been identified, most researchers study virus binding to more or less well-defined receptor analogues – natural and modified sialic acid-containing molecules or cells. The three most commonly used assays are discussed below.

Generation of specific receptor determinants using sialyltransferases

The introduction of this method by Paulson and colleagues was an important advance in research on the receptor specificity of viruses that bind to sialylglycoconjugates (see Paulson, 1985; Paulson and Rogers, 1987, for a review). In this method, erythrocytes are first treated with *V.cholerae* sialidase to abolish virus binding. The cells are then incubated with CMP-activated sialic acids and specific sialyltransferases to generate sialyloligosaccharides of defined sequence and sialic acid content. Initially, four purified mammalian sialyltransferases were used to generate the following sequences.

Neu5Ac(α2-6)Gal(β1-4)GlcNAc- (I);   Neu5Ac(α2-3)Gal(β1-3)GalNAc (II);
Neu5Ac(α2-3)Gal(β1-3/4)GlcNAc- (III); Neu5Ac(α2-6)[Gal(β1-3)]GalNAcαThr/Ser (IV).

Modified erythrocytes carrying these sequences were used for the hemagglutination assay and for virus adsorption studies. These experiments revealed for the first time that influenza viruses could significantly differ in their recognition of sequences and that receptor specificity correlated with the species of virus origin (Carrol et al., 1981, Rogers and Paulson, 1983). As sequences (I) and (II) appeared more suitable for discriminating influenza viruses from different hosts, the first two sialyltransferases continued to be used in the later studies (for examples, see Higa et al., 1985; Anders et al., 1986; Daniels et al., 1987; Rogers and D'Souza, 1989; Connor et al., 1994; Ito et al., 1998).

The same methodology was also applied to soluble glycoproteins (Matrosovich et al., 2003) and to live cells in tissue culture (Carrol and Paulson, 1985). The latter elegant study demonstrated that the ability of influenza viruses to infect cells depends on the presence of specific receptor determinants.

Binding to gangliosides

Unlike most sialylglycoproteins, gangliosides contain only one oligosaccharide chain per molecule. They can be purified to homogeneity, and the structure of individual species can be determined (Ledeen and Yu, 1982; Nagai and Iwamori, 1995). Moreover, the hydrophobic ceramide component easily anchors gangliosides in various assay media for binding studies.

Three major techniques to assess influenza virus binding to gangliosides have been used. Suzuki and his group incorporated gangliosides into sialidase-treated chicken erythrocytes and evaluated the ability of distinct ganglioside species to restore virus-mediated hemagglutination and hemolysis at low pH (Suzuki et al., 1985, 1986, 1989; Nobusawa et al., 1991). They revealed that attachment of viruses to gangliosides depends on the virus strain, the amount of the ganglioside adsorbed, the molecular species of sialic acid, and the structure of the asialic oligosaccharide portion of the ganglioside.

In the so-called TLC-overlay technique, the viruses bind to separated glycolipids directly on thin-layer chromatograms. This method was first devised to study the glycolipid receptor of cholera toxin (Magnani et al., 1980) and was later developed for analyses of the binding of viruses, bacteria, and cells (Hansson et al., 1984, Karlsson and Stromberg, 1987). Suzuki and colleagues adapted the TLC-overlay assay to analyze the binding of influenza A and B viruses to individual structurally defined gangliosides (Suzuki et al., 1992, 1996, 1997; Xu et al., 1994; Ito et al., 1997) (see Suzuki, 1994 for a review). This technique was also used to study the binding of human influenza viruses to complex mixtures of gangliosides from human and animal tissues (Müthing et al., 1993; Matrosovich et al., 1996; Müthing, 1996; Miller-Podraza et al., 1998, 2000; Gambaryan et al., 2002). In these experiments, the slowest moving, most polar gangliosides with the longest complex oligosaccharide chains revealed markedly higher receptor-binding activity compared to the more abundant, simple ganglioside species with shorter chains (3–7 sugars).

In the microwell adsorption technique, the virus binds to gangliosides adsorbed in the wells of plastic microtiter plates either alone or mixed with auxiliary phospholipids (Müthing et al., 1993; Matrosovich et al., 1996, 1997; Gambaryan et al., 1998b, 1999). The microwell assay requires less material and is easier to perform than the TLC-overlay assay. It is, therefore, more suitable for the simultaneous comparative analysis of a large number of influenza viruses strains or for the estimation of binding affinity from serial dilution of gangliosides.

Two ganglioside species, sialyl-3-paragloboside (3'SPG) and sialyl-6-paragloboside (6'SPG), which contain terminal Neu5Ac(α2-3)Gal- and Neu5Ac(α2-6)Gal-moieties, respectively, attached to a neolacto core [Gal(β1-4)GlcNAc(β1-3)Gal(β1-4)Glc-] have been found to be particularly useful for determining the ability of influenza viruses to recognize the Sia-Gal linkage (Suzuki et al., 1986; Nobusawa et al., 1991; Suzuki, 1994; Ito et al., 1997; Suzuki et al., 2000).

Binding to sialosides, sialylglycopolymers, and sialylglycoproteins

Measuring the binding affinity of the virus for structurally defined low molecular mass monovalent sialic acid glycosides (sialosides) permits the modelling of HA interactions with the terminal sialyloligosaccharide moieties of natural receptors. Synthetic analogs modified at distinct positions of the sialic acid moiety can also be used for structure-activity analysis (Sauter et al., 1989; Toogood et al., 1991; Sauter et al., 1992; Kelm et al., 1992; Matrosovich et al., 1993; Machytka et al., 1993; Itoh et al., 1995). A direct binding assay based on the perturbation of the NMR-spectrum of sialic acid in complex with HA was developed (Sauter et al., 1989, 1992; Hanson et al., 1992; Machytka et al., 1993). However, two less time- and material consuming competitive assays have since been developed and are more widely used. These assays rely on the ability of monovalent sialosides to block virus-mediated hemagglutination (Pritchett et al., 1987; Kelm et al., 1992) or to inhibit virus binding to peroxidase-labeled fetuin in a solid phase assay analogous to a competitive ELISA (Gambaryan and Matrosovich, 1992; Matrosovich et al., 1993).

Binding experiments with monovalent sialosides provide valuable information about HA-receptor interactions, but cannot account for the effects of presentation of the sialyloligosaccharide moiety in the context of the receptor macromolecule as a whole and for polyvalency of virus-receptor interactions. In addition, these experiments require large amounts of expensive sialosides because of the low binding affinity of the virus for monovalent receptors. Both drawbacks have been overcome with the generation of polyvalent sialylglycopolymers that are prepared by anchoring multiple copies of natural sialyloligosaccharides such as 3'SL, 6'SL, and 6'SLN to a hydrophilic polymeric carrier (Bovin et al., 1993; Bovin, 1998). The binding of sialylglycopolymers to the virus can be measured using the fetuin binding inhibition assay (Gambaryan et al., 1997). Alternatively, the binding of biotin-labeled sialylglycopolymers to solid-phase immobilized viruses can be assayed directly using streptavidin-peroxidase (Matrosovich et al., 2000). Both assays are quantitative, that is, allowing estimation of association constants for virus-receptor complexes. Simultaneous assaying of the virus affinity for monovalent sialosides and for corresponding sialylglycopolymers has proven to be useful for characterizing the receptor-binding properties of influenza viruses from different hosts (Gambaryan et al., 1997, 1998ab; Matrosovich et al., 1998, 2000).

Although natural sialylglycoproteins contain structurally different and heterogeneous oligosaccharide chains, few glycoproteins have been used in studies of the receptor-binding phenotypes of influenza viruses. Equine and pig α2-macroglobulins contain predominantly Sia(α2-6)Gal-terminated N-linked oligosaccharides (Hanaoka et al., 1989; Ryan-Poirrier and Kawaoka, 1993), and bind with a higher affinity to human influenza viruses than to avian viruses (Connor et al., 1994; Gambaryan et al, 1999; Rogers et al., 1983). Hen egg ovomucin contains predominantly Sia(α2-3)Gal-terminated oligosaccharides and binds to avian but not to human viruses (Gambaryan et al., 1999). Two peroxidase-labeled sialylglyco-proteins, pig macroglobulin and ovomucin, were successfully used to distinguish between the receptor binding phenotypes of avian and human viruses (Matrosovich et al., 1999, 2001). Rat α2-macroglubulin has been shown to be a potent inhibitor of influenza C virus-mediated hemagglutination (Herrler et al., 1985).

RECEPTOR SPECIFICITY OF INFLUENZA VIRUSES FROM DIFFERENT HOSTS

Aquatic birds as a source of influenza viruses in other species

The primary natural reservoir of influenza A viruses are wild aquatic birds, which harbor all currently known 15 HA and 9 NA antigenic subtypes (reviewed by Webster et al., 1992; Alexander, 2000; Horimoto and Kawaoka, 2001). Occasionally, these viruses transmit to sea mammals, land-based poultry, horses, swine, and humans, and cause infections of various severities. Shortly after interspecies transfer, the viruses usually die out because of poor fitness in their new hosts. However, on rare occasions, they adapt to efficiently replicate and transmit in the new species and continue to circulate for a prolonged period of time forming a stable host-specific virus lineage. All known lineages of influenza A viruses in land-based birds and mammals originated from the viruses of wild aquatic birds (Figure 5).

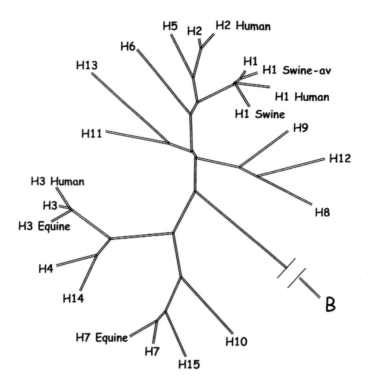

Figure 4.5 Relationships between influenza viruses of aquatic birds and mammals based on the complete HA amino acid sequences of representative virus strains. The tree was generated using neighbor-joining method implemented in the PHYLIP 3.572 package (Felsenstein, 1989) and drawn using TreeView 1.6.6 program (Page, 1996). The avian virus HAs are depicted by their antigenic subtype only; for the mammalian viruses, host species are additionally indicated: H1, A/duck/Wiskonsin/259/80; H1 Human, A/ Suita/1/89; H1 Swine, 'classical' swine virus A/swine/Nebraska/1/92; H1 Swine-av, 'avian-like' swine virus A/ Schleswig-Holstein/1/93; H2, A/mallard/New York/6750/78; H2 Human, A/Berkley/1/68; H3, A/duck/Hokkaido/33/80; H3 Human, A/Vienna/47/96; H3 Equine, A/equine/Kentucky/1/94; H4, A/duck/Czechoslovakia/56; H5, A/duck/Hong Kong/342/78; H6, A/shearwater/Australia/1/72; H7, A/duck/Hong Kong/293/78; H7 Equine, A/equine/New Market/1/77, H8, A/turkey/Ontario/6118/68; H9, A/turkey/Wisconsin/1/66; H10, A/chicken/Germany/N/49; H11, A/duck/England/1/56; H12, A/duck/Alberta/60/76, H13, A/gull/Maryland/704/77; H14, A/mallard/Gurjev/263/82; H15, A/duck/Australia/341/83; B, B/Lee/40.

Mechanisms of the influenza virus adaptation to a new host and selective pressures on the HA receptor specificity

The lack of proofreading/repair and postreplicative error correction mechanisms in RNA viruses contributes to replication errors of 1 per 10^3 to 10^4 nucleotides (Drake, 1993). Therefore, each round of influenza virus replication results in a mixed population with many variants, described as quasispecies (Holland et al., 1992; Domingo et al., 1996). In a population of influenza viruses containing more than 10^5 particles, variants with mutations in any amino acid of the HA can potentially be present, as long as they are not lethal. Most variants offer no competitive advantage over the dominant virus, however, transmission of the virus to a new host species changes the selective pressure and results in the expansion of the 'best fit' minor variants for adaptation to the new environment.

Major selective pressures on virus receptor specificity in a given host species are likely the patterns of sialic acid receptors on the target cells and on the soluble extracellular inhibitors. The involvement of these two pressures in the selection of virus variants with optimal receptor-binding phenotypes has been demonstrated under experimental conditions (for examples, see Carrol and Paulson, 1985; Matrosovich et al., 1998; Rogers et al., 1983; Ryan-Porrier and Kawaoka, 1991); the relative roles of these pressures in natural virus hosts are not yet well defined.

Receptor specificity of duck viruses

Wild ducks appear to be particularly important for the perpetuation of influenza viruses in nature (Webster et al., 1992). In ducks, the viruses preferentially replicate in the cells that line the intestinal tract and cause no major disease signs or symptoms (Slemons and Easterday, 1977; Webster et al., 1978). Duck viruses are excreted in high concentrations in the feces and are transmitted by the fecal-oral route through contaminated water (reviewed by Webster et al., 1992; Webster, 1997). The avirulent nature of influenza viruses in ducks and their efficient mode of transmission via fecal material in water supplies results most likely from the adaptation of the virus to this host over many centuries, creating the natural reservoir that maintains the virus without endangering its host.

Rogers and Paulson (1983) were the first to report that H3 avian viruses differentiate between types of Sia-Gal glycosidic linkage. Subsequent studies on the viruses of twelve HA subtypes confirmed that avian viruses bind to α2-3-linked sialic acids much more strongly than to α2-6-linked ones (Rogers and D'Souza, 1989; Nobusawa et al., 1991; Connor et al., 1994; Gambaryan et al., 1997, Matrosovich et al., 1997, 1999, 2000). Histochemical analysis of duck intestinal epithelial cells using linkage-specific lectins from *Maackia amurensis* (MAA, recognizes 2-3-linkage) and *Sambucus nigra* (SNA, recognizes 2-6-linkage) revealed the presence of Neu5Ac(α2-3)Gal-terminated sequences and no detectable expression of Neu5Ac(α2-6)Gal (Ito et al., 1998). Human viruses do not bind to plasma membranes isolated from duck intestinal cells, confirming a lack of Neu5Ac(α2-6)Gal-terminated receptors in duck intestine (Gambaryan et al., 2002). Thus, the receptor specificity of duck viruses correlates with the predominance of Neu5Ac(α2-3)Gal residues on the tissues in which the virus replicates in ducks.

Comparison of the amino acid sequences of avian and human virus HAs (Matrosovich et al., 1997) shows a high level of conservation of the avian RBS (Figure 2 and 6). The RBS of avian viruses apparently formed before the divergence of individual subtypes and was preserved during HA evolution in aquatic and land-based birds. Six more amino acids are conserved in the receptor-binding site of most avian viruses (Ala$_{138}$, Glu$_{190}$, Leu$_{194}$, Gly$_{225}$, Gln$_{226}$, and Gly$_{228}$) than in the binding site of human viruses. These amino acids either directly participate in hydrogen bond formation and van der Waals interactions with the Neu5Ac(α2-3)Gal moiety (Glu$_{190}$, Leu$_{194}$, Gly$_{225}$, and Gln$_{226}$), or make direct contacts with such amino acids (Ala$_{138}$ and Gly$_{228}$). Mutations at any of these positions decrease avian virus binding to the 3-linked galactose moiety of the receptor and lower virus affinity for Neu5Ac(α2-3)Gal-containing sialylglycopolymers (Matrosovich et al., 2000). Conservation of these amino acids thus appears to be required for the fit of the minimum energy *trans* conformer of the Neu5Ac(α2-3)Gal moiety to the avian HA RBS (see Figure 3).

The role of structural features of receptors other than the Sia-Gal-linkage in duck virus binding remains undefined. Avian influenza A viruses vary significantly in their recognition of Neu5Gc and usually bind more strongly to Neu5Ac-containing receptors than to homologous Neu5Gc-containing receptors (Higa et al., 1985; Nobusawa et al., 1991; Matrosovich et al., 1997; Ito et al., 2000). This feature argues against the importance of Neu5Gc binding for virus replication in aquatic birds. However, Ito et al. (2000) observed an apparent correlation between the virus affinity for NeuGc-containing receptors and its ability to replicate in duck intestine. These authors also detected expression of Neu5Gc-moieties in the epithelial cells of duck colon. A more recent study revealed that duck viruses bind to Neu5Ac(α2-3)Gal(β1-3)GalNAc-containing receptor analogs with significantly greater affinity than to Neu5Ac(α2-3)Gal(β1-4)GlcNAc-containing ones, indicating that these viruses discriminate between the core structures of oligosaccharides (Gambaryan et al., 2003). This specificity could be based on the different orientation of the (β1-3)- and (β1-4)-linked oligosaccharide cores and hence their different interactions with the rim of the RBS.

Low receptor-binding affinity can restrict influenza virus replication in ducks
Duck influenza viruses do not replicate efficiently in humans (Beare and Webster, 1991), and human influenza viruses do not replicate efficiently in ducks (Webster et al., 1978; Hinshaw et al, 1983). The host-range restriction of influenza viruses is a polygenic trait, with virus HA, nucleoprotein, matrix protein, neuraminidase and polymerase genes being important host-range determinants (for reviews, see Klenk and Rott, 1988; Horimoto and Kawaoka, 2001; Baigent and McCauley, 2003). The contribution of HA to host range restriction was first demonstrated by Hinshaw et al. (1983), who found that a reassortant virus harbouring the HA gene from the human virus A/Udorn/307/72 (H3N2) and the rest of its genes from A/mallard/NY/6750/78 (H2N2) virus did not grow in duck intestine. The acquisition by this reassortant virus of two mutations in its HA, namely Leu→Gln at position 226 and Gly→Ser at position 228, allowed its replication in the intestinal tract of ducks; however, the single mutation Leu$_{226}$→Gln$_{226}$ was not sufficient for such replication (Naeve et al., 1984; Vines et al., 1998). The substitution Leu$_{226}$→Gln$_{226}$ in the human virus HA is primarily responsible for binding to the Neu5Ac(α2-3)Gal-containing receptor, while the second mutation, Ser$_{228}$→Gly$_{228}$, enhances the binding affinity (Matrosovich et al., 2000). Thus, replication of the virus in ducks requires a greater affinity for the target cells in duck intestine than that provided by native human virus HA or by human HA with a single Leu$_{226}$→Gln$_{226}$ substitution.

Figure 4.6 (*opposite*) Partial HA amino acid sequences (HA1 positions 90-260) of 15 HA subtypes of avian influenza A viruses. The avian virus strains and their abbreviations are the same as in the Figure 5; the sequences are listed according to their homology. Amino acids conserved among at least 12 of 15 sequences are shaded. Top line shows the H3 numbering; amino acid positions that are absent in the H3 HA are indicated by 'A'. RBS line shows position of the amino acid with respect to the HA RBS. Star indicates that amino acid is within 15 Å distance from the C$_2$ atom of the sialic acid in the X31 virus HA complex with sialyloligosaccharides (Figure 4). 'R' indicates that amino acid residue contacts either sialic acid or penultimate 3-linked galactose of LSTa. The figure was generated by using GeneDoc 2.6 software (Nicholas et al., 1997).

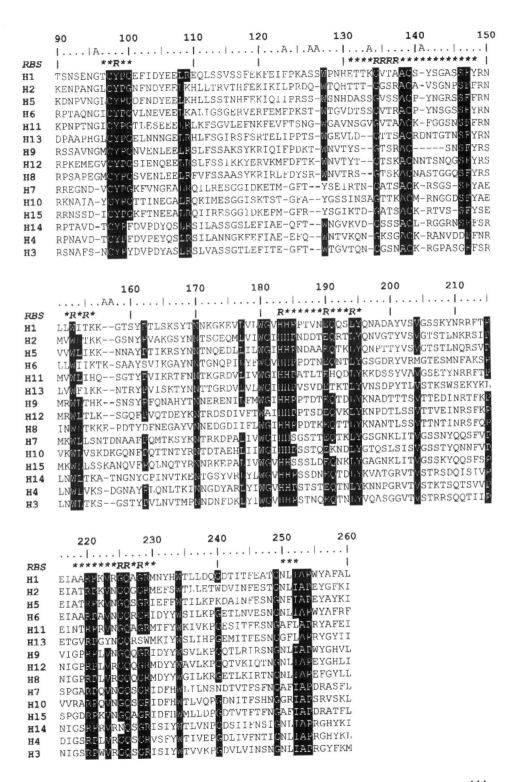

Viruses of other avian species

Data on receptor specificity of avian influenza viruses isolated from non-duck species are limited. Virus strains of several HA subtypes isolated from gulls and shorebirds preferentially bind to Neu5Ac(α2-3)Gal-terminated receptors (Connor et al., 1994; Gambaryan et al., 1997; Matrosovich et al., 1997, 1999, 2000), that is, behave similar to duck viruses. However, only some gull viruses can infect ducks (Kawaoka et al., 1988), suggesting some host-range restriction between avian species. In particular, viruses with H13 HA are widely distributed among gulls and shorebirds but until recently were never isolated from ducks. H13 HAs differ from all other avian virus HAs by two conserved amino acid substitutions at positions 228 and 229 in the receptor-binding site (Figure 6). These viruses differ from duck viruses by their relatively weak binding to 3'SL caused by weak interactions with the 3-linked Gal moiety (Matrosovich et al., 1997). In addition, unlike duck viruses, H13 gull viruses do not discriminate between Neu5Ac(α2-3)Gal(β1-3)GalNAc- and Neu5Ac(α2-3)Gal(β1-4)GlcNAc-containing receptors (Yamnikova et al., 2003). Because H13 viruses appear to be well adapted to gulls and shorebirds, it seems that the receptor specificity of the best-fit viruses in ducks differs from that of gulls and shorebirds.

Swine viruses and a variety of avian influenza viruses have been isolated from turkeys. Of two turkey virus strains tested by Nobusawa et al. (1991), one (A/Ty/Ontario/6118/68, H8N4) displayed typical duck virus-like receptor specificity, whereas the other, A/Ty/Wiskonsin/66 (H9N2), bound to Neu5Ac(α2-6)Gal-containing ganglioside more strongly than to a Neu5Ac(α2-3)Gal-containing analog. Two H1N1 classical swine virus strains isolated from turkeys also preferentially recognize Neu5Ac(α2-6)Gal determinants typical of swine and human viruses (Rogers and D'Souza, 1989). Thus at least some avian species support the replication of influenza viruses with human virus-like receptor specificity. This notion gained direct support in studies on the H9N2 virus lineage, which recently emerged in poultry in Asia (Lin et al., 2000; Matrosovich et al., 2001; Ha et al., 2002). H9N2 viruses from Hong Kong harbor Leu$_{226}$ instead of Gln, and some carry additional mutations at HA positions 183, 190, and 225, which are conserved in the HAs of other avian viruses. In addition, the neuraminidase of poultry H9N2 viruses has mutations in its hemadsorbing site, a characteristic resembling that of human H2N2 and H3N2 viruses but not shared by other avian viruses. These H9N2 viruses display receptor specificity similar to that of human H3N2 viruses and preferentially bind to Neu5Ac(α2-6)Gal-terminated receptors, which fits with the detection of Neu5Ac(α2-6)Gal–moieties in chicken tissues and with other differences observed between the sialic acid receptors of ducks and chickens (Feldmann et al., 2000; Gambaryan et al., 2002). H9N2 viruses with human virus-like specificity have been isolated from a variety of land-based poultry in live bird markets (for example, chicken, quail, pheasant, pigeon). The avian or mammalian species responsible for the generation of viruses with this receptor phenotype have yet to be identified.

Avian influenza is usually a mild or non-symptomatic disease. However, occasionally, the circulation of low-pathogenic aquatic bird viruses with H5 and H7 subtype HA in domestic poultry results in the selection of highly pathogenic viruses. One of the major features of these viruses is that their HA can be cleaved by ubiquitous proteases of the host, permitting virus replication in different tissues (for reviews, see Klenk and Rott, 1988; Klenk and Garten, 1994; Ito and Kawaoka, 1998; Steinhauer, 1999). The influenza outbreak in Hong Kong in 1997 was caused by direct transmission of such highly pathogenic chicken H5N1 viruses to humans (Claas et al., 1998; Subbarao et al., 1998) and prompted investigations of the receptor specificity of chicken strains. H5N1 chicken viruses, like duck viruses, bind to Neu5Ac(α2-3)Gal-containing receptors but not to Neu5Ac(α2-6)Gal (Matrosovich et al., 1999). However, H5 chicken viruses differ from closely related H5 duck and gull viruses by their lower affinity for soluble receptor analogs and chicken erythrocytes and by a lower neuraminidase (NA) activity. In addition, unlike most duck viruses, the 1997 H5N1 viruses from Hong Kong bind to Neu5Ac(α2-3)Gal(β1-4)GlcNAc-containing receptor analogs more strongly than to Neu5Ac(α2-3)Gal(β1-3)GalNAc-containing ones (Gambaryan et al., 2003). Analysis of the HA and NA amino acid sequences of H5 and H7 viruses from different avian species revealed that chicken viruses differ from duck viruses by additional N-linked glycans at the top of their HA and by large deletions in the NA stalk (Matrosovich et al., 1999; Banks et al., 2001). Both changes occurred independently in different lineages of chicken viruses, suggesting their functional role in the virus adaptation from duck to chickens. A carbohydrate at position 156 of the HA of H5N1 viruses from Hong Kong decreases the affinity of the virus for chicken erythrocytes, whereas the deletion in NA slows virus elution from these cells (Matrosovich et al., 1999).

Domestic land-based birds thus represent distinct avian hosts of influenza viruses, with the receptor specificity of poultry viruses being substantially different from that of duck viruses.

Equine and seal viruses

Equine influenza is associated with two stable virus lineages, H7N7 and H3N8 (Figure 5) (see Mumford and Chambers, 1998 for a review). H3 equine viruses have even more strict receptor specificity than H3 duck viruses. They agglutinate Neu5Ac(α2-3)Gal-containing erythrocytes but, unlike some duck viruses, they do not agglutinate erythrocytes with Neu5Ac(α2-6)Gal-terminated receptors (Rogers and Paulson, 1983; Connor et al., 1994). Equine tracheal tissues are rich in Neu5Gc-containing receptors, and equine viruses can bind to both Neu5Ac and Neu5Gc (Suzuki et al., 2000). Equine viruses have little if any affinity for 4-O-acetylated sialic acid (Matrosovich et al., 1998), which is also abundant in this species, suggesting that binding to Neu4, $5Ac_2$ reduces virus fitness in horses. This phenomenon is likely explained by the fact that Neu4, $5Ac_2$ resists cleavage by viral NA, thus increasing virus sensitivity to neutralization by sialic acid-containing inhibitors, making virus progeny more prone to self-aggregation and limiting their release from infected cells (for recent reviews on the functional interplay between HA and NA and on inhibitors of influenza viruses, see Wagner et al., 2002; Matrosovich and Klenk, 2003).

Influenza A viruses isolated from seals appear to originate from the avian reservoir without establishing stable lineages in this species (for review, see Hinshaw, 1998). In 1991, H3N3 avian-like viruses were isolated from the lung tissue of seals that had died of pneumonia along the Cape Code peninsula of Massachusetts (Callan et al., 1995). Of four viruses from this outbreak, three viruses were indistinguishable from the H3 duck viruses in their binding to monovalent and polyvalent receptor analogs (Matrosovich et al., 2000). One of the seal viruses displayed a lower binding affinity for Neu5Ac(α2-3)Gal-containing receptors with no increase in binding to Neu5Ac(α2-6)Gal-specific receptors. These data suggested that either a selective pressure against viruses with avian receptor specificity is low (or absent) in seals, or that the animals were only recently infected with the avian virus, and selection had not yet occurred. Osterhaus et al. (2000) reported transmission of influenza B viruses from humans to harbor seals. The receptor-binding phenotype of these viruses was not specified, and it remains unclear whether human viruses can replicate and transmit in seals without altering receptor specificity.

Swine viruses

Swine is susceptible to infection by both avian and human influenza viruses under experimental conditions (Hinshaw et al., 1981; Kida et al., 1994) and in nature (reviewed by Scholtissek et al., 1998). As with other mammals, the introduction of viruses from other hosts only rarely results in the establishment of stable virus lineages. Included among such lineages currently recognized in pigs are the H1N1 'classical' and 'avian-like' lineages (Figure 5), and several 'human-like' H3N2 lineages. Classical swine H1N1 virus shares its immediate ancestor with human H1N1 virus and probably originated from birds in the second decade of the last century. The avian-like H1N1 virus was introduced from birds to European pigs at the end of 1970s. H3N2 human-like swine virus lineages originate from H3N2 viruses, which have been circulating in humans since the 1968 pandemic.

Because of the relative ease with which pigs become infected with avian and human influenza viruses, they are seen as 'mixing vessels', in which reassortment between human and avian viruses can result in the emergence of pandemic viruses (Scholtissek et al., 1985).

Rogers and D'Souza (1989) found that both classical swine and avian-like swine viruses agglutinate Neu5Ac(α2-6)Gal-containing erythrocytes but not Neu5Ac(α2-3)Gal-containing cells. The human virus-like receptor specificity of H1N1 swine viruses was confirmed in later studies (Gambaryan et al., 1997; Matrosovich et al., 1997; Ito et al., 1998; Suzuki et al., 1997). Few H3N2 swine viruses tested are similar in their receptor specificity to human viruses (Higa et al, 1985; Suzuki et al, 1997). Ito et al. (1998) detected the presence of both Sia(α2-6)Gal and Sia(α2-3)Gal determinants on the surface of pig respiratory epithelia, which may explain the human virus-like specificity of pig viruses and the susceptibility of pigs to both avian and human viruses.

The avian-like swine viruses were first isolated in 1979 (Scholtissek et al., 1983), and have continued to circulate in European pigs (Ludwig et al., 1995; Brown et al., 1997). Ito et al. (1998) and Matrosovich et al. (2000) found that the earliest available isolates from swine displayed a substantially enhanced affinity for Neu5Ac(α2-6)Gal-containing receptors relative to the closely related H1N1 avian viruses. However, unlike pandemic human H2 and H3 viruses, the early avian-like swine viruses retained the ability of their avian predecessors to bind Neu5Ac(α2-3)Gal. The virus affinity for Neu5Ac(α2-3)Gal-containing receptors gradually reduced during subsequent circulation in pigs, and the affinity of viruses isolated after 1985 is similar to that of classical swine viruses.

These data indicate that the zoonotic importance of swine may not be limited to providing a milieu for reassortment between avian and human viruses. Changes in the receptor-binding properties of avian HA in pigs could increase the likelihood of virus transmission to humans because of better recognition of human receptors and/or a lower sensitivity to human mucin inhibitors.

Analysis of amino acid substitutions in the HA of H1N1 avian-like swine viruses with respect to H1N1 avian viruses showed that substitutions at positions 190 (Glu→Asp) and 225 (Gly→Glu) were likely responsible for the initial changes in receptor specificity (Matrosovich et al., 2000). Of note, all five H1N1 human viruses from the 1918 'Spanish' influenza pandemic sequenced to date bear Asp_{190}; three of them also have Asp_{225} (Reid et al., 1999, 2003). These findings suggest that the 1918 pandemic viruses had swine virus-like receptor specificity and could bind to Sia(α2-6)Gal-containing receptors. Another typical mutation that occurs in the HAs of all H1N1 avian-like and classical swine viruses is Thr_{155}→Val_{155}/Ile_{155}. The side chain of the amino acid at position 155 participates in the formation of the pocket that accommodates the acyl substituent at the N_5 atom of Neu5Ac (see Figure 2). Therefore, the substitution Thr$_{155}$→Val$_{155}$/Ile$_{155}$ could increase the affinity of swine viruses for Neu5Gc-containing receptors, which are absent in birds and humans, but account for about half of the sialic acid species on the respiratory epithelia of pigs (Suzuki et al., 1997). Although this concept has not been proven experimentally, it is consistent with two reports concerning the ability of H1N1 and H3N2 swine viruses to bind Neu5Gc (Higa et al., 1985; Suzuki et al., 1997).

Human viruses

The evolution of influenza A viruses in humans includes antigenic shift and antigenic drift (Kilbourne, 1987; Murphy and Webster, 1996; Cox and Subbarao, 2000). Antigenic shift occurs when a virus with a new HA or new HA and NA genes emerges in humans and causes a pandemic because of the lack of immunity to these new strains. Pandemic viruses continue their evolution by point mutations that result in a gradual change in their HA and NA (antigenic drift). The three pandemics of the 20th century were caused by the introduction into humans of entire genomes or parts of virus genes from non-human hosts: 'Spanish' flu (H1N1 virus, 1918), 'Asian' flu (H2N2 strain, 1957), and 'Hong Kong' flu (H3N2, 1968). The fourth pandemic ('Russian' flu) in 1977 was caused by the reappearance of an H1N1 human virus that was genetically almost identical to the virus circulating in the 1950s. The H2N2 virus lineage disappeared in 1968, but descendents of the H3N2 and H1N1 pandemic viruses continue to circulate in the human population, together with type B viruses, which may have originated from an ancient introduction of an avian virus to humans (Webster et al., 1992).

Human influenza A and B viruses bind to 6'-sialyl(N-acetyllactosamine) but do not bind to Neu5Ac(α2-3Gal)-containing receptors

Studies of the binding of many human influenza A (H1, H2, and H3) and B virus strains to derivatized erythrocytes, gangliosides, sialylglycoproteins, and sialylglycopolymers have revealed that all human viruses have high affinity for receptor analogs that contain terminal Neu5Ac(α2-6)Gal moieties and typically bind weakly to Neu5Ac(α2-3)Gal-containing receptors (Rogers and Paulson, 1983; Suzuki et al., 1985, 1986; Rogers and D'Souza, 1989; Xu et al., 1994; Connor et al., 1994; Ito et al., 1997; Gambaryan et al., 1997, 1999; Matrosovich et al, 2000). In most of these studies, 2-6-linked analogs were represented by the neolacto or type II sequence, Neu5Ac(α2-6)Gal(β1-4)GlcNAcβ1, [6'-sialyl(N-acetyllactosamine), 6'SLN], in which the galactose residue is linked to the penultimate GlcNAc by a 1-4 linkage. Suzuki et al. (1992) and Xu et al. (1994) reported that human influenza A and B viruses also bind to lacto (type I) sequences Neu5Ac(α2-6)Gal(β1-3)GlcNAcβ1, and some data on virus binding to 2-6-linked analogs in few other contexts have been published (Pritchett et al., 1987; Sauter et al., 1989; Gambaryan et al., 1995). However, no studies have reported a systematic comparison of different core sequences recognized by human influenza viruses. Presently, 6'SLN appears to be the most probable receptor determinant used by human influenza A and B viruses in nature, because it is abundant in natural sialylglycoconjugates and binds with high affinity to all human virus strains. In early studies on egg-adapted human viruses, some virus strains exhibited anomalous behavior by binding to Neu5Ac(α2-3)Gal-containing receptor analogs with comparable or higher affinity than to 6'SLN-containing receptors (Rogers and D'Souza, 1989; Connor et al., 1994; Xu et al., 1996). This inconsistency results from changes in the HAs introduced during egg-adaptation (see page 118). Influenza A and B viruses isolated and passaged solely in MDCK cells have very weak, if any, binding to Neu5Ac(α2-3)Gal-containing macromolecules (Gambaryan et al., 1997; Ito et al., 1997).

Using specific lectins and sections of human trachea, Baum and Paulson (1990) and Couceiro et al. (1993) revealed predominant expression of 2-6-linked sialic acids on the apical surface of tracheal epithelium. The staining for 2-3-linked sialic acids appeared to co-localize with secretory granules of mucus-secreting cells in agreement with the biochemical data on the abundance of Neu5Ac(α2-3)Gal-moieties and the deficiency of Neu5Ac(α2-6)Gal/GalNAc-residues in human respiratory mucins (Breg et al., 1987; Scharfman et al., 1995). These results suggested that receptor specificity of human influenza viruses may be maintained by simultaneous selective pressures that result from the predominant expression of Neu5Ac(α2-6)Gal-containing receptors on cells and from the predominant expression of Neu5Ac(α2-3)Gal-containing determinants on extracellular mucin inhibitors.

Although human influenza A and B viruses share a high binding affinity for 6'SLN-terminated receptors, they exhibit distinct type, subtype-, and lineage-specific patterns of binding to monovalent receptor analogs (Matrosovich et al., 1993, 1997, 2000; Gambaryan et al., 1995, 1999). For example, type B viruses typically bind sialic acid glycosides much more strongly than free sialic acid, which is indicative of favorable interactions with both 2-3-linked and 2-6-linked asialic moieties. H1N1 human viruses are distinct from the H3N2 influenza A viruses isolated from 1968 to the early 1990s and from type B viruses in that they bind 6'SLN with a higher affinity than 6'SL [Neu5Ac(α2-6)Gal(β1-4)Glc] (Gambaryan et al., 1997). The ability of H1N1 viruses to discriminate between 6'SLN and 6'SL suggests that these viruses could bind to the N-acetamide group of the GlcNAc residue of 6'SLN. Interestingly, this property can be completely destroyed by a single mutation at position 190 of the H1 subtype of HA (Gambaryan et al., 1999). Moreover, H3N2 viruses that have been circulating in humans since 1992 have Asp_{190} instead of Glu_{190} and, like H1N1 viruses, bind 6´SLN more strongly than 6´SL (Mochalova et al., 2003).

Receptor specificity and pandemics

Reassortant viruses that carried HA (H3N2, 1968) or both HA and NA (H2N2, 1957) derived from avian influenza viruses caused two human pandemics in the last century (reviewed by Webster et al., 1992; Cox and Subbarao, 2000). Studies on the receptor specificity of the earliest virus isolates from these pandemics revealed that they differ from closely related avian viruses in their binding to 6'SLN-containing receptors and have a low affinity for 2-3-linked receptor determinants (Rogers and Paulson, 1983; Suzuki et al., 1986; Connor et al., 1994; Matrosovich et al., 2000). The 'Spanish' pandemic of 1918 likely originated from an avian virus that was transmitted to humans as a whole (Webster et al., 1992; Taubenberger et al., 1997). Until recently, the 1918 human viruses were not available for receptor binding studies; however, comparison of their HA sequences (Reid et al., 1999) with those of human and swine H1N1 viruses suggest that the pandemic viruses of 1918 had human virus-like receptor specificity (Reid et al., 1999, 2003; Matrosovich et al., 2000). This hypothesis was proven recently correct when recombinant viruses containing the HA and NA genes of the 1918 strain have been were produced by reverse genetics and tested for their binding to NeuAc(α2-6)Gal and NeuAc(α2-3)Gal sialylglycopolymers (Kobasa et al., 2004). Thus, the acquisition of 6'SLN recognition and a decrease in binding to Neu5Ac(α2-3)Gal occurred relatively soon after the introduction of avian H1, H2 and H3 HAs into humans, at least, by the time these epidemics in humans occurred and the viruses were isolated. By contrast, the H5N1 chicken viruses that infected humans in Hong Kong in 1997 (Claas et al., 1998; Subbarao et al., 1998) maintained the avian-like receptor specificity (Matrosovich et al., 1999). The cases of H5N1 influenza in humans resulted from direct chicken-to-human introduction of the viruses with no evidence of efficient human-to-human transmission. Thus, although avian influenza virus can infect humans and even cause a fatal disease without a significant change in its receptor specificity, alterations in the receptor specificity appear necessary for effective virus transmission among humans.

For H1N1 viruses, the substitution $Glu_{190} \rightarrow Asp_{190}$ either alone or together with the substitution $Gly_{225} \rightarrow Asp_{225}$ appears to be essential for avian virus HA to acquire human-virus-like receptor specificity (Matrosovich et al., 1997, 2000; Reid et al., 1999, 2003). This hypothesis agrees with the fact that substitutions at positions 190 and 225 (often with reversion to the avian virus consensus sequence) are commonly observed during egg-adaptation of contemporary human H1N1 influenza viruses and have a most profound effect on virus recognition of the Sia-Gal linkages (Gambaryan et al., 1997, 1999).

Few amino acid substitutions separate avian H2 and H3 viruses from their corresponding earliest human isolates (Bean et al., 1992; Connor et al., 1994; Klimov et al., 1996); they included mutations at positions 226 (Qln→Leu) and 228 (Gly→Ser). A comparison of the receptor specificity of laboratory versus field virus isolates that differ at these HA positions indicates that the substitution $Qln_{226} \rightarrow L_{226}$ in both H3 and H2 HAs is critical for virus binding to 6-linked sialic acids (Rogers et al., 1983, 1985; Matrosovich et al. 1993, 1997, 2000; Vines et al., 1998), and that the additional mutation $Gly_{228} \rightarrow Ser_{228}$ strengthens this binding (Matrosovich et al., 2000). Interestingly, some H2 human viruses isolated in 1957 had L_{226} but maintained the avian virus-like Gly_{228} (Klimov et al., 1996). These viruses may represent the earliest step of adaptation of the avian H2 HA to humans.

Alterations of the RBS during circulation of influenza viruses in humans

The receptor-binding sites of human influenza viruses are much more diverse than those of avian influenza viruses (Matrosovich et al., 1997). This divergence stems, in part, from the different substitutions acquired by pandemic viruses during interspecies transmission (see above). Further divergence occurs during subsequent circulation of the virus in the human population due to accumulation of point mutations in HA, which allow the virus to evade neutralization by antibodies. Several major HA antigenic epitopes are located in close proximity to the receptor-binding site (see reviews by Wiley and Skehel, 1987; Wilson and Cox, 1990), and some neutralizing antibodies completely overlap the sialic acid binding pocket (Bizebard et al., 1995). It is not surprising, therefore, that antibody-escape mutations in HA often affect receptor specificity (see, for example, Underwood et al., 1987).

Bush et al. (1999) studied the evolution of an H3 subtype of human HA and found that 18 amino acid positions in the HA1 HA subunit were under positive selection for changes. Remarkably, 5 of these positions (135, 138, 190, 194, and 226) reside in the sialic acid binding pocket (see Figure 2). Five other positions (145, 156, 158, 186, and 193) surround the pocket (Figure 4), and substitutions at these positions can affect the recognition of the asialic portions of the receptors (Gambaryan et al., 1998b, 1999; Matrosovich et al., 1998, 2000; Kaverin et al., 2000).

Variations in the receptor-binding characteristics of human influenza viruses of the same evolutionary lineage have not been studied in detail, although several authors have noticed the evolution of the receptor specificity of H3N2 human virus strains isolated between 1968 and 1987 (Underwood, 1985; Anders et al., 1986; Matrosovich et al., 1991,

1993; Ryan-Poirrier et al., 1998). In the early 1990s, the H3N2 viruses rapidly acquired multiple substitutions in the RBS (including at positions 190 and 226) that substantially changed their receptor-binding characteristics. In particular, these viruses lost the ability to agglutinate chicken erythrocytes (Nobusawa et al., 2000; Medeiros et al., 2001; Mochalova et al., 2003).

Changes in receptor specificity during egg-adaptation of human viruses

Before tissue culture systems were available, embryonated chicken eggs were the most popular substrate for propagating influenza viruses because of their historical use, availability, and good virus yield. The allantoic cavity of eggs is typically used to propagate influenza viruses; however, as first recognized by Burnet and Bull (1943), egg-grown viruses differ considerably from the original virus isolates. Burnet and Bull noticed that the original (O-type) viruses present in human throat washes grew in the amniotic but not in the allantoic cavity of eggs. Adaptation of the viruses to grow in the allantoic cavity always resulted in the selection of derivative (D) virus with altered patterns of agglutination of chicken and guinea pig erythrocytes (the O-D change). Studies in the 1980s identified the molecular basis of the O-D change as the result of amino acid substitutions around the HA RBS (reviewed by Robertson, 1993). Several variants with different substitutions can be derived from a single human specimen, although usually only one or occasionally two substitutions are found in each variant. Viruses that belong to different HA types and subtypes acquire distinct (type- and subtype- and strain specific) substitutions during egg-adaptation. Unlike egg-adapted variants, human viruses isolated in Madin-Darby canine kidney (MDCK) cell cultures are usually homogeneous with unaltered HA sequences (Katz et al., 1990; Robertson et al., 1990; 1991; Katz and Webster, 1992).

Ito et al. (1997) found that the cells of chicken embryo chorio-allantoic membrane (CAM), contain Neu5Ac(α2-3)Gal-moieties, but do not express Neu5Ac(α2-6)Gal, whereas the cells of the amniotic membrane and MDCK cells express both types of Neu5Ac-Gal determinants. Gambaryan et al. (1997) demonstrated that non-egg-adapted human influenza A and B viruses do not bind to Neu5Ac(α2-3)Gal-containing receptors, whereas all their egg-adapted variants acquire this ability. All egg-adapted viruses had increased affinity for receptors on CAM cells compared to their non-adapted parents and neither egg-adapted variant displayed a decreased affinity for inhibitors present in allantoic fluid (Gambaryan et al., 1999). These data suggest that eggs select receptor-binding variants of human viruses that have increased affinity for Neu5Ac(α2-3)Gal -containing receptors on the target cells of CAM. By contrast, the cells of the amniotic membrane and MDCK cells appear to express sufficient amount of Neu5Ac(α2-6)Gal-determinants to permit relatively unrestricted growth of human viruses.

Amino acid substitutions in the HA of egg-adapted viruses increase their binding to CAM cells by several different mechanisms (Gambaryan et al., 1997, 1999), including: i) enhancing HA atomic interactions with the Neu5Ac(α2-3)Gal-moiety, ii) decreasing steric interference with more distant parts of the Neu5Ac(α2-3Gal)-containing receptors, and iii) enhancing ionic interactions with cells due to mutations that increase the positive charge of the HA molecule.

One mechanism of egg-adaptation involves the loss of the N-linked carbohydrates at the HA head at positions 163 of H1 HA, 246 of H3 HA, and 187 of type B HA, which are adjacent to the 'left upper' rim of the receptor-binding site of the neighbor HA monomer.

According to the structural model of Eisen et al. (1997), these carbohydrates could interfere with binding to Neu5Ac(α2-3)Gal-containing receptors by overlapping with their asialic portions (see Figure 4). Indeed, the loss of N-glycans had been shown to have a marginal effect on virus affinity for the low molecular weight receptor analog 3'SL, and for Neu5Ac(α2-6)Gal-terminated macromolecular receptors, but a much greater effect on virus binding to Neu5Ac(α2-3)Gal-containing macromolecules and CAM cells (Gambaryan et al. 1997, 1999; Matrosovich et al., 1998).

In addition to enhancing the affinity for Neu5Ac(α2-3)Gal-terminated receptors, substitutions in the HA of egg-adapted viruses generally decrease the affinity of the virus for 6'SLN-containing receptors (Gambaryan et al., 1999). Hence, the egg-adaptation of human viruses appears to inevitably change their fine receptor specificity and could impede their replication in humans. Indeed, in the course of experimental infection of humans with egg-adapted viruses, selection of revertants with the original HA consensus sequence has been observed (Gubareva et al, 2001).

Because egg-adaptation often changes the antigenicity of influenza viruses and can compromise surveillance studies and vaccine production, different cell lines have been used for the primary isolation and propagation of human influenza viruses. While some of these cell lines (e.g., LLC-MK2, MRC-5, Vero, primary chick kidney cells) are similar to MDCK cell in that they appear to be permissive to non-egg-adapted human influenza viruses (Govorkova et al., 1996; Katz and Webster, 1992, Romanova et al., 2003), others can select receptor-binding mutants, as was shown with BHK-21 cells (Govorkova et al., 1999). Furthermore, even in the case of MDCK cells, clinical human viruses appear not to fit completely to receptors on these cells, as receptor variants with amino substitutions in the HA have been shown to occasionally displace the natural virus (Robertson, 1993; Günther et al., 1993; Robertson et al., 1995; Romanova et al., 2003).

RECEPTOR SPECIFICITY AND PATHOGENICITY

Although data are accumulating on the role of receptor specificity in virus host range, the effects of sialic acid-binding on influenza virus virulence and pathogenicity have been less well studied. This section discusses the studies to date.

Cellular tropism

The ciliated epithelium of the respiratory tract consists of several distinct cell types with different functions (Jeffery and Li, 1997). Late in the infectious process, influenza viruses are believed to infect many different types of airway epithelial cells (Tateno et al., 1966; Ebisawa et al., 1969), however, neither the initial target cells of the virus attack, nor specific cell types essential for virus replication have been defined. Using differentiated cultures of human airway epithelium, Matrosovich et al. (2004) observed that during the course of a single-cycle infection, human viruses preferentially infect non-ciliated cells, whereas avian viruses and egg-adapted human virus variant with avian-virus-like receptor specificity mainly infect ciliated cells. This pattern correlated with the predominant localization of 2-6-linked sialic acids on non-ciliated cells and of 2-3-linked sialic acids on ciliated cells. These findings suggest that differences in replication and pathogenicity of human and avian viruses in humans may be related to the differential cellular tropism of the viruses.

Spread of virus progeny

Kilbourne and colleagues showed that egg-grown 'classical' swine H1N1 influenza viruses isolated from pigs, turkeys, and humans contain a mixture of L and H variants that differ by their antigenic properties and replicative characteristics (Kilbourne, 1988ab and references therein). L variants are low yielding in eggs and produced small clear plaques in MDCK cells, whereas H variants have the opposite characteristics and outgrow L variants on passaging in the laboratory. These differences are mediated by the HA gene and involve amino acid substitutions around the RBS (positions 156 and/or 158) that increase the negative charge of the HA molecule. Subsequent studies by Gambaryan et al. (1998b) on the L and H variants of A/NJ/11/76 (H1N1) virus revealed the reason for the low yield of the L variant in eggs and MDCK cells: excessive affinity for the target cells, which caused the progeny of the L virus to remain predominantly cell-associated such that its accumulation in the allantoic or culture fluid was slower than that of the low-affinity H variant.

This study highlights the potential for selection of high-yielding low-affinity variants with HA mutations during passaging of influenza viruses in the laboratory because of investigator bias towards choosing of a virus harvest with a higher hemagglutination titer or picking of a bigger plaque. On the other hand, a higher virus affinity appear to be required for virus survival under the more stringent conditions of natural infection compared to in the laboratory hosts, as demonstrated by the outgrowth of H variant by L variant in swine (Kilbourne et al., 1988b).

HA-mediated interactions with leukocytes

Polyclonal T-cell independent B-cell activation

Polyclonal B-cell activation is used by the viruses to counteract the efforts of the host to establish a specific humoral immune response and can be implicated in the pathogenesis of influenza virus-induced autoimmunity (reviewed by Cash et al., 1996).

Influenza A virus-induced polyclonal B-cell activation was first described by Butchko et al. (1978) and by Anders et al. (1984) who observed murine B-lymphocyte proliferation in response to treatment with UV-inactivated viruses. In addition to inducing the proliferation of mature B-cells, the viruses also stimulated a high rate of immunoglobulin synthesis and arrested the growth of immature B-cells (Rott and Cash, 1994). Virus-mediated B-lymphocyte activation involves interactions between the viral HA and the B-cell surface sialylglyco-conjugates. Thus, the mitogenic activity of the influenza virus reassortants correlates with the HA gene (Rott and Cash, 1994). Purified HA micelles prepared from detergent-disrupted virus are mitogenic (Poumbourios et al., 1987). Pretreatment of spleen cells with bacterial sialidase reduces their response to influenza viruses, whereas inhibition of the viral NA enhances the response (Anders et al., 1986).

Presentation of virus antigens to helper T-cells

T-helper cells participate in both humoral and cytotoxic T-cell responses against influenza virus infection and immunization (see Thomas et al., 1998; Stevenson and Doherty, 1998 for a review). Eisenlohr et al. (1987) studied stimulation of T-cell hybridomas specific for the HA and internal proteins of the PR/8/34 virus using B-cell lymphoma cells as antigen-presenting cells (APC). They found that the virus particles associated rapidly with APC, and that pretreatment of APC with sialidase or anti-HA antibodies abolished binding and reduced the efficiency of stimulation for both HA and the other viral proteins. These results suggested that the concentration of virus antigen on the APC due to HA binding to sialic acid-containing receptors greatly increases the efficiency of antigen presentation.

Given that both polyclonal activation of B-cells and stimulation of T-helper cells involve influenza virus interactions with cellular sialylglycoconjugates, the immune response to these viruses can be dependent on their receptor specificity. This may explain the marked differences in the immune response in mice elicited by two experimental influenza virus vaccines that differed by a single amino acid (Glu_{156} or Lys_{156}) in the vicinity of HA receptor-binding site (Kodihalli et al., 1995). The vaccine with Lys_{156} induced antibodies that were predominantly of the immunoglobulin M isotype and were nonprotective.

Deactivation of neutrophils

Secondary bacterial infections are the most common cause of mortality during influenza epidemics (Kilbourne, 1987). An enhanced susceptibility to bacterial superinfections can partially result from the ability of the virus to cause phagocyte dysfunction (reviewed by Abramson and Mills, 1988; Hartshorn and Tauber, 1988; Abramson and Wheeler, 1994).

Influenza viruses induce at least two early responses in neutrophils. Within seconds to minutes of exposure to the virus, the cell is activated to generate a respiratory burst (Daigneault et al. 1992). At the same time, the influenza virus impairs the normal bactericidal functions of neutrophils by depressing their secretory, oxidative, and chemotactic responses (Abramson et al., 1986; Cassidy et al., 1988 and references therein). The virus also stimulates apoptosis of neutrophils and potentiates the apoptotic effects of bacteria (Colamussi et al., 1999). Both the initial respiratory burst and the deactivation of neutrophils are mediated by the polyvalent virus binding to the sialylglycoconjugates on the surface of the cells. Thus, deactivation of neutrophils is not dependent on the fusion activity of the viral HA, and does not require internalization of virus particles (Cassidy et al., 1989). Sialic acid-binding lectins from *Limulus polyphemus* or *Limax flavus* cause neutrophil deactivation similarly to that seen with influenza virus or with solubilized virus glycoproteins in the form of aggregated rosettes or lipid-containing virosomes (Cassidy et al., 1989). Both activation of the respiratory burst and the deactivation of neutrophils are inhibited by preincubation of the virus with sialyllactose, by preincubation of cells with purified HA trimers, and by desialylation of the neutrophil surface with bacterial sialidase. The addition of ganglioside G_{T1b} to desialyzed cells totally restores virus binding but does not reverse the inhibition of activation/deactivation (Daigneault et al. 1992), suggesting that these effects may be mediated through a specific sialic acid-containing receptor, presumably, sialylglycoprotein (Daigneault et al. 1992; Hartshorn et al., 1995). Sialophorin (CD43) may be one such receptor (Rothwell and Wright, 1994; Abramson and Hudnor, 1995), although there are other neutrophil membrane proteins to which influenza virus can bind and independently mediate the functional depression of neutrophils (Hartshorn et al., 1995). Neutrophil gangliosides with long complex carbohydrate chains could also be candidate receptors involved in influenza virus-mediated dysfunction of neutrophils, as these sialylglycolipids bind influenza viruses in vitro with a much higher affinity than the more abundant simple ganglioside species (Müthing et al., 1993; Müthing, 1996; Matrosovich et al., 1996; Miller-Podraza et al., 1998, 2000).

Recognition of virus-infected cells by NK cells.

Natural killer (NK) cells are bone marrow-derived lymphocytes that lyse virus-infected and tumor cells spontaneously without antigen stimulation. This activity is mediated in part by a family of triggering receptors on the NK cells. A functional interaction between two such receptors, NKp46 and NKp44, with the HA of influenza viruses was recently identified (Mandelboim et al., 2001; Arnon et al., 2004). This interaction critically depends on the sialylation of the NK receptors, although other elements of the NK proteins are also involved. The receptors bind to HAs of several distinct human influenza A (H1N1 and H3N2) and influenza B viruses, suggesting that this phenomenon may represent a general mechanism by which NK cells recognize and destroy cells infected with influenza viruses.

HA-dependent enhancement of bacterial adhesion

Humans infected with influenza virus are at increased risk of developing local and systemic infections with a variety of bacteria (for review, see Abramson and Wheeler, 1994). Several mechanisms promote secondary bacterial infections in influenza, including impairment of the function of cells involved in immune defences (see above) and alterations in the epithelia leading to increased bacterial binding.

Sanford et al. (1978) showed that group B *Streptococcus* and *Streptococcus sanguis* adhere to MDCK cells infected with influenza A/NWS/33 (H1N1) virus. Antiserum to the virus completely blocks adhesion, suggesting that this enhanced adherence is due to the action of viral proteins. Jones and Menna (1982) studied the adherence of intranasally administered type 1a group B streptococci to the tracheal tissue of influenza virus-infected mice. Virus (A/PR/8/34, H1N1) infection led to a 120-fold increase in the adherence of bacteria relative to that observed in mock-infected mice, whereas prior intranasal exposure to virus-specific antiserum reduced the adherence by more than 90%. Further studies on the binding of several human influenza A viruses to different strains of group B *Streptococcus* have revealed that virus binding to bacterial serotypes Ia, Ic, and III and some Ib serotype strains can be abolished by desialylation of bacteria with *V.cholerae* sialidase (Sanford et al., 1980). The authors concluded that at least some streptococci adhere to the virus and virus-infected cells via virus HA binding to bacterial sialylglycoconjugates.

Besides streptococci, terminal sialyloligosaccharide residues can be found in the lipooligosaccharides (LOS) of mucosal Gram-negative bacteria, including members of the genera *Neisseria* and *Haemophilus* (see Preston et al., 1996 for a review on LOS) and in fungi (Alviano et al., 1999). This suggests that the HA-mediated enhancement of microbial adhesion may be a more common mechanism of stimulation of secondary infections by influenza viruses than it is currently recognized.

Receptor specificity affects the severity of experimental virus infection

Like humans, ferrets express high amounts of Neu5Ac(α2-6)Gal-terminated glycoconjugates on the apical surface of their tracheal epithelium (Leigh et al., 1995). To study the effect of virus receptor specificity on infection, Leigh et al (1995) inoculated ferrets intranasally with either wild-type human virus A/Memphis/102/72 (H3N2) or with its horse serum-resistant variant (HS) that contains the mutation $L_{226} \rightarrow Q_{226}$, which changes the receptor specificity from preferential Neu5Ac(α2-6)Gal recognition to preferential binding to Neu5Ac(α2-3)Gal-containing receptors. Although both viruses replicated equally well in the nasal turbinates of ferrets, virus was recovered from the lungs of fewer HS-infected ferrets (5/12 vs 11/12) and at lower titers than from the wild-type virus group. The authors suggested that a relatively impaired replication of HS in the lower respiratory tract of ferrets could be connected to a poor virus fit of the virus to the cellular receptors/intercellular inhibitors in these tissues. Over the 5 days after inoculation, HS-infected ferrets had a lower mean elevation in body temperature, greater weight gain, and less sneezing than the group infected with the wild-type virus. It may be that these symptoms were reduced in the HS-infected animals simply because of the reduced HS-virus replication in the lower respiratory tract. Alternatively, a difference in receptor specificity between the two viruses may have affected their interactions with epithelial cells and leukocytes, and/or promoted the release of different cytokines and inflammatory mediators.

Oxford et al. (1990) studied experimental infection in groups of volunteers who were inoculated intranasally with four influenza B viruses that originated from the same clinical virus isolate but had different passage history in man, human embryo trachea, and eggs. The HA of one egg-derived virus differed from the other three viruses by the loss of a glycosylation site at Asn_{187}, the typical egg-adaptation change known to affect receptor specificity. Unlike the three other viruses, which produced disease in 20%–30% of the volunteers, the egg-derived receptor-binding mutant produced no illness. Attenuation of virulence in human volunteers exposed to the same egg-adapted type B virus was also reported by Zuckerman et al. (1994).

CONCLUDING REMARKS

During the last years, significant effort has been devoted to improving our understanding of influenza virus interactions with sialylglycoconjugate receptors at the molecular level. Atomic interactions between HA and sialic acid have been resolved in great detail by X-ray analysis and by binding studies using synthetic and natural sialic acid analogs. Virus recognition of specific types of sialic acid-galactose linkage appears to depend on several factors. First, in addition to binding to the sialic acid moiety, the HA can interact with one or two penultimate sugars, avian virus HA binding to 3-linked Gal, and human virus HA binding to 6-linked N-acetyllactosamine. Second, the asialic portions of Sia(α2-3)Gal and Sia(α2-6)Gal-terminated receptors have different spatial orientation and, therefore, can either bind to, or sterically interfere with, different regions on the periphery of the receptor-binding site. Such interactions may play an essential role in virus polyvalent binding to soluble macromolecular inhibitors and to target cells. Namely, the binding can be markedly affected not only by amino acid substitutions in the receptor-binding site, but also by mutations outside the sialic-acid binding pocket, among them, substitutions that change the N-linked glycosylation or electrostatic charge of the HA globular head.

Recent studies have highlighted a correlation between the receptor-binding characteristics and biological properties of influenza viruses, such as host-range and pathogenicity. Human, swine, horse, and avian influenza viruses all have distinct receptor-binding specificity. Moreover, it appears that viruses of different avian species do not necessarily have identical receptor-binding characteristics, as duck, gull, and land-based poultry viruses differ in this respect. The host-specific receptor-binding phenotypes of viruses appear to be maintained by many selective pressures, among them, the availability of receptors on target cells and neutralization by competitive inhibitors. The relative contribution of these pressures in different host species and in distinct target tissues is unknown. Although it is believed that these pressures may create the barrier for the interspecies transmission of influenza viruses and, by this virtue, limit the emergence of new influenza outbreaks and pandemics, further studies are needed to estimate the magnitude of such restrictions in different avian and mammalian species.

Sialylglycoconjugates are ubiquitous on the surface of most avian and mammalian cells. Therefore, in addition to mediating infection in susceptible cells, influenza viruses can bind to a variety of other cell types leading to important biological effects, such as polyclonal activation of B-lymphocytes, presentation of viral antigens to helper T-lymphocytes, deactivation of neutrophils, and stimulation of inflammatory responses. Any correlation between the receptor-binding characteristics of the viruses and their ability to cause these effects has yet to be fully investigated. The range of possible variations of receptor-binding activity of human influenza viruses during antigenic drift has not been systematically studied, but some evidence indicates that it could be significant. Whether this variation could contribute to strain-dependent variation in influenza virus pathogenicity is not known. These issues represent attractive targets for future studies that could ultimately create new approaches to influenza prophylaxis and treatment.

Acknowledgements

We apologise to all researchers whose work was not cited due to limited space. We thank Susan Watson for editing of the manuscript. We acknowledge support of our own work from the Deutsche Forschungsgemeinschaft (SFB 286 and SFB 593); Fonds der Chemischen Industrie; Roche Products Ltd.; VIRGIL European Network of Excellence on Antiviral Drug Resistance, EU; National Institute of Allergy and Infectious Diseases Public Health Service, USA; CREST (Japan Science and Technology Agency); and the Ministry of Education, Culture, Sports, Science, and Technology of Japan.

References

Ablan, S., Rawat, S. S. , Blumenthal, R. , and Puri, A. (2001). Entry of influenza virus into a glycosphingolipid-deficient mouse skin fibroblast cell line. Arch. Virol. *146*, 2227–2238.

Abramson, J. S., Wheeler, J. G., Parce, J. W., Rowe, M. J., Lyles, D. S., Seeds, M., and Bass, D. A. (1986). Suppression of endocytosis in neutrophils by influenza A virus in vitro. J. Infect. Dis. *154*, 456–463.

Abramson, J. S. and Mills, E. L. (1988). Depression of neutrophil function induced by viruses and its role in secondary microbial infections. Rev. Infect. Dis. *10*, 326–341.

Abramson, J. S. and Wheeler, J. G. (1994). Virus-induced neutrophil dysfunction, role in the pathogenesis of bacterial infections. Pediatr. Infect. Dis. J. *13*, 643–652.

Abramson, J. S. and Hudnor, H. R. (1995). Role of the sialophorin (CD43) receptor in mediating influenza A virus-induced polymorphonuclear leukocyte dysfunction. Blood *85*, 1615–1619.

Alexander, D. J. (2000). A review of avian influenza in different bird species. Vet. Microbiol. *74*, 3–13.

Alviano, C. S., Travassos, L. R., and Schauer, R. (1999). Sialic acids in fungi, a minireview. Glycoconj. J. *16*, 545–554.

Anders, E. M., Scalzo, A. A., and White, D. O. (1984). Influenza viruses are T cell-independent B cell mitogens. J. Virol. *50*, 960–963.

Anders, E. M., Scalzo, A. A., Rogers, G. N., and White, D. O. (1986). Relationship between mitogenic activity of influenza viruses and the receptor-binding specificity of their hemagglutinin molecules. J. Virol. *60*, 476–482.

Arnon, T. I., Achdout, H., Lieberman, N., Gazit, R., Gonen-Gross, T., Katz, G., Bar-Ilan, A., Bloushtain, N., Lev, M., Joseph, A., Kedar, E., Porgador, A., and Mandelboim, O. (2004). The mechanisms controlling the recognition of tumor- and virus-infected cells by NKp46. Blood *103*, 664–672.

Aytay, S. and Schulze, I. T. (1991). Single amino acid substitutions in the hemagglutinin can alter the host range and receptor binding properties of H1 strains of influenza A virus. J. Virol. *65*, 3022–3028.

Baigent, S. J. and McCauley, J. W. (2003). Influenza type A in humans, mammals and birds, determinants of virus virulence, host-range and interspecies transmission. Bioessays *25*, 657–671.

Banks, J., Speidel, E. S., Moore, E., Plowright, L., Piccirillo, A., Capua, I., Cordioli, P., Fioretti, A., and Alexander, D. J. (2001). Changes in the haemagglutinin and the neuraminidase genes prior to the emergence of highly pathogenic H7N1 avian influenza viruses in Italy. Arch. Virol. *146*, 963–973.

Baum, L. G. and Paulson, J. C. (1990). Sialyloligosaccharides of the respiratory epithelium in the selection of human influenza virus receptor specificity. Acta Histochemica – Supplement-Band XL, S, 35–38.

Bean, W. J., Schell, M., Katz, J., Kawaoka, Y., Naeve, C., Gorman, O., and Webster, R. G. (1992). Evolution of the H3 influenza virus hemagglutinin from human and nonhuman hosts. J. Virol. *66*, 1129–1138.

Beare, A. S. and Webster, R. G. (1991). Replication of avian influenza viruses in humans. Arch. Virol. *119*, 37–42.

Bergelson, L. D., Bukrinskaya, A. G., Prokazova, N. V., Shaposhnikova, G. I., Kocharov, S. L., Shevchenko, V. P., Kornilaeva, G. V., and Fomina-Ageeva, E. V. (1982). Role of gangliosides in reception of influenza virus. Eur. J. Biochem. *128*, 467–474.

Bizebard, T., Gigant, B., Rigolet, P., Rasmussen, B., Diat, O., Bosecke, P., Wharton, S. A., Skehel, J. J., and Knossow, M. (1995). Structure of influenza virus haemagglutinin complexed with a neutralizing antibody. Nature *376*, 92–94.

Bovin, N. V., Korchagina, E. Y., Zemlyanukhina, T. V., Byramova, N. E., Galanina, O. E., Zemlyakov, A. E., Ivanov, A. E., Zubov, V. P., and Mochalova, L. V. (1993). Synthesis of polymeric neoglycoconjugates based on N-substituted polyacrylamides. Glycoconj. J. *10*, 142–151.

Bovin, N. V. (1998). Polyacrylamide-based glycoconjugates as tools in glycobiology. Glycoconj. J. *15*, 431–446.

Breg, J., van Halbeek, H., Vliegenthart, J. F., Lamblin, G., Houvenaghel, M. C., and Roussel, P. (1987). Structure of sialyl-oligosaccharides isolated from bronchial mucus glycoproteins of patients (blood group O) suffering from cystic fibrosis. Eur. J. Biochem. *168*, 57–68.

Breg, J., Kroon-Batenburg, L. M., Strecker, G., Montreuil, J., and Vliegenthart, J. F. (1989). Conformational analysis of the sialyl alpha(2–3/6)N-acetyllactosamine structural element occurring in glycoproteins, by two-dimensional NOE 1H-NMR spectroscopy in combination with energy calculations by hard-sphere exo-anomeric and molecular mechanics force- field with hydrogen-bonding potential. Eur. J. Biochem. *178*, 727–739.

Brown, I. H., Ludwig, S., Olsen, C. W., Hannoun, C., Scholtissek, C., Hinshaw, V. S., Harris, P. A., McCauley, J. W., Strong, I., and Alexander, D. J. (1997). Antigenic and genetic analyses of H1N1 influenza A viruses from European pigs. J. Gen. Virol. *78*, 553–562.

Bukrinskaia, A. G., Kornilaeva, G. V., Vorkunova, N. K., Timofeeva, N. G., and Shaposhnikova, G. I. (1982). Gangliosides – specific receptors for the influenza virus. Vopr. Virusol. *27*, 661–666.

Bukrinskaya, A. G. (1982). Penetration of viral genetic material into host cell. Adv. Virus Res. *27*, 141–204.

Burnet, F. M. and Bull, D. R. (1943). Changes in influenza virus associated with adaptation to passage in chick embryous. Aust. J. Exp. Biol. Med. Sci. *21*, 55–69.

Bush, R. M., Bender, C. A., Subbarao, K., Cox, N. J., and Fitch, W. M. (1999). Predicting the evolution of human influenza A. Science *286*, 1921–1925.

Butchko, G. M., Armstrong, R. B., Martin, W. J., and Ennis, F. A. (1978). Influenza A viruses of the H2N2 subtype are lymphocyte mitogens. Nature *271*, 66–67.

Callan, R. J., Early, G., Kida, H., and Hinshaw, V. S. (1995). The appearance of H3 influenza viruses in seals. J. Gen. Virol. *76*, 199–203.

Carroll, S. M., Higa, H. H., and Paulson, J. C. (1981). Different cell-surface receptor determinants of antigenically similar influenza virus hemagglutinins. J. Biol. Chem. *256*, 8357–8363.

Carroll, S. M. and Paulson, J. C. (1985). Differential infection of receptor-modified host cells by receptor-specific influenza viruses. Virus Res. *3*, 165–179.

Cash, E., Charreire, J., and Rott, O. (1996). B-cell activation by superstimulatory influenza virus hemagglutinin, a pathogenesis for autoimmunity? Immunol. Rev. *152*, 67–88.

Cassidy, L. F., Lyles, D. S., and Abramson, J. S. (1988). Synthesis of viral proteins in polymorphonuclear leukocytes infected with influenza A virus. J. Clin. Microbiol. *26*, 1267–1270.

Cassidy, L. F., Lyles, D. S., and Abramson, J. S. (1989). Depression of polymorphonuclear leukocyte functions by purified influenza virus hemagglutinin and sialic acid-binding lectins. J. Immunol. *142*, 4401–4406.

Claas, E. C. J., Osterhaus, A. D. M. E., Vanbeek, R., Dejong, J. C., Rimmelzwaan, G. F., Senne, D. A., Krauss, S., Shortridge, K. F., and Webster, R. G. (1998). Human influenza A H5N1 virus related to a highly pathogenic avian influenza virus. Lancet *351*, 472–477.

Colamussi, M. L., White, M. R., Crouch, E., and Hartshorn, K. L. (1999). Influenza A virus accelerates neutrophil apoptosis and markedly potentiates apoptotic effects of bacteria. Blood *93*, 2395–2403.

Connor, R. J., Kawaoka, Y., Webster, R. G., and Paulson, J. C. (1994). Receptor specificity in human, avian, and equine H2 and H3 influenza virus isolates. Virology *205*, 17–23.

Couceiro, J. N., Paulson, J. C., and Baum, L. G. (1993). Influenza virus strains selectively recognize sialyloligosaccharides on human respiratory epithelium; the role of the host cell in selection of hemagglutinin receptor specificity. Virus Res. *29*, 155–165.

Cox, N. J. and Subbarao, K. (2000). Global epidemiology of influenza, past and present. Annu. Rev. Med. *51*, 407–421.

Crecelius, D. M., Deom, C. M., and Schulze, I. T. (1984). Biological properties of a hemagglutinin mutant of influenza virus selected by host cells. Virology *139*, 164–177.

Daigneault, D. E., Hartshorn, K. L., Liou, L. S., Abbruzzi, G. M., White, M. R., Oh, S. K., and Tauber, A. I. (1992). Influenza A virus binding to human neutrophils and cross-linking requirements for activation. Blood *80*, 3227–3234.

Daniels, P. S., Jeffries, S., Yates, P., Schild, G. C., Rogers, G. N., Paulson, J. C., Wharton, S. A., Douglas, A. R., Skehel, J. J., and Wiley, D. C. (1987). The receptor-binding and membrane-fusion properties of influenza virus variants selected using anti-haemagglutinin monoclonal antibodies. EMBO J. *6*, 1459–1465.

Deom, C. M. and Schulze, I. T. (1985). Oligosaccharide composition of an influenza virus hemagglutinin with host-determined binding properties. J. Biol. Chem. *260*, 14771–14774.

Deom, C. M., Caton, A. J., and Schulze, I. T. (1986). Host cell-mediated selection of a mutant influenza A virus that has lost a complex oligosaccharide from the tip of the hemagglutinin. Proc. Natl. Acad. Sci. U. S. A *83*, 3771–3775.

Domingo, E., Escarmis, C., Sevilla, N., Moya, A., Elena, S. F., Quer, J., Novella, I. S., and Holland, J. J. (1996). Basic concepts in RNA virus evolution. FASEB J. *10*, 859–864.

Drake, J. W. (1993). Rates of spontaneous mutation among RNA viruses. Proc. Natl. Acad. Sci. U. S. A *90*, 4171–4175.

Ebisawa, I. T., Kitamoto, O., Takeuchi, Y., and Makino, M. (1969). Immunocytologic study of nasal epithelial cells in influenza. Am. Rev. Respir. Dis. *99*, 507–515.

Eisen, M. B., Sabesan, S., Skehel, J. J., and Wiley, D. C. (1997). Binding of the influenza A virus to cell-surface receptors, structures of five hemagglutinin-sialyloligosaccharide complexes determined by X-ray crystallography. Virology *232*, 19–31.

Eisenlohr, L. C., Gerhard, W., and Hackett, C. J. (1987). Role of receptor-binding activity of the viral hemagglutinin molecule in the presentation of influenza virus antigens to helper T cells. J. Virol. *61*, 1375–1383.

Feldmann, A., Schafer, M. K., Garten, W., and Klenk, H.-D. (2000). Targeted infection of endothelial cells by avian influenza virus A/FPV/Rostock/34 (H7N1) in chicken embryos. J. Virol. *74*, 8018–8027.

Felsenstein, J. (1989). PHYLIP – phylogeny inference package (Version 3. 2). Cladistics *5*, 164–166.

Gambaryan, A. S. and Matrosovich, M. N. (1992). A solid-phase enzyme-linked assay for influenza virus receptor-binding activity. J. Virol. Meth. *39*, 111–123.

Gambaryan, A. S., Piskarev, V. E., Yamskov, I. A., Sakharov, A. M., Tuzikov, A. B., Bovin, N. V., Nifant'ev, N. E., and Matrosovich, M. N. (1995). Human influenza virus recognition of sialyloligosaccharides. FEBS Lett. *366*, 57–60.

Gambaryan, A. S., Tuzikov, A. B., Piskarev, V. E., Yamnikova, S. S., Lvov, D. K., Robertson, J. S., Bovin, N. V., and Matrosovich, M. N. (1997). Specification of receptor-binding phenotypes of influenza virus isolates from different hosts using synthetic sialylglycopolymers, non-egg-adapted human H1 and H3 influenza A and influenza B viruses share a common high binding affinity for 6'-sialyl(N-acetyllactosamine). Virology *232*, 345–350.

Gambaryan, A. S., Marinina, V. P., Tuzikov, A. B., Bovin, N. V., Rudneva, I. A., Sinitsyn, B. V., Shilov, A. A., and Matrosovich, M. N. (1998a). Effects of host-dependent glycosylation of hemagglutinin on receptor-binding properties of H1N1 human influenza A virus grown in MDCK cells and in embryonated eggs. Virology *247*, 170–177.

Gambaryan, A. S., Matrosovich, M. N., Bender, C. A., and Kilbourne, E. D. (1998b). Differences in the biological phenotype of low-yielding (L) and high-yielding (H) variants of swine influenza virus A/NJ/11/76 are associated with their different receptor-binding activity. Virology *247*, 223–231.

Gambaryan, A. S., Robertson, J. S., and Matrosovich, M. N. (1999). Effects of egg-adaptation on the receptor-binding properties of human influenza viruses. Virology *258*, 232–239.

Gambaryan, A., Webster, R., and Matrosovich, M. (2002). Differences between influenza virus receptors on target cells of duck and chicken. Arch. Virol. *147*, 1197–1208.

Gambaryan, A. S., Tuzikov, A. B., Bovin, N. V., Yamnikova, S. S., Lvov, D. K., Webster, R. G., and Matrosovich, M. N. (2003). Differences between influenza virus receptors on target cells of duck and chicken and receptor specificity of the 1997 H5N1 chicken and human influenza viruses from Hong Kong. Avian. Dis. *47*, 1154–1160.

Gamblin, S. J., Haire, L. F., Russell, R. J., Stevens, D. J., Xiao, B., Ha, Y., Vasisht, N., Steinhauer, D. A., Daniels, R. S., Elliot, A., Wiley, D. C., and Skehel, J. J. (2004). The structure and receptor binding properties of the 1918 influenza hemagglutinin. Science *303*, 1838–1842.

Garten, W. and Klenk, H.-D. (1999). Understanding influenza virus pathogenicity. Trends Microbiol. *7*, 99–100.

Glick, G. D., Toogood, P. L., Wiley, D. C., Skehel, J. J., and Knowles, J. R. (1991). Ligand recognition by influenza virus. The binding of bivalent sialosides. J. Biol. Chem. *266*, 23660–23669.

Govorkova, E. A., Murti, G., Meignier, B., de Taisne, C., and Webster, R. G. (1996). African green monkey kidney (Vero) cells provide an alternative host cell system for influenza A and B viruses. J. Virol. *70*, 5519–5524.

Govorkova, E. A., Matrosovich, M. N., Tuzikov, A. B., Bovin, N. V., Gerdil, C., Fanget, B., and Webster, R. G. (1999). Selection of receptor-binding variants of human influenza A and B viruses in baby hamster kidney cells. Virology *262*, 31–38.

Gubareva, L. V., Matrosovich, M. N., Brenner, M. K., Bethell, R. C., and Webster, R. G. (1998). Evidence for zanamivir resistance in an immunocompromised child infected with influenza B virus. J. Infect. Dis. *178*, 1257–1262.

Gubareva, L. V., Kaiser, L., Matrosovich, M. N., Soo-Hoo, Y., and Hayden, F. G. (2001). Selection of influenza virus mutants in experimentally infected volunteers treated with oseltamivir. J. Infect. Dis. *183*, 523–531.

Günther, I., Glatthaar, B., Doller, G., and Garten, W. (1993). A H1 hemagglutinin of a human influenza A virus with a carbohydrate- modulated receptor binding site and an unusual cleavage site. Virus Res. *27*, 147–160.

Ha, Y., Stevens, D. J., Skehel, J. J., and Wiley, D. C. (2002). H5 avian and H9 swine influenza virus haemagglutinin structures, possible origin of influenza subtypes. EMBO J. *21*, 865–875.

Ha, Y., Stevens, D. J., Skehel, J. J., and Wiley, D. C. (2003). X-ray structure of the hemagglutinin of a potential H3 avian progenitor of the 1968 Hong Kong pandemic influenza virus. Virology *309*, 209–218.

Hanaoka, K., Pritchett, T. J., Takasaki, S., Kochibe, N., Sabesan, S., Paulson, J. C., and Kobata, A. (1989). 4-O-acetyl-N-acetylneuraminic acid in the N-linked carbohydrate structures of equine and guinea pig alpha 2-macroglobulins, potent inhibitors of influenza virus infection. J. Biol. Chem. *264*, 9842–9849.

Hanson, J. E., Sauter, N. K., Skehel, J. J., and Wiley, D. C. (1992). Proton nuclear magnetic resonance studies of the binding of sialosides to intact influenza virus. Virology *189*, 525–533.

Hansson, G. C., Karlsson, K.-A., Larson, G., Stromberg, N., Thurin, J., Orvell, C., and Norrby, E. (1984). A novel approach to the study of glycolipid receptors for viruses. Binding of Sendai virus to thin-layer chromatograms. FEBS Lett. *170*, 15–18.

Hartshorn, K. L. and Tauber, A. I. (1988). The influenza virus-infected phagocyte. A model of deactivation. Hematol. Oncol. Clin. North. Am. *2*, 301–315.

Hartshorn, K. L., Liou, L. S., White, M. R., Kazhdan, M. M., Tauber, J. L., and Tauber, A. I. (1995). Neutrophil deactivation by influenza A virus. Role of hemagglutinin binding to specific sialic acid-bearing cellular proteins. J. Immunol. *154*, 3952–3960.

Herrler, G., Geyer, R., Muller, H. P., Stirm, S., and Klenk, H.-D. (1985). Rat alpha 1 macroglobulin inhibits hemagglutination by influenza C virus. Virus Res. *2*, 183–192.

Herrler, G. and Klenk, H.-D. (1987). The surface receptor is a major determinant of the cell tropism of influenza C virus. Virology *159*, 102–108.

Herrler, G., Hausmann, J., and Klenk, H.-D. (1995). Sialic acid as receptor determinant of ortho- and paramyxoviruses. In, Biology of the sialic acids. A. Rosenberg, ed. Plenum, New York. pp. 315–336.

Higa, H. H., Rogers, G. N., and Paulson, J. C. (1985). Influenza virus hemagglutinins differentiate between receptor determinants bearing N-acetyl-, N-glycollyl-, and N, O-diacetylneuraminic acids. Virology *144*, 279–282.

Hinshaw, V. S., Webster, R. G., Easterday, B. C., and Bean, W. J., Jr. (1981). Replication of avian influenza A viruses in mammals. Infect. Immun. *34*, 354–361.

Hinshaw, V. S., Webster, R. G., Naeve, C. W., and Murphy, B. R. (1983). Altered tissue tropism of human-avian reassortant influenza viruses. Virology *128*, 260–263.

Hinshaw, V. S. (1998). Influenza in other species (seal, whale, and mink). In, Textbook of influenza. K. G. Nicholson, R. G. Webster, and A. J. Hay, eds. Blackwell Science, London. pp. 163–167.

Holland, J. J., de la Torre, J. C., and Steinhauer, D. A. (1992). RNA virus populations as quasispecies. Curr. Top. Microbiol. Immunol. *176*, 1–20.

Horimoto, T. and Kawaoka, Y. (2001). Pandemic threat posed by avian influenza A viruses. Clin. Microbiol. Rev. *14*, 129–149.

Inkster, M. D., Hinshaw, V. S., and Schulze, I. T. (1993). The hemagglutinins of duck and human H1 influenza viruses differ in sequence conservation and in glycosylation. J. Virol. *67*, 7436–7443.

Ito, T., Suzuki, Y., Takada, A., Kawamoto, A., Otsuki, K., Masuda, H., Yamada, M., Suzuki, T., Kida, H., and Kawaoka, Y. (1997). Differences in sialic acid-galactose linkages in the chicken egg amnion and allantois influence human influenza virus receptor specificity and variant selection. J. Virol. *71*, 3357–3362.

Ito, T., Couceiro, J. N. S. S., Kelm, S., Baum, L. G., Krauss, S., Castrucci, M. R., Donatelli, I., Kida, H., Paulson, J. C., Webster, R. G., and Kawaoka, Y. (1998). Molecular basis for the generation in pigs of influenza A viruses with pandemic potential. J. Virol. *72*, 7367–7373.

Ito, T. and Kawaoka, Y. (1998). Avian influenza. In, Textbook of influenza. K. G. Nicholson, R. G. Webster, and A. J. Hay, eds. Blackwell Science, London. pp. 126–136.

Ito, T., Suzuki, Y., Suzuki, T., Takada, A., Horimoto, T., Wells, K., Kida, H., Otsuki, K., Kiso, M., Ishida, H., and Kawaoka, Y. (2000). Recognition of N-glycolylneuraminic acid linked to galactose by the alpha2, 3 linkage is associated with intestinal replication of influenza A virus in ducks. J. Virol. *74*, 9300–9305.

Itoh, M., Hetterich, P., Isecke, R., Brossmer, R., and Klenk, H.-D. (1995). Suppression of influenza virus infection by an N-thioacetylneuraminic acid acrylamide copolymer resistant to neuraminidase. Virology *212*, 340–347.

Jeffery, P. K. and Li, D. (1997). Airway mucosa, secretory cells, mucus and mucin genes. Eur. Respir. J *10*, 1655–1662.

Jones, W. T. and Menna, J. H. (1982). Influenza type A virus-mediated adherence of type 1a group B streptococci to mouse tracheal tissue in vivo. Infect. Immun. *38*, 791–794.

Karlsson, K.-A. and Stromberg, N. (1987). Overlay and solid-phase analysis of glycolipid receptors for bacteria and viruses. Methods Enzymol. *138*, 220–232.

Karlsson, K.-A. (1989). Animal glycosphingolipids as membrane attachment sites for bacteria. Annu. Rev. Biochem. *58*, 309–350.

Katz, J. M., Wang, M., and Webster, R. G. (1990). Direct sequencing of the HA gene of influenza (H3N2) virus in original clinical samples reveals sequence identity with mammalian cell-grown virus. J. Virol. *64*, 1808–1811.

Katz, J. M. and Webster, R. G. (1992). Amino acid sequence identity between the HA1 of influenza A (H3N2) viruses grown in mammalian and primary chick kidney cells. J. Gen. Virol. *73*, 1159–1165.

Kaverin, N. V., Matrosovich, M. N., Gambaryan, A. S., Rudneva, I. A., Shilov, A. A., Varich, N. L., Makarova, N. V., Kropotkina, E. A., and Sinitsin, B. V. (2000). Intergenic HA-NA interactions in influenza A virus, postreassortment decrease of the receptor-binding affinity due to charged amino acid substitutions in the hemagglutinin of different subtypes. Virus Res. *66*, 123–129.

Kawaoka, Y., Chambers, T. M., Sladen, W. L., and Webster, R. G. (1988). Is the gene pool of influenza viruses in shorebirds and gulls different from that in wild ducks? Virology *163*, 247–250.

Keil, W., Niemann, H., Schwarz, R. T., and Klenk, H.-D. (1984). Carbohydrates of influenza virus. V. Oligosaccharides attached to individual glycosylation sites of the hemagglutinin of fowl plague virus. Virology *133*, 77–91.

Keil, W., Geyer, R., Dabrowski, J., Dabrowski, U., Niemann, H., Stirm, S., and Klenk, H.-D. (1985). Carbohydrates of influenza virus. Structural elucidation of the individual glycans of the FPV hemagglutinin by two-dimensional 1H n. m. r. and methylation analysis. EMBO J. *4*, 2711–2720.

Kelm, S., Paulson, J. C., Rose, U., Brossmer, R., Schmid, W., Bandgar, B. P., Schreiner, E., Hartmann, M., and Zbiral, E. (1992). Use of sialic acid analogues to define functional groups involved in binding to the influenza virus hemagglutinin. Eur. J. Biochem. *205*, 147–153.

Kelm, S. and Schauer, R. (1997). Sialic acids in molecular and cellular interactions. Int. Rev. Cytol. *175*, 137–240.

Kida, H., Ito, T., Yasuda, J., Shimizu, Y., Itakura, C., Shortridge, K. F., Kawaoka, Y., and Webster, R. G. (1994). Potential for transmission of avian influenza viruses to pigs. J. Gen. Virol. *75*, 2183–2188.

Kiessling, L. L. and Pohl, N. L. (1996). Strength in numbers, non-natural polyvalent carbohydrate derivatives. Chem. Biol. *3*, 71–77.

Kilbourne, E. D. (1987). Influenza. Plenum Pub Corp., New York

Kilbourne, E. D., Taylor, A. H., Whitaker, C. W., Sahai, R., and Caton, A. J. (1988a). Hemagglutinin polymorphism as the basis for low- and high-yield phenotypes of swine influenza virus. Proc. Natl. Acad. Sci. U. S. A *85*, 7782–7785.

Kilbourne, E. D., Easterday, B. C., and McGregor, S. (1988b). Evolution to predominance of swine influenza virus hemagglutinin mutants of predictable phenotype during single infections of the natural host. Proc. Natl. Acad. Sci. U. S. A 85, 8098–8101.

Klenk, E., Fallard, H., and Lempfrid, H. (1955). The enzymatic activity of influenza virus. Hoppe-Seyler's Z. Physiol. Chem. 301, 235–246.

Klenk, H.-D. and Rott, R. (1988). The molecular biology of influenza virus pathogenicity. Adv. Virus Res. 34, 247–281.

Klenk, H.-D. and Garten, W. (1994). Host cell proteases controlling virus pathogenicity. Trends Microbiol. 2, 39–43.

Klenk, H.-D., Wagner, R., Heuer, D., and Wolff, T. (2002). Importance of hemagglutinin glycosylation for the biological functions of influenza virus. Virus Res. 82, 73–75.

Klimov, A. I., Bender, C. A., Hall, H. E., and Cox, N. J. (1996). Evolution of human influenza A (H2N2) viruses. In, Options for the control of influenza III. L. E. Brown, A. W. Hampson, and R. G. Webster, eds. Elsevier Science B. V. pp. 546–552.

Kobasa, D., Takada, A., Shinya, K., Hatta, M., Halfmann, P., Theriault, S., Suzuki, H., Nishimura, H., Mitamura, K., Sugaya, N., Usui, T., Murata, T., Maeda, Y., Watanabe, S., Suresh, M., Suzuki, T., Suzuki, Y., Feldmann, H., and Kawaoka, Y. (2004). Enhanced virulence of influenza A viruses with the haemagglutinin of the 1918 pandemic virus. Nature 431, 703–707.

Kodihalli, S., Justewicz, D. M., Gubareva, L. V., and Webster, R. G. (1995). Selection of a single amino acid substitution in the hemagglutinin molecule by chicken eggs can render influenza A virus (H3) candidate vaccine ineffective. J. Virol. 69, 4888–4897.

Lamb, R. A. and Krug, R. M. (1996). Orthomyxoviridae, The viruses and their replication. In, Fields Virology. B. N. Fields, D. M. Knipe, and P. M. Howley, eds. Lippencott-Raven Publishers, Philadelphia. pp. 1353–1397.

Ledeen, R. W. and Yu, R. K. (1982). Gangliosides, structure, isolation, and analysis. Meth. Enzymol. 83, 139–191.

Leigh, M. W., Connor, R. J., Kelm, S., Baum, L. G., and Paulson, J. C. (1995). Receptor specificity of influenza virus influences

severity of illness in ferrets. Vaccine 13, 1468–1473.

Levinson, B., Pepper, D., and Belyavin, G. (1969). Substituted sialic acid prosthetic groups as determinants of viral hemagglutination. J. Virol. 3, 477–483.

Lin, Y. P., Shaw, M., Gregory, V., Cameron, K., Lim, W., Klimov, A., Subbarao, K., Guan, Y., Krauss, S., Shortridge, K., Webster, R., Cox, N., and Hay, A. (2000). Avian-to-human transmission of H9N2 subtype influenza A viruses, relationship between H9N2 and H5N1 human isolates. Proc. Natl. Acad. Sci. U. S. A 97, 9654–9658.

Lonberg-Holm, K. and Philipson, L. (1974). Early interaction between animal viruses and cells. Monogr. Virol. 9, 1–148.

Ludwig, S., Stitz, L., Planz, O., Van, H., Fitch, W. M., and Scholtissek, C. (1995). European swine virus as a possible source for the next influenza pandemic? Virology 212, 555–561.

Machytka, D., Kharitonenkov, I., Isecke, R., Hetterich, P., Brossmer, R., Klein, R. A., Klenk, H.-D., and Egge, H. (1993). Methyl alpha-glycoside of N-thioacetyl-D-neuraminic acid, a potential inhibitor of influenza A virus. A 1H NMR study. FEBS Lett. 334, 117–120.

Magnani, J. L., Smith, D. F., and Ginsburg, V. (1980). Detection of gangliosides that bind cholera toxin, direct binding of 125I-labeled toxin to thin-layer chromatograms. Anal. Biochem. 109, 399–402.

Mandelboim, O., Lieberman, N., Lev, M., Paul, L., Arnon, T. I., Bushkin, Y., Davis, D. M., Strominger, J. L., Yewdell, J. W., and Porgador, A. (2001). Recognition of haemagglutinins on virus-infected cells by NKp46 activates lysis by human NK cells. Nature 409, 1055–1060.

Martin, J., Wharton, S. A., Lin, Y. P., Takemoto, D. K., Skehel, J. J., Wiley, D. C., and Steinhauer, D. A. (1998). Studies of the binding properties of influenza hemagglutinin receptor-site mutants. Virology 241, 101–111.

Masuda, H., Suzuki, T., Sugiyama, Y., Horiike, G., Murakami, K., Miyamoto, D., Jwa Hidari, K. I., Ito, T., Kida, H., Kiso, M., Fukunaga, K., Ohuchi, M., Toyoda, T., Ishihama, A., Kawaoka, Y., and Suzuki, Y.

(1999). Substitution of amino acid residue in influenza A virus hemagglutinin affects recognition of sialyl-oligosaccharides containing N-glycolylneuraminic acid. FEBS Lett. *464*, 71–74.

Matrosovich, M. N. (1989). Towards the development of antimicrobial drugs acting by inhibition of pathogen attachment to host cells, a need for polyvalency. FEBS Lett. *252*, 1–4.

Matrosovich, M. N., Mochalova, L. V., Marinina, V. P., Byramova, N. E., and Bovin, N. V. (1990). Synthetic polymeric sialoside inhibitors of influenza virus receptor-binding activity. FEBS Lett. *272*, 209–212.

Matrosovich, M. N., Gambaryan, A. S., Reizin, F. N., and Chumakov, M. P. (1991). Recognition by human A and B influenza viruses of 8- and 7-carbon analogues of sialic acid modified in the polyhydroxyl side chain. Virology *182*, 879–882.

Matrosovich, M. N., Gambaryan, A. S., and Chumakov, M. P. (1992). Influenza viruses differ in recognition of 4-O-acetyl substitution of sialic acid receptor determinant. Virology *188*, 854–858.

Matrosovich, M. N., Gambaryan, A. S., Tuzikov, A. B., Byramova, N. E., Mochalova, L. V., Golbraikh, A. A., Shenderovich, M. D., Finne, J., and Bovin, N. V. (1993). Probing of the receptor-binding sites of the H1 and H3 influenza A and influenza B virus hemagglutinins by synthetic and natural sialosides. Virology *196*, 111–121.

Matrosovich, M., Miller-Podraza, H., Teneberg, S., Robertson, J., and Karlsson, K.-A. (1996). Influenza viruses display high-affinity binding to human polyglycosylceramides represented on a solid-phase assay surface. Virology *223*, 413–416.

Matrosovich, M. N., Gambaryan, A. S., Teneberg, S., Piskarev, V. E., Yamnikova, S. S., Lvov, D. K., Robertson, J. S., and Karlsson, K.-A. (1997). Avian influenza A viruses differ from human viruses by recognition of sialyloligosaccharides and gangliosides and by a higher conservation of the HA receptor-binding site. Virology *233*, 224–234.

Matrosovich, M., Gao, P., and Kawaoka, Y. (1998). Molecular mechanisms of serum resistance of human influenza H3N2 virus and their involvement in virus adaptation in a new host. J. Virol. *72*, 6373–6380.

Matrosovich, M., Zhou, N., Kawaoka, Y., and Webster, R. (1999). The surface glycoproteins of H5 influenza viruses isolated from humans, chickens, and wild aquatic birds have distinguishable properties. J. Virol. *73*, 1146–1155.

Matrosovich, M., Suzuki, T., and Webster, R. (2000a). Influenza virus can infect cells lacking glycosphingolipids. Keystone Symposium 'Cell Biology of Virus Entry, Replication and Pathogenesis', Taos, New Mexico. p. 61.

Matrosovich, M., Tuzikov, A., Bovin, N., Gambaryan, A., Klimov, A., Castrucci, M. R., Donatelli, I., and Kawaoka, Y. (2000). Early alterations of the receptor-binding properties of H1, H2, and H3 avian influenza virus hemagglutinins after their introduction into mammals. J. Virol. *74*, 8502–8512.

Matrosovich, M. N., Krauss, S. and Webster, R. G. (2001). H9N2 influenza A viruses from poultry in Asia have human virus-like receptor specificity. Virology *281*, 156–162.

Matrosovich, M. and Klenk, H.-D. (2003). Natural and synthetic sialic-acid-containing inhibitors of influenza virus receptor binding. Rev. Med. Virol. *13*, 85–97.

Matrosovich, M., Matrosovich, T., Carr, J., Roberts, N. A., and Klenk, H.-D. (2003). Overexpression of the alpha–2, 6-sialyltransferase in MDCK cells increases influenza virus sensitivity to neuraminidase inhibitors. J. Virol. *77*, 8418–8425.

Matrosovich, M. N., Matrosovich, T. Y., Gray, T., Roberts, N. A., and Klenk, H.-D. (2004). Human and avian influenza viruses target different cell types in cultures of human airway epithelium. Proc. Natl. Acad. Sci. U. S. A *101*, 4620–4624.

Medeiros, R., Escriou, N., Naffakh, N., Manuguerra, J. C., and van der Werf, S. (2001). Hemagglutinin residues of recent human A(H3N2) influenza viruses that contribute to the inability to agglutinate chicken erythrocytes. Virology *289*, 74–85.

Miller-Podraza, H., Larsson, T., Nilsson, J., Teneberg, S., Matrosovich, M., and Johansson, L. (1998). Epitope dissection of receptor-active gangliosides with affinity for Helicobacter pylori and influenza virus. Acta Biochimica Polonica 45, 439–449.

Miller-Podraza, H., Johansson, L., Johansson, P., Matrosovich, M., and Karlsson, K.-A. (2000). A strain of human influenza A virus binds to extended but not short gangliosides as assayed by thin-layer chromatography overlay. Glycobiology 10, 975–982.

Mochalova, L., Gambaryan, A., Romanova, J., Tuzikov, A., Chinarev, A., Katinger, D., Katinger, H., Egorov, A., and Bovin, N. (2003). Receptor-binding properties of modern human influenza viruses primarily isolated in Vero and MDCK cells and chicken embryonated eggs. Virology 313, 473–480.

Mumford, J. A. and Chambers, T. M. (1998). Equine influenza. In, Textbook of influenza. K. G. Nicholson, R. G. Webster, and A. J. Hay, eds. Blackwell Science, London. pp. 146–162.

Murphy, B. R. and Webster, R. G. (1996). Orthomyxoviruses. In, Fields Virology. B. N. Fields, D. M. Knipe, and P. M. Howley, eds. Lippincott-Raven Publishers, Philadelphia. pp. 1397–1495.

Müthing, J., Unland, F., Heitmann, D., Orlich, M., Hanisch, F. G., Peter-Katalinic, J., Knauper, V., Tschesche, H., Kelm, S., and Schauer, R. (1993). Different binding capacities of influenza A and Sendai viruses to gangliosides from human granulocytes. Glycoconj. J. 10, 120–126.

Müthing, J. (1996). Influenza A and Sendai viruses preferentially bind to fucosylated gangliosides with linear poly-N-acetyllactosaminyl chains from human granulocytes. Carbohydr. Res. 290, 217–224.

Naeve, C. W., Hinshaw, V. S., and Webster, R. G. (1984). Mutations in the hemagglutinin receptor-binding site can change the biological properties of an influenza virus. J. Virol. 51, 567–569.

Nagai, Y. and Iwamori, M. (1995). Cellular biology of gangliosides. A. Rosenberg, ed. Plenum, New York. pp. 197–241.

Nicholas, K. B., Nicholas, H. B. Jr., and Deerfield, D. W. (1997). GeneDoc, Analysis and visualization of genetic variation. EMBNEW News 4, 14–14.

Nobusawa, E., Aoyama, T., Kato, H., Suzuki, Y., Tateno, Y., and Nakajima, K. (1991). Comparison of complete amino acid sequences and receptor-binding properties among 13 serotypes of hemagglutinins of influenza A viruses. Virology 182, 475–485.

Nobusawa, E., Ishihara, H., Morishita, T., Sato, K., and Nakajima, K. (2000). Change in receptor-binding specificity of recent human influenza A viruses (H3N2), a single amino acid change in hemagglutinin altered its recognition of sialyloligosaccharides. Virology 278, 587–596.

Ohuchi, M., Feldmann, A., Ohuchi, R., and Klenk, H.-D. (1995). Neuraminidase is essential for fowl plague virus hemagglutinin to show hemagglutinating activity. Virology 212, 77–83.

Ohuchi, M., Ohuchi, R., Feldmann, A., and Klenk, H.-D. (1997). Regulation of receptor binding affinity of influenza virus hemagglutinin by its carbohydrate moiety. J. Virol. 71, 8377–8384.

Ohuchi, M., Ohuchi, R., and Matsumoto, A. (1999). Control of biological activities of influenza virus hemagglutinin by its carbohydrate moiety. Microbiol. Immunol. 43, 1071–1076.

Osterhaus, A. D., Rimmelzwaan, G. F., Martina, B. E., Bestebroer, T. M., and Fouchier, R. A. (2000). Influenza B virus in seals. Science 288, 1051–1053.

Oxford, J. S., Schild, G. C., Corcoran, T., Newman, R., Major, D., Robertson, J., Bootman, J., Higgins, P., al-Nakib, W., and Tyrrell, D. A. (1990). A host-cell-selected variant of influenza B virus with a single nucleotide substitution in HA affecting a potential glycosylation site was attenuated in virulence for volunteers. Arch. Virol. 110, 37–46.

Page, R. D. M. (1996). TREEVIEW, An application to display phylogenetic trees on personal computers. Computer Applications in Biosciences 12, 357–358.

Paulson, J. C. (1985). Interactions of animal viruses with cell surface receptors. In, The receptors. Vol. 2. M. Conn, ed. Academic Press, Orlando, FL. pp. 131–219.

Paulson, J. C. and Rogers, G. N. (1987). Resialylated erythrocytes for assessment of the specificity of sialyloligosaccharide binding proteins. Meth. Enzymol. *138*, 162–168.

Poppe, L., Dabrowski, J., von der Lieth, C. W., Numata, M., and Ogawa, T. (1989). Solution conformation of sialosylcerebroside (GM4) and its NeuAc(alpha 2–3)Gal beta sugar component. Eur. J. Biochem. *180*, 337–342.

Poppe, L., Stuike-Prill, R., Meyer, B., and van Halbeek, H. (1992). The solution conformation of sialyl-alpha (2-6)-lactose studied by modern NMR techniques and Monte Carlo simulations. J. Biomol. NMR *2*, 109–136.

Poumbourios, P., Anders, E. M., Scalzo, A. A., White, D. O., Hampson, A. W., and Jackson, D. C. (1987). Direct role of viral hemagglutinin in B-cell mitogenesis by influenza viruses. J. Virol. *61*, 214–217.

Preston, A., Mandrell, R. E., Gibson, B. W., and Apicella, M. A. (1996). The lipooligosaccharides of pathogenic gram-negative bacteria. Crit. Rev. Microbiol. *22*, 139–180.

Pritchett, T. J., Brossmer, R., Rose, U., and Paulson, J. C. (1987). Recognition of monovalent sialosides by influenza virus H3 hemagglutinin. Virology *160*, 502–506.

Pritchett, T. J. and Paulson, J. C. (1989). Basis for the potent inhibition of influenza virus infection by equine and guinea pig alpha 2-macroglobulin. J. Biol. Chem. *264*, 9850–9858.

Reid, A. H., Fanning, T. G., Hultin, J. V., and Taubenberger, J. K. (1999). Origin and evolution of the 1918 'Spanish' influenza virus hemagglutinin gene. Proc. Natl. Acad. Sci. U. S. A *96*, 1651–1656.

Reid, A. H., Janczewski, T. A., Lourens, R. M., Elliot, A.J., Daniels, R. S., Berry, C. L., Oxford, J. S., and Taubenberger, J. K. (2003). 1918 influenza pandemic caused by highly conserved viruses with two receptor-binding variants. Emerg. Infect. Dis. *9*, 1249–1253.

Reuter, G. and Gabius, H. J. (1996). Sialic acids structure-analysis-metabolism-occurrence-recognition. Biol. Chem. Hoppe-Seyler *377*, 325–342.

Robertson, J. S., Bootman, J. S., Nicolson, C., Major, D., Robertson, E. W., and Wood, J. M. (1990). The hemagglutinin of influenza B virus present in clinical material is a single species identical to that of mammalian cell-grown virus. Virology *179*, 35–40.

Robertson, J. S., Nicolson, C., Bootman, J. S., Major, D., Robertson, E. W., and Wood, J. M. (1991). Sequence analysis of the haemagglutinin (HA) of influenza A (H1N1) viruses present in clinical material and comparison with the HA of laboratory-derived virus. J. Gen. Virol. *72*, 2671–2677.

Robertson, J. S. (1993). Clinical influenza virus and the embryonated hen's eggs. Rev. Med. Virol. *3*, 97–106.

Robertson, J. S., Cook, P., Attwell, A. M., and Williams, S. P. (1995). Replicative advantage in tissue culture of egg-adapted influenza virus over tissue-culture derived virus, implications for vaccine manufacture. Vaccine *13*, 1583–1588.

Rogers, G. N. and Paulson, J. C. (1983). Receptor determinants of human and animal influenza virus isolates, differences in receptor specificity of the H3 hemagglutinin based on species of origin. Virology *127*, 361–373.

Rogers, G. N., Pritchett, T. J., Lane, J. L., and Paulson, J. C. (1983). Differential sensitivity of human, avian, and equine influenza A viruses to a glycoprotein inhibitor of infection, selection of receptor specific variants. Virology *131*, 394–408.

Rogers, G. N., Daniels, R. S., Skehel, J. J., Wiley, D. C., Wang, X. F., Higa, H. H., and Paulson, J. C. (1985). Host-mediated selection of influenza virus receptor variants. Sialic acid-alpha 2, 6Gal-specific clones of A/duck/Ukraine/1/63 revert to sialic acid-alpha 2, 3Gal-specific wild type in ovo. J. Biol. Chem. *260*, 7362–7367.

Rogers, G. N., Herrler, G., Paulson, J. C., and Klenk, H.-D. (1986). Influenza C virus uses 9-O-acetyl-N-acetylneuraminic acid as a high affinity receptor determinant for attachment to cells. J. Biol. Chem. *261*, 5947–5951.

Rogers, G. N. and D'Souza, B. L. (1989). Receptor binding properties of human and animal H1 influenza virus isolates. Virology *173*, 317–322.

Romanova, J., Katinger, D., Ferko, B., Voglauer, R., Mochalova, L., Bovin, N., Lim, W., Katinger, H., and Egorov, A. (2003). Distinct host range of influenza H3N2 virus isolates in Vero and MDCK cells is determined by cell specific glycosylation pattern. Virology *307*, 90–97.

Rosenberg, A., Howe, C., and Chargaff, E. (1956). Inhibition of influenza virus hemagglutination by a brain lipid fraction. Nature *177*, 234–235.

Rothwell, S. W. and Wright, D. G. (1994). Characterization of influenza A virus binding sites on human neutrophils. J. Immunol. *152*, 2358–2367.

Rott, O. and Cash, E. (1994). Influenza virus hemagglutinin induces differentiation of mature resting B cells and growth arrest of immature WEHI–231 lymphoma cells. J. Immunol. *152*, 5381–5391.

Rott, R. and Klenk, H.-D. (1987). Significance of viral glycoproteins for infectivity and pathogenicity. Zentralbl. Bakteriol. Mikrobiol. Hyg. [A] *266*, 145–154.

Ruigrok, R. W. H. (1998). Structure of influenza A, B and C viruses. In, Textbook of influenza. K. G. Nicholson, R. G. Webster, and A. J. Hay, eds. Blackwell Science, London. pp. 29–42.

Ryan-Poirier, K., Suzuki, Y., Bean, W. J., Kobasa, D., Takada, A., Ito, T., and Kawaoka, Y. (1998). Changes in H3 influenza A virus receptor specificity during replication in humans. Virus Res. *56*, 169–176.

Ryan-Poirier, K. A. and Kawaoka, Y. (1991). Distinct glycoprotein inhibitors of influenza A virus in different animal sera. J. Virol. *65*, 389–395.

Ryan-Poirier, K. A. and Kawaoka, Y. (1993). Alpha 2-macroglobulin is the major neutralizing inhibitor of influenza A virus in pig serum. Virology *193*, 974–976.

Sanford, B. A., Shelokov, A., and Ramsay, M. A. (1978). Bacterial adherence to virus-infected cells, a cell culture model of bacterial superinfection. J. Infect. Dis. *137*, 176–181.

Sanford, B. A., Smith, N., Shelokov, A., and Ramsay, M. A. (1980). Adherence of influenza A viruses to group B Streptococci. J. Infect. Dis. *141*, 496–506.

Sauter, N. K., Bednarski, M. D., Wurzburg, B. A., Hanson, J. E., Whitesides, G. M., Skehel, J. J., and Wiley, D. C. (1989). Hemagglutinins from two influenza virus variants bind to sialic acid derivatives with millimolar dissociation constants, a 500-MHz proton nuclear magnetic resonance study. Biochemistry *28*, 8388–8396.

Sauter, N. K., Hanson, J. E., Glick, G. D., Brown, J. H., Crowther, R. L., Park, S. J., Skehel, J. J., and Wiley, D. C. (1992). Binding of influenza virus hemagglutinin to analogs of its cell-surface receptor, sialic acid, analysis by proton nuclear magnetic resonance spectroscopy and X-ray crystallography. Biochemistry *31*, 9609–9621.

Scharfman, A., Lamblin, G., and Roussel, P. (1995). Interactions between human respiratory mucins and pathogens. Biochem. Soc. Trans. *23*, 836–839.

Schauer, R. (1982). Chemistry, metabolism, and biological functions of sialic acids. Adv. Carbohydr. Chem. Biochem. *40*, 131–234.

Schauer, R., Kelm, S., Schröder, C., and Miller, E. (1995). Biochemistry and role of sialic acids. In, Biology of the sialic acids. A. Rosenberg, ed. Plenum, New York. pp. 7–67.

Schauer, R. and Kamerling, J. P. (1997). Chemistry, biochemistry and biology of sialic acids. In, Glycoproteins II (New Comprehensive Biochemistry, vol. 29b). J. Montreuil, J. F. G. Vliegenthart, and H. Schachter, eds. Elsevier Science B. V. pp. 243–402.

Scholtissek, C., Burger, H., Bachmann, P. A., and Hannoun, C. (1983). Genetic relatedness of hemagglutinins of the H1 subtype of influenza A viruses isolated from swine and birds. Virology *129*, 521–523.

Scholtissek, C., Burger, H., Kistner, O., and Shortridge, K. F. (1985). The nucleoprotein as a possible major factor in determining host specificity of influenza H3N2 viruses. Virology *147*, 287–294.

Scholtissek, C., Hinshaw, V. S., and Olsen, C. W. (1998). Influenza in pigs and their role as the intermediate host. In, Textbook of influenza. K. G. Nicholson, R. G. Webster, and A. J. Hay, eds. Blackwell Science, London. pp. 137–145.

Schulze, I. T. (1997). Effects of glycosylation on the properties and functions of influenza virus hemagglutinin. J. Infect. Dis. 176 Suppl *1*, S24-S28.

Skehel, J. J. and Wiley, D. C. (2000). Receptor binding and membrane fusion in virus entry, the influenza hemagglutinin. Annu. Rev. Biochem. *69*, 531–569.

Slemons, R. D. and Easterday, B. C. (1977). Type-A influenza viruses in the feces of migratory waterfowl. J. Am. Vet. Med. Assoc. *171*, 947–948.

Spaltenstein, A. and Whitesides, G. M. (1991). Polyacrylamides bearing pendant α–sialoside groups strongly inhibit agglutination of erythrocytes by influenza virus. J. Amer. Chem. Soc. *113*, 686–687.

Steinhauer, D. A. (1999). Role of hemagglutinin cleavage for the pathogenicity of influenza virus. Virology *258*, 1–20.

Stevens, J., Corper, A. L., Basler, C. F., Taubenberger, J. K., Palese, P., and Wilson, I. A. (2004). Structure of the uncleaved human H1 hemagglutinin from the extinct 1918 influenza virus. Science *303*, 1866–1870.

Stevenson, P. G. and Doherty, P. C. (1998). Cell-mediated response to influenza viruses. In, Textbook of influenza. K. G. Nicholson, R. G. Webster, and A. J. Hay, eds. Blackwell Science, London. pp. 278–290.

Suarez, D. L., Garcia, M., Latimer, J., Senne, D., and Perdue, M. (1999). Phylogenetic analysis of H7 avian influenza viruses isolated from the live bird markets of the Northeast United States. J. Virol. *73*, 3567–3573.

Subbarao, K., Klimov, A., Katz, J., Regnery, H., Lim, W., Hall, H., Perdue, M., Swayne, D., Bender, C., Huang, J., Hemphill, M., Rowe, T., Shaw, M., Xu, X. Y., Fukuda, K., and Cox, N. (1998). Characterization of an avian influenza A (H5N1) virus isolated from a child with a fatal respiratory illness. Science *279*, 393–396.

Suzuki, T., Sometani, A., Yamazaki, Y., Horiike, G., Mizutani, Y., Masuda, H., Yamada, M., Tahara, H., Xu, G., Miyamoto, D., Oku, N., Okada, S., Kiso, M., Hasegawa, A., Ito, T., Kawaoka, Y., and Suzuki, Y. (1996). Sulphatide binds to human and animal influenza A viruses, and inhibits the viral infection. Biochem. J. *318*, 389–393.

Suzuki, T., Horiike, G., Yamazaki, Y., Kawabe, K., Masuda, H., Miyamoto, D., Matsuda, M., Nishimura, S. I., Yamagata, T., Ito, T., Kida, H., Kawaoka, Y., and Suzuki, Y. (1997). Swine influenza virus strains recognize sialylsugar chains containing the molecular species of sialic acid predominantly present in the swine tracheal epithelium. FEBS Lett. *404*, 192–196.

Suzuki, Y., Matsunaga, M., and Matsumoto, M. (1985). N-Acetylneuraminyllactosylcera-mide, GM3-NeuAc, a new influenza A virus receptor which mediates the adsorption-fusion process of viral infection. Binding specificity of influenza virus A/Aichi/2/68 (H3N2) to membrane-associated GM3 with different molecular species of sialic acid. J. Biol. Chem. *260*, 1362–1365.

Suzuki, Y., Nagao, Y., Kato, H., Matsumoto, M., Nerome, K., Nakajima, K., and Nobusawa, E. (1986). Human influenza A virus hemagglutinin distinguishes sialyloligosaccharides in membrane-associated gangliosides as its receptor which mediates the adsorption and fusion processes of virus infection. Specificity for oligosaccharides and sialic acids and the sequence to which sialic acid is attached. J. Biol. Chemistry *261*, 17057–17061.

Suzuki, Y., Kato, H., Naeve, C. W., and Webster, R. G. (1989). Single-amino-acid substitution in an antigenic site of influenza virus hemagglutinin can alter the specificity of binding to cell membrane-associated gangliosides. J. Virol. *63*, 4298–4302.

Suzuki, Y., Nakao, T., Ito, T., Watanabe, N., Toda, Y., Xu, G., Suzuki, T., Kobayashi, T., Kimura, Y., and Yamada, A. (1992). Structural determination of gangliosides that bind to influenza A, B, and C viruses by an improved binding assay, strain-specific receptor epitopes in sialo-sugar chains. Virology *189*, 121–131.

Suzuki, Y. (1994). Gangliosides as influenza virus receptors. Variation of influenza viruses and their recognition of the receptor sialo-sugar chains. Progr. Lipid Res. *33*, 429–457.

Suzuki, Y., Ito, T., Suzuki, T., Holland, R. E., Jr., Chambers, T. M., Kiso, M., Ishida, H., and Kawaoka, Y. (2000). Sialic acid species as a determinant of the host range of influenza A viruses. J. Virol. *74*, 11825–11831.

Tateno, I., Kitamoto, O., and Kawamura, A., Jr. (1966). Diverse immunocytologic findings of nasal smears in influenza. N. Engl. J. Med. *274*, 237–242.

Taubenberger, J. K., Reid, A. H., Krafft, A. E., Bijwaard, K. E., and Fanning, T. G. (1997). Initial genetic characterization of the 1918 'Spanish' influenza virus. Science *275*, 1793–1796.

Thomas, D. B., Patera, A. C., Graham, C. M., and Smith, C. A. (1998). Antibody-mediated immunity. In, Textbook of influenza. K. G. Nicholson, R. G. Webster, and A. J. Hay, eds. Blackwell Science, London. pp. 267–277.

Toogood, P. L., Galliker, P. K., Glick, G. D., and Knowles, J. R. (1991). Monovalent sialosides that bind tightly to influenza A virus. J. Med. Chem. *34*, 3138–3140.

Tuzikov, A. B., Byramova, N. E., Bovin, N. V., Gambaryan, A. S., and Matrosovich, M. N. (1997). Monovalent and polymeric 5N-thioacetamido sialosides as tightly-bound receptor analogs of influenza viruses. Antiviral Res. *33*, 129–134.

Underwood, P. A. (1985). Receptor binding characteristics of strains of the influenza Hong Kong subtype, using a periodate sensitivity test. Arch. Virol. *84*, 53–61.

Underwood, P. A., Skehel, J. J., and Wiley, D. C. (1987). Receptor-binding characteristics of monoclonal antibody-selected antigenic variants of influenza virus. J. Virol. *61*, 206–208.

Varki, A. (1992). Diversity in the sialic acids. Glycobiology *2*, 25–40.

Varki, A. (1997). Sialic acids as ligands in recognition phenomena. FASEB J. *11*, 248–255.

Veit, M., Herrler, G., Schmidt, M. F., Rott, R., and Klenk, H.-D. (1990). The hemagglutinating glycoproteins of influenza B and C viruses are acylated with different fatty acids. Virology *177*, 807–811.

Veit, M., Kretzschmar, E., Kuroda, K., Garten, W., Schmidt, M. F., Klenk, H. D., and Rott, R. (1991). Site-specific mutagenesis identifies three cysteine residues in the cytoplasmic tail as acylation sites of influenza virus hemagglutinin. J. Virol. *65*, 2491–2500.

Vines, A., Wells, K., Matrosovich, M., Castrucci, M. R., Ito, T., and Kawaoka, Y. (1998). The role of influenza A virus hemagglutinin residues 226 and 228 in receptor specificity and host range restriction. J. Virol. *72*, 7626–7631.

Wagner, R., Wolff, T., Herwig, A., Pleschka, S., and Klenk, H.-D. (2000). Interdependence of hemagglutinin glycosylation and neuraminidase as regulators of influenza virus growth, a study by reverse genetics. J. Virol. *74*, 6316–6323.

Wagner, R., Matrosovich, M., and Klenk, H.-D. (2002). Functional balance between haemagglutinin and neuraminidase in influenza virus infections. Rev. Med. Virol. *12*, 159–166.

Webster, R. G., Yakhno, M., Hinshaw, V. S., Bean, W. J., and Murti, K. G. (1978). Intestinal influenza, replication and characterization of influenza viruses in ducks. Virology *84*, 268–278.

Webster, R. G., Bean, W. J., Gorman, O. T., Chambers, T. M., and Kawaoka, Y. (1992). Evolution and ecology of influenza A viruses. Microbiological Reviews *56*, 152–179.

Webster, R. G. (1997). Influenza virus, transmission between species and relevance to emergence of the next human pandemic. Arch. Virol. Suppl. *13*, 105–113.

Weis, W., Brown, J. H., Cusack, S., Paulson, J. C., Skehel, J. J., and Wiley, D. C. (1988). Structure of the influenza virus haemagglutinin complexed with its receptor, sialic acid. Nature *333*, 426–431.

Weis, W. I. (1997). Cell-surface carbohydrate recognition by animal and viral lectins. Curr. Opin. Struct. Biol. *7*, 624–630.

Wiley, D. C. and Skehel, J. J. (1987). The structure and function of the hemagglutinin membrane glycoprotein of influenza virus. Annu. Rev. Biochem. *56*, 365–394.

Wilson, I. A. and Cox, N. J. (1990). Structural basis of immune recognition of influenza virus hemagglutinin. Annu. Rev. Immunol. *8*, 737–771.

Xu, G., Suzuki, T., Tahara, H., Kiso, M., Hasegawa, A., and Suzuki, Y. (1994). Specificity of sialyl-sugar chain mediated recognition by the hemagglutinin of human influenza B virus isolates. J. Biochem. *115*, 202–207.

Xu, G., Horiike, G., Suzuki, T., Miyamoto, D., Kumihashi, H., and Suzuki, Y. (1996). A novel strain, B/Gifu/2/73, differs from other influenza B viruses in the receptor binding specificities toward sialo-sugar chain linkage. Biochem. Biophys. Research Commun. *224*, 815–818.

Yamnikova, S. S., Gambaryan, A. S., Tuzikov, A. B., Bovin, N. V., Matrosovich, M. N., Fedyakina, I. T., Grinev, A. A., Blinov, V. M., Lvov, D. K., Suarez, D. L., and Swayne, D. E. (2003). Differences between HA receptor-binding sites of avian influenza viruses isolated from Laridae and Anatidae. Avian. Dis. *47*, 1164–1168.

Zuckerman, M. A., Cox, R. J., and Oxford, J. S. (1994). Attenuation of virulence in influenza B viral infection of volunteers. J. Infect. *28*, 41–48.

CHAPTER 5

Dendritic cells: induction and regulation of the adaptive immune response to influenza virus infection

Kevin L. Legge and Thomas J. Braciale

Abstract:
Clearance of respiratory viruses, like influenza virus, from the respiratory tract requires induction of an adaptive immune response. Initiation of adaptive immunity to foreign pathogens, like influenza virus, is thought to be mediated by dendritic cells. Dendritic cells perform this function by first sensing the invader in peripheral sites, maturing, and then migrating to the draining regional lymph nodes where they interact with and activate naïve T and B cells. In this review, we will highlight and discuss what role dendritic cells may play in the induction and regulation of the adaptive immune response to pulmonary influenza virus infections, and how the interaction of influenza virus with dendritic cells may influence the shape of the developing adaptive immune response.

DENDRITIC CELLS: INDUCTION OF CD8+ T CELL IMMUNITY TO INFLUENZA VIRUS

Infection of the respiratory tract with influenza virus leads to both innate and adaptive immune responses that are targeted to control and eliminate the viral infection. The adaptive arm of the anti-influenza response is mediated by neutralizing antibody producing B cells (Gerhard et al., 1997), CD4[+] helper T cells (Brown et al., 2004; Doherty et al., 1997; Wong and Pamer, 2003), and CD8[+] cytotoxic T lymphocytes (CTL) that target and kill virally-infected cells (Doherty and Christensen, 2000; Doherty et al., 1997; Topham et al., 1997; Wong and Pamer, 2003). Since T cell immunity requires presentation of peptide antigens by MHC molecules on the surface of antigen presenting cells (APC) (Banchereau et al., 2000; Germain and Jenkins, 2004; Guermonprez et al., 2002; Itano and Jenkins, 2003; Norbury et al., 2002; Pamer and Cresswell, 1998; Rock and Goldberg, 1999), early studies on the nature of the induction of the adaptive immune response to influenza virus focused on the type(s) of APC involved in triggering T cell immunity. *In vitro* exposure of purified dendritic cells (DC), but not macrophages, to infectious influenza virus was shown to be sufficient to stimulate a primary virus-specific CTL response from lymph node T cells (Bhardwaj et al., 1994; Macatonia et al., 1989; Nonacs et al., 1992). Furthermore, the influenza-infected DC were potent inducers of influenza-specific CD8[+] T cell responses – and could represent as little as 0.1–1% of the cultured cells and still trigger significant CTL responses without the need for addition of exogenous cytokines (Bhardwaj et al., 1994; Macatonia et al., 1989). These results suggested that DC might be the primary cell responsible for induction of influenza-specific T cell immunity.

Experiments subsequent to those described above demonstrated that presentation of viral antigen by DC to naïve CTL in these *in vitro* culture systems required that influenza be infectious (Bhardwaj et al., 1994; Macatonia et al., 1989; Nonacs et al., 1992). These results suggested that presentation of MHC class I-dependent influenza virus antigen utilized the endogenous pathway, but not the alternative cross-presentation pathway of antigen presentation (Heath and Carbone, 2001; Pamer and Cresswell, 1998; Rock and Goldberg, 1999). The lack of cross-presentation of the influenza antigen was not due to poor uptake of inactivated virus by DC, as incubation of DC with inactivated virus did stimulate CD4[+] T cell responses (Nonacs et al., 1992). These results suggested that effective *in vivo* induction of influenza-specific CD8[+] T cells might require direct infection of respiratory dendritic cells (rDC), and were consistent with earlier studies that showed that infectious, but not inactivated, influenza virus was a potent inducer of *in vivo* CTL responses (Braciale and Yap, 1978).

The involvement of *in vivo* cross-presented influenza virus antigen in the activation of naïve influenza-specific CD8[+] T cells following natural infection has been difficult to assess. Subsequent studies have also used inactivated virus to drive naïve CTL activation. However, these studies have also shown either a low-level of expression of NS1 (a non-structural influenza virus protein whose expression suggests direct influenza virus infection) in DC (Bender et al., 1995) or have led to changes in the immunodominance hierarchy of the CD8[+] T cell response (Cho et al., 2003) suggesting that administration of inactivated virus does not mimic responses to natural influenza infection. Furthermore inactivated influenza virus does not induce maturation and migration of rDC (Legge and Braciale, 2003). Such maturation and migration of rDC would be needed to drive activation of influenza-specific naïve CD8[+] T cells in the draining lymph nodes (Albert et al., 2001; Larsson et al., 2000; Wilson et al., 2003), as encounter of naïve T cells with influenza antigen-pulsed immature DC does not drive activation, but rather tolerance to viral antigen (Dhodapkar and Steinman, 2002; Dhodapkar et al., 2001). Evidence showing that rDC uptake of influenza-driven apoptotic cells *in vivo* in the respiratory tract after natural influenza infection would, however, support the rationale for a role for cross-presentation of influenza virus antigen in the stimulation of naïve influenza-specific T cell responses. In fact, influenza virus infection of the respiratory tract does lead to apoptosis of virally-infected respiratory epithelium (Brydon et al., 2003; Mori et al., 1995; Technau-Ihling et al., 2001) and uptake of apoptotic bodies coupled with inflammation allows the presentation of apoptosis-derived antigen and maturation of DC (Banchereau et al., 2000; Gallucci et al., 1999; Guermonprez et al., 2002; Sauter et al., 2000). Furthermore, human DC can take up and cross-present viral antigen *in vitro* from cultures of influenza-infected apoptotic cells (Albert et al., 1998). Therefore, it remains to be seen what role, if any, cross-presentation of influenza viral antigen by rDC following natural infections may play in the induction of CD8[+] influenza-specific T cell immunity.

Many studies have suggested that the induction of influenza-specific CD8[+] T cells by DC does not require CD4[+] T cell help (Bender et al., 1995; Larsson et al., 2000; Oh et al., 2001). This property has been attributed to the direct infection (Bhardwaj et al., 1994; Macatonia et al., 1989; Nonacs et al., 1992) and licensing of the DC (Ridge et al., 1998) by the influenza virus infection. Acquisition of virus antigen through cross-presentation most likely would not allow licensing of the DC. Interestingly, when mice lacking CD4[+] T cells are infected with influenza virus, they mount comparable primary effector influenza-specific CD8[+] T cell responses, yet have significantly diminished memory responses when compared to wild-type mice (Belz et al., 2002). A previously described consequence of CD8$^+$ T cell induction without CD4$^+$ T cell help has been the inhibition of memory responses (Behrens et al., 2004; Wherry and Ahmed, 2004). When the above results are considered together with data that demonstrates that only a fraction of DC within the draining lymph nodes express the non-structural influenza virus protein, NS1 (suggesting direct infection) (Legge and Braciale, unpublished) they suggest that the CD8$^+$ T cell response to influenza may be induced in the lymph nodes by mixed populations of DC that have acquired antigen – either through direct infection or cross-presentation. The ratio (dependent upon the degree of exposure to virus, i.e. inoculum size and viral titer) and localization of these two populations within the lymph nodes may in turn drive differences in the development of the CD8$^+$ T cell response. Alternatively, the degree of viral infection of DC (i.e. MOI) may be responsible for the differences in the ability of DC to drive effective effector and memory CD8$^+$ T cell development. Ultimately the answer to this question awaits careful single cell analysis of both viral gene content and protein in DC isolated from the lymph nodes following influenza virus infection.

DENDRITIC CELL MIGRATION TO THE REGIONAL LYMPH NODES FOLLOWING INFLUENZA VIRUS INFECTION

In order to initiate the CD8$^+$ response to influenza, naïve influenza-specific CD8$^+$ T cell precursors must encounter influenza-antigen bearing cells in the draining lymph nodes (Banchereau et al., 2000; Guermonprez et al., 2002; Norbury et al., 2002). As pulmonary macrophages, B cells, and DC all contain viral antigen following intranasal (i.n.) infection with influenza virus (Hamilton-Easton and Eichelberger, 1995) and could therefore present viral antigen, the question remained as to which cell population trafficked to the regional lymph nodes and triggered naïve T cells. Examination of purified APC populations from the lung-draining mediastinal lymph nodes after influenza virus infection showed that DC were greatly enriched in influenza-infected cells (Hamilton-Easton and Eichelberger, 1995). Furthermore the purified DC (but not macrophages or B cells) were able to present antigen to and activate influenza-specific T cell hybridomas (Hamilton-Easton and Eichelberger, 1995).

These results along with related results demonstrating an increase in DC numbers in the regional lymph nodes following pulmonary infections (McWilliam et al., 1994) suggested that rDC may be sensitive to respiratory virus infections in the lungs and in response mature/migrate to the regional lymph nodes to initiate subsequent virus-specific T cell responses. Indeed, adoptive transfer of mature antigen-bearing (dye-tagged) splenic and bone marrow derived DC into the lungs demonstrated that DC could traffic from the lungs to the regional lymph nodes and then trigger CD8$^+$ T cells (Havenith et al., 1993; Lambrecht et al., 2000; Ritchie et al., 2000). Yet these experiments did not directly address whether rDC, which are immature prior to influenza infection (Calder et al., 2004; Holt et al., 1992; Legge and Braciale, 2003) and are phenotypically distinct from splenic and bone marrow derived DC (Legge and Braciale, unpublished) (Belz et al., 2004b; Calder et al., 2004; Vermaelen and Pauwels, 2003), likewise differ in their function. Direct labeling of the respiratory tract by i.n. administration of the intravital dye, CFSE showed that 95% of the cells that trafficked from the lungs to the lung draining lymph nodes following influenza virus infection were rDC (the remaining 5% were macrophages)(Legge and Braciale, 2003). These rDC migrated into the T cell areas in the lymph nodes near the HEV and were in close proximity to CD8β$^+$ T cells (Legge and Braciale, 2003) (Legge and Braciale, unpublished). Together these results suggest that the rDC acquire influenza

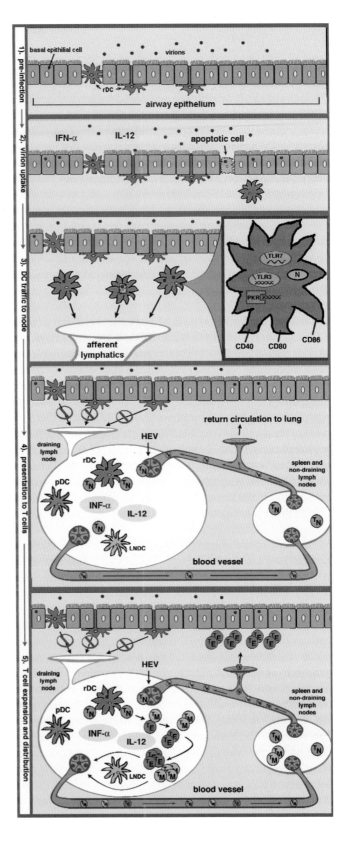

Figure 5.1 Induction of adaptive immunity by dendritic cells following pulmonary influenza virus infection.

Panel 1 Respiratory dendritic cells (RDC) reside within the respiratory epithelial layer and the underlying submucosa (as well as on alveolar surfaces [not shown]) prior to the onset of influenza virus infection.

Panel 2 Immature RDC, respiratory epithelial cells, and alveolar macrophages (not shown) can become infected after encounter with influenza virus resulting in local production of Type 1 interferon and IL-12 along with the induction of apoptosis of epithelial cells as a result of the virus infection.

Panel 3 RDC activate and mature after influenza infection via TLR3, TLR7/8 and PKR dependent pathways and/or as a result of uptake of apoptotic cell bodies (*Insert:* TLR3,TLR7/8 and PKR signaling eventually results in DC activation, maturation [including the upregulation of the costimulatory molecules CD40, CD80, CD86 expression], and migration of the RDC from the respiratory tract through the afferent lymphatics to the draining lymph nodes).

Panel 4 After 24–48 hrs of infection, the RDC remaining in the lungs become refractory to migration from the respiratory tract – while RDC that have already migrated to the lymph node, localize near high endothelial venules (HEV) within the lymph nodes where they encounter naïve virus-specific CD8$^+$ T lymphocytes (T$_N$) which enter the draining lymph nodes through the HEV. RDC then present viral antigen to virus-specific T$_N$ which triggers T cell activation/differentiation. Influx of activated RDC into the draining lymph nodes results in the local production of both Type 1 interferon and IL-12, which can support T cell activation/differentiation. CD8α^+ lymph node resident DC (LNDC) may also served as antigen presenting cells to T$_N$ via direct infection, or more likely via uptake of viral antigen derived from infected migrant RDC and/or from migrant RDC that have taken up apoptotic material containing viral antigen in the respiratory tract.

Panel 5 T$_N$ differentiate into effector (T$_E$) and memory (T$_M$) T cells which can then leave the draining lymph nodes through the efferent lymphatics and circulatory system to localize to the site of virus infection (i.e. the infected lungs). During activation/differentiation virus-specific T cells migrate through the lymph node and encounter LNDC which may regulate T cell expansion by triggering T cell apoptosis.

virus antigen in the lungs, respond to the influenza antigen acquisition by migrating to the regional lymph nodes, and, once there, use the antigen to trigger and activate naïve influenza-specific T cells.

Recent evidence suggested that rDC traffic from the lungs to the draining lymph nodes was limited to the early (i.e. first 24–48 hrs) stages of influenza virus infection (Legge and Braciale, 2003). This result suggested that the interaction of antigen-carrying DC with naive T cells may also be limited to the early stages of influenza virus infection (Legge and Braciale, 2003). Purification of the rDC, $CD8\alpha^+$ lymph node-resident DC, $CD8\alpha^-$ lymph node-resident DC, and lymph node-resident plasmacytoid subpopulations of DC from the lung draining lymph nodes showed that the presentation of influenza viral antigen to T cells after influenza virus infection was confined to the rDC and $CD8\alpha^+$ lymph node-resident subpopulations of dendritic cells (Belz et al., 2004b). The rDC isolated from the lymph nodes most likely acquired antigen in the respiratory tract from direct influenza virus infection, as some rDC in the lung-draining lymph nodes expressed a non-structural protein from influenza virus, NS1 [Legge and Braciale, unpublished]. However, the source of the influenza antigen in the $CD8\alpha^+$ DC is less evident. Both $CD8\alpha^-$ and plasmacytoid dendritic cells are infectable *in vitro* with influenza virus and will subsequently present antigen to $CD8^+$ T cells (Fonteneau et al., 2003; Kronin et al., 2001) – yet they do not present antigen in the lymph nodes after i.n. influenza virus infection. This suggests that the $CD8\alpha^+$ lymph node-resident dendritic cells do not acquire antigen through direct influenza virus infection within the lymph nodes. $CD8\alpha^+$ dendritic cells have shown the ability to cross-present viral antigen (Belz et al., 2004a; Smith et al., 2003). Therefore, the $CD8\alpha^+$ dendritic cells within the lymph nodes following influenza virus infections may acquire antigen from the migrating rDC, and then subsequently participate in the activation of naïve influenza-specific $CD8^+$ T cells. In support of the hypothesis that rDC serve as the source of antigen for transfer to $CD8\alpha^+$ DC, blockade of rDC migration to the lymph node (by i.n. CpG instillation) blocks influenza antigen presentation by $CD8\alpha^+$ DC in the lymph nodes (Belz et al., 2004b; Legge and Braciale, 2003). Since $CD8\alpha^+$ DC are know to acquire antigen through cross-presentation and the isolation of the subpopulations of DC from lymph nodes currently requires the dissociation of the lymph nodes by enzymatic digestion (and hence allows the mixing of populations of cells that may be architecturally sequestered), the transfer of influenza antigen between the DC populations *in vivo* and the possible role that each of these DC sub-populations (i.e. rDC and $CD8\alpha^+$ DC) play in activating naïve T cells with regard to lymph node localization and order of encounter with naïve T cells remains to be explored.

The experiments described above examining of the kinetics of rDC migration to the regional lymph nodes following influenza virus infection also yielded a second unexpected finding. Following instillation of influenza virus and many inflammatory mediators (e.g. polyI:C, CpG) to the lungs, the rDC that remain in the lungs and do not mature/migrate to the regional lymph nodes become refractory (~18–24 hrs post instillation/infection) to the initial stimulus as well as to subsequent differing maturation/migration stimuli (Legge and Braciale, 2003). The induction of this rDC refractory state halts rDC migration to the regional lymph nodes and has a profound effect on the generation of immunity to challenge infection – as mice given CpG or polyI:C 18 hrs prior to infection (to induce the refractory state) mount greatly inhibited $CD8^+$ T cell responses to challenge influenza virus infection (Legge and Braciale, 2003). A major complication that increases mortality after respiratory influenza virus infection is bacterial super-infections (Hament et al., 1999). The induction of the rDC refractory state and decreased T cell immunity to secondary or challenge infections suggests that a loss of DC maturation and migration to the lymph nodes may play a role in increasing susceptibility to bacterial super-infections following influenza virus infections.

RECRUITMENT OF DC TO THE LUNGS AFTER INFLUENZA VIRUS INFECTION

DC isolated from various immune organs have multiple phenotypes that have been associated with differing ability to shape adaptive immune responses to pathogens (Banchereau et al., 2000; Guermonprez et al., 2002). Prior to influenza virus infection the DC populations found in the lungs appear to be surprisingly homogenous. rDC in the lungs are overwhelmingly predominated by $CD8\alpha^-CD4^-CD11b^{lo}$ DC (Legge and Braciale, unpublished). However, after influenza virus infection, there is substantial recruitment of $CD8\alpha^+$ and $CD8\alpha^-$ DC into the lungs (Legge and Braciale, unpublished) (Legge and Braciale, 2003; McWilliam et al., 1997; McWilliam et al., 1996). These cells do not traffic into the draining lymph nodes (Belz et al., 2004b; Legge and Braciale, 2003), and therefore, do not appear to participate in the induction of adaptive immunity. However, recruitment of these DC into the lungs has been shown to augment sensitivity to respiratory antigen (Upham and Stumbles, 2003; Yamamoto et al., 2000) and has been implicated in both asthma and allergic diseases (Holt et al., 1999; Julia et al., 2002; Lambrecht et al., 2000; Lambrecht et al., 1998; Upham and Stumbles, 2003; Yamamoto et al., 2000). The DC recruited to the lungs after i.n. challenge with respiratory viruses acquire respiratory antigen, localize in close proximity to activated T cells (Byersdorfer and Chaplin, 2001; Yamamoto et al., 2000), and drive Th2 polarization in the airway mucosa (Constant et al., 2002; Holt et al., 1997; Holt et al., 1999; Yamamoto et al., 2000). Furthermore, the recruited DC are retained in the lungs for extended periods after infection (De Heer et al., 2004; Julia et al., 2002; Yamamoto et al., 2000) and can secrete chemokines to selectively attract Th2 memory cells (Gonzalo et al., 1999; Imai et al., 1999). Therefore, while a current role for the recruitment of these DC into the lungs following influenza virus infection is unknown, given their ability to acquire respiratory antigen and present them in the lungs to activated T cells, it is likely that, in addition to driving allergic diseases, they may also enhance and shape pulmonary immunity to influenza virus.

DC RESPONSE TO INFLUENZA VIRUS INFECTION

Both murine and human DC are activated and mature following encounter with inflammatory cytokines (e.g. TNFα and IL-1) and microbial products (e.g. LPS) (Banchereau et al., 2000; Cumberbatch and Kimber, 1995). Likewise, following influenza virus infection rDC respond by undergoing maturation [i.e. upregulation of costimulatory ligands (CD40, CD80, CD86) and MHC class II], and migrate to the regional lymph nodes (Legge and Braciale, 2003). The source for this activation of rDC appears to be, at least in part, due to exposure of the DC to viral dsRNA upon influenza virus infection. *In vitro* exposure of DC to dsRNA leads to the upregulation of both costimulatory/adhesion molecules and cell surface levels of MHC class I, thereby increasing the DC antigen presenting capacity (Cella et al., 1999b). Introduction of polyI:C (a dsRNA analog) to the respiratory tract mimics the dsRNA stimulus provided by infectious influenza virus and likewise leads to *in vivo* rDC maturation/migration to the draining lymph nodes (Legge and Braciale, 2003).

Experiments using GM-CSF/IL-4 derived human peripheral blood DC have shown that dsRNA may also confer an additional benefit to the DCs – namely the upregulation of low levels of IFNα and the subsequent IFNα-autocrine-driven expression of MxA (Cella et al., 1999b). MxA is a protein which has been shown to protect cells from the cytopathic effects of many viruses and is thought to be responsible for the lack of productive influenza virus infection except at high multiplicities of infection (i.e. 25MOI) (Oh et al., 2000)) seen in DC (Pavlovic et al., 1995; Pavlovic et al., 1992). Experiments using murine bone marrow-derived DC (BM-DC) have further demonstrated that production of IFNα by DCs is tightly controlled by influenza virus NS1 protein (Diebold et al., 2003). While direct influenza infection of BM-DC leads to low level IFNα production, the cytoplasmic introduction of polyI:C (by liposomes) instead triggers high-level type I IFN expression (Diebold et al., 2003). Influenza virus NS1 protein is known to bind to and sequester viral dsRNA and is expressed at high levels in BM-DC infected with influenza virus (Diebold et al., 2003). These results suggest that, given the high level NS1 expression by BM-DC after influenza virus infection, NS1 might sequester influenza viral dsRNA resulting in suppression of IFNα production in influenza-infected BM-DC. Indeed, *in vitro* infection of BM-DC with a NS1-deleted mutant influenza virus (ΔNS1) results in high level IFNα production (like that seen with the cytoplasmic polyI:C), and yields a corresponding poor level of replication of the virus (Diebold et al., 2003).

Recognition of influenza virus dsRNA by DCs appears to be primarily mediated by PKR, a serine-threonine kinase whose triggering ultimately leads to inhibition of protein synthesis and restriction of virus replication and IFNα production (Williams, 2001). However, TLR3 has also been shown to mediate the induction of type I IFN and IL-12p40 by DC in response to dsRNA (Fujimoto et al., 2004; Lund et al., 2004). Therefore, the pathway of dsRNA recognition in DCs following influenza virus infection remains complex. In favor of the PKR recognition pathway dsRNA is recognized by TLR3-deficient BM-DC, but not by wild-type DC treated with 2-aminopurine, a PKR inhibitor or by DC from mice lacking the PKR dsRNA binding domain (Diebold et al., 2003). However, DC from PKR-deficient mice still make IFNα to dsRNA, and the incubation of DC with IFNα results in upregulation of TLR3. Therefore, due to the high level expression of type 1 IFN in the respiratory tract after influenza virus infection, TLR3 on rDC (and type 1 IFN operating on TLR3 expression) may trigger rDC activation/maturation via recognition of dsRNA generated within cells during influenza infection of the respiratory tract.

In addition to the recognition of inflammatory mediators (TNFα/IFNα) and microbial products (dsRNA), dendritic cells also appear to process antigen, mature, and migrate to the draining lymph nodes following interactions with virus-induced apoptotic cells (Qu et al., 2003). Infection of the respiratory tract with influenza virus leads to apoptosis of the pulmonary epithelium (Brydon et al., 2003; Mori et al., 1995; Technau-Ihling et al., 2001). Influenza induced-apoptotic cells have been shown to be potent stimulators of *in vitro* CTL responses (Albert et al., 1998). DC that are activated in this manner appear to acquire antigen through cross-presentation – as *in vitro* incubation of influenza-apoptotic cells with neutralizing anti-HA antibody or filtered supernatant from influenza apoptotic cells does not lead to transfer of antigen to DC (Albert et al., 1998). Furthermore, transfer of antigen from the influenza-apoptotic cells to the DC is sensitive to blockade with Z-VAD-CHO, an irreversible inhibitor of caspase activity (Albert et al., 1998).

Therefore, DC may respond to influenza virus infections of the respiratory tract through direct infection of the DC, interactions with viral dsRNA, uptake of infected apoptotic cells, or reaction to viral induced inflammatory cytokines, or a combination thereof. These differences in recognition may have an important impact on the regulation of DC function (i.e. antigen presentation, co-stimulation, DC migration and localization) and may play an important role in the education of the adaptive T cell immune response due to recognition-dependent differential programming of the ability of responding DC to produce inflammatory mediators (like IL-12p40 and IFNα) that can drive naïve CD8[+] T cell activation.

EFFECT OF INFLUENZA NA PROTEIN ON DC FUNCTION

Influenza virus infection of DC leads to upregulation of costimulatory molecules, MHC class I and II, adhesion molecules, and viral proteins. While influenza virus infection of DC is unproductive, except at high MOI (Oh et al., 2000), increases in the number of virions that infect individual DC leads to overall higher levels of virus protein expression and influenza virus-induced changes in DC function (Oh et al., 2000). At low MOI, influenza virus NA expression on the surface of DC allows for the removal of sialic acids from the surface of the DC (Oh and Eichelberger, 1999; Oh et al., 2000). Previous studies have shown that sialic acids can inhibit intercellular interactions (Bagriacik and Miller, 1999; Shimamura et al., 1994) and expression of influenza NA on DC (at low MOI) allows greater interactions of the costimulatory molecules CD80, CD86 and the adhesion ICAM-1 with their ligands on T cells (Oh and Eichelberger, 1999; Oh et al., 2000). These interactions facilitate increased interaction between the infected DC and the T cells leading to enhanced allogenic and influenza-specific T cell activation, IL-2, and IFNγ production (Oh and Eichelberger, 1999; Oh and Eichelberger, 2000; Oh et al., 2000). Infection of DC at higher MOI, however, leads instead to NA cleavage of TGFβ and production of IL-4 and IL-10 by the T cells (Oh and Eichelberger, 2000; Oh et al., 2000). The MOI-induced changes in T cell polarization appear to be related to release of influenza virus particles from the high MOI-infected DC (Oh et al., 2000). The presence of influenza NA protein on the surface of these particles further desialylates the surface to the DC leading to DC-DC clusters that can exclude T cells (Oh et al., 2000). The mechanism regulating the change from non-productive to productive influenza virus infection in the DC, as well as, the changes in the polarization potential of the DC remains unknown. However, it is tempting to speculate that increases in MOI may allow influenza virus to overcome (possibly by overproduction of viral proteins) the regulatory mechanisms that prevent virus production (i.e. possibly MxA) and alter the cytokine expression of the DC.

CYTOKINE PRODUCTION AND REGULATION
OF ADAPTIVE IMMUNITY

While the production of IL-12 and IFNα by DC is regulated, as described above, by the type of the encounter between influenza virus and DC, the production of these cytokines is important in shaping the developing immune response (Banchereau et al., 2000; Monteiro et al., 1998; Price et al., 2000; Qu et al., 2003). Secretion of IFNα, in conjunction with IL-6, from DC in response to influenza virus augments B cell activation/differentiation into influenza-specific plasma cells. This increase in plasma cells in turn raises the level of secretion of anti-influenza-specific IgG (Jego et al., 2003). Furthermore, mice deficient in STAT1[-/-] have inhibited IFN signaling and shift their CD4[+] influenza-specific T cell responses from Th1 to Th2, and exhibit enhanced disease following influenza virus infection (Durbin et al., 2000). IL-12 production in response to influenza virus infection has been shown to enhance influenza-specific CD8[+] T cell proliferation, IFNγ production, and cytolytic responses (Bhardwaj et al., 1996; Monteiro et al., 1998). Therefore, the regulated enhanced production of pro-inflammatory cytokines following influenza virus infection appears to have beneficial effects on the development of adaptive immunity to influenza virus infections.

These results suggest that perturbations in the level of cytokines produced by rDC, in response to different levels of influenza virus infection, may regulate the magnitude and shape of the T cell response to influenza virus infections.

INFLUENZA VIRUS ENHANCEMENT OF CO-ADMINISTERED ANTIGEN

Since influenza virus infection enhances the rDC maturation and migration to the regional lymph nodes, it would be expected that influenza infections would increase immunity to i.n. co-administered inert antigen. Indeed, when LPS-free ovalbumin (OVA) was administered in conjunction with respiratory influenza virus infection, the CD4[+] and CD8[+] T cell response to OVA was altered from a tolerogenic and type II cytokine response to that of a type I cytokine response including high level production of IL-2 and IFNγ (Brimnes et al., 2003). Furthermore, a protective T cell response to OVA was generated as the mice demonstrated enhanced recovery to challenge infection with a recombinant OVA-gene expressing vaccinia virus (Brimnes et al., 2003). Therefore, the activation of DC by influenza virus appears to increase activation and responses to bystander respiratory antigen.

PLASMACYTOID DC

In addition to role that myeloid DC play in the immune response to influenza virus infection, recent studies have shown that a second type of DC, the plasmacytoid dendritic cell (pDC) may also contribute to the immune response to influenza virus (Cella et al., 2000; Fonteneau et al., 2003). Previously referred to as plasmacytoid lymphocytes, plasmacytoid monocytes, or IFN-producing cells (IPC) due to their production of large amounts of type I IFN upon interaction with virus (Asselin-Paturel et al., 2001; Cella et al., 1999a; De Heer et al., 2004; Kadowaki et al., 2000), these cells were shown to primarily localize to the blood and to cluster around HEV in inflamed lymph nodes upon inflammation (Cella et al., 2000; Cella et al., 1999a; Fonteneau et al., 2003; Kadowaki et al., 2000; O'Keeffe et al., 2002). However, theses APC were under-appreciated as DC until CD4[+]CD11c[-] (human) cells (Kadowaki et al., 2000) were shown to behave as pre-dendritic cells (preDC) and to become functional DC upon culture with IL-3, CD40L, viral, and bacterial stimuli (Cella et al., 2000; Cella et al., 1999a; Grouard et al., 1997; Kadowaki et al., 2000; O'Keeffe et al., 2002). Subsequent studies in mice have identified a corresponding blood-resident CD11c[lo]B220[+] preDC that differentiates into a functional CD8α[+]DEC-205[-] pDCs (De Heer et al., 2004; Gilliet et al., 2002; O'Keeffe et al., 2002) upon stimulation.

The initial focus on pDC following influenza virus infection centered upon pDC as a source for type I IFN upon encounter with virus (Kadowaki et al., 2000). The presence of pDC in small numbers in the lungs prior to infection (De Heer et al., 2004)(Legge and Braciale, unpublished) and the recruitment of pDC to the lungs upon influenza infection (Legge and Braciale, unpublished) suggests that pDC may contribute [in addition to and independent of NK cells (Abb et al., 1984)] to the important protective anti-viral role that type I IFN plays in the lungs following influenza virus infection (Price et al., 2000). pDC make type I IFN following recognition of influenza virus by at least two mechanisms: 1) the recognition of virus dsRNA through the methods described above for conventional DCs (Price et al., 2000; Williams, 2001), and 2) the recognition of viral genomic ssRNA (Diebold et al., 2004; Lund et al., 2004). Unlike conventional DC, where IFNα production is suppressed by influenza NS1-dsRNA sequestration (Diebold et al., 2003), infection/exposure of pDC to influenza virus (live or killed) yields significant IFNα production (Diebold et al., 2004; Lund et al., 2004). IFNα production is triggered by the interaction of ssRNA with TLR7 within acidic endosomes through a MyD88-dependent signaling pathway since pDC deficient in either TLR7 or MyD88 failed to produce IFNα and IL-12p40 in response to infection with multiple different RNA viruses including influenza (Diebold et al., 2004; Lund et al., 2004). The ssRNA pathway of influenza-driven IFNα production is dependent upon an acidic endosome as an inhibitor of endosomal acidification, chlorquine, blocks IFNα production in a dose-dependant manner (Lund et al., 2004). These results suggest that some particles of influenza are degraded by endosomal proteases leading to exposure of viral genomic RNA to TLR7, triggering the maturation of the pDC and production of IFNα (Diebold et al., 2004).

The involvement of pDC in induction of adaptive immunity to influenza had initially been largely overlooked, due to the fact that immature pDC were such poor stimulators of naïve T cells (Fonteneau et al., 2003). However, when purified pDC were matured *in vitro* by influenza virus, they were found to be as efficient as conventional CD11c$^+$ myeloid DC in expanding allogenic naïve T cells (Cella et al., 2000; Fonteneau et al., 2003). Furthermore, pDC exposed to influenza virus appeared more resistant to influenza virus infection (less influenza protein expression and apoptosis) and expressed CCR7 and responded to CCL19 (Fonteneau et al., 2003) suggesting that they may traffic, as described above, to lymph nodes. Subsequent studies have shown that blood purified pDC exposed to influenza virus induce *in vitro* proliferation of influenza-specific CTL and CD4 (Th1) T cell clones equivalent to that induced by conventional DC (Fonteneau et al., 2003). Together these results suggest that, following influenza virus infection, pDC could migrate to the HEV in lymph nodes where they would be better equipped (due to increased resistance to infection) than conventional myeloid DC (which can be productively infected at high MOI) to activate naive influenza-specific T cells. In acute influenza virus infections, where virus does not become systemic, this process (i.e. pDC participation in naïve T cell activation) appears unlikely as pDC only represent ~1.5 % of pulmonary DCs (De Heer et al., 2004), do not migrate to the lymph nodes from the lungs upon i.n. influenza virus infection (Belz et al., 2004b; Legge and Braciale, 2003), and do not appear to present influenza-derived antigen within the lymph nodes (Belz et al., 2004b). Therefore, a role for pDC in induction of adaptive T cell immunity to acute influenza virus infections remains in question. Systemic spread of influenza virus to the blood would alternatively lead to encounter with the preDC-pDC precursors and could involve pDC in the induction of the adaptive T cell response to influenza virus.

References.

Abb, J., Abb, H., and Deinhardt, F. (1984). Relationship between natural killer (NK) cells and interferon (IFN) alpha-producing cells in human peripheral blood. Studies with a monoclonal antibody with specificity for human natural killer cells. Immunobiology. *167*, 359–364.

Albert, M. L., Jegathesan, M., and Darnell, R. B. (2001). Dendritic cell maturation is required for the cross-tolerization of CD8$^+$ T cells. Nat Immunol. *2*, 1010–1017.

Albert, M. L., Sauter, B., and Bhardwaj, N. (1998). Dendritic cells acquire antigen from apoptotic cells and induce class I-restricted CTLs. Nature. *392*, 86–89.

Asselin-Paturel, C., Boonstra, A., Dalod, M., Durand, I., Yessaad, N., Dezutter-Dambuyant, C., Vicari, A., O'Garra, A., Biron, C., Briere, F., and Trinchieri, G. (2001). Mouse type I IFN-producing cells are immature APCs with plasmacytoid morphology. Nat Immunol. *2*, 1144–1150.

Bagriacik, E. U., and Miller, K. S. (1999). Cell surface sialic acid and the regulation of immune cell interactions, the neuraminidase effect reconsidered. Glycobiology. 9, 267–275.

Banchereau, J., Briere, F., Caux, C., Davoust, J., Lebecque, S., Liu, Y. J., Pulendran, B., and Palucka, K. (2000). Immunobiology of dendritic cells. Annu Rev Immunol. 18, 767–811.

Behrens, G., Li, M., Smith, C. M., Belz, G. T., Mintern, J., Carbone, F. R., and Heath, W. R. (2004). Helper T cells, dendritic cells and CTL Immunity. Immunol Cell Biol. 82, 84–90.

Belz, G. T., Smith, C. M., Eichner, D., Shortman, K., Karupiah, G., Carbone, F. R., and Heath, W. R. 2004a. Cutting edge, conventional CD8 alpha$^+$ dendritic cells are generally involved in priming CTL immunity to viruses. J Immunol. 172, 1996–(2000).

Belz, G. T., Smith, C. M., Kleinert, L., Reading, P., Brooks, A., Shortman, K., Carbone, F. R., and Heath, W. R. 2004b. Distinct migrating and nonmigrating dendritic cell populations are involved in MHC class I-restricted antigen presentation after lung infection with virus. Proc Natl Acad Sci USA. 101, 8670–8675.

Belz, G. T., Wodarz, D., Diaz, G., Nowak, M. A., and Doherty, P. C. (2002). Compromised influenza virus-specific CD8($^+$)-T-cell memory in CD4($^+$)-T-cell-deficient mice. J Virol. 76, 12388–12393.

Bender, A., Bui, L. K., Feldman, M. A., Larsson, M., and Bhardwaj, N. (1995). Inactivated influenza virus, when presented on dendritic cells, elicits human CD8$^+$ cytolytic T cell responses. J Exp Med. 182, 1663–1671.

Bhardwaj, N., Bender, A., Gonzalez, N., Bui, L. K., Garrett, M. C., and Steinman, R. M. (1994). Influenza virus-infected dendritic cells stimulate strong proliferative and cytolytic responses from human CD8$^+$ T cells. J Clin Invest. 94, 797–807.

Bhardwaj, N., Seder, R. A., Reddy, A., and Feldman, M. V. (1996). IL-12 in conjunction with dendritic cells enhances antiviral CD8$^+$ CTL responses in vitro. J Clin Invest. 98, 715–722.

Braciale, T. J., and Yap, K. L. (1978). Role of viral infectivity in the induction of influenza virus-specific cytotoxic T cells. J Exp Med. 147, 1236–1252.

Brimnes, M. K., Bonifaz, L., Steinman, R. M., and Moran, T. M. (2003). Influenza virus-induced dendritic cell maturation is associated with the induction of strong T cell immunity to a coadministered, normally nonimmunogenic protein. J Exp Med. 198, 133–144.

Brown, D. M., Roman, E., and Swain, S. L. (2004). CD4 T cell responses to influenza infection. Semin Immunol. 16, 171–177.

Brydon, E. W., Smith, H., and Sweet, C. (2003). Influenza A virus-induced apoptosis in bronchiolar epithelial (NCI-H292) cells limits pro-inflammatory cytokine release. J Gen Virol. 84, 2389–2400.

Byersdorfer, C. A., and Chaplin, D. D. (2001). Visualization of early APC/T cell interactions in the mouse lung following intranasal challenge. J Immunol. 167, 6756–6764.

Calder, C. J., Liversidge, J., and Dick, A. D. (2004). Murine respiratory tract dendritic cells, isolation, phenotyping and functional studies. J Immunol Methods. 287, 67–77.

Cella, M., Facchetti, F., Lanzavecchia, A., and Colonna, M. (2000). Plasmacytoid dendritic cells activated by influenza virus and CD40L drive a potent TH1 polarization. Nat Immunol. 1, 305–310.

Cella, M., Jarrossay, D., Facchetti, F., Alebardi, O., Nakajima, H., Lanzavecchia, A., and Colonna, M. 1999a. Plasmacytoid monocytes migrate to inflamed lymph nodes and produce large amounts of type I interferon. Nat Med. 5, 919–923.

Cella, M., Salio, M., Sakakibara, Y., Langen, H., Julkunen, I., and Lanzavecchia, A. 1999b. Maturation, activation, and protection of dendritic cells induced by double-stranded RNA. J Exp Med. 189, 821–829.

Cho, Y., Basta, S., Chen, W., Bennink, J. R., and Yewdell, J. W. (2003). Heat-aggregated noninfectious influenza virus induces a more balanced CD8($^+$)-T-lymphocyte immunodominance hierarchy than infectious virus. J Virol. 77, 4679–4684.

Constant, S. L., Brogdon, J. L., Piggott, D. A., Herrick, C. A., Visintin, I., Ruddle, N. H., and Bottomly, K. (2002). Resident lung antigen-presenting cells have the capacity to promote Th2 T cell differentiation in situ. J Clin Invest. *110*, 1441–1448.

Cumberbatch, M., and Kimber, I. (1995). Tumour necrosis factor-alpha is required for accumulation of dendritic cells in draining lymph nodes and for optimal contact sensitization. Immunology. *84*, 31–35.

De Heer, H. J., Hammad, H., Soullie, T., Hijdra, D., Vos, N., Willart, M. A., Hoogsteden, H. C., and Lambrecht, B. N. (2004). Essential role of lung plasmacytoid dendritic cells in preventing asthmatic reactions to harmless inhaled antigen. J Exp Med. *200*, 89–98.

Dhodapkar, M. V., and Steinman, R. M. (2002). Antigen-bearing immature dendritic cells induce peptide-specific CD8($+$) regulatory T cells *in vivo* in humans. Blood. *100*, 174–177.

Dhodapkar, M. V., Steinman, R. M., Krasovsky, J., Munz, C., and Bhardwaj, N. (2001). Antigen-specific inhibition of effector T cell function in humans after injection of immature dendritic cells. J Exp Med. *193*, 233–238.

Diebold, S. S., Kaisho, T., Hemmi, H., Akira, S., and Reis e Sousa, C. (2004). Innate antiviral responses by means of TLR7-mediated recognition of single-stranded RNA. Science. *303*, 1529–1531.

Diebold, S. S., Montoya, M., Unger, H., Alexopoulou, L., Roy, P., Haswell, L. E., Al-Shamkhani, A., Flavell, R., Borrow, P., and Reis e Sousa, C. (2003). Viral infection switches non-plasmacytoid dendritic cells into high interferon producers. Nature. *424*, 324–328.

Doherty, P. C., and Christensen, J. P. (2000). Accessing complexity, the dynamics of virus-specific T cell responses. Annu Rev Immunol. *18*, 561–592.

Doherty, P. C., Topham, D. J., Tripp, R. A., Cardin, R. D., Brooks, J. W., and Stevenson, P. G. (1997). Effector CD4$^+$ and CD8$^+$ T-cell mechanisms in the control of respiratory virus infections. Immunol Rev. *159*, 105–117.

Durbin, J. E., Fernandez-Sesma, A., Lee, C. K., Rao, T. D., Frey, A. B., Moran, T. M.,

Vukmanovic, S., Garcia-Sastre, A., and Levy, D. E. (2000). Type I IFN modulates innate and specific antiviral immunity. J Immunol. *164*, 4220–4228.

Fonteneau, J. F., Gilliet, M., Larsson, M., Dasilva, I., Munz, C., Liu, Y. J., and Bhardwaj, N. (2003). Activation of influenza virus-specific CD4$^+$ and CD8$^+$ T cells, a new role for plasmacytoid dendritic cells in adaptive immunity. Blood. *101*, 3520–3526.

Fujimoto, C., Nakagawa, Y., Ohara, K., and Takahashi, H. (2004). Polyriboinosinic polyribocytidylic acid [poly(I,C)]/TLR3 signaling allows class I processing of exogenous protein and induction of HIV-specific CD8$^+$ cytotoxic T lymphocytes. Int Immunol. *16*, 55–63.

Gallucci, S., Lolkema, M., and Matzinger, P. (1999). Natural adjuvants, endogenous activators of dendritic cells. Nat Med. *5*, 1249–1255.

Gerhard, W., Mozdzanowska, K., Furchner, M., Washko, G., and Maiese, K. (1997). Role of the B-cell response in recovery of mice from primary influenza virus infection. Immunol Rev. *159*, 95–103.

Germain, R. N., and Jenkins, M. K. (2004). *In vivo* antigen presentation. Curr Opin Immunol. *16*, 120–125.

Gilliet, M., Boonstra, A., Paturel, C., Antonenko, S., Xu, X. L., Trinchieri, G., O'Garra, A., and Liu, Y. J. (2002). The development of murine plasmacytoid dendritic cell precursors is differentially regulated by FLT3-ligand and granulocyte/macrophage colony-stimulating factor. J Exp Med. *195*, 953–958.

Gonzalo, J. A., Pan, Y., Lloyd, C. M., Jia, G. Q., Yu, G., Dussault, B., Powers, C. A., Proudfoot, A. E., Coyle, A. J., Gearing, D., and Gutierrez-Ramos, J. C. (1999). Mouse monocyte-derived chemokine is involved in airway hyperreactivity and lung inflammation. J Immunol. *163*, 403–411.

Grouard, G., Rissoan, M. C., Filgueira, L., Durand, I., Banchereau, J., and Liu, Y. J. (1997). The enigmatic plasmacytoid T cells develop into dendritic cells with interleukin (IL)–3 and CD40-ligand. J Exp Med. *185*, 1101–1111.

Guermonprez, P., Valladeau, J., Zitvogel, L., Thery, C., and Amigorena, S. (2002). Antigen presentation and T cell stimulation by dendritic cells. Annu Rev Immunol. *20*, 621–667.

Hament, J. M., Kimpen, J. L., Fleer, A., and Wolfs, T. F. (1999). Respiratory viral infection predisposing for bacterial disease, a concise review. FEMS Immunol Med Microbiol. *26*, 189–195.

Hamilton-Easton, A., and Eichelberger, M. (1995). Virus-specific antigen presentation by different subsets of cells from lung and mediastinal lymph node tissues of influenza virus-infected mice. J Virol. *69*, 6359–6366.

Havenith, C. E., van Miert, P. P., Breedijk, A. J., Beelen, R. H., and Hoefsmit, E. C. (1993). Migration of dendritic cells into the draining lymph nodes of the lung after intratracheal instillation. Am J Respir Cell Mol Biol. *9*, 484–488.

Heath, W. R., and Carbone, F. R. (2001). Cross-presentation in viral immunity and self-tolerance. Nat Rev Immunol. *1*, 126–134.

Holt, P. G., Macaubas, C., Cooper, D., Nelson, D. J., and McWilliam, A. S. (1997). Th-1/Th-2 switch regulation in immune responses to inhaled antigens. Role of dendritic cells in the aetiology of allergic respiratory disease. Adv Exp Med Biol. *417*, 301–306.

Holt, P. G., Oliver, J., McMenamin, C., and Schon-Hegrad, M. A. (1992). Studies on the surface phenotype and functions of dendritic cells in parenchymal lung tissue of the rat. Immunology. *75*, 582–587.

Holt, P. G., Stumbles, P. A., and McWilliam, A. S. (1999). Functional studies on dendritic cells in the respiratory tract and related mucosal tissues. J Leukoc Biol. *66*, 272–275.

Imai, T., Nagira, M., Takagi, S., Kakizaki, M., Nishimura, M., Wang, J., Gray, P. W., Matsushima, K., and Yoshie, O. (1999). Selective recruitment of CCR4-bearing Th2 cells toward antigen-presenting cells by the CC chemokines thymus and activation-regulated chemokine and macrophage-derived chemokine. Int Immunol. *11*, 81–88.

Itano, A. A., and Jenkins, M. K. (2003). Antigen presentation to naive CD4 T cells in the lymph node. Nat Immunol. *4*, 733–739.

Jego, G., Palucka, A. K., Blanck, J. P., Chalouni, C., Pascual, V., and Banchereau, J. (2003). Plasmacytoid dendritic cells induce plasma cell differentiation through type I interferon and interleukin 6. Immunity. *19*, 225–234.

Julia, V., Hessel, E. M., Malherbe, L., Glaichenhaus, N., O'Garra, A., and Coffman, R. L. (2002). A restricted subset of dendritic cells captures airborne antigens and remains able to activate specific T cells long after antigen exposure. Immunity. *16*, 271–283.

Kadowaki, N., Antonenko, S., Lau, J. Y., and Liu, Y. J. (2000). Natural interferon alpha/beta-producing cells link innate and adaptive immunity. J Exp Med. *192*, 219–226.

Kronin, V., Fitzmaurice, C. J., Caminschi, I., Shortman, K., Jackson, D. C., and Brown, L. E. (2001). Differential effect of CD8($^+$) and CD8(-) dendritic cells in the stimulation of secondary CD4($^+$) T cells. Int Immunol. *13*, 465–473.

Lambrecht, B. N., De Veerman, M., Coyle, A. J., Gutierrez-Ramos, J. C., Thielemans, K., and Pauwels, R. A. (2000). Myeloid dendritic cells induce Th2 responses to inhaled antigen, leading to eosinophilic airway inflammation. J Clin Invest. *106*, 551–559.

Lambrecht, B. N., Salomon, B., Klatzmann, D., and Pauwels, R. A. (1998). Dendritic cells are required for the development of chronic eosinophilic airway inflammation in response to inhaled antigen in sensitized mice. J Immunol. *160*, 4090–4097.

Larsson, M., Messmer, D., Somersan, S., Fontenau, J. F., Donahoe, S. M., Lee, M., Dunbar, P. R., Cerundolo, V., Julkunen, I., Nixon, D. F., and Bhardwaj, N. (2000). Requirement of mature dendritic cells for efficient activation of influenza A-specific memory CD8$^+$ T cells. J Immunol. *165*, 1182–1190.

Legge, K. L., and Braciale, T. J. (2003). Accelerated migration of respiratory dendritic cells to the regional lymph nodes is limited to the early phase of pulmonary infection. Immunity. *18*, 265–277.

Lund, J. M., Alexopoulou, L., Sato, A., Karow, M., Adams, N. C., Gale, N. W., Iwasaki, A., and Flavell, R. A. (2004). Recognition of single-stranded RNA viruses by Toll-like receptor 7. Proc Natl Acad Sci USA. *101*, 5598–5603.

Macatonia, S. E., Taylor, P. M., Knight, S. C., and Askonas, B. A. (1989). Primary stimulation by dendritic cells induces antiviral proliferative and cytotoxic T cell responses *in vitro*. J Exp Med. *169*, 1255–1264.

McWilliam, A. S., Marsh, A. M., and Holt, P. G. (1997). Inflammatory infiltration of the upper airway epithelium during Sendai virus infection, involvement of epithelial dendritic cells. J Virol. *71*, 226–236.

McWilliam, A. S., Napoli, S., Marsh, A. M., Pemper, F. L., Nelson, D. J., Pimm, C. L., Stumbles, P. A., Wells, T. N., and Holt, P. G. (1996). Dendritic cells are recruited into the airway epithelium during the inflammatory response to a broad spectrum of stimuli. J Exp Med. *184*, 2429–2432.

McWilliam, A. S., Nelson, D., Thomas, J. A., and Holt, P. G. (1994). Rapid dendritic cell recruitment is a hallmark of the acute inflammatory response at mucosal surfaces. J Exp Med. *179*, 1331–1336.

Monteiro, J. M., Harvey, C., and Trinchieri, G. (1998). Role of interleukin-12 in primary influenza virus infection. J Virol. *72*, 4825–4831.

Mori, I., Komatsu, T., Takeuchi, K., Nakakuki, K., Sudo, M., and Kimura, Y. (1995). *In vivo* induction of apoptosis by influenza virus. J Gen Virol.*76*, 2869–2873.

Nonacs, R., Humborg, C., Tam, J. P., and Steinman, R. M. (1992). Mechanisms of mouse spleen dendritic cell function in the generation of influenza-specific, cytolytic T lymphocytes. J Exp Med. *176*, 519–529.

Norbury, C. C., Malide, D., Gibbs, J. S., Bennink, J. R., and Yewdell, J. W. (2002). Visualizing priming of virus-specific CD8[+] T cells by infected dendritic cells *in vivo*. Nat Immunol. *3*, 265–271.

Oh, S., Belz, G. T., and Eichelberger, M. C. (2001). Viral neuraminidase treatment of dendritic cells enhances antigen-specific CD8([+]) T cell proliferation, but does not account for the CD4([+]) T cell independence of the CD8([+]) T cell response during

influenza virus infection. Virology. *286*, 403–411.

Oh, S., and Eichelberger, M. C. (1999). Influenza virus neuraminidase alters allogeneic T cell proliferation. Virology. *264*, 427–435.

Oh, S., and Eichelberger, M. C. (2000). Polarization of allogeneic T-cell responses by influenza virus-infected dendritic cells. J Virol. *74*, 7738–7744.

Oh, S., McCaffery, J. M., and Eichelberger, M. C. (2000). Dose-dependent changes in influenza virus-infected dendritic cells result in increased allogeneic T-cell proliferation at low, but not high, doses of virus. J Virol. *74*, 5460–5469.

O'Keeffe, M., Hochrein, H., Vremec, D., Caminschi, I., Miller, J. L., Anders, E. M., Wu, L., Lahoud, M. H., Henri, S., Scott, B., et al. (2002). Mouse plasmacytoid cells, long-lived cells, heterogeneous in surface phenotype and function, that differentiate into CD8([+]) dendritic cells only after microbial stimulus. J Exp Med. *196*, 1307–1319.

Pamer, E., and Cresswell, P. (1998). Mechanisms of MHC class I – restricted antigen processing. Annu Rev Immunol. *16*, 323–358.

Pavlovic, J., Arzet, H. A., Hefti, H. P., Frese, M., Rost, D., Ernst, B., Kolb, E., Staeheli, P., and Haller, O. (1995). Enhanced virus resistance of transgenic mice expressing the human MxA protein. J Virol. *69*, 4506–4510.

Pavlovic, J., Haller, O., and Staeheli, P. (1992). Human and mouse Mx proteins inhibit different steps of the influenza virus multiplication cycle. J Virol. *66*, 2564–2569.

Price, G. E., Gaszewska-Mastarlarz, A., and Moskophidis, D. (2000). The role of alpha/beta and gamma interferons in development of immunity to influenza A virus in mice. J Virol. *74*, 3996–4003.

Qu, C., Moran, T. M., and Randolph, G. J. (2003). Autocrine type I IFN and contact with endothelium promote the presentation of influenza A virus by monocyte-derived APC. J Immunol. *170*, 1010–1018.

Ridge, J. P., Di Rosa, F., and Matzinger, P. (1998). A conditioned dendritic cell can be a temporal bridge between a CD4[+] T-helper and a T-killer cell. Nature. *393*, 474–478.

Ritchie, D. S., Hermans, I. F., Lumsden, J. M., Scanga, C. B., Roberts, J. M., Yang, J., Kemp, R. A., and Ronchese, F. (2000). Dendritic cell elimination as an assay of cytotoxic T lymphocyte activity *in vivo*. J Immunol Methods. *246*, 109–117.

Rock, K. L., and Goldberg, A. L. (1999). Degradation of cell proteins and the generation of MHC class I-presented peptides. Annu Rev Immunol. *17*, 739–779.

Sauter, B., Albert, M. L., Francisco, L., Larsson, M., Somersan, S., and Bhardwaj, N. (2000). Consequences of cell death, exposure to necrotic tumor cells, but not primary tissue cells or apoptotic cells, induces the maturation of immunostimulatory dendritic cells. J Exp Med. *191*, 423–434.

Shimamura, M., Shibuya, N., Ito, M., and Yamagata, T. (1994). Repulsive contribution of surface sialic acid residues to cell adhesion to substratum. Biochem Mol Biol Int. *33*, 871–878.

Smith, C. M., Belz, G. T., Wilson, N. S., Villadangos, J. A., Shortman, K., Carbone, F. R., and Heath, W. R. (2003). Cutting edge, conventional CD8 alpha[+] dendritic cells are preferentially involved in CTL priming after footpad infection with herpes simplex virus-1. J Immunol. *170*, 4437–4440.

Technau-Ihling, K., Ihling, C., Kromeier, J., and Brandner, G. (2001). Influenza A virus infection of mice induces nuclear accumulation of the tumorsuppressor protein p53 in the lung. Arch Virol. *146*, 1655–1666.

Topham, D. J., Tripp, R. A., and Doherty, P. C. (1997). CD8[+] T cells clear influenza virus by perforin or Fas-dependent processes. J Immunol. *159*, 5197–5200.

Upham, J. W., and Stumbles, P. A. (2003). Why are dendritic cells important in allergic diseases of the respiratory tract? Pharmacol Ther. *100*, 75–87.

Vermaelen, K., and Pauwels, R. (2003). Accelerated airway dendritic cell maturation, trafficking, and elimination in a mouse model of asthma. Am J Respir Cell Mol Biol. *29*, 405–409.

Wherry, E. J., and Ahmed, R. (2004). Memory CD8 T-cell differentiation during viral infection. J Virol. *78*, 5535–5545.

Williams, B. R. (2001). Signal integration via PKR. Sci STKE. *2001*, RE2.

Wilson, N. S., El-Sukkari, D., Belz, G. T., Smith, C. M., Steptoe, R. J., Heath, W. R., Shortman, K., and Villadangos, J. A. (2003). Most lymphoid organ dendritic cell types are phenotypically and functionally immature. Blood. *102*, 2187–2194.

Wong, P., and Pamer, E. G. (2003). CD8 T cell responses to infectious pathogens. Annu Rev Immunol. *21*, 29–70.

Yamamoto, N., Suzuki, S., Shirai, A., Suzuki, M., Nakazawa, M., Nagashima, Y., and Okubo, T. (2000). Dendritic cells are associated with augmentation of antigen sensitization by influenza A virus infection in mice. Eur J Immunol. *30*, 316–326.

CHAPTER 6

Quantitative and qualitative characterization of the CD8+ T cell response to influenza virus infection

Nicole L. La Gruta and Peter C. Doherty

Abstract

The CD8+ T cell response to influenza virus infection is critical for the efficient clearance of viral infection. The advent of tetramer staining has, over the last 5–10 years, enabled accurate quantitation of the epitope specific CD8+ T cell response to influenza virus infection, and has revealed much about immunodominance hierarchies and the kinetics of individual epitope specific responses. More recently, particular interest has been paid to the quality of CD8+ T cell responses, such as cytokine production and cytolytic ability, since T cell function must be a key factor in determining the efficacy of the response. Here, we describe recent advances in the characterization of both magnitude and quality of the CD8+ T cell response to influenza virus infection. These studies may also serve as a model to elucidate general mechanisms of CD8+ T cell-mediated viral clearance.

INTRODUCTION

Though the pathogenesis of influenza A virus pneumonia in the mouse is not strictly equivalent to the infection process observed in humans, or the various avian and other vertebrate species that maintain influenza in nature, the analytical power of genetically modified mouse model systems and the availability of analytical reagents has allowed rigorous dissection of the cellular immune response to these viruses. Indications are that the conclusions reached are applicable to both vaccine development strategies and to the understanding of protection and immunopathology in the human disease.

CD8+ T CELLS IN INFLUENZA A VIRUS INFECTION

Following influenza virus infection of mice, CD8+ T cells, through the production of inflammatory cytokines and their ability to kill virus-infected cells, are considered key mediators of viral clearance, and are crucial for the efficient elimination of virus (Bender et al., 1992; Doherty, 1996). Evidence for the importance of CD8+ T cells in clearing influenza viral infection comes from studies demonstrating delayed viral clearance in mice lacking CD8+ T cells (Bender et al., 1992), as well as studies demonstrating that the significant immune pressure from CD8+ T cells can result in virus escape mutants and thus drive antigenic evolution of influenza viruses (Price et al., 2000).

Most recent studies of influenza virus immunity have used either C57Bl/6J (B6, H2^b) or BALB/c (H2^d) mice. The majority of the work described in this chapter focuses on the C57Bl/6J system due to the larger number of identified viral epitopes and the more extensive characterization of CD8+ T cell responses in this system. Respiratory infection of naïve B6 mice with influenza causes an acute pneumonia and subsequent viral clearance by d10 after infection (Doherty and Christensen, 2000), with no evidence of viral antigen persistence (Allan et al., 1990). Investigation of the secondary CD8+ T cell response to influenza infection necessitates the use of heterologous virus strains such as A/PR8/34 (PR8, H1N1) and A/HKx31 (HKx31, H3N2), which share the same six internal gene products, from which the CD8+ T cell epitopes are derived, but express distinct surface hemagglutinin (H) and neuraminidase (N) proteins at which the neutralizing antibody response is directed (Kilbourne, 1969). Thus, priming of mice with one virus and challenging with the other circumvents the complication of neutralization of the challenge inoculum and enables analysis of secondary virus-specific CD8+ T cell responses (Belz et al., 2000b; Belz et al., 2001; Flynn et al., 1998; La Gruta et al., 2004). The CD8+ T cell response of B6 mice to influenza virus infection is directed at multiple epitopes, six of which have been identified for H2D^b and H2K^b, collectively (Chen et al., 2001b). The most dominant responses are directed at epitopes derived from the nucleoprotein (NP$_{366-374}$, H2D^b) (Falk et al., 1991; Townsend et al., 1986) and acid polymerase (PA$_{224-233}$, H2D^b) (Belz et al., 2000b) with other identified responses directed at the basic polymerase subunit 1 (PB1$_{703-711}$, H2K^d), non-structural protein 2 (NS2$_{114-121}$, H2K^d), matrix protein 1 (M1$_{128-135}$, H2K^b), and a recently identified protein, PB1-F2, which originates from an 87 amino acid alternative open reading frame within the PB1 gene (PB1-F2$_{62-70}$, H2D^b) (Chen et al., 2001a).

CIRCULATION OF INFLUENZA-SPECIFIC CD8+ T CELLS FOLLOWING ACTIVATION

Following initial activation and proliferation of naïve virus-specific CD8+ T cells in the mediastinal lymph node (MLN) 3–4 days post-infection (Lawrence and Braciale, 2004; Tripp et al., 1995b), activated T cells migrate to the infected lung, where they undergo further proliferative expansion between d5–7 (Lawrence et al., 2005). Studies have shown that the capacity of influenza-specific CD8+ T cells to mediate protective effects against infection strongly depends on their ability to localize to the lungs, and specifically to infected airway epithelium (Cerwenka et al., 1999a; Cerwenka et al., 1999b). Further, retention of activated effector CD8+ T cells in the lungs has been shown to be dependent on their expression of LFA-1, which is thought to promote T cell binding to vascular endothelium and lung parenchymal cells (Thatte et al., 2003). Interestingly, although antigen is required to specifically activate CD8+ T cells, the localization of these cells to the lungs has been shown to be antigen independent and mediated by localized inflammation (Ely et al., 2003; Topham et al., 2001). Activated cells exiting the draining lymph node can also disseminate throughout lymphoid tissues including spleen and distal lymph nodes (Lawrence and Braciale, 2004). In fact, analysis of the tissue diaspora for the influenza-specific recall response revealed widely dispersed epitope specific CD8+ T cells in tissues such as lung, spleen, bone marrow, blood, liver and non-draining lymph nodes (Marshall et al., 2001) and the distribution of these CD8+ T cell populations was maintained out to at least d120 after infection. A recent study has suggested that the retention of these memory influenza-specific CD8+ T cells in non-lymphoid tissues, including the lung, is due to expression of VLA-1 by the T cells which mediates attachment to the extracellular matrix (Ray et al., 2004).

QUANTITATION OF THE CD8+ T CELL RESPONSE TO INFLUENZA A VIRUS INFECTION

Epitope hierarchies in primary and secondary immune responses

Following primary infection with virus, CD8+ T cell responses to the D^bNP_{366} and D^bPA_{224} epitopes are roughly equidominant, although it seems that the D^bPA_{224}-specific response peaks 1–2 days earlier than the D^bNP_{366} response which is maximal around d9–10 after infection (Belz et al., 2001; Wiley et al., 2001), with all other responses being similarly low. Conversely, in a secondary virus infection, the D^bNP_{366}-specific response dominates over all others constituting up to 80% of the total virus-specific CD8+ T cell response, and ranging from 5 to10 fold greater in magnitude than the next largest D^bPA_{224}-specific response (Belz et al., 2000a; Belz et al., 2001; Flynn et al., 1998; Flynn et al., 1999). This differential immunodominance exhibited between primary and secondary virus encounters was initially suggested to be a consequence of differential antigen presentation by dendritic cells (DCs) and non-DCs, where DCs present both D^bNP_{366} and D^bPA_{224}, while only D^bNP_{366} is efficiently presented by non-DCs (Crowe et al., 2003). Thus in a primary infection, where DCs are essential for the stimulation of naïve CD8+ T cells, D^bNP_{366} and D^bPA_{224}-specific responses are equivalent, while following secondary challenge when it is argued that recall of memory CD8+ T cells may be achieved by both DCs and non-DCs, a distinct advantage is conferred to the D^bNP_{366}-specific response. This explanation of the changing D^bNP_{366} and D^bPA_{224}-specific epitope hierarchies may prove to be too simplistic, however, as it fails to address two key findings. Firstly, even within the DC population, D^bNP_{366} is presented at significantly higher levels than D^bPA_{224} (Chen et al., 2004), and consequently one would anticipate dominance of the D^bNP_{366}-specific response following both primary and secondary infection, if antigen presentation levels were the sole determinant of immunodominance. Secondly, a recent study has suggested that DCs are essential, not only in a primary response, but also to achieve sizeable recall responses and are likely to be the major cell type driving T cell stimulation following secondary virus challenge (Zammit et al., 2005). Thus, it is probable that the changing D^bNP_{366} and D^bPA_{224}-specific immunodominance hierarchies between primary and secondary responses are more than likely due to a combination of antigen presentation and other factors associated with the epitope-specific CD8+ T cells themselves.

Comparison of the spleen and the site of infection show that while the virus-specific CD8+ T cell numbers in spleen vastly outnumber those in the lung parenchyma or the inflammatory population obtained by bronchoalveolar lavage (BAL) (Andreansky et al., 2005; La Gruta et al., 2004; Marshall et al., 2001; Turner et al., 2003; Webby et al., 2003), this is due to the greater size and cellularity of the organ. The percentage of CD8+ T cells which are virus epitope-specific is, in fact, far higher at the site of infection indicating their enrichment in the pneumonic lung (Belz et al., 2000a; Belz et al., 2001; Flynn et al., 1998; Marshall et al., 2001).

Quantitative compensation in the influenza virus-specific CD8+ T cell response

Considering that a defined, but limited number of epitopes have been identified in the B6 model of influenza A virus infection, this provides an ideal system for investigating the effect of epitope-specific CD8+ T cell responses on one other. Studies from our laboratory using influenza A viruses effectively lacking in either or both of the native NP_{366} or PA_{224} epitopes (via an N5Q mutation which prevents binding to $H2D^b$) showed that although no compensatory expansion in known epitope-specific responses was apparent following primary infection, assays using virus infected stimulators were able to demonstrate total compensation in the CD8+ T cell compartment (Andreansky et al., 2005). Similarly, following secondary challenge, total compensation was also observed which in this case was evidenced by compensatory expansion in, in particular, the subdominant epitope-specific populations (Andreansky et al., 2005; Webby et al., 2003). Thus, these results suggest that, at least in some circumstances, alternative epitope-specific CD8+ T cells are capable of restoring the magnitude of the total virus-specific response when immunodominant responses are removed and it seems possible that at least some of these responses are directed at as yet unidentified epitopes (Zhong et al., 2003). The capacity of individual epitope-specific responses to compensate functionally for one another is discussed in a later section.

Quantitation of influenza virus-specific memory CD8+ T cells

After the acute phase (d10) of the primary CD8+ T cell response to virus infection, D^bNP_{366} and D^bPA_{224}-specific CD8+ T cells in the spleen contract at a similarly steady rate over the ensuing 10–20 days (Flynn et al., 1999; La Gruta et al., 2004; Wiley et al., 2001). The size of the D^bNP_{366} and D^bPA_{224}-specific CD8+ T cell populations appears then to stabilize at equivalent levels, at approximately 10-fold fewer than at the acute timepoint, from approximately 40–50 days after infection out to at least d570, when the number of memory cells is still readily detectable (Belz et al., 2000b; Hogan et al., 2001a; Kedzierska, K. and Stambas, J, unpublished data). Importantly, the transition of effector influenza specific CD8+ T cells into memory is dependent on CD4+ T cells as demonstrated by a significantly reduced magnitude and recall of primed influenza-specific CD8+ T cells in MHC class II⁻ (IA^b-/-) mice (Belz et al., 2002; Doherty and Christensen, 2000; Tripp et al., 1995b).

Similar kinetics are observed for the contraction into and maintenance of memory following secondary virus challenge, with the numbers of virus-specific memory cells remaining steady out to at least 120 days after challenge (Flynn et al., 1999; Marshall et al., 2001). However, presumably due to the huge dominance of the D^bNP_{366}- over the D^bPA_{224}-specific response at the acute timepoint (d8) (5–10-fold), this difference is carried into memory resulting in a significantly larger number of D^bNP_{366}- relative to D^bPA_{224}-specific memory cells (Flynn et al., 1999; Hou et al., 1994; Marshall et al., 2001). Further, the number of secondary memory cells even at very late timepoints (>d100) is significantly higher than those detected before secondary challenge (Flynn et al., 1999; Marshall et al., 2001). Therefore, it is clear that influenza-specific memory levels are significantly increased by a second viral exposure, however tertiary challenge has been shown to have little effect on virus-specific CD8+ T cell numbers (Christensen et al., 2000).

QUALITATIVE ANALYSIS OF INFLUENZA VIRUS-SPECIFIC CD8+ T CELLS

Characteristics of epitope-specific CD8+ T cell repertoires

Analysis of CDR3β usage in T cell populations has long been employed to provide information on the spectrum of TCRs used, as well as the identification of unique characteristics associated with epitope-specific responses which may provide information on the functional output of these cells. Antigen specific CD8+ T cell responses are sometimes biased towards usage of a particular TCR Vα (Aebischer et al., 1990; Davis et al., 1995; Mikszta et al., 1999), but are more commonly determined by the TCR Vβ profile (Acha-Orbea et al., 1988; Aebischer et al., 1990; Argaet et al., 1994; Cose et al.,

1995; Edouard et al., 1993; Stewart-Jones et al., 2003; Trentin et al., 1996; Urban et al., 1988). Characterization of TCRβ usage in D^bNP_{366} and D^bPA_{224}-specific $CD8^+$ T cell populations induced by influenza infection has revealed preferential usage of Vβ8.3 (30–50%) (Belz et al., 2000a; Deckhut et al., 1993; Flynn et al., 1998; Kedzierska et al., 2004) and Vβ7 (30–60%) (Belz et al., 2000a; Turner et al., 2003), respectively. Within the dominant Vβ+ populations, D^bNP_{366} and D^bPA_{224} specific CD8+ T cell repertoires have been extensively characterized to reveal several unique features; CD8+ D^bPA_{224}+Vβ7+ cells are highly diverse, containing predominantly 'private' (specific to individual mice) TCRβ sequences, which utilise a variety of Jβ elements and have a predominant CDR3β length of 6 aa (Deckhut et al., 1993; Kedzierska et al., 2004; Turner et al., 2004), while CD8+D^bNP_{366}+Vβ8.3+ cells have limited diversity with a high frequency of 'public' (shared between different individuals) TCRβ sequences using Jβ2.2 and having a modal CDR3β length of 9 aa (Turner et al., 2003; Turner et al., 2004) (Table 1). The obviously distinct repertoire characteristics observed between D^bNP_{366}- and D^bPA_{224}-specific CD8+ T cell repertoires has been suggested to be a consequence of the structure of the presented epitope, with the more 'featureless' NP_{366} peptide less able to recruit a highly diverse range of T cell precursors relative to the 'prominent' PA_{224} peptide (Turner et al., 2005).

As an extension of this analysis, TCR CDR3β sequences from relatively high avidity D^bNP_{366}- and D^bPA_{224}-specific CD8+ T cells (identified as those able to bind tetramer under conditions of limited tetramer availability) were analysed and compared to the total tetramer binding population. Interestingly, partitioning of TCR CDR3β sequences was observed in the high avidity D^bPA_{224}-specific, but not the D^bNP_{366}-specific, population relative to the total tetramer binding population (Kedzierska et al, 2005). Thus, TCR affinity makes a greater contribution to the range of overall avidities in the D^bPA_{224}-specific relative to the D^bNP_{366}-specific population. These data probably also reflect that D^bPA_{224}-specific TCRs encode a broader range of TCR affinities than D^bNP_{366}-specific TCRs which is concordant with the limited D^bNP_{366}-specific TCRβ diversity.

Production of cytokines and cytotoxic mediators

Intensive analysis of cytokine production by influenza-specific CD8+ T cells has revealed several interesting phenomena. Firstly, it was observed upon analysis of CD8+ T cell production of IFN-γ, TNF-α, and IL-2 that all epitope-specific populations exhibited a subsetting, or hierarchical, profile of cytokine production (Belz et al., 2001; La Gruta et al., 2004) (Figure 1A-C). Thus, while the majority of tetramer+ cells produce IFN-γ, the TNF-α producing population is a subset of these IFN-γ+ cells, while those cells producing IL-2 are almost exclusively found within the IFN-γ+ TNF-α+ subset. Such a pattern of cytokine production may be suggestive of a progressive differentiation, or acquisition, of particular cytokine profiles.

Following on from this observation, qualitative differences were observed between epitope-specific populations, as comparison of the D^bNP_{366}- and D^bPA_{224}-specific $CD8^+$ T cell populations indicated that a substantially higher proportion of D^bPA_{224}-specific $CD8^+$ T cells (as measured by IFN-γ production) also produced TNF-α and IL-2, relative to the D^bNP_{366}-specific population (Belz et al., 2001; La Gruta et al., 2004) (Figure 1B). This has been correlated with a relatively higher avidity of the D^bPA_{224}-specific population and a greater propensity for D^bPA_{224}-specific cells to undergo apoptosis in vitro, relative to the D^bNP_{366}-specific set (La Gruta et al., 2004). Interestingly, the proportion of cells producing multiple cytokines increases over time such that the majority of long term

Table 6.1 Characteristics of D^bNP_{366}- and D^bPA_{224}-specific CD8+ T cell repertoires

	D^bNP_{366}	D^bPA_{224}
Mice analyzed	10	12
TCRs sequenced	976	1,630
Modal CDR3β length	9	6
Different sequences	45	241
'Public' sequences	3	0
Sequences per mouse	7.9 ± 2.5	$20.6 \pm 3.8*$

Table reproduced from Turner et al. (2005) and summarizes data from Turner et al. (2003) and Kedzierska et al. (2004). *p<0.0001.

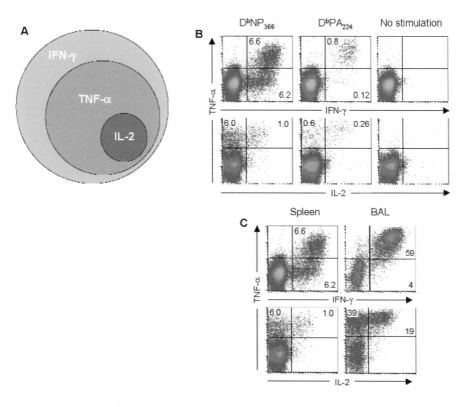

Figure 1 Cytokine profiles of influenza-specific CD8+ T cells in the B6 model of infection. Schematic representation of subsetting observed in influenza specific CD8+ T cell production of IFN-γ, TNF-α, and IL-2 (**A**). Secondary immune spleen and lung airway CD8+ T cell production of IFN-γ, TNF-α, and IL-2 is shown following short term in vitro stimulation in the presence or absence of influenza epitopes. After stimulation, cells are stained with fluorescent antibodies to CD8α, IFN-γ, TNF-α, and IL-2. The IFN-γ > TNF-α > IL-2 hierarchy is evident in both spleen and BAL (**B,C**). A larger proportion of CD8+ D^bPA_{224}-specific, compared to D^bNP_{366}-specific splenocytes (as determined by production of IFN-γ), produce TNF-α and IL-2 (**B**). Analysis of D^bNP_{366}-specific cells reveals a higher proportion of epitope-specific CD8+ T cells in the BAL compared to the spleen. Further, a higher proportion of epitope-specific (IFN-γ+) CD8+ T cells in the BAL produce TNF-α and IL-2, compared to the spleen (**C**).

memory cells produce all three cytokines upon in vitro restimulation, again suggesting differentiation to a multiple cytokine phenotype, or possibly preservation of cells expressing a differentiated profile.

Comparison of CD8+ T cells recovered from the spleen, lungs (airways and tissue), and MLN after intranasal influenza infection has also revealed qualitative differences in cytokine profiles (Baumgarth et al., 1997; Baumgarth and Kelso, 1996a; Johnson et al., 2003; La Gruta et al., 2004). Initial studies showed that bulk CD8+ T cells from the lung tissue of acutely infected animals had the capacity to produce higher levels of IFN-γ and IL-2, compared to CD8+ T cells from MLN, following short term in vitro anti-CD3 stimulation as measured by cytokine levels in culture supernatant (Baumgarth et al., 1997; Baumgarth and Kelso, 1996a). Later experiments using intracellular cytokine staining showed that a higher proportion of epitope-specific CD8+ T cells (as determined by production of IFN-γ following short term in vitro stimulation with characterized D^b restricted viral peptides) from the lung airways were able to produce TNF-α and IL-2, and in higher amounts, compared to those isolated from the spleen (Figure 1C) or lung parenchyma (La Gruta et al., 2004; La Gruta, NL, unpublished data). Surprisingly, the proportion of IFN-γ$^+$ cells producing both TNF-α and IL-2 was found to be the lowest in cells from the lung parenchyma relative to all other tissues studied (i.e. spleen, lung airways and MLN), while the proportion of epitope-specific triple cytokine producers in the MLN was almost as high as seen in the airways. Thus, though the number of epitope-specific cells in various tissues must influence the overall level of cytokine production, these data suggest that cells derived from the lung airways after infection may be the most functionally activated (or differentiated), closely followed by those from the MLN. The fact that the BAL and lung populations exhibit distinct cytokine profiles probably reflects the distribution of antigenic epitopes in the infected respiratory tract. CD8$^+$ T cells that are found in the lung airways are more likely to interact with the luminal surface of the virus-infected alveolar epithelium while the lymphocytes in the lung parenchyma may have less exposure to high levels of antigen.

The importance of IFN-γ to the CD8+ T cell response to influenza infection has been studied in mice with a targeted disruption in the IFN-γ gene or in mice treated with neutralizing antibodies to IFN-γ (Baumgarth and Kelso, 1996b; Graham et al., 1993). A significant decrease in the level of cellular infiltrate in the lungs was observed in mice lacking IFN-γ (Baumgarth and Kelso, 1996b) which may be a consequence of an accumulation (or trapping) of epitope-specific cells in the spleen (Badovinac et al., 2000).

Analysis of the expression of cytotoxic mediators by virus epitope-specific CD8+ T cells has revealed that perforin and granzymes A and B are widely expressed in CD8+ T cells from lungs, spleen and mediastinal lymph node (draining the lung) with the expression peaking at around 5–10 days after infection (Johnson et al., 2003; Lawrence et al., 2005). Interestingly, analysis of individual CD8+ epitope-specific cells by single cell PCR revealed that the frequencies of epitope-specific cells expressing perforin or granzyme B were similar in all three tissues, while granzyme A was more frequently expressed in lung and spleen than MLN, and granzyme C was expressed at a relatively low frequency in all tissues (Johnson et al., 2003).

Cytotoxicity of influenza virus specific CD8+ T cells

Limited data on the direct ex vivo cytotoxicity of influenza virus-specific CD8+ T cells has shown a significantly higher degree of specific lysis (i.e. lytic ability measured on a per cell basis) by $D^b NP_{366}$-specific CD8+ T cells extracted from the BAL compared to the spleen (Marshall et al., 2001). Similarly, following short term in vitro stimulation, BAL cells showed potent cytotoxicity compared to the minimal cytotoxic activity by cells from

MLN (Allan et al., 1990). In the BALB/c model of influenza virus infection measurements of direct ex vivo cytotoxicity following primary infection with influenza A/Mem71 virus demonstrated low to undetectable levels of cytotoxicity by virus-specific cells derived from MLN or spleen, though substantial IFN-γ independent killing was observed for cells from both lung parenchyma and airways, with slightly better killing observed in cells from the airwarys (Baumgarth and Kelso, 1996a; Baumgarth and Kelso, 1996b; Johnson et al., 2003). This enhanced cytotoxicity of virus specific cells extracted from the lung airways correlates with an apparently more differentiated cytokine profile and presumably higher activation state of these cells (see above paragraph).

Qualitative contributions of individual influenza epitope-specific CD8+ T cell responses

The relative functional importance of individual epitope specific CD8+ T cell responses has been elucidated using influenza viruses lacking either or both of the immunodominant NP_{366} and PA_{224} epitopes (HK-NP, HK-PA, HK-NP/PA) (Andreansky et al., 2005; Webby et al., 2003). Only infection of mice with the HK-NP virus revealed a slight delay in viral clearance after primary infection, however lung pathology was more pronounced for all three of the mutant viruses compared to wt virus (Webby et al., 2003). Interestingly, if a compensatory role of antibody is removed from the equation, increased mortality was observed in mice infected with all three of the mutant viruses compared to wt virus (Andreansky et al., 2005). Thus, despite the fact that some evidence for quantitative compensation can be found following infection with these epitope-deficient viruses (see above paragraph), this is insufficient to reconstitute the functionality of the wt response, potentially suggesting qualitative superiority of the D^bNP_{366}- and D^bPA_{224}-specific responses over other subdominant responses.

Qualitative analysis of respiratory CD8+ T cell memory

Recent studies have described heterogeneity within T cell memory populations with regard to surface phenotype, localization and functionality (Flynn et al., 1998; Hogan et al., 2001a; Masopust et al., 2001; Murali-Krishna et al., 1998; Reinhardt et al., 2001; Sallusto et al., 2004; Sallusto et al., 1999; Tripp et al., 1995a; Usherwood et al., 1999; Wherry et al., 2003). While CD62Lhi or 'central' memory cells tend to localize to secondary lymphoid tissues, the CD62Llo or 'effector' memory cells are found in both spleen and peripheral tissues (Sallusto et al., 2004; Sallusto et al., 1999; Wherry et al., 2003). These CD62Llo memory cells have been postulated, due to their localization in peripheral tissues, to provide an early response to antigen rechallenge (Hogan et al., 2001a; Hogan et al., 2001b), while CD62Lhi memory are thought to be more efficient at mediating long term protective immunity due to their persistence and enhanced proliferative potential (Sallusto et al., 1999; Wherry et al., 2003).

In the B6 model of influenza infection there is no current data on the differential functions of these CD8+ memory T cell populations although intensive work in this area is currently underway. It is clear, however, that these populations exist following influenza A infection (Flynn et al., 1998; Hogan et al., 2001a; Tripp et al., 1995a) and in fact persistence of highly functional effector memory CD8+ T cells in the lung airways and upper respiratory tract for several months after viral clearance has been described and shown to correlate with protective immunity (Hogan et al., 2001a; Wiley et al., 2001).

Investigation of the contributions of CD62Llo and hi memory populations to CD8+ T cell recall in a respiratory model has recently been performed using Sendai virus (Roberts et al., 2005; Roberts and Woodland, 2004). In this respiratory infection model it was found that the CD62Llo population contributed significantly to recall if challenged relatively soon after memory establishment, functioning at least as well as the CD62Lhi memory cells (Roberts and Woodland, 2004). However, with increased periods of time after initial memory establishment, the CD62Lhi memory cells made a significantly greater contribution to the recall response than the CD62Llo cells (Roberts et al., 2005). Thus, it appears that at least in this respiratory infection model, the CD62Llo and hi memory populations behave in a fashion that is in concordance with earlier studies.

Analysing the CD8+ T cell response to influenza infection is clearly of value as it has provided much insight into the fundamental aspects of anti-viral CD8+ T cell immunity. In particular, the ability to compare multiple concurrent viral epitope-specific responses has demonstrated an enormous potential for diversity within virus-specific CD8+ T cell responses. Future research may enable the correlation of such diverse characteristics of CD8+ T cell responses with functional outcomes, in order to provide a more complete understanding of the implications for the human anti-viral response and potential vaccine development.

Acknowledgements
The authors would ike to thank Stephen Turner, Katherine Kedzierska, John Stambas and Misty Jenkins for helpful discussion.

References

Acha-Orbea, H., Mitchell, D. J., Timmermann, L., Wraith, D. C., Tausch, G. S., Waldor, M. K., Zamvil, S. S., McDevitt, H. O., and Steinman, L. (1988). Limited heterogeneity of T cell receptors from lymphocytes mediating autoimmune encephalomyelitis allows specific immune intervention. Cell 54, 263–273.

Aebischer, T., Oehen, S., and Hengartner, H. (1990). Preferential usage of V alpha 4 and V beta 10 T cell receptor genes by lymphocytic choriomeningitis virus glycoprotein-specific H-2Db-restricted cytotoxic T cells. Eur. J. Immunol. 20, 523–531.

Allan, W., Tabi, Z., Cleary, A., and Doherty, P. C. (1990). Cellular events in the lymph node and lung of mice with influenza. Consequences of depleting CD4+ T cells. J. Immunol. 144, 3980–3986.

Andreansky, S. S., Stambas, J., Thomas, P. G., Xie, W., Webby, R. J., and Doherty, P. C. (2005). Consequences of immunodominant epitope deletion for minor influenza virus-specific CD8+-T-cell responses. J. Virol. 79, 4329–4339.

Argaet, V. P., Schmidt, C. W., Burrows, S. R., Silins, S. L., Kurilla, M. G., Doolan, D. L., Suhrbier, A., Moss, D. J., Kieff, E., Sculley, T. B., and Misko, I. S. (1994). Dominant selection of an invariant T cell antigen receptor in response to persistent infection by Epstein-Barr virus. J. Exp. Med. 180, 2335–2340.

Badovinac, V. P., Tvinnereim, A. R., and Harty, J. T. (2000). Regulation of antigen-specific CD8+ T cell homeostasis by perforin and interferon-gamma. Science 290, 1354–1358.

Baumgarth, N., Egerton, M., and Kelso, A. (1997). Activated T cells from draining lymph nodes and an effector site differ in their responses to TCR stimulation. J. Immunol. 159, 1182–1191.

Baumgarth, N., and Kelso, A. (1996a). Functionally distinct T cells in three compartments of the respiratory tract after influenza virus infection. Eur. J. Immunol. 26, 2189–2197.

Baumgarth, N., and Kelso, A. (1996b). In vivo blockade of gamma interferon affects the influenza virus-induced humoral and the local cellular immune response in lung tissue. J. Virol. 70, 4411–4418.

Belz, G. T., Stevenson, P. G., and Doherty, P. C. (2000a). Contemporary analysis of MHC-related immunodominance hierarchies in the CD8+ T cell response to influenza A viruses. J. Immunol. 165, 2404–2409.

Belz, G. T., Wodarz, D., Diaz, G., Nowak, M. A., and Doherty, P. C. (2002). Compromised influenza virus-specific CD8(+)-T-cell memory in CD4(+)-T-cell-deficient mice. J. Virol. 76, 12388–12393.

Belz, G. T., Xie, W., Altman, J. D., and Doherty, P. C. (2000b). A previously unrecognized H-2D(b)-restricted peptide prominent in the primary influenza A virus-specific CD8(+) T-cell response is much less apparent following secondary challenge. J. Virol. 74, 3486–3493.

Belz, G. T., Xie, W., and Doherty, P. C. (2001). Diversity of epitope and cytokine profiles for primary and secondary influenza a virus-specific CD8+ T cell responses. J. Immunol. 166, 4627–4633.

Bender, B. S., Croghan, T., Zhang, L., and Small, P. A., Jr. (1992). Transgenic mice lacking class I major histocompatibility complex-restricted T cells have delayed viral clearance and increased mortality after influenza virus challenge. J. Exp. Med. 175, 1143–1145.

Cerwenka, A., Morgan, T. M., and Dutton, R. W. (1999a). Naive, effector, and memory CD8 T cells in protection against pulmonary influenza virus infection: homing properties rather than initial frequencies are crucial. J. Immunol. 163, 5535–5543.

Cerwenka, A., Morgan, T. M., Harmsen, A. G., and Dutton, R. W. (1999b). Migration kinetics and final destination of type 1 and type 2 CD8 effector cells predict protection against pulmonary virus infection. J. Exp. Med. 189, 423–434.

Chen, W., Calvo, P. A., Malide, D., Gibbs, J., Schubert, U., Bacik, I., Basta, S., O'Neill, R., Schickli, J., Palese, P., et al. (2001a). A novel influenza A virus mitochondrial protein that induces cell death. Nat. Med. 7, 1306–1312.

Chen, W., Norbury, C. C., Cho, Y., Yewdell, J. W., and Bennink, J. R. (2001b). Immunoproteasomes shape immunodominance hierarchies of antiviral CD8(+) T cells at the levels of T cell repertoire and presentation of viral antigens. J. Exp. Med. 193, 1319–1326.

Chen, W., Pang, K., Masterman, K. A., Kennedy, G., Basta, S., Dimopoulos, N., Hornung, F., Smyth, M., Bennink, J. R., and Yewdell, J. W. (2004). Reversal in the immunodominance hierarchy in secondary CD8+ T cell responses to influenza A virus: roles for cross-presentation and lysis-independent immunodomination. J. Immunol. 173, 5021–5027.

Christensen, J. P., Doherty, P. C., Branum, K. C., and Riberdy, J. M. (2000). Profound protection against respiratory challenge with a lethal H7N7 influenza A virus by increasing the magnitude of CD8(+) T-cell memory. J. Virol. 74, 11690–11696.

Cose, S. C., Kelly, J. M., and Carbone, F. R. (1995). Characterization of diverse primary herpes simplex virus type 1 gB-specific cytotoxic T-cell response showing a preferential V beta bias. J. Virol. 69, 5849–5852.

Crowe, S. R., Turner, S. J., Miller, S. C., Roberts, A. D., Rappolo, R. A., Doherty, P. C., Ely, K. H., and Woodland, D. L. (2003). Differential antigen presentation regulates the changing patterns of CD8+ T cell immunodominance in primary and secondary influenza virus infections. J. Exp. Med. 198, 399–410.

Davis, M. M., McHeyzer-Williams, M., and Chien, Y. H. (1995). T-cell receptor V-region usage and antigen specificity. The cytochrome c model system. Ann. N. Y Acad. Sci. 756, 1–11.

Deckhut, A. M., Allan, W., McMickle, A., Eichelberger, M., Blackman, M. A., Doherty, P. C., and Woodland, D. L. (1993). Prominent usage of V beta 8.3 T cells in the H-2Db-restricted response to an influenza A virus nucleoprotein epitope. J. Immunol. 151, 2658–2666.

Doherty, P. C. (1996). Cytotoxic T cell effector and memory function in viral immunity. Curr. Top Microbiol. Immunol. *206*, 1–14.

Doherty, P. C., and Christensen, J. P. (2000). Accessing complexity: the dynamics of virus-specific T cell responses. Annu. Rev. Immunol. *18*, 561–592.

Edouard, P., Thivolet, C., Bedossa, P., Olivi, M., Legrand, B., Bendelac, A., Bach, J. F., and Carnaud, C. (1993). Evidence for a preferential V beta usage by the T cells which adoptively transfer diabetes in NOD mice. Eur. J. Immunol. *23*, 727–733.

Ely, K. H., Cauley, L. S., Roberts, A. D., Brennan, J. W., Cookenham, T., and Woodland, D. L. (2003). Nonspecific recruitment of memory CD8(+) T cells to the lung airways during respiratory virus infections. J. Immunol. *170*, 1423–1429.

Falk, K., Rotzschke, O., Stevanovic, S., Jung, G., and Rammensee, H. G. (1991). Allele-specific motifs revealed by sequencing of self-peptides eluted from MHC molecules. Nature *351*, 290–296.

Flynn, K. J., Belz, G. T., Altman, J. D., Ahmed, R., Woodland, D. L., and Doherty, P. C. (1998). Virus-specific CD8+ T cells in primary and secondary influenza pneumonia. Immunity *8*, 683–691.

Flynn, K. J., Riberdy, J. M., Christensen, J. P., Altman, J. D., and Doherty, P. C. (1999). In vivo proliferation of naive and memory influenza-specific CD8(+) T cells. Proc. Natl. Acad. Sci. USA. *96*, 8597–8602.

Graham, M. B., Dalton, D. K., Giltinan, D., Braciale, V. L., Stewart, T. A., and Braciale, T. J. (1993). Response to influenza infection in mice with a targeted disruption in the interferon gamma gene. J. Exp. Med. *178*, 1725–1732.

Hogan, R. J., Usherwood, E. J., Zhong, W., Roberts, A. A., Dutton, R. W., Harmsen, A. G., and Woodland, D. L. (2001a). Activated antigen-specific CD8+ T cells persist in the lungs following recovery from respiratory virus infections. J. Immunol. *166*, 1813–1822.

Hogan, R. J., Zhong, W., Usherwood, E. J., Cookenham, T., Roberts, A. D., and Woodland, D. L. (2001b). Protection from respiratory virus infections can be mediated by antigen-specific CD4(+) T cells that persist in the lungs. J. Exp. Med. *193*, 981–986.

Hou, S., Hyland, L., Ryan, K. W., Portner, A., and Doherty, P. C. (1994). Virus-specific CD8+ T-cell memory determined by clonal burst size. Nature *369*, 652–654.

Johnson, B. J., Costelloe, E. O., Fitzpatrick, D. R., Haanen, J. B., Schumacher, T. N., Brown, L. E., and Kelso, A. (2003). Single-cell perforin and granzyme expression reveals the anatomical localization of effector CD8+ T cells in influenza virus-infected mice. Proc. Natl. Acad. Sci. USA.

Kedzierska, K., Turner, S. J., and Doherty, P. C. (2004). Conserved T cell receptor usage in primary and recall responses to an immunodominant influenza virus nucleoprotein epitope. Proc. Natl. Acad. Sci. USA. *101*, 4942–4947.

Kilbourne, E. D. (1969). Future influenza vaccines and the use of genetic recombinants. Bull World Health Organ *41*, 643–645.

La Gruta, N. L., Turner, S. J., and Doherty, P. C. (2004). Hierarchies in cytokine expression profiles for acute and resolving influenza virus-specific CD8+ T cell responses: correlation of cytokine profile and TCR avidity. J. Immunol. *172*, 5553–5560.

Lawrence, C. W., and Braciale, T. J. (2004). Activation, differentiation, and migration of naive virus-specific CD8+ T cells during pulmonary influenza virus infection. J. Immunol. *173*, 1209–1218.

Lawrence, C. W., Ream, R. M., and Braciale, T. J. (2005). Frequency, specificity, and sites of expansion of CD8+ T cells during primary pulmonary influenza virus infection. J. Immunol. *174*, 5332–5340.

Marshall, D. R., Turner, S. J., Belz, G. T., Wingo, S., Andreansky, S., Sangster, M. Y., Riberdy, J. M., Liu, T., Tan, M., and Doherty, P. C. (2001). Measuring the diaspora for virus-specific CD8+ T cells. Proc. Natl. Acad. Sci. USA. *98*, 6313–6318.

Masopust, D., Vezys, V., Marzo, A. L., and Lefrancois, L. (2001). Preferential localization of effector memory cells in nonlymphoid tissue. Science *291*, 2413–2417.

Mikszta, J. A., McHeyzer-Williams, L. J., and McHeyzer-Williams, M. G. (1999). Antigen-driven selection of TCR In vivo: related TCR alpha-chains pair with diverse TCR beta-chains. J. Immunol. *163*, 5978–5988.

Murali-Krishna, K., Altman, J. D., Suresh, M., Sourdive, D. J., Zajac, A. J., Miller, J. D., Slansky, J., and Ahmed, R. (1998). Counting antigen-specific CD8 T cells: a reevaluation of bystander activation during viral infection. Immunity 8, 177–187.

Price, G. E., Ou, R., Jiang, H., Huang, L., and Moskophidis, D. (2000). Viral escape by selection of cytotoxic T cell-resistant variants in influenza A virus pneumonia. J. Exp. Med. *191*, 1853–1867.

Ray, S. J., Franki, S. N., Pierce, R. H., Dimitrova, S., Koteliansky, V., Sprague, A. G., Doherty, P. C., de Fougerolles, A. R., and Topham, D. J. (2004). The collagen binding alpha1beta1 integrin VLA-1 regulates CD8 T cell-mediated immune protection against heterologous influenza infection. Immunity *20*, 167–179.

Reinhardt, R. L., Khoruts, A., Merica, R., Zell, T., and Jenkins, M. K. (2001). Visualizing the generation of memory CD4 T cells in the whole body. Nature *410*, 101–105.

Roberts, A. D., Ely, K. H., and Woodland, D. L. (2005). Differential contributions of central and effector memory T cells to recall responses. J. Exp. Med. *202*, 123–133.

Roberts, A. D., and Woodland, D. L. (2004). Cutting edge: effector memory CD8+ T cells play a prominent role in recall responses to secondary viral infection in the lung. J. Immunol. *172*, 6533–6537.

Sallusto, F., Geginat, J., and Lanzavecchia, A. (2004). Central memory and effector memory T cell subsets: function, generation, and maintenance. Annu. Rev. Immunol. *22*, 745–763.

Sallusto, F., Lenig, D., Forster, R., Lipp, M., and Lanzavecchia, A. (1999). Two subsets of memory T lymphocytes with distinct homing potentials and effector functions. Nature *401*, 708–712.

Stewart-Jones, G. B., McMichael, A. J., Bell, J. I., Stuart, D. I., and Jones, E. Y. (2003). A structural basis for immunodominant human T cell receptor recognition. Nat. Immunol. *4*, 657–663.

Thatte, J., Dabak, V., Williams, M. B., Braciale, T. J., and Ley, K. (2003). LFA-1 is required for retention of effector CD8 T cells in mouse lungs. Blood *101*, 4916–4922.

Topham, D. J., Castrucci, M. R., Wingo, F. S., Belz, G. T., and Doherty, P. C. (2001). The role of antigen in the localization of naive, acutely activated, and memory CD8(+) T cells to the lung during influenza pneumonia. J. Immunol. *167*, 6983–6990.

Townsend, A. R., Rothbard, J., Gotch, F. M., Bahadur, G., Wraith, D., and McMichael, A. J. (1986). The epitopes of influenza nucleoprotein recognized by cytotoxic T lymphocytes can be defined with short synthetic peptides. Cell *44*, 959–968.

Trentin, L., Zambello, R., Facco, M., Sancetta, R., Cerutti, A., Milani, A., Tassinari, C., Crivellaro, C., Cipriani, A., Agostini, C., and Semenzato, G. (1996). Skewing of the T-cell receptor repertoire in the lung of patients with HIV-1 infection. Aids *10*, 729–737.

Tripp, R. A., Hou, S., and Doherty, P. C. (1995a). Temporal loss of the activated L-selectin-low phenotype for virus-specific CD8+ memory T cells. J. Immunol. *154*, 5870–5875.

Tripp, R. A., Sarawar, S. R., and Doherty, P. C. (1995b). Characteristics of the influenza virus-specific CD8+ T cell response in mice homozygous for disruption of the H-2lAb gene. J. Immunol. *155*, 2955–2959.

Turner, S. J., Diaz, G., Cross, R., and Doherty, P. C. (2003). Analysis of clonotype distribution and persistence for an influenza virus-specific CD8+ T cell response. Immunity *18*, 549–559.

Turner, S. J., Kedzierska, K., Komodromou, H., La Gruta, N. L., Dunstone, M. A., Webb, A. I., Webby, R., Walden, H., Xie, W., McCluskey, J., et al. (2005). Lack of prominent peptide-major histocompatibility complex features limits repertoire diversity in virus-specific CD8(+) T cell populations. Nat. Immunol. *6(4)*, 382–389

Turner, S. J., Kedzierska, K., La Gruta, N. L., Webby, R., and Doherty, P. C. (2004). Characterization of CD8+ T cell repertoire diversity and persistence in the influenza A virus model of localized, transient infection. Semin Immunol. *16*, 179–184.

Urban, J. L., Kumar, V., Kono, D. H., Gomez, C., Horvath, S. J., Clayton, J., Ando, D. G., Sercarz, E. E., and Hood, L. (1988). Restricted use of T cell receptor V genes in murine autoimmune encephalomyelitis raises possibilities for antibody therapy. Cell *54*, 577–592.

Usherwood, E. J., Hogan, R. J., Crowther, G., Surman, S. L., Hogg, T. L., Altman, J. D., and Woodland, D. L. (1999). Functionally heterogeneous CD8(+) T-cell memory is induced by Sendai virus infection of mice. J. Virol. *73*, 7278–7286.

Webby, R. J., Andreansky, S., Stambas, J., Rehg, J. E., Webster, R. G., Doherty, P. C., and Turner, S. J. (2003). Protection and compensation in the influenza virus-specific CD8+ T cell response. Proc. Natl. Acad. Sci. USA. *100*, 7235–7240.

Wherry, E. J., Teichgraber, V., Becker, T. C., Masopust, D., Kaech, S. M., Antia, R., von Andrian, U. H., and Ahmed, R. (2003). Lineage relationship and protective immunity of memory CD8 T cell subsets. Nat. Immunol. *4*, 225–234.

Wiley, J. A., Hogan, R. J., Woodland, D. L., and Harmsen, A. G. (2001). Antigen-specific CD8(+) T cells persist in the upper respiratory tract following influenza virus infection. J. Immunol. *167*, 3293–3299.

Zammit, D. J., Cauley, L. S., Pham, Q. M., and Lefrancois, L. (2005). Dendritic cells maximize the memory CD8 T cell response to infection. Immunity *22*, 561–570.

Zhong, W., Reche, P. A., Lai, C. C., Reinhold, B., and Reinherz, E. L. (2003). Genome-wide characterization of a viral cytotoxic T lymphocyte epitope repertoire. J. Biol. Chem. *278*, 45135–45144.

CHAPTER 7

M2 and neuraminidase inhibitors: anti-influenza activity, mechanisms of resistance, and clinical effectiveness

Larisa V. Gubareva and Frederick G. Hayden

Abstract

Antivirals have an important role in the treatment and prevention of influenza infections. This chapter describes the antiviral activity, mechanisms of action and resistance, clinical efficacy, and consequences of antiviral resistance for two available classes of anti-influenza drugs. Amantadine and rimantadine target the M2 protein of influenza A viruses; single mutations in the trans-membrane domain of M2 confer high-level resistance to this drug class. Therapeutic use is frequently associated with emergence of drug-resistant variants; such variants are transmissible from person-to-person and pathogenic. The approved neuraminidase inhibitors (zanamivir and oseltamivir) and related investigative drugs are potent, specific inhibitors of influenza A and B viruses. Resistance emerges in vitro due to point mutations in hemagglutinin that alter cellular receptor binding or in viral neuraminidase that alter drug binding. Zanamivir and oseltamivir are highly effective for prophylaxis of influenza A and B infections; early therapeutic use reduces illness duration, lower respiratory complications, and in the case of oseltamivir, hospitalizations. Resistant variants with neuraminidase mutations have be infrequently isolated from adults and more commonly from pediatric patients treated with oseltamivir. Available evidence indicates that the relative efficiency of resistance emergence, the biologic fitness of resistant variants, and their transmissibility varies for two drug classes and for specific drugs within the neuraminidase inhibitor class. These differences have important implications for their clinical use.

INTRODUCTION

Chemoprophylaxis and chemotherapy with specific anti-influenza antivirals are important strategies for influenza management and provide adjuncts to prevention by immunization. The potential advantages of using antiviral agents include retention of activity independent of frequent antigenic changes, rapid onset of protective action in contrast to immunization, prophylactic activity that supplements that provided by immunization, and therapeutic activity in established illness. If available in sufficient quantities, these drugs could play a major role in responding to an influenza pandemic or major epidemic threat, particularly in the absence of an effective vaccine (Hayden, 2001). Modeling studies indicate that wide-scale use of oseltamivir treatment in an influenza pandemic could markedly reduce morbidity and mortality, healthcare utilization, and hospitalizations (van Genugten et al., 2003; Gani et al., 2005; Hayden, 2001). However, current supplies and production capacity are very limited, so that stockpiling of drugs is essential to ensure adequate availability.

The available agents have important differences in mechanism of action and spectrum, frequency and clinical-epidemiologic importance of drug-resistant variants, pharmacology and ease of administration, and side effect profiles. These drugs target either the M2 protein of influenza A viruses (amantadine, rimantadine) or the neuraminidase (NA) of influenza A and B viruses (oseltamivir, zanamivir). The M2 inhibitors are associated with dose-related central nervous system and gastrointestinal side effects, oseltamivir with gastrointestinal side effects, and inhaled zanamivir infrequently with bronchospasm. The NA inhibitors and rimantadine are superior to amantadine in regard to need for individual prescribing, tolerance monitoring, and frequency of serious side effects. The M2 inhibitors are more likely than the NA inhibitors to have clinically significant issues with emergence and spread of drug-resistant influenza viruses. The clinical pharmacology and tolerance profiles of both M2 and NA inhibitors have been reviewed extensively (WHO, 2004; Hayden and Aoki, 2005; Hayden and Aoki, 1999) and this chapter focuses on their antiviral activity, resistance profiles and implications of resistance emergence, and clinical effectiveness.

M2 ION CHANNEL INHIBITORS

Two M2 ion channel inhibitors, amantadine and rimantadine, are in clinical use for influenza indications. Amantadine (1-adamantanamine hydrochloride) is composed of a unique tricyclic 10-carbon ring structure with a primary amine group on the superior pole (Figure 1). It was shown to have anti-influenza A virus activity in vitro and in humans in the early 1960s (Davies et al., 1964; Jackson et al., 1963) Initially approved for prevention of influenza A due to H2N2 subtype viruses in 1966 in the United States, the indications were extended for prevention and therapy of infections due to all influenza A subtypes a decade later. Rimantadine (alpha-methyl-1-adamantanemethylamine hydrochloride) is a closely related derivative that shares the same hydrocarbon structure but incorporates a carbon with a methyl group between the nitrogen and adamantane ring (Figure 1). Rimantadine was approved for influenza indications in 1993 in the United States. Both compounds retain antiviral activity after long-term storage (>25 years at ambient temperature) (Scholtissek and Webster, 1998).

Antiviral activity

Mechanism of action
The specific target of low amantadine and rimantadine concentrations ($\leq 1\mu g$ per ml) is the ion channel function of the M2 protein. This 96 amino acid-long (the *N*-terminal methionine is cleaved) integral membrane protein contains three domains: a 23-residue *N*-terminal extracellular domain, a 19-residue transmembrane domain, and a 54-residue cytoplasmic tail (Lamb et al., 1985). It is expressed as a homotetramer in the plasma membranes of influenza A virus-infected cells and is incorporated as a minor component into influenza A virions (Zebedee and Lamb, 1988). Models of the structure of the M2 ion channel indicate that the transmembrane domain forms a left-handed alpha-helical coiled-coil containing a central ion-conducting pore through the axis of symmetry (Kovacs and Cross, 1997; Pinto et al., 1997; Zhong et al., 1998). The M2 channel is opened (activated) when its *N*-terminal ectodomain is exposed to a low pH environment of the endosomal lumen (Pinto et al., 1992; Chizhmakov et al., 1996). The models predict occlusion of the ion-conducting pore by the inward facing histidine residue 37 (Pinto et al., 1992; Wang et al., 1995). The side chain of tryptophane 41 acts as the gate that opens and closes the pore (Tang et al., 2002).

The M2 inhibitors affect two steps in the virus replication cycle, virus uncoating and virus maturation. An influx of protons into the virion interior is required for dissociation of the viral

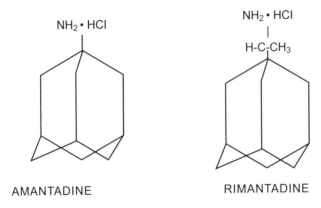

Figure 7.1 Chemical structures of amantadine and rimantadine.

M1 (matrix) and ribonucleoproteins (RNPs) during virus entry into the cell (Hay, 1992; Hay, 1996; Chizhmakov et al., 1996). The drugs bind to the transmembrane domain of the M2 and inhibit uncoating of the viral genome thus preventing the RNPs import into the nucleus (Martin and Helenius, 1991; Bukrinskaya et al., 1982a; Bukrinskaya et al., 1982b). The effect of amantadine and rimantadine on inhibition of virus uncoating is general to all strains of influenza A virus. The ion channel activity of the M2 is also activated during transport through the exocytic pathway. The M2 activity raises the luminal pH of the trans Golgi network that prevents conformational change of the meta-stable form of the hemagglutinin (HA)(Sakaguchi et al., 1996; Sugrue et al., 1990; Takeuchi and Lamb, 1994). The drugs affect the maturation of virulent avian H5 and H7 subtype viruses, the HAs of which are cleaved intracellularly (Hay et al., 1985; Ruigrok et al., 1991). These drugs also raise endosomal-lysosomal pH and may cause inhibition of virus-mediated membrane fusion at high concentrations. Although amantadine accumulates intracellularly in lysosomes, its anti-influenza effect is rapidly lost upon removal from cell culture medium (Richman et al., 1981).

Spectrum and potency

Amantadine and rimantadine specifically inhibit the *in vitro* replication of influenza A viruses at low concentrations (≤ 1 μg per ml), ones that are achievable clinically after oral administration in humans. Amantadine concentrations that inhibit plaque formation in Madin-Darby canine kidney cells (MDCK) by 50 per cent range from 0.2 to 0.4 μg per ml for clinical isolates of influenza A viruses including H1N1, H2N2, and H3N2 subtypes (Hayden et al., 1980; Browne et al., 1983). Rimantadine shows comparable or greater activity in plaque assays and is more active in other in vitro assays (Burlington et al., 1982; Belshe et al., 1989). Amantadine and rimantadine inhibit the replication of a recombinant influenza virus possessing the 1918 M gene at low concentrations (Tumpey et al., 2002). High concentrations (10 to 50μg per ml) have *in vitro* inhibitory activity against a range of other viruses (eg, influenza B, rubella, paramyxoviruses, arenaviruses, flaviviruses, and rabies). Amantadine concentrations of 50 μg/ml are usually cytotoxic and those ≥ 100 μg/ml consistently toxic for most cell types after several days of in vitro exposure.

Both prophylactic and therapeutic activities of amantadine have been demonstrated in experimental influenza A virus infection of animals (Hayden, 1986a). Systemic rimantadine protects mice from lethal infection by a reassortant virus possessing the 1918 M gene (Tumpey et al., 2002), and amantadine is active against lethal influenza viral encephalitis following olfactory bulb inoculation of a neurovirulent strain (Mori et al., 2002). In animals, aerosol delivery to the respiratory tract has greater efficacy than systemic administration (Fenton et al., 1977; Walker et al., 1976). In several animal model studies rimantadine has somewhat greater antiviral activity than amantadine against influenza A viruses (Tsunoda et al., 1965; Schulman, 1968). In mice, rimantadine administration reduces mortality, pulmonary viral titers, and the ability of mice to transmit infection to uninfected cage mates more than comparable doses of amantadine(Schulman, 1968).

Rimantadine and amantadine exhibit enhanced antiviral effects when combined with ribavirin, interferon, or neuraminidase inhibitors in vitro (Madren et al., 1995; Hayden et al., 1984; Hayden, 1996b; Govorkova et al., 2004). In experimental murine influenza, combinations of amantadine and ribavirin may show enhanced antiviral activity and survival compared to monotherapy (Galegov et al., 1977; Wilson et al., 1980; Hayden, 1986b), and combinations of rimantadine and oseltamivir are more effective in avian influenza A/H9N2 and H5N1 subtype virus infections (Leneva et al., 2000; Gorokova et al., 2004). Whether combinations can prevent development of resistance to the anti-influenza action of amantadine and rimantadine remains to be established.

Resistance
Mechanisms
Amantadine susceptibility is mediated by the M gene (segment 7) encoding the M1 and M2 protein, although the HA gene affects the sensitivity of certain strains (Hay et al., 1985; Lubeck et al., 1978; Scholtissek and Faulkner, 1979). Resistant viruses have point mutations in their M gene and corresponding single amino acid substitutions in the transmembrane domain of the M2 protein (positions 26, 27, 30, 31, or 34) (Hay, 1996). The frequency with which different mutations emerge is subtype dependent. For human H3N2 subtype viruses a Ser31Asn mutation predominates, whereas Val27Ala is more common in H1N1 subtype viruses (Saito et al., 2003; Shiraishi et al., 2003). Resistance is readily selected by growth in the presence of the drugs in vitro or *in vivo* (Hay, 1996; Hayden, 1996a; Herlocher et al., 2002), and resistant variants exist as subpopulations in nondrug-exposed virus pools (approximately 10^{-4}). Resistance is high level ($\geq$30–100 fold reductions in in vitro susceptibility)(Abed et al., 2005) and confers cross-resistance to other M2 inhibitors.

It has been proposed that virus develops resistance by two alternative pathways to avoid blockage of its channel: (1) a conventional route in which the channel no longer binds the blocker and, hence, the blocker cannot exert its inhibitory function; and (2) a novel mechanism in which binding of the blocker is retained, yet the function of the protein is not affected. In the amantadine-resistant mutants that have lost the ability to bind amantadine, the mutations introduce a larger amino acid (e.g., Ser31Asn, Ala30The) (Astrahan et al., 2004). The drug therefore can no longer bind due to steric hindrance, or chemical incompatibility. In the amantadine-resistant mutants that retain drug binding, the mutations introduce a smaller amino acid (e.g., Val27Gly or Ala, Ile27Ser or Thr). Thus, it is possible that in these mutants, amantadine can still bind the channel, but since the pore is large, the channel is no longer blocked. Recent studies indicate that the binding site for amantadine is located near residues 30 and 31, while the site in which the disruption of the water molecule file is alleviated is located in the vicinity of residue 27 (Astrahan et al., 2004).

Mutants resistant to amantadine are also cross-resistant to rimantadine. However, BL-1743 (2-[3-azaspiro (5, 5) undecanol]-2-imidazoline), another inhibitor of the ion M2 channel activity, interacts somewhat differently than amantadine with the residues of the transmembrane domain (Tu et al., 1996). For example, the mutations at residues 27, 30, 31, and 34 induce cross-resistance to amantadine and BL-1743, whereas the mutation Leu38Phe has a differential effect; it confers resistance to amantadine but retains susceptibility to BL-1743 (Tu et al., 1996). A mutation Ile35Val induces a several hundred-fold reduction in sensitivity to BL-1743 and only a 10-fold reduction in sensitivity to amantadine. Mutations in the transmembrane domain located closer to the outer leaflet of the lipid bilayer (extracellular side) have similar properties with respect to amantadine and BL-1743, whereas those closer to the inner leaflet (cytoplasmic side) can exhibit major differences in resistance profiles (Tu et al., 1996).

The structural data indicate that the amino acid residues whose replacements generate resistance to amantadine (27, 30, 34) face the pore of the channel. The amino acid 31 is partially in the protein-protein interface and partially in the pore (Shuck et al., 2000). The sites that are most restricted along the pore are in the vicinity of the two residues implicated in channel activation and gating, H37 and W41, respectively. Despite the existing variability of the transmembrane M2 protein sequences among the viruses detected in avian species and swine, no mutations have been detected at residues 34, 36, 37, and 42 (more than 400 sequences from the Influenza Sequence Database (Macken et al., 2001) were analyzed). The mutants containing substitutions at residues 35, 40, and 41 are also extremely rare (Table 1).

In addition, resistance in certain avian viruses may be due to mutations in viral HA that increase acid stability, so that inhibition of M2 function by amantadine does not lead to low pH-inactivation of HA (Steinhauer et al., 1991). As yet this mechanism of resistance has not been recognized in human influenza viruses.

Frequency and duration

All contemporary pandemic strains of influenza A and almost all viruses recovered from persons not exposed to these drugs or to someone receiving an M2 inhibitor remain susceptible (Hayden, 1996a). However, resistance has been found in human influenza H1N1viruses from the pre-amantadine era (Bean, 1992) and from contemporary swine (Marozin et al., 2002) and recently avian H5N1 viruses (Hien et al., 2004) infecting humans. Earlier surveys of community isolates found approximately 1–3% frequencies of resistance (Zeigler et al., 1999). However, a recent survey found increasing frequencies of M2 inhibitor resistance among community isolates of A/H3N2 viruses that reached over 70% in China and Hong Kong in 2004 and exceeded 10% in the United States, Europe and other countries in 2005 (Bright et al., 2005).

Resistant variants also emerge as early as 2–3 days after initiating treatment of influenza and are detectable in approximately 30% of treated children and adults (Hall et al., 1987; Hayden et al., 1989; Hayden et al., 1991). Detailed studies employing molecular cloning techniques have detected resistant variants in up to 80% of amantadine-treated children, often with multiple resistance genotypes (Shiraishi et al., 2003).

Immunocompetent persons generally stop shedding resistant variants within 10 days, but prolonged shedding, sometimes extending weeks to months after cessation of drug, has been documented in highly immunocompromised persons (Klimov et al., 1995; Englund et al., 1998). In such persons, most isolates recovered after 3 days of therapy are M2 inhibitor resistant.

Table 7.1 Amino acid variance identified in influenza A viruses M2 transmembrane domain

	25	26	27	28	29	30	31	32	33	34	35	36	37	38	39	40	41	42	43
Mutations detected in swine and avian viruses	H, L	I, F	G, F, T, I, A	I, A, T, D	I, G, T, V	S	K, N	F, V, T	T, V		V*		F, P, M	F, P, M	M, V	W**	G***	F, I	F, I
Residue position	25	26	27	28	29	30	31	32	33	34	35	36	37	38	39	40	41	42	43
M2 TM	P	L	V	V	A	A	S	I	I	G	I	L	H	L	I	L	W	I	L
Mutations in human viruses associated with resistance		S, I	F		V	G	N, T, V	R	M	E	(T)	V		(F)					

420 sequences of avian and swine influenza A viruses from the Database of the National Laboratories, Los Alamos. (resistance to BL-1743)

This table includes both laboratory selected and clinically detected variants.

34E – also in an avian amantadine-resistant virus

* – detected in a single isolate A/turkey/New Jersey/98 (H7N2).

** – detected in a single isolate A/aquatic bird/Hong Kong/M603/90 (H11N1).

*** – detected in a single isolate A/duck/Hong Kong/p50/97 (H11N9).

Consequences

Most M2 inhibitor-resistant viruses appear to be readily transmissible and fully pathogenic. In experimentally infected chickens, amantadine treatment causes rapid selection of drug-resistant virus that remains infectious and lethal for contact birds receiving amantadine, although combined administration of inactivated vaccine and amantadine to contact birds is protective (Webster et al., 1985). Resistant variants are usually comparable to their drug-susceptible parents in replication, transmissibility in the absence of selective drug pressure, and virulence in avian models (Bean et al., 1989). Resistant human H3N2 subtype isolates retain virulence in ferrets (Sweet et al., 1991) and cause typical illness in humans (Hayden et al., 1989). Recombinant H1N1 subtype viruses with resistance mutations also show no impairment in replication in vitro and are at least as virulent as wild-type virus in mice (Abed et al., 2005).

Illness may be prolonged by 1–2 days in immunocompetent adults or children shedding resistant virus while receiving M2 inhibitor therapy (Hall et al., 1987; Hayden et al., 1991). Immunocompromised hosts have experienced prolonged illness and sometimes progressive lower respiratory tract disease in association with emergence of resistant virus. Severe infections due to infection by resistant variants have also been documented in the institutionalized elderly (Houck et al., 1995).

Transmission of resistant viruses from treated persons to close contacts in households (Hayden et al., 1989) or nursing homes causes failures of drug prophylaxis (Mast et al., 1991; Houck et al., 1995). Multiple institutional outbreaks managed with amantadine have continued because of the emergence and transmission of M2 inhibitor-resistant variants (Houck et al., 1995; Bowles et al., 2002; Lee et al., 2000). The occurrence of new influenza illnesses more than a few days after start of M2 inhibitor chemoprophylaxis should raise concern about drug resistance. Of note, NA inhibitors are effective alternatives for managing closed population outbreaks due to M2 inhibitor-resistant strains (Lee et al., 2000; Bowles et al., 2002). The potential for epidemic spread of resistant variants in closed (Stilianakis et al., 1998) or open communities and the degree of selective drug pressure necessary to induce such an event are uncertain. However, recent wide-spread community circulation of M2 inhibitor-resistant variants in certain Asian countries may relate in part to amantadine use fostered by over-the-counter availability (Bright et al., 2005).

Clinical effectiveness
Prophylaxis

The clinical usefulness of amantadine and rimantadine as antiviral agents is limited to the prevention and treatment of influenza A virus infections. In experimentally challenged volunteers, prophylactic administration of amantadine is 15–40% effective in reducing the frequency of infection and 50–90% effective in reducing the frequency of influenza A virus illness (Treanor and Hayden, 1998). Prophylaxis in experimental influenza also reduces viral replication and influenza-specific antibody responses in blood and nasal secretions (Reuman et al., 1989a; Reuman et al., 1989b). No clear dose-related antiviral effects have been found over a dose range of 50–200 mg daily, but doses of 100 mg daily provide approximately 75% protection against illness (Sears and Clements, 1987; Reuman et al., 1989a).

Seasonal prophylaxis

When taken daily for seasonal prophylaxis in open populations of children and adults, protective efficacy against epidemic illness ranges from 70–100%. Rimantadine is comparable to amantadine in preventing influenza A virus infection and illness (Zlydnikov et al., 1981; Dolin et al., 1982; WHO, 1985). One comparative six week seasonal prophylaxis trial in students found that equivalent doses (100 mg bid) of amantadine and rimantadine provided 91% and 85% protection, respectively, against illness due to influenza A virus (Dolin et al., 1982).Termination of prophylaxis before influenza activity has ceased is sometimes associated with early post-treatment failures of drug prophylaxis (Payler and Purdham, 1984; Muldoon et al., 1976). Efficacy against laboratory documented infection is lower than against illness, but subclinical seroconversions during chemoprophylaxis may protect against subsequent reinfection. When seasonal prophylaxis has been given to school-aged children, the frequency of influenza A virus infection in other family members is reduced (Crawford et al., 1988).

The minimally effective dose for preventing natural influenza illness has not been determined for either drug. Doses of 100 mg per day of amantadine (Payler and Purdham, 1984; Rose, 1983) or rimantadine (Brady et al., 1990) appear to be effective for chemoprophylaxis and also reduce the risk of drug side effects. Studies in the former Soviet Union suggested that a rimantadine dose of 50 mg per day was effective for prophylaxis (Zlydnikov et al., 1981). Protection from chemoprophylaxis appears to be additive to that provided by specific antibody induced by prior infection or immunization (Quilligan et al., 1966; Galbraith et al., 1969b; Smorodintsev et al., 1970b), but protection has also been found in sero-susceptible populations experiencing infection by novel influenza A viruses (Monto et al., 1979; Pettersson et al., 1980; Nafta et al., 1970; Smorodintsev et al., 1970a). However, efficacy against pandemic strains has been lower averaging approximately 60–70% across studies (Hayden, 2001). Amantadine at 100 mg daily provided 63% protection against illness during the A/Hong Kong/68(H3N2) pandemic in one trial in the former Soviet Union (Smorodintsev et al., 1970a).

Nosocomial/ institutional prophylaxis

Prophylaxis can protect against influenza A illness in hospitals, chronic care facilities, and nursing homes, and mass chemoprophylaxis in institutional outbreaks has been temporally associated with cessation of influenza activity in the majority of instances (Arden et al., 1988; Tamblyn, 2001) Chemoprophylaxis for at least 10 days or until 7 days after onset of the last case has been advised (Gravenstein and Davidson, 2002). Because of its lower risk of CNS adverse effects and lesser dependence on renal function for elimination, rimantadine would be preferred to amantadine in elderly patients.

Amantadine administration to patients hospitalized during a community outbreak provided complete protection against nosocomial influenza in one study (O'Donoghue et al., 1973). In immunized elderly nursing home residents, administration of rimantadine (100 mg bid) during an outbreak period significantly improved protection against illness (75 per cent efficacy) compared to placebo (WHO, 1985). Rimantadine prophylaxis at doses of 100 or 200 mg daily for up to 8 weeks in immunized nursing home residents found lower risk of influenza-like illness (58% efficacy) but did not reduce influenza infection (Monto et al., 1995). One study comparing chemoprophylaxis with oral rimantadine and inhaled zanamivir in nursing home residents during 3 influenza seasons and found an additional protective efficacy of 68% in the zanamivir recipients, in part because of the frequent occurrence of illness due to M2 inhibitor resistance in rimantadine recipients (Gravenstein et al., 2000; Gravenstein and Davidson, 2002).

Post-exposure prophylaxis

Post-exposure prophylaxis has been useful in limiting the spread of infection within households. When taken for 10 days by family contacts after onset of illness in an ill index case, amantadine and rimantadine reduced the risk of influenza A illness by 70–100% (Galbraith et al., 1969a; Bricaire et al., 1990). In contrast, negligible protection of contacts has been observed in two studies involving concurrent treatment of ill index cases, including one trial conducted during the 1968 H3N2 pandemic (Galbraith et al., 1969b; Hayden et al., 1989). Prophylaxis failures were probably due to transmission of drug-resistant virus from treated index cases to household contacts (Hayden et al., 1989).

Treatment

In acute uncomplicated influenza of previously healthy, young adults, early treatment (within 48 hours of symptom onset) at doses of 200 mg per day reduces the duration of fever and systemic complaints by 1–2 days, virus titers in upper respiratory secretions, and duration of functional limitation (VanVoris et al., 1981; Hayden and Monto, 1986; Hayden et al., 1991; Hayden and Aoki, 1999). One comparison of aspirin and amantadine treatment found that aspirin-treated patients became afebrile more rapidly but experienced significantly higher rates of aspirin side effects and slower overall symptomatic improvement than amantadine recipients (Younkin et al., 1983). Similar therapeutic benefit in regard to resolution of fever and symptoms occurs in elderly nursing home residents treated with rimantadine (Betts et al., 1987). In H3N2 subtype infections, abnormalities in peripheral airway flow but not bronchial hyperreactivity improve more rapidly in amantadine-treated patients (Little et al., 1976; Little et al., 1978), but similar studies have not been conducted with rimantadine. Antiviral treatment generally does not reduce the humoral immune responses to influenza virus, although virus-specific nasal IgG and IgA responses may be reduced, perhaps because of lower antigenic stimulation (Clover et al., 1991).

No large prospective studies have been conducted to determine if amantadine prevents or treats the pulmonary complications of influenza. Unlike neuraminidase inhibitors, rimantadine does not prevent middle ear pressure abnormalities after experimental influenza (Skoner et al., 1999). Rimantadine is also much less effective than oseltamivir in a murine model of sequential influenza virus and pneumococal infection (McCullers, 2004). One placebo-controlled study in 80 nursing home residents found more rapid symptom improvement and fever resolution with early rimantadine treatment (Betts et al., 1987) but no differences in duration of viral shedding or the occurrence of pneumonia (2 rimantadine, 4 placebo) and deaths (2 rimantadine, 1 placebo). In nursing home residents one retrospective study found a reduced risk of influenzal illness-related sequelae in amantadine-treated patients, particularly those with prior influenza immunization, compared to no treatment (Libow et al., 1996). Compared to no treatment, retrospective studies found that early M2 inhibitor treatment appeared to reduce antibiotic use (37% vs 65%) and the risk of pneumonia (16% vs 48%) in institutionalized elderly (Bowles et al., 2002) and risk of progression to pneumonia (35% vs 76%) in high-risk adults with leukemia or hematopoietic stem cell transplantation with influenza (La Rosa et al., 2001). One uncontrolled study of high amantadine doses (400 to 550 mg per day) found a 55 per cent survival rate in 11 patients with primary influenza viral pneumonia (Couch and Jackson, 1976), and combinations of amantadine and oseltamivir have been used to treat individual patients with severe influenza (Ison et al., 2005a).

In children studies have found variable clinical benefits with early M2 inhibitor treatment. Amantadine treatment (50 to 150 mg per day for 7 days) of children with influenza

A/H3N2 infection was associated with decreased duration of fever (Kitamoto, 1971). In children with influenza A H3N2 subtype infection, rimantadine treatment (6.6 mg/ kg/day, up to 150 mg/day for those <9 years, and 200 mg/day for older) for 5 days was associated with lower symptom burden, fever, and viral titers during the first 2–3 days of treatment compared to acetaminophen, but rimantadine-treated children had more prolonged shedding of influenza virus by an average of one day (Hall et al., 1987). Another study of children with predominately H1N1 subtype infection found reduced frequencies of viral shedding but no significant clinical benefit of rimantadine compared to acetaminophen (Thompson et al., 1987). Amantadine has been used in treatment of influenza A-associated encephalopathy with uncertain benefit (Sugaya et al., 2002).

Treatment of ill index cases in the household, usually children, with an M2 inhibitor appeared to reduce the risk of infection to close contacts by approximately 30% in one study (Couch et al., 1986). In contrast, antiviral treatment of ill children combined with prophylaxis for their household contacts has been associated with prophylaxis failures due to probable transmission of drug-resistant viruses (Hayden et al., 1989; Galbraith et al., 1969b). The optimal duration of therapy in children is uncertain, and the possibility that shorter treatment courses might provide therapeutic benefit and reduce the frequency of drug-resistant virus emergence requires study.

NEURAMINIDASE INHIBITORS

Several potent, highly selective influenza neuraminidase (NA) inhibitors have been developed in part through analysis of the NA crystal structure (von Itzstein et al., 1993). The first clinically effective inhibitor zanamivir, a dehydrated sialic acid derivative in which a guanidinyl group replaces the hydroxyl at the C4 position (Figure 2), was approved for influenza treatment in 1999 in the United States. Oseltamivir phosphate, the prodrug of oseltamivir carboxylate (formerly GS4071), has a cyclohexene ring with replacement of the polar glycerol with a lipophilic side chain but without a guanidinyl group. This oral NA inhibitor also entered clinical practice in 1999. Several other investigational NA inhibitors including peramivir (formerly BCX-1812, RWJ-270201), A-315675, and the prodrug R-118958 are also sialic acid analogues that confer potent NA inhibition in vitro and animals (Sidwell and Smee, 2002; Kati et al., 2002; Yamashita et al., 2003).

No direct clinical comparisons between NA inhibitors or, in influenza A infections, between NA inhibitors and M2 inhibitors have been published to date, except for one nursing home prophylaxis trial (Gravenstein et al., 2000). Consequently, it is unproven whether there are important differences in clinical efficacy, although reductions in influenza-associated complications have been established with NA inhibitors (Kaiser et al., 2000b; Kaiser et al., 2003), and this antiviral class is associated with lesser problems with antiviral resistance compared to M2 inhibitors and has an antiviral spectrum including influenza B viruses.

Antiviral activity
Mechanism of action
Influenza HA and NA play essential roles during viral entry and release from host cells. The principle action of NA is to remove terminal sialic acid residues from receptors that are recognized by HA on progeny virions, cells, or respiratory secretions. Destruction of these receptors promotes virion release from infected cells and spread within the respiratory tract (Gubareva et al., 2000). In addition, neuraminidase has a role in initial viral infection

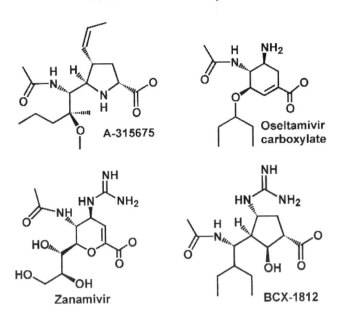

Figure 7.2 Chemical structures of approved (oseltamivir, zanamivir) and investigational (BCX-1812 or peramivir, A-315675) neuraminidase inhibitors.

one of which is preventing inactivation of virus by respiratory mucins (Matrosovich et al., 2004b). The catalytic site of viral NA is highly conserved across influenza A and B viruses, whereas influenza C viruses lack this glycoprotein. NA inhibitors (NAIs) bind in the active enzyme site and block its function, so that virion release from infected cells and spread with the respiratory tract are inhibited.

In general, zanamivir and oseltamivir carboxylate are slow, tight binders of NA, exhibiting time-dependent inhibition. These inhibitors reversibly inhibit influenza NAs and also display slow-binding characteristics, such that inhibition increases over time. For oseltamivir slow binding is actually a consequence of slow dissociation of the inhibitor from the NA (Kati et al., 1998).

All of these agents are highly selective for influenza NA and generally inhibit NAs from other pathogens or mammalian cells at 10^6-fold or higher concentrations (McClellan and Perry, 2001). High zanamivir concentrations appear to inhibit parainfluenza virus NA activity in vitro (Greengard et al., 2000). Cellular cytotoxicity is generally observed only at very high (millimolar) concentrations.

Spectrum and Potency

Oseltamivir carboxylate and zanamivir inhibit NA enzyme activity at low nanomolar concentrations and inhibit viral replication in cell culture at approximately 10–100 fold higher concentrations (reviewed in (Gubareva et al., 2000; McClellan and Perry, 2001)). They are generally more active against influenza A than B NAs. The range of inhibitory concentrations in cell culture is broad (1, 000 fold or more) and depends heavily on the assay method. Influenza viruses exhibit a wide range of the susceptibility to NA inhibitors in MDCK cell culture (Woods et al., 1993), most likely due to varying requirements for the NA activity during virus release. These drugs are inhibitory for range of nine NAs present in influenza viruses in nature (Gubareva et al., 1995; Govorkova et al., 2001). Among recent clinical isolates, oseltamivir carboxylate is somewhat more active against N2 than zanamivir but less active against B NAs (McKimm-Breschkin et al., 2003). Oseltamivir and zanamivir are inhibitory for the N1 derived from the 1918 pandemic virus, and a virus possessing this NA was inhibited by oseltamivir in cell culture and experimental murine infection (Tumpey et al., 2002); oseltamivir is active in vitro and in animal models against the avian H5N1 and H9N2 subtype viruses that have caused human disease (Leneva et al., 2000). However, recent murine studies indicate that higher doses and more prolonged administration of oseltamivir is required for inhibition of a 2004 H5N1 clinical isolate compared to one from 1997 (Yen et al., 2005). Oral oseltamivir and topical zanamivir are active in murine and ferret models of influenza by routes of administration that reflect their clinical pharmacology. However, the inhibitory effects of oral oseltamivir and peramivir on the viral titers in the ferret respiratory tract were modest (Sweet et al., 2002; Mendel et al., 1998). In a murine model, the antiviral effect is also virus strain-dependent (Sidwell et al., 2001). For M2 inhibitor susceptible viruses, NAIs and M2 inhibitors show enhanced antiviral activity in vitro and in animal models of influenza A virus infection (Govorkova et al., 2004; Leneva et al., 2000). However, it is unknown if combinations of agents might be able to exert more potent antiviral effects and provide greater clinical benefits in human infections.

Resistance
Mechanisms
To evaluate the potential for the development of resistance to NA inhibitors during the clinical use, it is important first to understand the genetic basis and molecular mechanisms of resistance to this new class of drugs. The emergence of resistant variants has been studied by passage of laboratory strains in tissue culture in the presence of drug (McKimm-Breschkin, 2000; Cheam et al., 2004). This approach allows characterization of the emerged resistant variants in phenotypic (enzyme activity and virus growth inhibition assays) and genotypic assays. Analysis of the escape mutants selected in cell culture has revealed two basic mechanisms of resistance to NA inhibitors: 1) reduced binding efficiency between HA and its receptor and 2) reduced binding efficiency between NA and its inhibitor.

Resistance to NA inhibitors in cell culture arises from point mutations in either the HA or NA genes and corresponding single amino acid substitutions in the related viral glycoprotein. These genotypic changes can be best explained in the context of the balance between the HA receptor-binding and the NA receptor-destroying activities. One of resistance emergence scenarios is shown in Figure 3. To assure the optimal virus propagation, the HA and NA activities of the virus must be balanced (Wagner et al., 2002). By reducing viral NA activity, the drug alters this balance leading to aggregation of progeny virions at the infected-cell surface thus limiting virus spread.

HA mutations. HA mutations decrease binding to sialic acid-bearing receptors and thus dependence on NA activity, so that drug-mediated inhibition of NA is less effective in blocking viral release. Such HA mutants are drug resistant in cell culture-based phenotypic assays but drug sensitive in the enzyme inhibition assay. Since HA mutants display drug resistance in cell culture that is enzyme-independent, they exhibit cross-resistance to all NA inhibitors (McKimm-Breschkin, 2000; Baum et al., 2003).

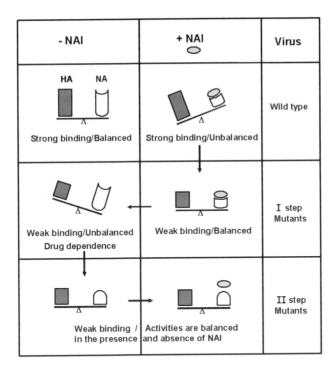

Figure 7.3 A schematic representation of possible resistance emergence patterns to neuraminidase inhibitors.

At low concentrations of NA inhibitor in cell culture media, the HA mutant yields may be increased (partial drug dependence), most likely due to a better balance of the mutant HA and wild type NA activities (Gubareva et al., 1996; McKimm-Breschkin et al., 1996; Barnett et al., 1999). Normal NA activity would decrease the efficiency of adsorption of the virus to the cell by premature destruction of receptors. In the presence of drug inhibiting NA activity, a variant containing the altered HA would show a greater ability to adsorb to cells, and thus more plaques would be formed than without drug in the media. Because of the HA's ability to compensate for the loss of the NA activity can be affected by receptor structure of the host cell (Gubareva et al., 1998; Hughes et al., 2001; Matrosovich et al., 2003), it is critical to conduct selection studies and assess resistance of the emerged variants in the same cell culture system.

Oligosaccharide chains attached in the vicinity of the receptor-binding site play important role in modulation of the HA binding to receptors (Ohuchi et al., 1997). When virus propagation takes place in the presence of NA inhibitor, the oligosaccharide chains remain sialylated. For viruses that contain oligosaccharide chains in vicinity to the receptor-binding site, the sialylation of the HA may provide a mechanism of NA-independence. (Mishin et al., 2005). Emergence of the NA inhibitor-resistant mutants of influenza A viruses lacking the NA activity due to a large internal deletion in the NA gene has been reported (Nedyalkova et al., 2002).

The HA mutants usually retain full susceptibility to NA inhibitors in animal models of influenza (Zambon and Hayden, 2001; Abed et al., 2002). HA mutations have been rarely recognized in clinical isolates to date, and it is not known whether they alone might confer clinically significant resistance to NAIs (Zambon and Hayden, 2001). Such mutations in HA sometimes result in reduced viral infectivity.

NA mutations. Further passage in the presence of NA inhibitor typically leads to emergence of the mutants carrying amino acid substitutions in the NA active site. The NA mutants selected in cell culture exhibit resistance in plaque and enzyme activity inhibition assays. In general, drug-resistant variants with NA mutations replicate as efficient as wild-type virus in the presence and in the absence of the drug (Staschke et al., 1995; Gubareva et al., 1997; Baum et al., 2003). Resistance in NA typically results from single amino acid changes in the active enzyme site that alter drug binding. These mutations are associated with approximately 30- to over 1, 000-fold lower susceptibility in NA inhibition assays. The highly conserved NA active site is believed to be a result of the virus optimizing its ability to propagate in susceptible hosts. X-ray crystallography and molecular modeling studies provide structural explanations of the mechanism of resistance to NA inhibitors (Smith et al., 1923; Baum et al., 2003; Varghese et al., 1995; Varghese et al., 1998; Molla et al., 2002). The amino acids directly interacting with substrate (R118, D151, R152, R224, E276, R292, R371, and Y406 based on the N2 amino acid residue numbering) are referred as 'catalytic' or functional residues. Additionally, the residues E119, W178, I222, R156, S179, D198 (N in N7 and N9 subtypes), E227, H274, E277, N294, and E425 provide a scaffold (framework) for the 'catalytic' residues (Varghese and Colman, 1991; Burmeister et al., 1992; Air and Laver, 1989). There is a certain degree of uncertainty in assigning of an amino acid residue to the NA active site (Colman, 1989; Burmeister et al., 1992). For example, Asp151 has been considered as one of the key residues in catalysis (Taylor and von Itzstein, 1994), however recent detection of natural isolates lacking the asparagine at this position challenges this role (McKimm-Breschkin et al., 2003). Therefore, it is possible that participation of certain amino acid residues in the catalysis may vary in structurally diverse NAs.

For convenience, the mutations in the NA active site detected in the viruses selected after drug exposure are described in the context of functional residues (R152 and R292) versus framework residues (E119, D198, and H274). The mutations at functional residues reduce the enzyme's specific activity and are accompanied by cross-resistance to inhibitors of this class in an enzyme inhibition assay. In contrast, framework mutations are often associated with retention of susceptibility to other NA inhibitors (Gubareva et al., 2002; Molla et al., 2002; Gubareva, 2004). The effects of specific framework mutations depend on drug and on virus type and subtype. For example, the mutation E119G in the enzyme of A/N2 subtype confers resistance to zanamivir but not to oseltamivir, whereas the mutation E119V produces an opposite effect (Gubareva, 2004; Ives et al., 2002). A Glu119Asp in an influenza B NA confers high-level (32, 000-fold) zanamivir and moderate (105-fold) oseltamivir resistance (Cheam et al., 2004).

The influence of the volume occupied by the amino acid side chain for the resistance profile has been reported for mutants carrying substitutions at the framework residue 274 (Wang et al., 2002). Replacement of His with Tyr or Phe in the enzyme of N1 subtype results in the high-level oseltamivir resistance but these mutants are susceptible to zanamivir (Wang et al., 2002; Gubareva et al., 2001). The mutation H274Y identified in the enzyme of influenza B confers resistance to oseltamivir and peramivir but not to zanamivir (Baum et al., 2003). However, this mutation does not alter oseltamivir-susceptibility of the N2 enzyme (Wang et al., 2002). Moreover, replacement of framework residue Asp198 with Asn confers resistance to oseltamivir in the type B enzyme (Gubareva, 2004), yet Asn is present at this position in the drug-susceptible enzymes of N7 and N9 subtypes (Colman, 1989). Interestingly, an influenza B virus with reduced susceptibility to oseltamivir and zanamivir has been recovered from an untreated patient (Hurt et al., 2004).

It is important to recognize that the amino acid residues beyond the catalytic site can affect virus susceptibility to NA inhibitors. For example, a replacement of Arg152 with Lys in the NA active site of influenza B/Memphis/20/96 virus was accompanied by at least 70-fold increase in the IC_{50} values against zanamivir(Gubareva et al., 1998) but the same substitution caused only a 5-fold increase in the influenza B/Beijing/1/87 virus genetic background. (Jackson et al., 2005). The lack of cross-resistance between some NA inhibitors may prove important if clinical resistance to a specific NA inhibitor emerges.

Importantly, mutations in NA active site are usually associated with loss of enzyme stability or activity in vitro (Jackson et al., 2005) and reduced viral fitness in animal models (reviewed in (Zambon and Hayden, 2001; McKimm-Breschkin, 2000)). Experimental murine infections have been used to evaluate resistance and virulence of the mutants selected in MDCK cells after drug exposure (McKimm-Breschkin, 2000; Sidwell and Smee, 2000). The value of NAI-resistance testing in a mouse model is limited due to a lack of information on the structure of the murine receptors for influenza. Human influenza viruses are shown to preferably replicate in the human respiratory tract epithelial cells that expressed NeuAc-alpha2, 6-Gal-receptors (Baum and Paulson, 1990; Matrosovich et al., 2004a). Ferrets, whose respiratory tract epithelial cells share the receptor similarity with that of the human (Leigh et al., 1995), provide a more suitable model for the assessment of the properties of the mutants selected in humans treated with NAIs (Gubareva et al., 1998; Herlocher et al., 2002; Herlocher et al., 2003).

Detection

Phenotypic monitoring for NA inhibitor-resistant variants is technically challenging in part because most conventional cell culture based assays are not reliable (reviewed in (Tisdale, 2000; Zambon and Hayden, 2001)). A recently described MDCK cell line that expresses an enzyme to enhance surface expression of alpha-2, 6 linked sialic acid receptors appears to overcome this limitation (Matrosovich et al., 2003) but more data are needed. The currently recommended assays to detect phenotypic NA resistance are one of several NA enzyme inhibition assays (Wetherall et al., 2003). The susceptibility of isolates, particularly those with NA mutations, are highly dependent on assay conditions including enzyme concentration, buffers, and substrate (Gubareva et al., 2002). Sequence analysis of the NA gene can detect known resistance mutations and provide genotypic evidence of resistance. Following oseltamivir treatment, the most commonly observed mutations in clinical isolates have been Arg292Lys and less often Glu119Val in N2 and His274Tyr in N1 neuraminidases (Roberts, 2001). The genotypic assays report mutations known to be associated with drug resistance and therefore can only interrogate those mutations that are already known to confer resistance.

For drug resistance monitoring therefore, it would be prudent to extend the sequence examination to the residues that are not highly conserved or do not directly interact with the substrate. For convenience, the 24 residues that have been implicated in the NA catalysis, substrate binding or framework are listed: R118, E119, L134, D151*, R152, R156, W178, S179, D198*, I222, R224, E227, D243, A250*, E276, E277, R292, D293, N294, T365, R371, Y406, W409, E425 (N2 numbering) (Colman, 1989; Burmeister et al., 1992). The residues marked with a star sign (*) are not highly conserved. An alignment of amino acid sequences of NAs that belong to different antigenic (sub)types could be found in (Colman, 1989).

Frequency and duration

Murine models have been used to evaluate the potential of NAIs to select drug resistant mutants (Nedyalkova et al., 2002). No emergence of oseltamivir-resistance was detected in mice infected with A/H3N2 virus after 5 days of twice daily treatment (Sidwell and Smee, 2000). A second passage of the virus through mice treated with oseltamivir also yielded virus that did not differ in its sensitivity to oseltamivir. These data indicate that mutants resistant to NAI do not emerge readily in a mouse model. Nevertheless, the emergence of the mutants resistant to NAIs due to substitutions in the HA receptor-binding site and in the NA active site (Arg292Lys) has been detected in NAI -treated mice infected with A/H2N2 virus (Ison et al., 2005b).

No de novo resistance to NA inhibitors has been found in almost all studies isolates collected before the introduction of the drugs into clinical practice (McKimm-Breschkin et al., 2003). Resistance emergence due to NA mutations occurs uncommonly during treatment with oseltamivir and has not been detected during prophylactic use to date (Jackson et al., 2000; Hayden et al., 2004). In oseltamivir-treated immunocompetent adults and children participating in controlled trials, the frequency of post-treatment viruses showing NA resistance was higher in children (approximately 4% of those treated) than in adults (approximately 0.4%) (Whitley et al., 2001; Roberts, 2001). Most resistant variants have been detected on days 4–6 after starting therapy and are often present in mixtures with wild-type virus. One recent report found that resistant viral clones, usually predominately over wild-type ones, were detectable in 18% of young children treated with oseltamivir (Kiso et al., 2004). During surveillance for oseltamivir-resistant variants has been conducted by the WHO-affiliated Neuraminidase Inhibitor Susceptibility Network, low frequencies (<1%) of resistant variants have been detected among community isolates during the first three years since introduction of the drug into clinical practice (Monto et al., 2005).

No clinical isolates with phenotypic resistance to zanamivir have been detected in clinical trials to date. Among 41 paired isolates from persons receiving intranasal zanamivir, no changes were found in zanamivir susceptibility in vitro, and sequence analysis found no mutations in NA or HA that have been associated with resistance to NAIs in vitro (Barnett et al., 2000). Resistant variants have been detected in highly immunocompromised hosts and may be associated with prolonged viral shedding (Weinstock et al., 2003). One pediatric bone marrow transplant recipient had persistent virus replication despite treatment with inhaled zanamivir for influenza B viral pneumonia; this child had emergence first of an HA mutation that that conferred reduced receptor binding and altered antigenicity and later of an NA mutation (Arg152Lys) that was associated with reduced zanamivir susceptibility and reduced viral fitness (Gubareva et al., 1998).

Consequences

The appearance of oseltamivir-resistant NA variants has not been linked to altered clinical course in immunocompetent children or adults to date (Jackson et al., 2000). In volunteers experimentally infected with an A/H1N1 virus, emergence of viruses with His274Tyr mutation was associated with rebound in nasal viral titers (Gubareva et al., 2001). Prolonged shedding of oseltamivir-resistant virus and associated illness have been observed uncommonly in highly immunocompromised hosts, and variants resistant to both M2 inhibitors and oseltamivir have been seen in such patients (Ison et al., 2005a; Weinstock et al., 2003).

In general NA mutations are associated with reduced infectivity, replication, and pathogenicity in animal models of influenza (Carr et al., 2002; McKimm-Breschkin, 2000; Zambon and Hayden, 2001; Jackson et al., 2000; Herlocher et al., 2002; Gubareva et al., 1997; Gubareva et al., 1998). The oseltamivir-resistant Arg292Lys N2 mutant, although not the His274Tyr N1 or Glu119Val N2 mutants, appear less transmissible in ferrets (Herlocher et al., 2004). No evidence for emergence or transmission of oseltamivir or zanamivir-resistant variants has been found in household-based studies in which NA inhibitors were used for both treatment of ill index cases and post-exposure prophylaxis of contacts (Hayden et al., 2000a; Hayden et al., 2004). A mathematical modeling study predicted that community circulation of oseltamivir resistant variants would be extremely unlikely, even if substantial drug were used for treatment (defined as 40% of symptomatic persons receiving oseltamivir) because of the reduced fitness and transmissibility of such variants (Ferguson et al., 2003). A recent survey of nearly 1,200 influenza A(H3N2) isolates collected during the 2003–4 season in Japan, during which about 5% of the population received oseltamivir treatment, found four (~ 0.4% frequency) with oseltamivir resistance (NISN, 2005). These results suggest that there might have been a very low frequency of person-person transmission of resistant variants in a season with high drug utilization.

Clinical effectiveness

Multiple controlled trials, primarily enrolling previously healthy adults and children have established that oseltamivir and inhaled zanamivir are effective for both prevention and treatment of influenza A and B virus infections (reviewed in Cooper et al., 2003; Fleming, 2003; Moscona, 2005). Early treatment of acute, uncomplicated influenza is associated with both symptomatic benefit and in some instances complications reduction. However, published studies provide limited information about the therapeutic efficacy of these drugs in those hospitalized with severe influenza or in high-risk populations, including immuno-compromised hosts, elderly institutionalized, pregnant women, or infants (McClellan and Perry, 2001).

Prophylaxis

The prophylactic efficacy of NA inhibitors has been established during placebo-controlled studies involving both long-term, seasonal and short-term dosing regimens. Oral oseltamivir in doses of 100 mg once or twice daily was highly effective in protecting against viral recovery and illness after experimental influenza A inoculation (Hayden et al., 1999b), whereas in experimental influenza B, doses of 75 mg once or twice daily showed significant antiviral effects but did not reduce infection (Hayden et al., 2000b). Intranasal zanamivir was highly protective against influenza infection and illness experimentally induced by intranasal virus exposure and also reduced associated otologic abnormalities (Walker et al., 1997). Intravenous zanamivir 600 mg twice daily for 5 days was well-tolerated and prevented infection, illness, and associated nasal cytokine and chemokine responses in experimental influenza (Fritz et al., 1999; Calfee et al., 1999). However, subsequent trials in naturally occurring disease found that inhaled but not intranasal zanamivir is protective against natural influenza (Kaiser et al., 2000a). Inhaled zanamivir is not approved currently for chemoprophylaxis in the United States or in most other countries.

Seasonal prophylaxis

Seasonal prophylaxis studies in otherwise healthy, non-immunized adults receiving oseltamivir for 6-weeks found that doses of 75 mg once daily reduced the risk of influenza infection by 50%, of influenza illness by 76%, of influenza illness with fever by 90%, and of culture proven influenza illness by 100% (Hayden et al., 1999a). The risk of influenza illness was reduced by 84% in one center with higher attack rates, and no greater efficacy was apparent with twice daily dosing. Once daily inhalation of zanamivir is also highly effective in protecting against naturally occurring influenza illness. One 4 weeks seasonal prophylaxis study in adults, mostly non-immunized and young, found that inhaled zanamivir 10 mg once daily reduced the likelihood of influenza infection by 31%, of proven influenza illness by 67%, and of influenza illness with fever (≥37.8°C) by 84% (Monto et al., 1999a). One uncontrolled study indicated that prophylaxis for 8 weeks was well-tolerated and effective in highly immunocompromised children (Chik et al., 2005). Oseltamivir has been used for protection against avian influenza (Koopmans et al., 2004), but whether the currently approved doses are effective for prevention of avian influenza infections has not been rigorously assessed.

Nosocomial/ institutional prophylaxis

In mostly immunized nursing home residents, 6 weeks prophylaxis with oseltamivir 75 mg once daily reduced the risk of proven influenza illness by 92% and of influenza-related complications (Peters et al., 2001). Both oseltamivir and inhaled zanamivir have been used with apparent success for control of influenza A and B outbreaks in immunized nursing home populations, including those ongoing in association with M2 inhibitor resistance (Parker et al., 2001; Bowles et al., 2002). In nursing home residents, 2-weeks of inhaled zanamivir was superior to oral rimantadine in preventing influenza A infection, in part because of the high frequency of rimantadine resistance found (Gravenstein et al., 2000). Zanamivir has also been employed successfully to terminate nursing home outbreaks, including those occurring in setting of influenza B and M2 inhibitor-resistant influenza A (Lee et al., 2000).

Post-exposure prophylaxis

Both oral oseltamivir and inhaled zanamivir have been shown to be effective for post-contact prophylaxis in household-based studies. When used for post-exposure prophylaxis in household contacts, a 7 day regimen of oseltamivir (75 mg once daily) reduced influenza illness by 90% reduction in contacts aged 13 years and older (Welliver et al., 2001). One non-blinded trial in which index cases were treated with oseltamivir and family contacts were randomized to observation or to post-exposure prophylaxis for 10 days, found that symptomatic influenza developed less often in families using post-exposure prophylaxis (68% reduction) compared to index case treatment alone (Hayden et al., 2004). Oseltamivir was protective against influenza illness in children aged 1–12 years old, although the protective efficacy appeared to be lower in young children than adults.

Post-exposure prophylaxis with zanamivir once daily for 10 days reduced the risk of influenza illness by 82% in healthy household contacts aged 5 years or older (Monto et al., 2003). Similarly, in households where both treatment of the ill index case and once daily prophylaxis for 10 days in healthy family members down to age 5 years were administered, found inhaled zanamivir reduced the likelihood of influenza illness in healthy contacts by 79% (Hayden et al., 2000a). Of note, this strategy had failed with oral rimantadine in part due to the rapid emergence and transmission of drug-resistant influenza viruses (Hayden et al., 1989).

Treatment

In experimentally induced influenza A/H1N1, early oseltamivir administration (28 hours post infection) significantly reduced virus titers, illness, and nasal pro-inflammatory cytokine (IL-6, TNF-alpha, IFN-gamma) levels in the upper respiratory tract (Hayden et al., 1999b). Doses of 75 and 150 mg twice daily showed antiviral effects in experimental influenza B (Hayden et al., 2000b).

In acute influenza patients treated within 36–48 hours of symptom onset, oseltamivir reduced the median times to alleviation of illness by 0.4 day in high-risk adults, 1.2 days in otherwise healthy adults, and 1.5 days in children compared to placebo (reviewed in (Cooper et al., 2003)). In previously healthy adults, early treatment was associated with approximate 2–3 day reductions in time to resuming usual activities (Treanor et al., 2000). Earlier initiation of treatment has been associated with both greater antiviral effects (Boivin et al., 2003) and clinical effectiveness with over 3 day reductions in the predicted disability period, if treatment was initiated within 12 hours (Aoki et al., 2003). Treatment in children aged 1–12 years reduced overall antibiotic use by 24% and by over 40% for new acute media diagnoses (Whitley et al., 2001). In adolescents and adults, including some persons with underlying risk conditions, treatment reduced the incidence of influenza-related lower respiratory complications leading to antibiotics by 55% and all-cause hospitalizations by an average of 59% (Kaiser et al., 2003). Oseltamivir treatment provides clinical benefits in influenza B but not in influenza-like illness due to other pathogens. Limited controlled data suggest that oseltamivir treatment is beneficial in elderly persons or those with underlying cardiopulmonary disease (McClellan and Perry, 2001; Martin et al., 2001). Retrospective analyses indicate that early treatment in outpatients, particularly children and adults aged 60 years and older, (Nordstrom et al., 2005) and of nursing home residents (Bowles et al., 2002) reduces the risk of pneumonia and hospitalization. In pediatric asthma patients with influenza, early treatment appeared to reduce asthma exacerbations and more rapidly improve spirometric abnormalities. (Johnston et al., 2005) Uncontrolled case series indicate that oseltamivir reduces the risk of complications in immunocompromised patients (Machado CM et al., 2004; Nichols et al., 2004); it has been used with uncertain benefit in treating established complications including encephalopathy (Straumanis et al., 2003). Late administration of oseltamivir at conventional doses has not been associated with reduced mortality in patients with avian H5N1 disease. (Hien et al., 2004; Chotpitayasunondh et al., 2005) Therapeutic use of oseltamivir generally does not inhibit the frequency or titer of influenza-specific humoral immune responses.

Treatment of acute influenza with both intranasal and inhaled zanamivir reduced upper respiratory virus levels, nasal symptoms, and complications risk to a somewhat greater extent than inhaled drug alone (Hayden et al., 1997; Kaiser et al., 2000b), but combined intranasal and inhaled dosing was not associated with substantially better overall clinical outcomes. Subsequent treatment studies were conducted primarily with orally inhaled zanamivir 10 mg administered twice daily by the approved Diskhaler[TM] device. Inhaled zanamivir shortens the duration of uncomplicated febrile influenza illness by about 1.5 days and by more in those treated within 30 hours of symptom onset (Hayden et al., 1997; MIST Study Group, 1998; Cooper et al., 2003). Therapeutic benefit has been observed in children aged 5–12 years (Mäkelä et al., 2000), and ambulatory patients with underlying risk conditions, but not in non-febrile influenza, those without influenza, or in those treated later in their illness.

Inhaled zanamivir reduces pharyngeal (Boivin et al., 2000) but not nasal viral titers (Hayden et al., 1997) viral titers. One analysis found that treatment benefits are greater (median 2.5–3 days reduced illness duration) in those with severe symptoms at entry, those aged over 50 years, and higher-risk patients (Monto et al., 1999b). In predominately otherwise healthy adults and adolescents, inhaled zanamivir treatment has been associated with a 40% reduction in lower respiratory tract events leading to antibiotic use and a 28% overall reduction in antibiotic prescriptions (Kaiser et al., 2000b). In a retrospective analysis of high-risk patients, over two-thirds of whom had chronic cardiopulmonary conditions, zanamivir treatment reduced illness duration, time to return to usual activities, and the incidence of complications leading to antibiotic use (Lalezari et al., 2001), and one prospective trial in patients with asthma or COPD found that zanamivir treatment reduced illness duration and complications leading to both antibiotics and change in respiratory medications (Murphy et al., 2000), although no differences in the frequency of FEV1 decreases or asthma adverse events were found. Zanamivir treatment has been associated with reduced absenteeism and healthcare resource use (Aoki et al., 2000) but not with reductions in hospitalization in studies to date (Kaiser et al., 2000b; Griffin et al., 2001). Therapeutic use does not inhibit the frequency or titer of influenza-specific humoral immune responses.

Data on zanamivir treatment in hospitalized patients or those with established complications are minimal. One study in 7 allogeneic stem cell transplant patients with influenza treated with zanamivir 10 mg twice daily until cessation of virus shedding; this took a median of 14 days and ranged up to 44 days (Johny et al., 2002). One controlled study of nebulized zanamivir given four times daily in combination with oral rimantadine found good tolerance but modest clinical or virologic effects in patients hospitalized with lower respiratory manifestations of influenza compared to a combination of rimantadine and placebo aerosol (Ison et al., 2003).

Whether treatment of ill persons reduces their risk of spreading infection to close contacts remains unstudied, although a modest reduction might be expected based on an earlier study of M2 inhibitors (Couch et al., 1986). However, in a household-based study in which all ill index cases received oseltamivir treatment, the frequency of secondary influenza illness in healthy contacts was over 60% higher in those observed compared to those receiving post-exposure prophylaxis (Hayden et al., 2004). This finding suggests that treatment of the ill index cases did not substantially reduce the likelihood of secondary transmission.

References

Abed, Y., Bourgault, A. M., Fenton, R. J., Morley, P. J., Gower, D., Owens, I. J., Tisdale, M., and Boivin, G. (2002). Characterization of 2 influenza A(H3N2) clinical isolates with reduced susceptibility to neuraminidase inhibitors due to mutations in the hemagglutinin gene. J. Infect. Dis. *186*, 1074–1080.

Abed, Y., Goyette, N., and Boivin, G. (2005). Generation and Characterization of Recombinant Influenza A (H1N1) Viruses Harboring Amantadine Resistance Mutations. Antimicrob. Agents Chemother. *49*, 556–559.

Air, G. M. and Laver, W. G. (1989). The neuraminidase of influenza virus. [Review] [56 refs]. Proteins. 6(4):341–56.

Aoki, F. Y., Fleming, D. M., Griffin, A. D., Lacey, L., and Edmundson, S. Impact of zanamivir treatment on productivity, health status and healthcare resource use in patients with influenza. Pharmacoeconomics 17[2], 187–195. 2000.

Aoki, F. Y., Macleod, M. D., Paggiaro, P., Carewicz, O., El Sawy, A., Wat, C., Griffiths, M., Waalberg, E., and Ward, P. (2003). Early administration of oral oseltamivir increases the benefits of influenza treatment. J. Antimicrob. Chemother. 51, 123–129.

Arden, N. H., Patriarca, P. A., Fasano, M. B., Lui, K. J., Harmon, M. W., Kendal, A. P., and Rimland, D. (1988). The roles of vaccination and amantadine prophylaxis in controlling an outbreak of influenza A (H3N2) in a nursing home. Arch. Intern. Med. 148, 865–868.

Astrahan, P., Kass, I., Cooper, M. A., and Arkin, I. T. (2004). A novel method of resistance for influenza against a channel-blocking antiviral drug. Proteins. 55(2):251–7.

Barnett, J., Cadman, A., Gor, D., Dempsey, M., Walters, M., Candlin, A., Tisdale, M., Morley, P. J., Owens, I. J., Fenton, R. J., Lewis, A. P., Claas, E. C., Rimmelzwaan, G. F., DeGroot, R., and Osterhaus, A. D. M. E. (2000). Zanamivir susceptibility monitoring and characterization of influenza virus clinical isolates obtained during phase II clinical efficacy studies. Antimicrob Agents Chemother 44, 78–87.

Barnett, J. M., Cadman, A., Burell, F. M., Madar, S. H., Lewis, A. P., Tisdale, M., and Bethell, R. (1999). In vitro selection and characterisation of influenza B/Beijing/1/87 isolates with altered susceptibility to zanamivir. Virology 265, 286–295.

Baum, E. Z., Wagaman, P. C., Ly, L., Turchi, I., Le, J., Bucher, D., and Bush, K. (2003). A point mutation in influenza B neuraminidase confers resistance to peramivir and loss of slow binding. Antiviral Res. 59, 13–22.

Baum, L. G. and Paulson, J. C. (1990). Sialyloligosaccharides of the respiratory epithelium in the selection of human influenza virus receptor specificity. Acta Histochemica – Supplementband. 40:35–8.

Bean, W. J. Unselected amantadine resistance in human influenza virus and its long-term survival in nature. American Society for Virology 11th Annual Mtg. 1992.

Bean, W. J., Threlkeld, S. C., and Webster, R. G. (1989). Biologic potential of amantadine-resistant influenza A virus in an avian model. J. Infect. Dis. 159, 1050–1056.

Belshe, R. B., Burk, B., Newman, F., Cerruti, R. L., and Sim, I. S. (1989). Resistance of influenza A virus to amantadine and rimantadine: results of one decade of surveillance. J. Infect. Dis. 159, 430–435.

Betts, R. F., Treanor, J. J., Graman, P. S., Bentley, D. W., and Dolin, R. (1987). Antiviral agents to prevent or treat influenza in the elderly. J. Resp. Dis. 8(Suppl), S56-S59.

Boivin, G., Coulombe, Z., and Wat, C. (2003). Quantification of the influenza virus load by real-time polymerase chain reaction in nasopharyngeal swabs of patients treated with oseltamivir. J. Infect. Dis. 188, 578–580.

Boivin, G., Goyette, N., Hardy, I., Aoki, F. Y., Wagner, A., and Trottier, S. (2000). Rapid antiviral effect of inhaled zanamivir in the treatment of naturally occurring influenza in otherwise healthy adults. J. Infect. Dis. 181, 1471–1474.

Bowles, S. K., Lee, W., Simor, A. E., Vearncombe, M., Loeb, M., Tamblyn, S., Fearon, M., Li, Y., and McGeer, A. (2002). Use of oseltamivir during influenza outbreaks in Ontario nursing homes, 1999–2000. J. Amer. Geriatrics Soc. 50, 608–616.

Brady, M. T., Sears, S. D., Pacini, D. L., Samorodin, R., DePamphilis, J., Oakes, M., Soo, W., and Clements, M. L. (1990). Safety and prophylactic efficacy of low-dose rimantadine in adults during an influenza A epidemic. Antimicrob Agents Chemother. 34, 1633–1636.

Bricaire, F., Hannoun, C., and Boissel, J. P. (1990). Prevention of influenza A: Effectiveness and tolerance of rimantadine hydrochloride. Presse Medicale 19, 69–72.

Bright, R. A., Medina, M. J., Xu, X. et al. (2005) Incidence of adamantadine resistance among influenza A (H3N2) viruses isolated worldwide from 1994 to 2005: a cause for concern. The Lancet. 366: 1175–81.

Browne, M. J., Moss, M. Y., and Boyd, M. R. (1983). Comparative activity of amantadine and ribavirin against influenza virus in vitro: possible clinical relevance. Antimicrob Agents Chemother. 23, 503–505.

Bukrinskaya, A. G., Vorkunova, N. K., Kornilayeva, G. V., Narmanbetova, R. A., and Vorkunova, G. K. (1982a). Influenza virus uncoating in infected cells and effect of rimantadine. J. Gen. Virol. 60, 49–59.

Bukrinskaya, A. G., Vorkunova, N. K., and Pushkarskaya, N. L. (1982b). Uncoating of a rimantadine-resistant variant of influenza virus in the presence of rimantadine. J. Gen. Virol. 60, 61–66.

Burlington, D. B., Meiklejohn, G., and Mostow, S. R. (1982). Anti-influenza A virus activity of amantadine hydrochloride and rimantadine hydrochloride in ferret tracheal ciliated epithelium. Antimicrob Agents Chemother. 21, 794–799.

Burmeister, W. P., Ruigrok, R. W., and Cusack, S. (1992). The 2. 2 A resolution crystal structure of influenza B neuraminidase and its complex with sialic acid. EMBO J. 11(1), 49–56.

Calfee, D. P., Peng, A. W., Cass, L., Lobo, M., and Hayden, F. G. (1999). Safety and efficacy of intravenous zanamivir in preventing experimental human influenza A virus infection. Antimicrob Agents Chemother 43, 1616–1620.

Carr, J., Ives, J., Kelly, L., Lambkin, R., Oxford, J., Mendel, D., Tai, L., and Roberts, N. (2002). Influenza virus carrying neuraminidase with reduced sensitivity to oseltamivir carboxylate has altered properties in vitro and is compromised for infectivity and replicative ability in vivo. Antiviral Res. 54, 79–88.

Cheam, A. L., Barr, I. G., Hampson, A. W., Mosse, J., and Hurt, A. C. (2004). In vitro generation and characterisation of an influenza B variant with reduced sensitivity to neuraminidase inhibitors. Antiviral Res. 63, 177–181.

Chik, K. W., Li, C. K., Chan, P. K. S., Shing, M. M. K., Lee, V., Tam, J. S., and Yuen, P. M. (2005). Oseltamivir prophylaxis during the influenza season in a paediatric cancer centre: prospective observational study. Hong Kong Med. J. 10, 103–106.

Chizhmakov, I. V., Geraghty, F. M., Ogden, D. C., Hayhurst, A., Antoniou, M., and Hay, A. J. (1996). Selective proton permeability and pH regulation of the influenza virus M2 channel expressed in mouse erythroleukaemia cells. J. Physiol. 494, 329–336.

Chotpitayasunondh, T., Ungchusak, K., Hanshaoworakul, W., Chunsuthiwat, S., Sawanpanyalert, P., Lok[hati, R., Lochindarat, S., Srisan, P., Suwan, P., Osotthanakorn, Y., Anantasetagood, T., Kanjanawasri, S., Tanupattarachai, S., Weerakul, J., Chaiwirattana, R., Maneerattanaporn, M., Poolsavatkitikool, R., Chokephaibulkit, K., Apisarnthanarak, A., and Dowell, S. (2005). Human disease from influenza A (H5N1), Thailand, 2004. Emerg. Infect. Dis. 11, 201–209.

Clover, R. D., Waner, J. L., Becker, L., and Davis, A. (1991). Effect of rimantadine on the immune response to influenza A infections. J. Med. Virol. 34, 68–73.

Colman, P. M. (1989). Influenza virus neuraminidase: Enzyme and antigen. In The Influenza Viruses, R. M. Krug, ed. (New York: Plenum Press), pp. 175–218.

Cooper, N. J., Sutton, A. J., Abrams, K. R., Wailoo, A., Turner, D., and Nicholson, K. G. (2003). Effectiveness of neuraminidase inhibitors in treatment and prevention of influenza A and B: systematic review and meta-analyses of randomised controlled trials. BMJ. 326, 1235–1239.

Couch, R. B. and Jackson, G. G. (1976). Antiviral agents in influenza – summary of Influenza Workshop VIII. J. Infect. Dis. 134, 516–527.

Couch, R. B., Kasel, J. A., Glezen, W. P., Cate, T. R., Six, H. R., Taber, L. H., Frank, A. L., Greenberg, S. B., Zahradnik, J. M., and Keitel, W. A. (1986). Influenza: its control in persons and populations. J. Infect. Dis. 153, 431–440.

Crawford, S. A., Clover, R. D., Abell, T. D., Ramsey, C. N. J., Glezen, P., and Couch, R. B. (1988). Rimantadine prophylaxis in children: a follow-up study. Pediatric Infect. Dis. J. *7*, 379–383.

Davies, W. L., Grunert, R. R., Haff, R. F., McGahen, J. W., Neumayer, E. M., Paulshock, M., Watts, J. C., Wood, T. R., Hermann, E. C., and Hoffmann, C. E. (1964). Antiviral Activity of 1-Adamantanamine (Amantadine). Science *144*, 862.

Dolin, R., Reichman, R. C., Madore, H. P., Maynard, R., Linton, P. N., and Webber-Jones, J. (1982). A controlled trial of amantadine and rimantadine in the prophylaxis of influenza A infection. New Eng. J. Med. *307*, 580–584.

Englund, J. A., Champlin, R. E., Wyde, P. R., Kantarjian, H., Atmar, R. L., Tarrand, J. J., Yousuf, H., Regnery, H., Klimov, A. I., Cox, N., and Whimbey, E. (1998). Common emergence of amantadine and rimantadine resistant influenza A viruses in symptomatic immunocompromised adults. Clin. Infect. Dis. *26*, 1418–1424.

Fenton, R. J., Bessell, C., Spilling, C. R., and Potter, C. W. (1977). The effects of peroral or local aerosol administration of 1-aminoadamantane hydrochloride (amantadine hydrochloride) on influenza infections of the ferret. J. Antimicrob. Chemother. *3*, 463–472.

Ferguson, N. M., Mallett, S., Jackson, H., Roberts, N., and Ward, P. (2003). A population-dynamic model for evaluating the potential spread of drug-resistant influenza virus infections during community-based use of antivirals. J. Antimicrob. Chemother. *51*, 977–990.

Fleming, D. M. (2003). Zanamivir in the treatment of influenza. Expert Opin. Pharmacother. *4*, 799–805.

Fritz, R. S., Hayden, F. G., Calfee, D. P., Cass, L., Peng, A. W., Alvord, G., Strober, W., and Straus, S. E. (1999). Nasal cytokine and chemokine responses in experimental influenza A virus infection: results of a placebo-controlled trial of intravenous zanamivir treatment. J. Infect. Dis. *180*, 586–593.

Galbraith, A. W., Oxford, J. S., Schild, G. C., and Watson, G. I. (1969b). Study of 1-adamantanamine hydrochloride used prophylactically during the Hong Kong influenza epidemic in the family environment. Bulletin of the World Health Organization *41*, 677–682.

Galbraith, A. W., Oxford, J. S., Schild, G. C., and Watson, G. I. (1969a). Protective effect of 1-adamantanamine hydrochloride on influenza A2 infections in the family environment: a controlled double- blind study. Lancet *2*, 1026–1028.

Galegov, G. A., Pushkarskaya, N. L., Obrosova-Serova, N. P., and Zhdanov, V. M. (1977). Combined action of ribavirin and rimantadine in experimental myxovirus infection. Experientia *33*, 905–906.

Gani, R., Hughes, H., Fleming, D., Griffin, T., Medlock, J., Leach, S. (2005). Potential impact of antiviral drug use during influenza pandemic. Emerging Infectious Diseases, 11(9): 1355–62.

Govorkova, E. A., Leneva, I. A., Goloubeva, O. G., Bush, K., and Webster, R. G. (2001). Comparison of efficacies of RWJ-270201, zanamivir, and oseltamivir against H5N1, H9N2, and other avian influenza viruses. Antimicrob Agents Chemother. *45*, 2723–2732.

Govorkova, E. A., Fang, H. B., Tan, M., and Webster, R. G. (2004). Neuraminidase inhibitor-rimantadine combinations exert additive and synergistic anti-influenza virus effects in MDCK cells. Antimicrob. Agents Chemother. *48*, 4855–4863.

Gravenstein, S. and Davidson, H. E. (2002). Current strategies for management of influenza in the elderly population. Clin. Infect. Dis. *35*, 729–737.

Gravenstein, S., Drinka, P., Osterweil, D., Schilling, M, McElhaney, J. E., Elliott, M, Hammond, J., Keene, O., Krause, P., and Flack, N. A multicenter prospective double-blind ramdomized controlled trial comparing the relative safety and efficacy of zanamivir to rimantadine for nursing home influenza outbreak control. Abstracts of the 40th Interscience Conference on Antimicrobial Agents and Chemotherapy, Toronto, Canada,

September 17–20, 2000, 270, Abst #1155. 2000.

Greengard, O., Poltoratskaia, N., Leikina, E., Zimmerberg, J., and Moscona, A. (2000). The anti-influenza virus agent 4-GU-DANA (zanamivir) inhibits cell fusion mediated by human parainfluenza virus and influenza virus HA. J. Virol. *74*, 11108–11114.

Griffin, A. D., Perry, A. S., and Fleming, D. M. (2001). Cost-effectiveness analysis of inhaled zanamivir in the treatment of influenza A and B in high-risk patients. Pharmacoeconomics *19*, 293–301.

Gubareva, L. V., Bethell, R. C., Penn, C. R., and Webster, R. G. (1996). *In vitro* characterization of 4-guanidino-Neu5Ac2en-resistant mutants of influenza A virus. In Options for the Control of Influenza III, L. E. Brown, A. W. Hampson, and R. G. Webster, eds. (Amsterdam, The Netherlands: Elsevier Science), pp. 753–760.

Gubareva, L. V., Hayden, F. G., and Kaiser, L. (2000). Influenza virus neuraminidase inhibitors. Lancet *355*, 827–835.

Gubareva, L. V., Kaiser, L., Matrosovich, M. N., Soo-Hoo, Y., and Hayden, F. G. (2001). Selection of influenza virus mutants in experimentally infected volunteers treated with oseltamivir. J. Infect. Dis. *183*, 523–531.

Gubareva, L. V., Matrosovich, M. N., Brenner, M. K., Bethell, R., and Webster, R. G. (1998). Evidence for zanamivir resistance in an immunocompromised child infected with influenza B virus. J. Infect. Dis. *178*, 1257–1262.

Gubareva, L. V., Penn, C. R., and Webster, R. G. (1995). Inhibition of replication of avian influenza viruses by the neuraminidase inhibitor 4-guanidino-2, 4-dideoxy-2, 3-dehydro-N-acetylneuraminic acid. Virology *212*, 323–330.

Gubareva, L. V., Robinson, M. J., Bethell, R. C., and Webster, R. G. (1997). Catalytic and framework mutations in the neuraminidase active site of influenza viruses that are resistant to 4-guanidino-Neu5Ac2en. J. Virol. *71*, 3385–3390.

Gubareva, L. V., Webster, R. G., and Hayden, F. G. (2002). Detection of influenza virus resistance to neuraminidase inhibitors by an enzyme inhibition assay. Antiviral Res. 53(1):47–61.

Gubareva, L. V. (2004). Molecular mechanisms of influenza virus resistance to neuraminidase inhibitors. Virus Res. *103*, 199–203.

Hall, C. B., Dolin, R., Gala, C. L., Markovitz, D. M., Zhang, Y. Q., Madore, P. H., Disney, F. A., Talpey, W. B., Green, J. L., Francis, A. B., and Pichichero, M. E. (1987). Children with influenza A infection: treatment with rimantadine. Pediatrics *80*, 275–282.

Hay, A. J. (1996). Amantadine and Rimantadine – Mechanisms. In Antiviral Drug Resistance, D. D. Richman, ed. John Wiley & Sons Ltd), pp. 43–58.

Hay, A. J. (1992). The action of adamantanamines against influenza A viruses: inhibition of the M2 ion channel protein. Seminars in Virol. *3*, 21–30.

Hay, A. J., Wolstenholme, A. J., Skehel, J. J., and Smith, M. H. (1985). The molecular basis of the specific anti-influenza action of amantadine. EMBO J. *4*, 3021–3024.

Hayden, F. G. (1996b). Combination antiviral therapy for respiratory virus infections. Antiviral Res. *29*, 45–48.

Hayden, F. G. (1986b). Combinations of antiviral agents for treatment of influenza virus infections. J. Antimicrob. Chemother. *18 Suppl B*, 177–183.

Hayden, F. G. (2001). Perspectives on antiviral use during pandemic influenza. Phil. Trans. Roy. Soc. Lond. *356*, 1877–1884.

Hayden, F. G. (1996a). Amantadine and rimantadine – clinical aspects. In Antiviral Drug Resistance, D. D. Richman, ed. John Wiley & Sons Ltd), pp. 59–77.

Hayden, F. G. (1986a). Animal models of influenza virus infection for evaluation of antiviral agents. In Experimental Models in Antimicrobial Chemotherapy, O. Zak and M. A. Sande, eds. (London: Academic Press), pp. 353–371.

Hayden, F. G. and Aoki, F. Y. (2005). Influenza neuraminidase inhibitors. In Antimicrobial therapy and vaccines, V. L. Yu, G. Edwards, P. S. McKinnon, C. Peloquin, and G. D. Morse, eds. (Pittsburg, PA: ESun Technologies, LLC), pp. 773–789.

Hayden, F. G. and Aoki, F. Y. (1999). Amantadine, Rimantadine, and Related Agents. In Antimicrobial Therapy and Vaccines, V. L. Yu, T. C. Merigan, N. J. White, and S. Barriere, eds. (Baltimore, MD: Williams & Wilkins), pp. 1344–1365.

Hayden, F. G., Atmar, R. L., Schilling, M., Johnson C, Poretz, D., Parr, D., Huson, L., Ward, P., and Mills, R. G. (1999a). Use of the selective oral neuraminidase inhibitor oseltamivir to prevent influenza. N. Engl. J. Med. 341, 1336–1343.

Hayden, F. G., Belshe, R., Villanueva, C., Lanno, R., Hughes, C., Small, I., Dutkowski, R., Ward, P., and Carr, J. (2004). Management of influenza in households: a prospective, randomized comparison of oseltamivir treatment with or without post-exposure prophylaxis. J. Infect. Dis. 189, 440–449.

Hayden, F. G., Belshe, R. B., Clover, R. D., Hay, A. J., Oakes, M. G., and Soo, W. (1989). Emergence and apparent transmission of rimantadine-resistant influenza A virus in families. New Eng. J. Med. 321, 1696–1702.

Hayden, F. G., Cote, K. M., and Douglas, R. G. J. (1980). Plaque inhibition assay for drug susceptibility testing of influenza viruses. Antimicrob Agents Chemother. 17, 865–870.

Hayden, F. G., Gubareva, L. V., Monto, A. S., Klein, T., Elliott, M., Hammond, J., Sharp, S., and Ossi, M. (2000a). Inhaled zanamivir for preventing influenza in families. New Eng. J. Med. 343, 1282–1289.

Hayden, F. G., Jennings, L., Robson, R., Schiff, G., Jackson, H., Rana, B., McClelland, G., Ipe, D., Roberts, N., and Ward, P. (2000b). Oral oseltamivir in human experimental influenza B infection. Antiviral Ther. 5, 205–213.

Hayden, F. G. and Monto, A. S. (1986). Oral rimantadine hydrochloride therapy of influenza A virus H3N2 subtype infection in adults. Antimicrob Agents Chemother. 29, 339–341.

Hayden, F. G., Osterhaus, A. D. M. E., Treanor, J. J., Fleming, D. M., Aoki, F. Y., Nicholson, K. G., Bohnen, A. M., Hirst, H. M., Keene, O., and Wightman, K. (1997). Efficacy and safety of the neuraminidase inhibitor zanamivir in the treatment of influenza virus infections. N. Engl. J. Med. 337, 874–879.

Hayden, F. G., Schlepushkin, A. N., and Pushkarskaya, N. L. (1984). Combined interferon-alpha 2, rimantadine hydrochloride, and ribavirin inhibition of influenza virus replication in vitro. Antimicrob Agents Chemother. 25, 53–57.

Hayden, F. G., Sperber, S. J., Belshe, R. B., Clover, R. D., Hay, A. J., and Pyke, S. (1991). Recovery of drug-resistant influenza A virus during therapeutic use of rimantadine. Antimicrob Agents Chemother. 35, 1741–1747.

Hayden, F. G., Treanor, J. J., Betts, R. F., Lobo, M., Esinhart, J. D., and Hussey, E. K. (1996). Safety and efficacy of the neuraminidase inhibitor GG167 in experimental human influenza. J. Amer. Med. Assoc. 275, 295–299.

Hayden, F. G., Treanor, J. J., Fritz, R. S., Lobo, M., Betts, R., Miller, M., Kinnersley, N., Mills, R. G., Ward, P., and Straus, S. E. (1999b). Use of the oral neuraminidase inhibitor oseltamivir in experimental human influenza. J. Amer. Med. Assoc. 282, 1240–1246.

Herlocher, M. L., Carr, J., Ives, J., Elias, S., Truscon, R., Roberts, N., and Monto, A. S. (2002). Influenza virus carrying an R292K mutation in the neuraminidase gene is not transmitted in ferrets. Antiviral Res. 54, 99–111.

Herlocher, M. L., Truscon, R., Elias, S., Yen, H. L., Roberts, N. A., Ohmit, S. E., and Monto, A. S. (2004). Influenza viruses resistant to the antiviral drug oseltamivir: transmission studies in ferrets. J. Infect. Dis. 190, 1627–1630.

Herlocher, M. L., Truscon, R., Fenton, R., Klimov, A., Elias, S., Ohmit, S. E., and Monto, A. S. (2003). Assessment of development of resistance to antivirals in the ferret model of influenza virus infection. J. Infect. Dis. 188, 1355–1361.

Hien, T. T., Liem, N. T., Dung, N. T., San, L. T., Mai, P. P., Chau, N., Suu, P. T., Dong, V. C., Mai, L. T. Q., Thi, N. T., Khoa, D. B., Phat, P. L., Truong, N. T., Long, H. T., Tung, C. V., Giang, L. T., Tho, N. D., Nga, L. H., Tien, N. T. K., San, L. H., Tuan, L. V., Dolecek, C.,

Thayaparan, T. T., de Jong, M., Schultsz, C., Cheng, P., Lim, W., and Horby, P. (2004). Avian influenza A (H5N1) in 10 patients in Vietnam. New Eng. J. Med. *350*, 1179–1188.

Houck, P., Hemphill, M., LaCroix, S., Hirsh, D., and Cox, N. (1995). Amantadine-resistant influenza A in nursing homes. Identification of a resistant virus prior to drug use. Arch. Intern. Med. *155*, 533–537.

Hughes, M. T., McGregor, M., Suzuki, T., Suzuki, Y., and Kawaoka, Y. (2001). Adaptation of influenza A viruses to cells expressing low levels of sialic acid leads to loss of neuraminidase activity. J. Virol. *75*, 3766–3770.

Hurt, A. C., McKimm-Breschkin, J. L., McDonald, M., Barr, I. G., Komadina, N., and Hampson, A. W. (2004). Identification of a human influenza type B strain with reduced sensitivity to neuraminidase inhibitor drugs. Virus Res. *103*, 205–211.

Ison, M. G., Gnann, J. W. Jr., Nagy-Agren, S., Treanor, J., Paya, C., Steigbigel, R., Elliott, M., Weiss, H. L., and Hayden, F. G. Safety and efficacy of nebulized zanamivir in hospitalized patients with serious influenza. Antiviral Ther. 8, 183–190. 2003.

Ison, M. G., Gubareva, L., Atmar, R. L., Treanor, J., and Hayden, F. G. (2005a). Recovery of drug-resistant influenza from immunocompromised patients: A case series. J. Infect. Dis. *In press*.

Ison, M. G., Mishin, V., Braciale, T. J., Hayden, F. G., and Gubareva, L. (2005b). Comparative activities of oseltamivir and A-322278 in immunocompetent and immunocompromised murine models for influenza virus infection. J. Infect. Dis. *in revision*.

Ives, J. A. L., Carr, J. A., Mendel, D. B., Tai, C. Y., Lambkin, R., Kelly, L., Oxford, J. S., Hayden, F. G., and Roberts, N. A. (2002). The H274Y mutation in the influenza A/H1N1 neuraminidase active site following oseltamivir phosphate treatment leave virus severely compromised both in vitro and in vivo. Antiviral Res. *55*, 307–317.

Jackson, D., Barclay, W., and Zurcher, T. (2005). Characterization of recombinant influenza B viruses with key neuraminidase inhibitor resistance mutations. J. Antimicrob. Chemother. *55*, 162–169.

Jackson, G. G., Muldoon, R. L., and Akers, L. W. (1963). Serological evidence for prevention of influenzal infection in volunteers by an anti-influenzal drug adamantanamine hydrochloride. Antimicrob Agents Chemother. *3*, 703–707.

Jackson, H. C., Roberts, N., Wang, Z., and Belshe, R. (2000). Management of influenza Use of new antivirals and resistance in perspective. Clin. Drug Invest *20*, 447–454.

Johnston, S., Ferrero, F., Garcia, M., and Dutkowski, R. (2005). Oral oseltamivir improves pulmonary function and reduces exacerbation frequency for influenza-infected children with asthma. Pediatric Infect. Dis. J. *24*, 225–232.

Johny, A. A., Clark, A., Price, N., Carrington, D., Oakhill, A., and Marks, D. I. (2002). The use of zanamivir to treat influenza A and B infection after allogeneic stem cell transplantation. Bone Marrow Transplantation. *29*, 113–115.

Kaiser, L., Henry, D., Flack, N., Keene, O., and Hayden, F. G. (2000a). Short-term treatment with zanamivir to prevent influenza: results of a placebo-controlled study. Clin. Infect. Dis. *30*, 587–589.

Kaiser, L., Keene, O. N., Hammond, J., Elliott, M., and Hayden, F. G. (2000b). Impact of zanamivir on antibiotics use for respiratory events following acute influenza in adolescents and adults. Arch. Intern. Med. *160*, 3234–3240.

Kaiser, L., Wat, C., Mills, T., Mahoney, P., Ward, P., and Hayden, F. (2003). Impact of oseltamivir treatment on influenza-related lower respiratory tract complications and hospitalizations. Arch. Intern. Med. *163*, 1667–1672.

Kati, W. M., Montgomery, D., Carrick, R., Gubareva, L., Maring, C., McDaniel, K., Steffy, K., Molla, A., Hayden, F., Kempf, D., and Kohlbrenner, W. (2002). In vitro characterization of A-315675, a highly potent inhibitor of A and B strain influenza virus neuraminidases and influenza virus replication. Antimicrobial Agents & Chemotherapy. *46*, 1014–1021.

Kati, W. M., Saldivar, A. S., Mohamadi, F., Sham, H. L., Laver, W. G., and Kohlbrenner, W. E. (1998). GS4071 is a slow-binding inhibitor of influenza neuraminidase from both A and B strains. Biochem. Biophys. Res. Commun. *244*, 408–413.

Kiso, M., Mitamura, K., Sakai-Tagawa, Y., Shiraishi, K., Kawakami, C., Kimura, K., Hayden, F. G., Sugaya, N., and Kawaoka, Y. (2004). Resistant influenza A viruses in children treated with oseltamivir: descriptive study. Lancet *364*, 759–765.

Kitamoto, O. (1971). Therapeutic effectiveness of amantadine hydrochloride in naturally occurring Hong Kong influenza. Jpn J. Tuberc Chest Dis. *17*, 1–17.

Klimov, A. I., Rocha, E., Hayden, F. G., Shult, P. A., Roumillat, L. F., and Cox, N. J. (1995). Prolonged shedding of amantadine-resistant influenzae A viruses by immunodeficient patients: detection by polymerase chain reaction-restriction analysis. J. Infect. Dis. *172*, 1352–1355.

Koopmans, M., Wilbrink, B., Conyn, M., Natrop, G., van der Nat, H., Vennema, H., Meijer, A., van Steenbergen, J., Fouchier, R., Osterhaus, A., and Bosman, A. (2004). Transmission of H7N7 avian influenza A virus to human beings during a large outbreak in commercial poultry farms in the Netherlands. The Lancet *363*, 587–593.

Kovacs, F. A. and Cross, T. A. (1997). Transmembrane four-helix bundle of influenza A M2 protein channel: structural implications from helix tilt and orientation. Biophys. J. *73*, 2511–2517.

La Rosa, A. M., Malik, S., Englund, J. A., Couch, R., Raad, I. I., Rolston, K. V., Jacobson, K. L., Kontoyiannis, D. P., and Whimbey, E. Influenza A in hospitalized adults with leukemia and hematopoietic stem call Transplant. (HSCT) recipients; risk factors for progression to pneumonia. Abstracts of the 39th Annual Meeting of the Infectious Diseases Society of America, San Francisco, CA, October 25–28, 2001, 111, Abst #418. 2001.

Lalezari, J. P., Elliott, M., and Keene, O. Zanamivir for the treatment of influenza A and B infection in high-risk patients. Arch. Internal Med. 161, 212–217. 2001.

Lamb, R. A., Zebedee, S. L., and Richardson, C. D. (1985). Influenza virus M2 protein is an integral membrane protein expressed on the infected-cell surface. Cell. 40(3):627–33.

Lee, C., Loeb, M., Phillips, A., Nesbitt, J., Smith, K., Fearon, M., McArthur, M. A., Mazzulli, T., Li, Y., and McGreer, A. (2000). Zanamivir use during transmission of amantadine-resistant influenza A in a nursing home. Infect. Control Hospit. Epidem. *21*, 700–704.

Leigh, M. W., Connor, R. J., Kelm, S., Baum, L. G., and Paulson, J. C. (1995). Receptor specificity of influenza virus influences severity of illness in ferrets. Vaccine. 13(15):1468–73.

Leneva, I. A., Roberts, N., Govorkova, E. A., Goloubeva, O. G., and Webster, R. G. (2000). The neuraminidase inhibitor GS4104 (oseltamivir phosphate) is efficacious against A/Hong Kong/156/97 (H5N1) and A/Hong Kong/1074/99 (H9N2) influenza viruses. Antiviral Res. *48*, 101–115.

Libow, L. S., Neufeld, R. R., Olson, E., Breuer, B., and Starer, P. (1996). Sequential outbreak of influenza A and B in a nursing home: efficacy of vaccine and amantadine. J. Amer. Geriatrics Soc. *44*, 1153–1157.

Little, J. W., Hall, W. J., Douglas, R. G. J., Hyde, R. W., and Speers, D. M. (1976). Amantadine effect on peripheral airways abnormalities in influenza. A study in 15 students with natural influenza A infection. Annal. Internal Med. *85*, 177–182.

Little, J. W., Hall, W. J., Douglas, R. G. J., Mudholkar, G. S., Speers, D. M., and Patel, K. (1978). Airway hyperreactivity and peripheral airway dysfunction in influenza A infection. Amer. Rev. Resp. Dis. *118*, 295–303.

Lubeck, M. D., Schulman, J. L., and Palese, P. (1978). Susceptibility of influenza A viruses to amantadine is influenced by the gene coding for M protein. J. Virol. *28*, 710–716.

Machado CM, Boas, L., Mendes, A., da Rocha, I., Sturaro, D., Dulley, F., and Pannuti, C. (2004). Use of oseltamivir to control influenza complications after bone marrow transplantation. Bone Marrow Transplan. *34*, 111–114.

Macken, C., Lu, H., Goodman, J., and Boykin, L. (2001). The value of a database in surveillance and vaccine selection. In Options for the Control of Influenza IV, N. Osterhaus and Cox & Hampson A. W., eds. (Amsterdam: Elsevier Science), pp. 103–106.

Madren, L. K., Shipman, C. Jr., and Hayden, F. G. (1995). In vitro inhibitory effects of combinations of anti-influenza agents. Antiviral Chemistry & Chemotherapy 6, 109–113.

Mäkelä, M. J., Pauksens, K., Rostila, R., Fleming, D. M., Man, C. Y., Keene, O. N., and Webster, A. (2000). Clinical efficacy and safety of the orally inhaled neuraminidase inhibitor zanamivir in the treatment of influenza: a randomized, double-blind, placebo-controlled European study. J. Infect. 40, 42–48.

Marozin, S., Gregory, V., Cameron, K., Bennett, M., Valette, M., Aymard, M., Foni, E., Barigazzi, G., Lin, Y., and Hay, A. (2002). Antigenic and genetic diversity among swine influenza A H1N1 and H1N2 viruses in Europe. J. Gen. Virol. 83, 735–745.

Martin, C., Mahoney, P., and Ward, P. Oral oseltamivir reduces febrile illness in patients considered at high risk of influenza complications. Osterhaus, A., Cox, N., and Hampson, A. Options for the Control of Influenza IV. 807–811. 2001. London, Excerpta Medica. International Congress Series 1219.

Martin, K. and Helenius, A. (1991). Nuclear transport of influenza virus ribonucleoproteins: the viral matrix protein (M1) promotes export and inhibits import. Cell 67, 117–130.

Mast, E. E., Harmon, M. W., Gravenstein, S., Wu, S. P., Arden, N. H., Circo, R., Tyszka, G., Kendal, A. P., and Davis, J. P. (1991). Emergence and possible transmission of amantadine-resistant viruses during nursing home outbreaks of influenza A (H3N2). Amer. J. Epidemiol. 134, 988–997.

Matrosovich, M., Matrosovich, T., Carr, J., Roberts, N. A., and Klenk, H. D. (2003). Overexpression of the alpha-2, 6-sialyltransferase in MDCK cells increases influenza virus sensitivity to neuraminidase inhibitors. J. Virol. 77, 8418–8425.

Matrosovich, M. N., Matrosovich, T. Y., Gray, T., Roberts, N. A., and Klenk, H. D. (2004a). Human and avian influenza viruses target different cell types in cultures of human airway epithelium. Proceedings of the National Academy of Sciences of the United States of America. 101(13):4620–4.

Matrosovich, M. N., Matrosovich, T. Y., Gray, T., Roberts, N. A., and Klenk, H. D. (2004b). Neuraminidase is important for the initiation of influenza virus infection in human airway epithelium. J. Virol. 78, 12665–12667.

McClellan, K. and Perry, C. M. (2001). Oseltamivir: a review of its use in influenza. Drugs. 61, 263–283.

McCullers, J. A. (2004). Effect of antiviral treatment on the outcome of secondary bacterial pneumonia after influenza. J. Infect. Dis. 190, 519–526.

McKimm-Breschkin, J., Trivedi, T., Hampson, A., Hay, A., Klimov, A., Tashiro, M., Hayden, F., and Zambon, M. (2003). Neuraminidase sequence analysis and susceptibilities of influenza virus clinical isolates to zanamivir and oseltamivir. Antimicrob Agents Chemother. 47, 2264–2272.

McKimm-Breschkin, J. L. (2000). Resistance of influenza viruses to neuraminidase inhibitors – a review. Antiviral Res. 47, 1–17.

McKimm-Breschkin, J. L., McDonald, M., Blick, T. J., and Colman, P. M. (1996). Mutation in the influenza virus neuraminidase gene resulting in decreased sensitivity to the neuraminidase inhibitor 4-guanidino-Neu5Ac2en leads to instability of the enzyme. Virology 225, 240–242.

Mendel, D. B., Tai, C. Y., Escarpe, P. A., Li, W., Sidwell, R. W., Huffman, J. H., Sweet, C., Jakeman, K. J., Merson, J., Lacy, S. A., Lew, W., Williams, M. A., Zhang, L., Chen, M. S., Bischofberger, N., and Kim, C. U. (1998). Oral administration of a prodrug of the influenza virus neuraminidase inhibitor GS4071 protects mice and ferrets against influenza infection. Antimicrob Agents Chemother. 42, 640–646.

Mishin, V., Novikov, D. V., Hayden, F. G., and Gubareva, L. (2005). Effect of hemagglutinin glycosylation on influenza virus susceptibility to neuraminidase inhibitors. J. Virol. 791 12416–12424.

MIST Study Group (1998). Randomized trial of efficacy and safety of inhaled zanamivir in treatment of influenza A and B virus infections. Lancet *352*, 1877–1881.

Molla, A., Kati, W., Carrick, R., Steffy, K., Shi, Y., Montgomery, D., Gusick, N., Stoll, V. S., Stewart, K. D., Ng, T. I., Maring, C., Kempf, D. J., and Kohlbrenner, W. (2002). In vitro selection and characterization of influenza A (A/N9) virus variants resistant to a novel neuraminidase inhibitor, A-315675. J. Virol. *76*, 5380–5386.

Monto, A. S., Gunn, R. A., Bandyk, M. G., and King, C. L. (1979). Prevention of Russian influenza by amantadine. J. Amer. Med. Assoc. *241*, 1003–1007.

Monto, A. S., Macken, C., McKimm-Breschkin, J., Hampson, A., Hay, A., Klimov, A., Tashiro, M., Webster, R., Aymard, M., Hayden, F. G., and Zambon, M. (2005). Influenza viruses resistant to the neuraminidase inhibitors detected during the first three years of their use. (*Submitted for publication.*)

Monto, A. S., Ohmit, S. E., Hornbuckle, K., and Pearce, C. L. (1995). Safety and efficacy of long-term use of rimantadine for prophylaxis of type A influenza in nursing homes. Antimicrob Agents Chemother. *39*, 2224–2228.

Monto, A. S., Robinson DP, Herlocher L, Hinson JM, Elliott, M., and Crisp, A. (1999a). Zanamivir in the prevention of influenza among healthy adults. J. Amer. Med. Assoc. *282*, 31–36.

Monto, A. S., Robinson, D. P., Griffin, A. D., and Edmundson, S. (2003). The effects of zanamivir on productivity in the prevention of influenza among healthy adults. J. Antimicrob. Chemother. *44*, 41 (P1).

Monto, A. S., Webster, A., and Keene, O. (1999b). Randomized, placebo-controlled studies of inhaled zanamivir in the treatment of influenza A and B: pooled efficacy analysis. J. Antimicrob. Chemother. *44*, 23–29.

Moscona, A. (2005) Neuraminidase Inhibitors for Influenza. The New England Journal of Medicine 2005 Sept 29; 353(13): 1363–73.

Mori, I., Liu, B., Hossain, M. J., Takakuwa, H., Daikoku, T., Nishiyama, Y., Naiki, H., Matsumoto, K., Yokochi, T., and Kimura, Y. (2002). Successful protection by amantadine hydrochloride against lethal encephalitis caused by a highly neurovirulent recombinant influenza A virus in mice. Virology. *303*, 287–296.

Muldoon, R. L., Stanley, E. D., and Jackson, G. G. (1976). Use and withdrawal of amantadine chemoprophylaxis during epidemic influenza A. Amer. Rev. Resp. Dis. *113*, 487–491.

Murphy, K., Eivindson, A., Pauksens, K., Stein, W. J., Tellier, G., Watts, R., Leophonte, P., Sharp, S. J., and Loeschel, E. (2000). Efficacy and safety of inhaled zanamivir for the treatment of influenza in patients with asthma or chronic obstructive pulmonary disease. Clin. Drug Invest *20*, 337–349.

Nafta, I., TTurcanu, A. G., Braun, I., Companetz, W., Simionescu, A., Birt, E., and Florea, V. (1970). Administration of amantadine for the prevention of Hong Kong influenza. Bull. World Health Organ. *42*, 423–427.

Nedyalkova, M. S., Hayden, F. G., Webster, R. G., and Gubareva, L. V. (2002). Accumulation of defective neuraminidase (NA) genes by influenza A viruses in the presence of NA inhibitors as a marker of reduced dependence on NA. J. Infect. Dis. 185(5):591–8.

Nichols, W. G., Guthrie, K. A., Corey, L., and Boeckh, M. (2004). Influenza infections after henatopoietic stem cell transplantation: Risk factors, mortality, and the effect of antiviral therapy. Clin. Infect. Dis. *39*, 1300–1305.

NISN (2005). Use of influenza antivirals during 2003–2004 and monitoring of neuraminidase inhibitor resistance. Wk. Epidemiol. Rec. *80*, 156.

Nordstrom, B. L., Sung, I., Suter, P., and Szneke, P. (2005). Risk of pneumonia and other complications of influenza-like illness in patients treated with oseltamivir. Curr. Med. Res. Opin. *21*, 761–768.

O'Donoghue, J. M., Ray, C. G., Terry, D. W. J., and Beaty, H. N. (1973). Prevention of nosocomial influenza infection with amantadine. Amer. J. Epidemiol. *97*, 276–282.

Ohuchi, M., Ohuchi, R., Feldmann, A., and Klenk, H. D. (1997). Regulation of receptor binding affinity of influenza virus hemagglutinin by its carbohydrate moiety. J. Virol. 71(11):8377–84.

Parker, R., Loewen, N., and Skowronski, D. Canada Communicable Disease Report. 37–40. 2001.

Payler, D. K. and Purdham, P. A. (1984). Influenza A prophylaxis with amantadine in a boarding school. Lancet *1*, 502–504.

Peters, P. H., Gravenstein, S., Norwood, P., DeBock, V., VanCouter, T., Gibbens, M., VonPlanta, T., and Ward, P. (2001). Long-term use of oseltamivir for the prophylaxis of influenza in a vaccinated frail older population. J. Amer. Geriatric Soc. *49*, 1–7.

Pettersson, R. F., Hellstrom, P. E., Penttinen, K., Pyhala, R., Tokola, O., Vartio, T., and Visakorpi, R. (1980). Evaluation of amantadine in the prophylaxis of influenza A (H1N1) virus infection: a controlled field trial among young adults and high-risk patients. J. Infect. Dis. *142*, 377–383.

Pinto, L. H., Holsinger, L. J., and Lamb, R. A. (1992). Influenza virus M2 protein has ion channel activity. Cell *69*, 517–528.

Pinto, L. H., Dieckmann, G. R., Gandhi, C. S., Papworth, C. G., Braman, J., Shaughnessy, M. A., Lear, J. D., Lamb, R. A., and DeGrado, W. F. (1997). A functionally defined model for the M2 proton channel of influenza A virus suggests a mechanism for its ionáselectivity. PNAS *94*, 11301–11306.

Quilligan, J. J. J., Hirayama, M., and Baernstein, H. D. J. (1966). The suppression of A2 influenza in children by the chemoprophylactic use of Amantadine. J. Pediat. *69*, 572–575.

Reuman, P. D., Bernstein, D. I., Keefer, M. C., Young, E. C., Sherwood, J. R., and Schiff, G. M. (1989a). Efficacy and safety of low dosage amantadine hydrochloride as prophylaxis for influenza A. Antiviral Res. *11*, 27–40.

Reuman, P. D., Bernstein, D. I., Keely, S. P., Young, E. C., Sherwood, J. R., and Schiff, G. M. (1989b). Differential effect of amantadine hydrochloride on the systemic and local

immune response to influenza A. J. Med. Virol. *27*, 137–141.

Richman, D. D., Yazaki, P., and Hostetler, K. Y. (1981). The intracellular distribution and antiviral activity of amantadine. Virology *112*, 81–90.

Roberts, N. (2001). Treatment of influenza with neuraminidase inhibitors: virological implications. Phil. Transact. Roy. Soc. *356*, 1895–1897.

Rose, H. J. (1983). Use of amantadine in influenza: a second report. J. Roy. College Gen. Pract. *33*, 651–653.

Ruigrok, R. W., Hirst, E. M., and Hay, A. J. (1991). The specific inhibition of influenza A virus maturation by amantadine: an electron microscopic examination. J. Gen. Virol. *72*, 191–194.

Saito, R., Sakai, T., Sato, I., Sano, Y., Oshitani, H., Sato, M., and Suzuki, H. (2003). Frequency of amantadine-resistant influenza A viruses during two seasons featuring cocirculation of H1N1 and H3N2. J. Clin. Microbiol. *41*, 2164–2165.

Sakaguchi, T., Leser, G. P., and Lamb, R. A. (1996). The ion channel activity of the influenza virus M2 protein affects transport through the Golgi apparatus. J. Cell Biol. *133*, 733–747.

Scholtissek, C. and Faulkner, G. P. (1979). Amantadine-resistant and -sensitive influenza A strains and recombinants. J. Gen. Virol. *44*, 807–815.

Scholtissek, C. and Webster, R. G. (1998). Long-term stability of the anti-influenza A compounds – amantadine and rimantadine. Antiviral Res. *38*, 213–215.

Schulman, J. L. (1968). Effect of 1-amantadine hydrochloride (amantadine HCl) and methyl-1-adamatanethylamine hydrochloride (rimantadine HCl) on transmission of influenza virus infection in mice (33222). Proc. Soc. Exp. Biol. Med. *128*.

Sears, S. D. and Clements, M. L. (1987). Protective efficacy of low-dose amantadine in adults challenged with wild-type influenza A virus. Antimicrob Agents Chemother. *31*, 1470–1473.

Shiraishi, K., Mitamura, K., Sakai-Tagawa, Y., Goto, H., Sugaya, N., and Kawaoka, Y. (2003). High frequency of resistant viruses harboring different mutations in amantadine-treated children with influenza. J. Infect. Dis. *188*, 57–61.

Shuck, K., Lamb, R. A., and Pinto, L. H. (2000). Analysis of the Pore Structure of the Influenza A Virus M2 Ion Channel by the Substituted-Cysteine Accessibility Method. J. Virol. *74*, 7755–7761.

Sidwell, R. W. and Smee, D. F. (2000). In vitro and in vivo assay systems for study of influenza virus inhibitors. [Review] [100 refs]. Antiviral Res. 48(1):1–16.

Sidwell, R. W. and Smee, D. F. (2002). Peramivir (BCX-1812, RWJ-270201): potential new therapy for influenza. Expert Opini. Invest. Drugs *11*, 859–869.

Sidwell, R. W., Smee, D. F., Huffman, J. H., Barnard, D. L., Bailey, K. W., Morrey, J. D., and Babu, Y. S. (2001). In vivo influenza virus-inhibitory effects of the cyclopentane neuraminidase inhibitor RJW-270201. Antimicrob. Agents Chemother. *45*, 749–757.

Skoner, D. P., Gentile, D. A., Patel, A., and Doyle, W. J. (1999). Evidence for cytokine mediation of disease expression in adults experimentally infected with influenza A virus. J. Infect. Dis. *180*, 10–14.

Smith, B. J., McKimm-Breshkin, J. L., McDonald, M., Fernley, R. T., Varghese, J. N., and Colman, P. M. (2002). Structural studies of the resistance of influenza virus neuramindase to inhibitors. J. Med. Chem. *45*, 2207–2212.

Smorodintsev, A. A., Karpuhin, G. I., Zlydnikov, D. M., Malyseva, A. M., Svecova, E. G., Burov, S. A., Hramcova, L. M., Romanov, J. A., Taros, L. J., Ivannikov, J. G., and Novoselov, S. D. (1970a). The prophylactic effectiveness of amantadine hydrochloride in an epidemic of Hong Kong influenza in Leningrad in 1969. Bull. World Health Organ. *42*, 865–872.

Smorodintsev, A. A., Zlydnikov, D. M., Kiseleva, A. M., Romanov, J. A., Kazantsev, A. P., and Rumovsky, V. I. (1970b). Evaluation of amantadine in artificially induced A2 and B influenza. J. Amer. Med. Assoc. *213*, 1448–1454.

Staschke, K. A., Colacino, J. M., Baxter, A. J., Air, G. M., Bansal, A., Hornback, W. J.,

Munroe, J. E., and Laver, W. G. (1995). Molecular basis for the resistance of influenza viruses to 4-guanidino-Neu5Ac2en. Virology *214*, 642–646.

Steinhauer, D. A., Wharton, S. A., Skehel, J. J., Wiley, D. C., and Hay, A. J. (1991). Amantadine selection of a mutant influenza virus containing an acid-stable hemagglutinin glycoprotein: evidence for virus- specific regulation of the pH of glycoprotein transport vesicles. Proc. Nat. Acad. Sci. USA. *88*, 11525–11529.

Stilianakis, N. I., Perelson, A. S., and Hayden, F. G. (1998). Emergence of drug resistance during an influenza epidemic: Insights from a mathematical model. J. Infect. Dis. 863–873.

Straumanis, J. P., Tapia, M., and King, J. (2003). Influenza B infection associated with encephalitis: treatment with oseltamivir. Pediatric Infect. Dis. J. *21*, 173–175.

Sugaya, N., Yoshikawa, T., Miura, M., Ishizuka, T., Kawakami, C., and Asano, Y. (2002). Influenza encephalopathy associated with infection with human herpesvirus 6 and/or human herpesvirus 7. Clin. Infect. Dis. *34*, 461–466.

Sugrue, R. J., Bahadur, G., Zambon, M. C., Hall-Smith, M., Douglas, A. R., and Hay, A. J. (1990). Specific structural alteration of the influenza haemagglutinin. EMBO J. *9*, 3469–3476.

Sweet, C., Hayden, F. G., Jakeman, K. J., Grambas, S., and Hay, A. J. (1991). Virulence of rimantadine-resistant human influenza A (H3N2) viruses in ferrets. J. Infect. Dis. *164*, 969–972.

Sweet, C., Jakeman, K. J., Bush, K., Wagaman, P. C., McKown, L. A., Streeter, A. J., sai-Krieger, D., Chand, P., and Babu, Y. S. (2002). Oral administration of cyclopentane neuraminidase inhibitors protects ferrets against influenza virus infection. Antimicrob. Agents Chemother. *46*, 996–1004.

Takeuchi, K. and Lamb, R. A. (1994). Influenza virus M2 protein ion channel activity stabilizes the native form of fowl plague virus hemagglutinin during intracellular transport. J. Virol. 68(2):911–9.

Tamblyn, S. E. Antiviral use during influenza outbreaks in long-term care facilities.

Osterhaus, A., Cox, N., and Hampson, A. Options for the control of influenza IV. [First], 817–822. 2001. New York, Excerpta Medica. International Congress Series 1219. 9-23-2000.

Tang, Y., Zaitseva, F., Lamb, R. A., and Pinto, L. H. (2002). The Gate of the Influenza Virus M2 Proton Channel Is Formed by a Single Tryptophan Residue. J. Biol. Chem. *277*, 39880–39886.

Taylor, N. R. and von Itzstein, M. (1994). Molecular modeling studies on ligand binding to sialidase from influenza virus and the mechanism of catalysis. J. Med. Chem. *37*, 616–624.

Thompson, J., Fleet, W., Lawrence, E., Pierce, E., Morris, L., and Wright, P. (1987). A comparison of acetaminophen and rimantadine in the treatment of influenza A infection in children. J. Med. Virol. *21*, 249–255.

Tisdale, M. (2000). Monitoring of viral susceptibility: new challenges with the development of influenza NA inhibitors. Rev. Med. Virol. *10*, 45–55.

Treanor, J. J. and Hayden, F. G. (1998). Volunteer Challenge Studies. In Textbook of Influenza, K. G. Nicholson, R. G. Webster, and A. J. Hay, eds. Blackwell Science Ltd.), pp. 517–537.

Treanor, J. J., Hayden, F. G., Vrooman, P. S., Barbarash, R. A., Bettis, R., Riff, D., Singh, S., Kinnersley, N., Ward, P., and Mills, R. G. (2000). Efficacy and safety of the oral neuraminidase inhibitor oseltamivir in treating acute influenza. J. Amer. Med. Assoc. *283*, 1016–1024.

Tsunoda, A., Maassab, H. F., Cochran, K. W., and Eveland, W. C. (1965). Antiviral activity of alpha-methyl-1-adamantanemethylamine hydrochloride. Antimicrob Agents Chemother. *5*, 553–560.

Tu, Q., Pinto, L. H., Luo, G., Shaughnessy, M. A., Mullaney, D., Kurtz, S., Krystal, M., and Lamb, R. A. (1996). Characterization of inhibition of M2 ion channel activity by BL-1743, an inhibitor of influenza A virus. J. Virol. *70*, 4246–4252.

Tumpey, T. M., Garcia-Sastre, A., Mikulasova, A., Taubenberger, J. K., Swayne, D. E., Palese, P., and Basler, C. F. (2002). Existing antivirals are effective against influenza viruses with genes from the 1918 pandemic virus. Proc. Natl. Acad. Sci. USA. *99*, 13849–13854.

van Genugten, M. L. L., Heijnen, M.-L. A., and Jager, J. C. (2003). Pandemic influenza and healthcare demand in the Netherlands: scenario analysis. Emerg. Infect. Dis. *9*, 531–537.

VanVoris, L. P., Betts, R. F., Hayden, F. G., Christmas, W. A., and Douglas, R. G. J. (1981). Successful treatment of naturally occurring influenza A/USSR/77 H1N1. J. Amer. Med. Assoc. *245*, 1128–1131.

Varghese, J. N. and Colman, P. M. (1991). Three-dimensional structure of the neuraminidase of influenza virus A/Tokyo/3/67 at 2. 2 A resolution. J. Mol. Biol. *221*, 473–486.

Varghese, J. N., Epa, V. C., and Colman, P. M. (1995). Three-dimensional structure of the complex of 4-guanidino-Neu5Ac2en and influenza virus neuraminidase. Protein Sci. *4*, 1081–1087.

Varghese, J. N., Smith, P. W., Sollis, S. L., Blick, T. J., Sahasrabudhe, A., McKimm-Breschkin, J. L., and Colman, P. M. (1998). Drug design against a shifting target: a structural basis for resistance to inhibitors in a variant of influenza virus neuraminidase. Structure. *6*, 735–746.

von Itzstein, M., Wu, W. Y., Kok, G. B., Pegg, M. S., Dyason, J. C., Jin, B., Van, P. T., Smythe, M. L., White, H. F., Oliver, S. W., and et, a. (1993). Rational design of potent sialidase-based inhibitors of influenza virus replication. Nature *363*, 418–423.

Wagner, R., Matrosovich, M., and Klenk, H. D. (2002). Functional balance between haemagglutinin and neuraminidase in influenza virus infections. Rev. Med. Virol. *12*, 159–166.

Walker, J. B., Hussey, E. K., Treanor, J. J., Montalvo, A., and Hayden, F. G. (1997). Effects of the neuraminidase inhibitor zanamivir on otologic manifestations of experimental human influenza. J. Infect. Dis. *176*, 1417–1422.

Walker, J. S., Stephen, E. L., and Spertzel, R. O. (1976). Small-particle aerosols of antiviral compounds in treatment of type A influenza pneumonia in mice. J. Infect. Dis. *133 Suppl*, A140–4.

Wang, C., Lamb, R. A., and Pinto, L. H. (1995). Activation of the M2 ion channel of influenza virus: a role for the transmembrane domain histidine residue. Biophys. J. *69*, 1363–1371.

Wang, M. Z., Tai, C. Y., and Mendel, D. B. (2002). Mechanism by which mutations at his274 alter sensitivity of influenza a virus n1 neuraminidase to oseltamivir carboxylate and zanamivir. Antimicrobial Agents & Chemotherapy. *46*, 3809–3816.

Webster, R. G., Kawaoka, Y., Bean, W. J., Beard, C. W., and Brugh, M. (1985). Chemotherapy and vaccination: a possible strategy for the control of highly virulent influenza virus. J. Virol. *55*, 173–176.

Weinstock, D. M., Gubareva, L. V., and Zuccotti, G. (2003). Prolonged shedding of multidrug-resistant influenza A virus in an immunocompromised patient. New England J. Med. *348*, 867–868.

Welliver, R., Monto, A. S., Carewicz, O., Schatteman, E., Hassman, M., Hedrick, J., Huson, L., Ward, P., and Oxford, J. S. (2001). Effectiveness of Oseltamivir in preventing influenza in household contacts. J. Amer. Med. Assoc. *285*, 748–754.

Wetherall, N. T., Trivedi, T., Zeller, J., Hodges-Savola, C., McKimm-Breschkin, J. L., Zambon, M., and Hayden, F. G. (2003). Evaluation of neuraminidase enzyme assays using different substrates to measure susceptibility of influenza virus clinical isolates to neuraminidase inhibitors: report of the neuraminidase inhibitor susceptibility network. J. Clin. Microbiol. *41*, 742–750.

Whitley, R. J., Hayden, F. G., Reisinger, K., Young, N., Dutkowski, R., Ipe, D., Mills, R. G., and Ward, P. (2001). Oral oseltamivir treatment of influenza in children. Pediatric Infect. Dis. J. *20*, 127–133.

WHO (1985). Current status of amantadine and rimantadine as anti-influenza-A agents: Memorandum from a WHO meeting. Bull. World Health Organ. *63*, 51–56.

WHO Guidelines on the use of vaccines and antivirals during influenza pandemics. http://www. who. int/csr/resources/ publications/influenza/en/11_29_01_A.pdf. 2004.

Wilson, S. Z., Knight, V., Wyde, P. R., Drake, S., and Couch, R. B. (1980). Amantadine and ribavirin aerosol treatment of influenza A and B infection in mice. Antimicrob Agents Chemother. *17*, 642–648.

Woods, J. M., Bethell, R. C., Coates, J. A., Healy, N., Hiscox, S. A., Pearson, B. A., Ryan, D. M., Ticehurst, J., Tilling, J., Walcott, S. M., and et, a. (1993). 4-Guanidino-2, 4-dideoxy-2, 3-dehydro-N-acetylneuraminic acid is a highly effective inhibitor both of the sialidase (neuraminidase) and of growth of a wide range of influenza A and B viruses in vitro. Antimicrob Agents Chemother. *37*, 1473–1479.

Yamashita, M., Ohno, A., Tomozawa, T., and Yoshida, S. *R-118958, a unique anti-influenza agent*: a prodrug form of R-125489, A novel inhibitor of influenza virus neuraminidase. Abstracts of the 43rd Interscience Conference on Antimicrobial Agents and Chemotherapy, Chicago, Illinois, USA, September 14–17, 2003. 2003.

Yen, H. L., Monto, A. S., Webster, R. G., and Govorkova, E. A. (2005). Virulence may determine the necessary duration and dosage of oseltamivir treatment for highly pathogenic A/Vietnam/1203/04 (H5N1) influenza virus in mice. J. Infect. Dis. *192*, 665–672.

Younkin, S. W., Betts, R. F., Roth, F. K., and Douglas, R. G. J. (1983). Reduction in fever and symptoms in young adults with influenza A/Brazil/78 H1N1 infection after treatment with aspirin or amantadine. Antimicrob Agents Chemother. *23*, 577–582.

Zambon, M. and Hayden, F. G. (2001). Position statement: global neuraminidase inhibitor susceptibility network. Antiviral Res. *49*, 147–156.

Zebedee, S. L. and Lamb, R. A. (1988). Influenza A virus M2 protein: monoclonal antibody restriction of virus growth and detection of M2 in virions. J. Virol. 62(8):2762–72.

Zhong, Q., Husslein, T., Moore, P., Newns, D., and Pattnaik, P. (1998). The M2 channel of influenza A virus: a molecular dynamics study. FEBS Letters *434*, 265–271.

Ziegler, T., Hemphill, M. L., Ziegler, M. L. et al. (1999). Low incidence of rimantadine resistance in field isolates of influenza A viruses. journal of Infectious diseases. 1999 Oct; 180(4): 935–9.

Zlydnikov, D. M., Kubar, O. I., Kovaleva, T. P., and Kamforin, L. E. (1981). Study of rimantadine in the USSR: a review of the literature. Rev. Infect. Dis. 3, 408–421.

CHAPTER 8

Influenza vaccines: current and future strategies

Jacqueline M. Katz, Sanjay Garg, Suryaprakash Sambhara

Abstract

Vaccination is the primary method for the prevention of influenza and its complications. The continual genetic and antigenic variation that influenza viruses undergo requires constant global surveillance to identify and select new variants with epidemic potential or novel viruses with pandemic potential for inclusion in vaccines. Two general types of influenza vaccines, inactivated or live attenuated vaccines, both grown in embryonated hen's eggs, are currently licensed for use. Inactivated vaccines induce immunity to infection in 70–90% of healthy adults <65 years of age when there is a good antigenic match between vaccine and circulating virus strain, but are generally less effective in older adults. Improved vaccines against epidemic influenza and effective vaccines against potential pandemic viruses are a public health priority. New strategies for influenza vaccines include altering the dose, site, or method of delivery of inactivated vaccines, the use of adjuvants or immunomodulators to enhance immune responses, or targeting viral proteins that may promote broader, cross-protective responses. Plasmid-based reverse genetics technology may provide a more rapid approach to the generation of candidate vaccine strains, and is essential for vaccine strains derived from highly pathogenic avian viruses. Cell culture-based vaccines may improve manufacturing capacity, particularly in the event of a newly emergent pandemic threat.

INTRODUCTION

Each year in the United States, influenza viruses infect 10–20% of the population. Infection rates for epidemic influenza are highest among children. However, serious illness, hospitalization, and death are greatest among elderly persons aged ≥65 years of age, children less than 2 years of age, or persons of any age who have certain chronic medical conditions (Centers for Diseases Control and Prevention, 2004). An average of approximately 114,000 influenza-related hospitalizations occur annually in the U.S. and this estimate may be even higher in years when influenza A H3N2 viruses predominate (Centers for Disease Control and Prevention, 2004). During the 1990s, epidemics of influenza caused an average of approximately 36,000 deaths annually in the U.S (Thompson et al., 2003). More than 90% of influenza-related deaths, which are primarily a result of pneumonia or exacerbation of cardiopulmonary disorders or other chronic diseases, occur in the elderly. Vaccination is the primary method for the prevention of influenza and its complications in the community. Annual vaccination with the currently licensed inactivated trivalent vaccine is recommended for persons at increased risk for complications from influenza, persons aged 50–64 years and those with frequent contact with high-risk groups (Table 1). This chapter will provide an overview of current influenza vaccines and focus on newer strategies that may yield improved influenza vaccines for the future.

Table 8.1 Recommendations for Use of Influenza Virus Vaccine, 2004*

Population group
Persons at increased risk for complications from influenza
• Persons aged ≥65 years
• Residents of long term care facilities or those that house persons with chronic medical conditions
• Persons with chronic pulmonary or cardiovascular disease, including asthma
• Persons with chronic metabolic disease, kidney disease, hemoglobinopathy or immunosuppression
• Children and teenagers receiving long-term aspirin therapy
• Women who will be pregnant during the influenza season
• Children aged 6 –23 months
Persons aged 50–64 years
Persons who have frequent contact with high-risk persons
• Healthcare workers
• Employees of long-term care facilities
• Persons who provide care to high-risk persons, including children aged 0–23 months
• Household contacts of high risk persons, including children aged 0–23 months

* Recommendations of the Centers for Disease Control and Prevention Advisory Committee on Immunization Practices, 2004

ANTIGENIC VARIABILITY AND VACCINE STRAIN SELECTION

Influenza viruses are unique among pathogens for which vaccines have been developed because of the continual emergence of antigenic variants which requires the annual re-formulation of influenza vaccines. Human influenza A and B viruses undergo a process of variation termed antigenic drift. Antigenic drift variants arise due to an accumulation of point mutations in the viral hemagglutinin (HA) and neuraminidase (NA) genes and selection of variants with amino acid mutations that enable the virus to escape neutralization by antibody acquired from previous infections or immunization. Antigenic drift leads to the emergence of new epidemic strains and is the force that drives the continual evolution of influenza viruses in humans. Currently, influenza A viruses with HA of the H1 or H3 subtypes and NA of the N1 or N2 subtype, as well as influenza B viruses cocirculate in humans. A second form of variation which occurs only rarely is termed antigenic shift, and occurs only in influenza A viruses of which there are 16 HA (H1-H16) and 9 NA (N1-N9) different subtypes. Antigenic shift occurs when a virus with a novel HA, with or without other accompanying genes derived from an influenza virus of avian or animal origin, appears in humans. Antigenic shift may result in a pandemic if the novel virus is capable of spreading from person-to-person among a serologically naïve and susceptible population. Vaccines for the pandemic situation will be considered separately, below.

Global virological and epidemiological surveillance conducted by an international World Health Organization (WHO) network of laboratories identifies new variants with epidemic potential, as well as unusual viruses with pandemic potential (Cox et al., 1994). Viruses with variant HA molecules are identified by antigenic and genetic analyses. If a variant is detected that also reacts poorly with human sera from individuals vaccinated with the existing vaccine, and epidemiological surveillance confirms geographic spread, this variant becomes a candidate for the new season's vaccine. Currently, vaccine strain selection occurs biannually; vaccine strain selection for the northern hemisphere occurs in February, while selection of vaccine strains for the southern hemisphere occurs in September. Once an influenza A wildtype variant is selected, a reassortant vaccine strain is generated and further characterized to confirm antigenic identity with the parental strain. The realization that novel epidemic strains may emerge from China or southern Asia has increased awareness of the importance of heightened surveillance in this area. Likewise, enhanced surveillance in southeast Asia is important for pandemic preparedness since the region has been implicated as an epicenter for the emergence of future pandemic strains.

CURRENT INFLUENZA VACCINES
Inactivated vaccines
Current inactivated influenza vaccines are grown in embryonated hen's eggs. Most influenza vaccines are split-vaccines produced from detergent-treated purified influenza virus, or surface-antigen vaccines containing purified HA and NA proteins. Whole-virus vaccines are nowadays used infrequently in annual influenza vaccines as they are associated with increased adverse reactions, especially in children. Annual influenza vaccines are trivalent, containing 15μg each of two influenza A subtypes (H1N1 and H3N2) and one influenza B strain and are delivered by the intramuscular route. The segmented viral genome of influenza allows for the preparation of influenza A vaccine strains by reassortment with a donor virus, A/Puerto Rico/8/34 (PR8), which confers the property of high growth in eggs. A comparable influenza B high growth donor strain is not available and thus a wildtype B virus strain is used. Immunity induced by inactivated influenza vaccines is based primarily on the induction of neutralizing serum antibody directed against the HA of the vaccine strain. The serum antibody induced by inactivated vaccine is strain-specific and antigenic differences between the vaccine and circulating strain may reduce the efficacy of the vaccine. Inactivated vaccines induce immunity to infection in 70–90% of healthy adults <65 years of age when there is a good antigenic match between vaccine and circulating virus strains (Bridges et al., 2000; Palache, 1997). While the majority of children and young adults develop levels of post-vaccination anti-HA− antibody considered to be protective (titers≥40), older adults and persons with chronic diseases develop lower anti-HA antibody responses following vaccination. In elderly individuals aged ≥65 years of age, the inactivated influenza vaccine is only 30–50% effective in preventing influenza illness (Govaert et al., 1994; Patriarca et al., 1985). Nevertheless, the inactivated vaccine is 50–80% effective in preventing influenza-related hospitalizations and deaths in this population (Nichol et al., 1998; Patriarca et al., 1985).

Live attenuated vaccines
A second type of vaccine, a live attenuated influenza vaccine (LAIV), was recently licensed in the U.S. and is approved for use in healthy persons aged 5–49 years. LAIVs have been widely used in the former Soviet Union and Russia in influenza vaccine prevention programs in children and working-age adults (Kendal, 1997; Rudenko et al., 2000). LAIVs are based on the concept of cold-adaptation, or the growth of human influenza viruses at less than optimal temperatures, resulting in attenuation of a donor strain. This strategy has been used by both U.S. and Russian investigators to generate highly stable master strains of influenza A and B viruses with cold-adapted (*ca*) and temperature sensitive (*ts*) phenotypes. Five mutations in the PB2, PB1 and NP genes were associated with the ts phenotype in *ca* A/Ann Arbor/6/60, the master donor strain used to produce influenza A virus components of the

U.S. licensed FluMist vaccine (Jin et al., 2003). These mutations in combination enable efficient replication of the master donor strain at lower temperatures (33°C) but essentially prevent replication at higher temperatures (39°C) resulting in an attenuated replication phenotype (Maassab et al., 1982). A similar set of *ts* attenuating mutations are found in the A/Leningrad/137/17/57 ca master donor stain used to generate the influenza A components of Russian ca vaccines (Klimov et al., 1992). Recombinant vaccine viruses that possess 6 internal genes from the master donor ca strains and the HA and NA genes from the wildtype epidemic strain are generated by traditional reassortment techniques. The resulting 6:2 reassortants possess the attenuation phenotypes of the donor strain and the desired antigenic properties of a given epidemic variant (Maassab et al., 1982; Murphy and Coelingh, 2002). Trivalent LAIVs contain influenza strains that are antigenically equivalent to those contained in the inactivated influenza vaccine. Doses of 10^7 50% tissue culture infectious doses ($TCID_{50}$) are typically safe and immunogenic in all age groups. Live vaccines delivered intranasally replicate to a limited extent in the upper respiratory tract and thus induce immunity more similar to that induced by natural infection than that induced by inactivated vaccines. LAIV are likely to induce mucosal antibody responses and cellular responses, which may contribute to high protective efficacy even against a variant epidemic strain that is antigenically distinct from the vaccine strain (Belshe et al., 2000). Nevertheless, like the current inactivated vaccines, there is a need to continually update LAIV with the HA and NA genes of currently circulating influenza strains.

Adjuvanted subunit vaccine

Although a spectrum of adjuvant formulations to improve the immunogenicity of inactivated influenza vaccines have been evaluated in preclinical studies, few such approaches have reached licensure. One of these is a subunit trivalent influenza vaccine formulated with MF59, an emulsion composed of 5% v/v squalene, 0.5% v/v polysorbate 80 and 0.5% v/v sorbitan trioleate, emulsified under high pressure to produce uniform droplets which is licensed in some European countries (Ott et al., 1995). MF59 adjuvanted vaccine was shown to significantly enhance serum hemagglutination-inhibition (HAI) antibody responses to influenza A/H3N2 and the influenza B vaccine component particularly in elderly individuals with chronic diseases (Banzhoff et al., 2003). Although local reactions were more frequent in individuals who received the MF59-adjuvanted vaccine compared with a non-adjuvanted vaccine, they were predominantly mild and transient (Frey et al., 2003). Another modification of the subunit vaccine approach is the virosomal influenza vaccine. Virosomes consist of influenza virus surface glycoproteins HA and NA released from the virus by detergent disruption and reconstituted into a phosholipid bilayer to form liposomes (Gluck et al., 2000; Gluck and Metcalfe, 2002; Zurbriggen, 2003). This approach is thought to present the HA and NA proteins to the host's immune response in a manner similar to that of the intact virus, resulting in augmented cellular responses (Bungener et al., 2002). The trivalent parenterally administered virosomal vaccine has been licensed in some European countries since 1996 (de Bruijn et al., 2004). More recently, the virosomal influenza vaccine was formulated for intranasal delivery (Gluck et al., 2000). A key difference in the formulation for intranasal delivery was the addition of a powerful mucosal adjuvant, heat labile enterotoxin from enterotoxigenic *Escherichia coli*, known as LT. Clinical trials demonstrated that virosomal vaccine plus LT elicited significant mucosal and serum antibody responses and was generally well tolerated (Durrer et al., 2003). However, after licensure and use in Switzerland, 46 cases of Bell's palsy were reported over a 7 month period and the vaccine was withdrawn from the market. A case-control study provided evidence to suggest a strong association between the inactivated intranasal influenza vaccine and Bell's palsy with a peak incidence occurring 31–60 days post vaccination (Mutsch et al., 2004). The LT adjuvanted intranasal inactivated influenza vaccine is no longer in clinical use.

NEW STRATEGIES FOR IMPROVED INFLUENZA VACCINES
Dosing, delivery and combinations of current vaccines

Current inactivated influenza vaccines contain 15μg of HA of each of the two influenza A and one influenza B virus components. Increasing the HA content of vaccines has been shown to substantially augment serum antibody titers in both healthy younger and elderly individuals (Keitel et al., 1994; Keitel et al., 1996). In young healthy adults with either low or high pre-existing serum antibody, the mean neutralizing antibody titers achieved 4 weeks post-vaccination were directly correlated with the vaccine dose administered which ranged from 15 to 405 μg of monovalent purified HA, while differences in local reactions were minimal (Keitel et al., 1994). Using a similar strategy in healthy persons aged >65 years, doses of up to 135 μg of purified HA vaccine or a split-product vaccine resulted in a doubling of the response rate and a 2- to 3-fold increase in mean serum antibody titer, although the higher doses of the split-product vaccine resulted in a higher rate of injection site discomfort. Nevertheless, the use of higher doses of purified vaccine product is one approach to improve the immunogenicity of vaccines, particularly in the elderly.

Other studies have explored the benefits of altering the site of delivery of inactivated vaccines. Intranasal administration of inactivated vaccine was shown to induce significant post-vaccination virus-specific IgA antibody responses in nasal wash specimens in studies conducted in either young, healthy adults or institutionalized elderly (Keitel et al., 2001; Muszkat et al., 2003). Boyce et al., (2000) investigated whether the addition of MF59 adjuvant to an intranasally administered subunit vaccine would enhance vaccine immunogenicity, but found no significant difference in nasal IgA and serum antibody responses in individuals that received the MF59-adjuvanted intranasal vaccine compared with unadjuvanted intranasal subunit vaccine. Although intranasal delivery of either inactivated vaccine or LAIV also induces serum HAI antibody responses, optimal serum antibody responses generally require parenteral immunization with inactivated vaccines. Since serum antibody responses are considered to be essential for optimal protection, vaccine strategies that elicit both serum antibody and nasal IgA responses are preferable. A combination of intranasal LAIV and intramuscular inactivated vaccine administered simultaneously may offer an optimal strategy for generating both nasal IgA antibody and serum HAI antibody responses (Keitel et al., 2001; Rudenko et al., 2000; Stepanova et al., 2002). Indeed, combined immunization with LAIV and inactivated vaccine was shown to provide greater protective efficacy against laboratory-documented influenza infection in elderly individuals compared with inactivated vaccine alone (Rudenko et al., 2000; Treanor et al., 1992).

Cell culture-grown influenza vaccines

Currently marketed influenza vaccines are produced in embryonated hen's eggs. Approximately 250 million doses are brought to market annually in over 100 countries. In addition to the tight timeline necessitated by the vaccine strain selection process, the supply of high quality fertilized eggs must be carefully timed well in advance to provide sufficient capacity for vaccine manufacture (Gerdil, 2003). Growth of virus for vaccines in a cell culture system has the potential advantage of vaccine production at any time of the year and hence a quicker response time in the event of a newly emergent epidemic variant or pandemic threat. Growth of virus in serum-free cell culture also minimizes the risk for microbiological contamination and would likely result in a vaccine with reduced endotoxin content compared with egg-derived vaccines. Influenza viruses grown in mammalian cell culture were shown to be more similar to those in human clinical specimens compared with their egg-grown counterparts (Katz et al., 1990; Schild et al., 1983). To be considered as a substrate for the commercial manufacture of human vaccines, a cell line must meet stringent criteria established by regulatory authorities, including a detailed history of derivation, extensive testing to document the lack of adventitious agents and demonstration of the lack of tumourogenicity in animals (Center for Biologics Evaluation and Research, 1987; Griffiths, 1999). Inactivated influenza vaccines grown in Madin Darby canine kidney (MDCK) cells, the most common mammalian cell substrate for influenza virus growth, or African green monkey kidney (Vero) cells have been licensed in some European countries but are not yet on the market. An MDCK cell-grown inactivated influenza vaccine was shown to be well-tolerated and the proportion of subjects that achieved a protective titer (HAI titer ≥ 40) were similar to that observed in recipients of egg-grown vaccine, although absolute geometric mean titers were higher in the latter (Palache, 1997). Similarly, Halperin et al., (2002) found no substantial differences in safety and immunogenicity between MDCK cell-grown and egg-grown vaccine in children, younger and elderly adults. Vero cell-derived influenza vaccines have been scaled up to fermenter batch production and highly purified whole virus vaccines have been generated (Kistner et al., 1999). In preclinical studies, Vero cell-derived inactivated vaccine was as immunogenic in the induction of humoral immunity and induced better cellular responses than traditional egg-grown vaccine (Bruhl et al., 2000). Another mammalian cell line, PER.C6, is also under consideration as a substrate for influenza vaccine production (Pau et al., 2001). The PER.C6 cell line is a human fetal retinoblast immortalized by transfection with an E1 minigene of adenovirus type 5. Efficient replication of egg-grown reference influenza A and B viruses in PER.C6 cells has been demonstrated.

Recombinant vaccine strategies and other influenza viral antigens

Recombinant DNA technology offers another approach for the production of influenza vaccines that would overcome the need for cultivation in eggs and the generation of high-growth influenza A virus reassortants. An HA gene may be rapidly cloned directly from a field isolate and expressed in insect cells by a recombinant baculovirus, a system known for high yield of expressed proteins. Clinical trials with recombinant monovalent H1 and H3 antigens expressed in insect cells demonstrated that doses of 15–45 µg induced neutralizing HA-specific antibody responses comparable to those elicited by subvirion licensed vaccines (Powers et al., 1995; Treanor et al., 1996). The highly purified HA protein vaccines also have the potential to be less reactogenic than traditional vaccines (Lakey et al., 1996; Powers et al., 1995). A dose escalation trial of trivalent recombinant HA influenza vaccine in an elderly population was recently completed and results from the study are pending (M. Cox, personal communication).

Current influenza vaccines are designed primarily to induce antibodies directed against the HA since only these antibodies can prevent infection. However, antibodies directed against the other influenza A transmembrane proteins, NA and the ion channel M2 protein can limit viral replication and ameliorate disease in animal models (Treanor et al., 1990; Webster et al., 1988). Alternate vaccine strategies based on the NA have long been considered and early studies indicated that NA-based vaccines could induce partial immunity in humans (Couch et al., 1974). Vaccination of mice with a baculovirus–expressed recombinant N2 protein reduced viral replication and morbidity as measured by weight loss in mice challenged with homologous virus (Kilbourne et al., 2004). Reduction in viral replication following a heterologous virus challenge was greater in mice vaccinated with a combination of recombinant H3 and N2 proteins compared with mice vaccinated with conventional monovalent H3N2 vaccine (Johansson, 1999). Baculovirus expressed NA vaccines have been evaluated in humans.

M2 protein is highly conserved among human influenza A virus subtypes, and is therefore considered a suitable candidate to elicit broad immunity against multiple influenza A subtypes. Parenteral vaccines based on a baculovirus-expressed recombinant M2 protein, a recombinant with the transmembrane region deleted for increased solubility, or a fusion protein representing the N terminal ectodomain of M2 combined with Hepatitis B virus core protein were all shown to protect mice by limiting pulmonary virus replication and preventing lethal disease induced by homologous or heterosubtypic viral challenge (Frace et al., 1999; Neirynck et al., 1999; Slepushkin et al., 1995). To augment the induction of anti-M2 antibody which is thought to be the effector of M2 immunity, (Mozdzanowska et al., 2003) expressed the M2 ectodomain as a synthetic multiple antigenic peptide (MAP) covalently linked to a helper T cell determinant. Intranasal immunization of mice with this construct induced immunity against infection in the upper and lower respiratory tract of mice. Although antibody responses to NA following infection and vaccination are well documented, responses to M2 following infection are less well characterized but may be less robust (Black et al., 1993). Nevertheless, the studies to date indicate that supplementation of HA-based vaccines with M2 antigen may provide a rational strategy to broaden the protective effect of influenza vaccines (Frace et al., 1999). On the other hand, a study in pigs indicated that immunization with M2 may actually lead to exacerbation of disease following subsequent influenza infection, indicating the need for some caution in future studies on vaccination with M2 (Heinen et al., 2002).

Reverse genetics technology

Recent advancements in reverse genetics techniques for negative sense viruses now allow for the generation of infectious influenza viruses entirely from cloned plasmid DNA. A set of 8 or 12 plasmids encoding influenza virion sense RNA and/or mRNA are used to cotransfect an appropriate cell line resulting in the rescue of infectious virus (Fodor et al., 1999; Hoffmann et al., 2000; Hoffmann et al., 2002a; Neumann et al., 1999). This technology has enormous potential to advance influenza vaccine technology, not only for the generation of inactivated and live attenuated reassortant vaccine strains for interpandemic influenza, but especially for the generation of vaccines against highly pathogenic avian influenza strains with pandemic potential. The application of reverse genetics for pandemic vaccines will be discussed separately below. The 12 plasmid system is based on co-transfection of mammalian cells with 8 plasmids encoding virion sense RNA (vRNA) under the control of a human Pol I promoter (Neumann and Kawaoka, 2002) and 4 plasmids encoding messenger RNAs (mRNA) encoding the RNP complex (PB1, PB2, PA and nucleoprotein gene products) under the control of a Pol II promoter. The 8 plasmid method further refined and simplified the process by engineering the influenza virus gene segments into a bi-directional or ambisense vector with Pol I and Pol II promoters flanking each gene segment (Hoffmann et al., 2000; Hoffmann et al., 2002a). The 8 and 12 plasmid systems have been applied to the generation of 6:2 reassortant viruses for use as vaccine seed

Table 8.2 Adjuvants, formulations, and delivery systems used with influenza vaccine in preclinical animal models – selected studies

Delivery system	References
Mucosal adjuvants	
Cholera toxin (CT)	Chen and Quinnan, Jr., 1989; Mbawuike and Wyde, 1993; Tamura et al., 1992
Heat-labile enterotoxin (LT)	Barackman et al., 1999; Tamura et al., 1994
Systemic adjuvants	
Monophosphorolipid -A	Baldridge et al., 2000; Mbawuike et al., 1996
Polymer-metal complexes	Mustafaev and Norimov, 1990
Non ionic block polymers	Deliyannis et al., 1998; Todd et al., 1998
PLGA particles	Chattaraj et al., 1999; Lemoine et al., 1999; Nixon et al., 1996
Esterified hyaluronic acid microspheres	Singh et al., 2001
Bifidobacterium breve +vaccine	Yasui et al., 1994; Yasui et al., 1999
Cytokine supplementation	Mbawuike et al., 1990; Pauksen et al., 2000; Taglietti, 1995
Bacterial Vectors	
Salmonella	Brett et al., 1991; Tite et al., 1990
Listeria	Ikonomidis et al., 1997
Viral Vectors	
Vaccinia and modified vaccinia	Itamura et al., 1990; Karupiah et al., 1992
TLR ligands	
Flagellin TLR5 ligand	Levi and Arnon, 1996; McEwen et al., 1992
LPS TLR 4 ligand	Tamizifar et al., 1995
Immunotargeting	
Delivering antigen to antigen presenting cells	Skea et al., 1993

viruses (Hoffmann et al., 2002a; Schickli et al., 2001). Plasmid systems to generate influenza B viruses have also been developed (Hoffmann et al., 2002b). Thus, technology is now available to routinely generate all three components of trivalent influenza virus vaccines using reverse genetics rather than traditional reassortment technology which can be cumbersome and sometimes, inefficient.

The ability to generate viruses with targeted mutations in the viral genome has provided further options for the development of vaccine strains with either additional determinants of attenuation or new strategies to produce replication-deficient influenza viruses as vaccine candidates. Using reverse genetics, Jin et al., (2004) demonstrated that the introduction of four mutations in the PB2 and PB1 genes responsible for the ts phenotype in the A/Ann Arbor/6/60 master donor ca strain into the PR8 virus, the donor of high-growth phenotype, further attenuated this strain with respect to replication in the upper and lower respiratory tract of ferrets. The modified high growth donor strain may provide an added level of safety in the manufacturing of inactivated influenza vaccines against avian viruses with pandemic potential (Jin et al., 2004; WHO, 2003).

Reverse genetics technology has also been used to introduce deletions in the influenza

A virus NS1 gene which has interferon antagonist activity (Garcia-Sastre et al., 1998). Viruses lacking substantial portions of the NS1 gene, although attenuated in their ability to replicate in BALB/c mice, were immunogenic and elicited a protective immune response against lethal challenge with the wildtype PR/8/34 virus (Talon et al., 2000). Plasmid-based reverse genetics has been used to generate replication-incompetent virus-like particles lacking the NS2 gene (Watanabe et al., 2002b). When administered intranasally to mice, the NS2-deficient virus induced virus-specific IgG and IgA in tracheal-lung washes and to a lesser extent in nasal washes, but no serum HAI antibody response. Nevertheless, mice were protected from subsequent challenge with a lethal influenza infection. Using a similar approach, the same group also generated an influenza A virus expressing an M2 protein that lacked the transmembrane (TM) region (residues 29–31) critical for ion channel activity (Watanabe et al., 2002a). However, only a modest decrease in virulence compared with the parental strain was achieved. Nevertheless, the M2 TM deletion may be considered as one of a number of attenuating mutations to be used in combination with others to generate future attenuated master donor strain that offer further enhancement of safety and stability. Along the same lines, a mutant virus that lacked most of the NA gene coding region induced serum anti-HA antibody and protection of mice against lethal challenge with homologous wildtype virus (Shinya et al., 2004). Reverse genetics also offers the prospect of developing a chimeric influenza A/B master donor strain, overcoming the current need to have separate influenza A and B attenuated master donor strains for LAIV or the lack of a high growth influenza B virus inactivated vaccine donor strain. Horimoto et al., (2003) generated chimeric viruses possessing an influenza B HA coding region flanked by one or both influenza A HA gene 3' or 5' noncoding regions on an influenza A background, but all viruses replicated far less efficiently than the parental influenza A strains. Further work is needed to develop a chimeric influenza A/B viruses that could be used as a universal vaccine donor strain.

VACCINE STRATEGIES THAT USE ADJUVANTS AND/OR ALTERNATE DELIVERY SYSTEMS

Identification of next generation adjuvants that can be safely combined with vaccine antigens to induce long-lasting humoral and cellular immune responses is a major goal of vaccine developers. A variety of plant, bacterial, or insect derived products, and pharmaceutical compounds, formulated with influenza vaccines and delivered by various routes, have shown much promise in animal studies. However, only few have been evaluated in Phase I/II clinical trials and alum remains the only adjuvant presently licensed for human use in the U.S. Because it is not possible to review all adjuvants/delivery systems used with influenza vaccine in preclinical animal models in detail, this review is restricted to adjuvants/delivery systems for influenza vaccines that have progressed to human Phase I/II trials. For other adjuvant/delivery strategies, we refer the reader to the selected studies which demonstrate the range of products investigated for influenza vaccines detailed in Table 2.

Proteosomes

Proteosomes are hydrophobic, membranous, multimolecular preparations of meningococcal outer membrane proteins that form nanoparticles ranging in size from 20–800 nm in aqueous solutions (Lowell et al., 1988). Porin B from *N. meningitidis* B strains, a major component of proteosomes, has been shown to activate antigen presenting cells (APC) by interacting with Toll-like receptor (TLR)2 and upregulating the expression of the costimulatory molecule CD86 on the surface of B cells and other APC (Massari et al., 2002). These interactions lead to the generation of an optimal priming environment to induce immune responses. Intranasal or intramuscular immunization of influenza subunit vaccines non-covalently formulated with proteosomes have been shown to induce potent antigen-specific mucosal and systemic immune responses in animal studies (Jones et al., 2003; Plante et al., 2001). Unlike the conventional vaccines, proteosome vaccines elicit antigen-specific sIgA responses delivered by the intranasal (i.n) or intramuscular (i.m) route. In a series of Phase I and II human trials, influenza proteosome vaccines have been well-tolerated and immunogenic. In a recently completed study in healthy subjects aged 18–64 years, influenza proteosome vaccine containing A/Panama/99 H3 HA conferred better protection than the conventional vaccine against clinical disease caused by the A/Fujian-like H3N2 drift variant virus (D. Burt, personal communication). Thus, proteosome vaccines show much potential as a next generation nasal influenza vaccine.

Immuno stimulating complexes (ISCOMs)

Immunostimulating complexes (ISCOMs) consist of the saponin Quil A from the tree *Quillaja saponaria* Molina which has adjuvant properties, cholesterol, phospholipids, and antigens. The components are incorporated into icosahedral particles of about 40 nm in size by hydrophobic interactions during assembly of the complex (Morein et al., 1984). A detailed technical review of the ISCOM technology can be found elsewhere (Johansson and Lovgren-Bengtsson, 1999). The enhanced immunogenicity of ISCOMs may correlate with enhanced expression of class II MHC and costimulatory molecules on professional APC. Moreover, ISCOMs deliver antigens for class I and class II presentation, induce either a predominant Th1 or a balanced Th1/Th2 response, and induce/recall cytotoxic T lymphocyte (CTL) responses (Behboudi et al., 1996; Mowat et al., 1993; Sambhara et al., 1998a).

A number of groups including ours have shown in animal models that influenza vaccines formulated as ISCOMs are highly immunogenic and induces long-lasting HAI titers and CTL responses (Coulter et al., 1998; Deliyannis et al., 1998; Rimmelzwaan et al., 1997; Sambhara et al., 1998b; Sundquist et al., 1988). Furthermore, an H1-ISCOMs-vaccine conferred cross-protection against challenge with a range of influenza A subtypes (H2N2, H3N2, H5N1 and H9N2) that have caused respiratory illness in humans (Sambhara et al., 2001). Immunogenicity and kinetic studies demonstrated that the ISCOM-formulated vaccine, even at 0.3 μg of HA, elicited HAI titers that were 8- to 16-fold higher than with the conventional vaccine at 3μg of HA.

In a Phase I clinical trial, conventional inactivated vaccine was compared with ISCOM-formulated influenza vaccine in healthy adults of 18–45 years of age. The ISCOM-formulated vaccine induced/recalled significant CTL activity against H1N1 and H3N2 viruses in 50–60% of vaccinated individuals, while conventional vaccine induced CTL activity in only 5% of vaccinated individuals (Ennis et al., 1999). Consistent with these data, results from another human trial of the ISCOM delivery system demonstrated that ISCOM-formulated vaccine induced a more rapid serum antibody titer rise against vaccine strains than did conventional vaccine (Rimmelzwaan et al., 2000). Enhanced T-cell proliferative responses and CTL responses occurred in a higher proportion of the individuals who received the ISCOM formulated vaccine compared with conventional vaccine. However, parenteral immunization of ISCOM formulated influenza vaccine have been associated with higher rate of local reactions compared with conventional vaccines and thus their use as an injectable interpandemic vaccine has not progressed. However, their use in the pandemic situations remains to be evaluated.

Chitosan-based vaccine delivery

Chitosan is a carbohydrate bipolymer derivative of chitin, a carbohydrate based material found in the exoskeletons of crustaceans and insects and in mushrooms has also been evaluated as an adjuvant for influenza vaccine delivery. Chitin and its derivatives have been shown to activate peritoneal macrophages in vitro, induce nitric oxide (NO) production and chemotaxis, and suppress tumour growth in syngeneic mice. Although the precise mechanism of action of chitosan on the immune system is not known, the ability to activate macrophages suggests that they may interact with pattern recognition receptors of the innate immune system to initiate an inflammatory response and create a microenvironment for the enhanced adaptive immune responses.

An influenza virus surface antigen vaccine formulated with chitosan and delivered i.n. to mice induced strong local and systemic antibody responses (Bacon et al., 2000). In a Phase I clinical trial, the immune response of volunteers immunized twice by the i.n. route with a chitosan-formulated vaccine was compared with those of individuals who received one dose of non-formulated vaccine by the i.m. route. The chitosan-formulated vaccine was well tolerated and greater than 40% of the volunteers who received the i.n. vaccine achieved a 4-fold or greater increase in HAI antibody titer against all three vaccine strains. Although the mean antibody titers induced by the i.n. chitosan formulated vaccine were generally lower, the number of volunteers who developed protective HAI titer of ≥ 40 was similar for both i.n. and i.m. administration protocols for at least two strains. Results from subsequent human trials are awaited (Illum et al., 2001).

Liposomes with or without adjuvants

Muramyl-dipeptides (MDP) have been shown to possess adjuvant activity in a number of preclinical studies with influenza vaccine (Byars et al., 1990; Hjorth et al., 1997). Cationic lipids which are the basis of liposomes have also been shown to possess adjuvant activity. A Phase I clinical study evaluated trivalent influenza vaccine containing A/Yamagata/120/86 (H1N1), A/Fukuoka/C29/86 (H3N2), and B/Nagasaki/1/87 formulated as liposomes together with MDP. The liposome/MDP-formulated vaccine induced higher HAI titers than unformulated vaccine against both influenza A components, while responses to the influenza B strain were comparable (Kaji et al., 1992). The liposome/MDP-formulated vaccine induced mild local reactions, but no systemic reactions.

A randomized, double-blinded study evaluated the immunogenicity of subvirion vaccine formulated as liposomes compared with control subvirion vaccine containing 15 mg dose of HA from A/Taiwan/1/86 (H1N1) virus in elderly seropositive volunteers. The liposome vaccine was well-tolerated and serologic responses were similar to those induced by the control vaccine. However, influenza A-specific CTL activity was enhanced to a greater extent by the liposome vaccine than by the control subvirion vaccine (Powers, 1997). Further studies are needed to assess the usefulness of liposome formulation without the addition of an adjuvant.

The cholesterol derivative, DC-Chol, can also form liposomes and was used to formulate a trivalent vaccine containing A/Texas/36/91 (H1N1), A/Johannesburg/33/94 (H3N2), and B/Harbin/7/94 (Guy et al., 2001). Subcutaneous or intranasal delivery of formulated vaccine in mice and intramuscular delivery in macaques induced enhanced humoral antibody responses compared with non-formulated vaccines and only formulated vaccine induced influenza NP-specific CTL responses, indicating that DC-Chol liposomes formulation facilitated class I presentation. Interaction of cationic lipids with the antigen is critical for the adjuvant activity. The results from these preclinical studies in animal models suggest that DC-Chol is a potential adjuvant for the delivery of influenza vaccine. Clinical trials in humans using this formulation were completed and the results are awaited.

Poly [di(carboxylatophenoxyl)phosphazene] (PCPP)

PCPP is a water soluble polymer with adjuvant activity that has been shown to augment humoral immune responses to trivalent influenza vaccine compared to non-adjuvanted vaccine in both young and aged mice (Payne et al., 1998). A Phase I clinical trial with PCPP-influenza vaccine was completed and the results are awaited.

Dehydroepiandrosterone (DHEA)

DHEA and its sulfated pro-hormone have been shown to possess immunomodulatory effects in animal models (Araneo et al., 1993). DHEA or DHEA sulfate (DHEAS) supplementation enhanced antibody responses against influenza vaccine and increased resistance to post-vaccination intranasal challenge with influenza virus in aged mice (Danenberg et al., 1995b; Danenberg et al., 1995a). In proof of concept studies in humans, the frequency of 4-fold rises and overall mean antibody HAI titers were greater in elderly individuals that received an oral form of DHEAS before influenza vaccination compared with individuals who did not receive DHEAS (Araneo et al., 1995). However, in other trials, similar enhancement of influenza-specific immune responses by DHEA was not observed (Ben-Yehuda et al., 1998). Additional studies are required to assess the usefulness of DHEA as an adjuvant for influenza vaccine.

TLR9 ligands

Oligodeoxynucleotides (ODN) containing unmethylated CpG motifs have been shown to bind to TLR9 on plasmacytoid DC and B cells (Chuang et al., 2002; Hemmi et al., 2000). CpG ODN activate monocytes and macrophages, skewing the priming environment towards a Th1 response, even in the presence of a Th2 adjuvant such as alum (Moldoveanu et al., 1998; Roman et al., 1997). A recent Phase I study in healthy volunteers aged 18–40 years seronegative for H1N1 and influenza B viruses evaluated i.m. delivery of a CpG ODN adjuvanted trivalent influenza vaccine containing a 1/10th antigen dose of A/Beijing/262/95, A/Sydney/5/97 and B/Harbin/7/94 viruses. The adjuvanted reduced-dose vaccine was well-tolerated and induced similar levels of serum HAI antibodies as the unadjuvanted full dose vaccine (Cooper et al., 2004). However, there was no enhancement of antibody titers when the adjuvanted vaccine was given at the full antigen dose. These results suggest that formulation of trivalent influenza vaccines with CpG ODN may allow for the use of reduced antigen dose without compromising the immunogenicity of the vaccine.

DNA vaccines

DNA vaccines have attracted much attention since they were first reported to induce protective immune responses (Ada and Ramshaw, 2003; Ulmer, 2002). DNA vaccines are non-infectious, non-replicating *E. coli*-derived plasmids containing transcription machinery that encodes only the protein(s) of interest. Unlike viral vectors such as adenovirus or vaccinia virus which induce immune response to vector derived proteins, DNA vaccines induce immune responses only to the protein of interest. This feature makes it possible to vaccinate individuals multiple times without induction of immune-dampening vector-specific responses.

Influenza virus DNA based vaccines induced protective and long-lasting HAI antibodies in rodents, ferrets and non-human primates and conferred protection against challenge with a homologous virulent strain in mice, chicken, and ferrets (Donnelly et al., 1995; Donnelly et al., 1997; Pertmer et al., 1995; Robinson et al., 1993; Ulmer et al., 1994; Ulmer et al., 1998; Webster et al., 1994). DNA vaccine against conserved internal proteins such as NP generated CTL responses and conferred heterosubtypic immunity against lethal challenge (Ulmer et al., 1993). Moreover, very small amounts of DNA are sufficient to induce long-lasting protective immune responses. Despite this extensive preclinical data which demonstrates the potential of DNA immunization as a future vaccine strategy, there is limited data available on the immunogenicity of influenza DNA vaccines in humans. The initial results of a Phase I study that evaluated intramuscular delivery of an influenza HA DNA vaccine trial were disappointing and results have not been formally released to assess the usefulness of this approach for next generation influenza vaccines. Further improvements in DNA vaccine techniques in the vector construction, improving transfection and translation efficiency, vector delivery modalities, and modulating priming environment to improve antigen processing and presentation may help develop second generation DNA vaccines for preventive and therapeutic immune intervention.

Transcutaneous vaccine delivery

Skin is a dynamic physical and functional barrier that maintains the physical and physiological integrity of the individual by preventing the entry and colonization of potential pathogens. The skin is endowed with well integrated innate and adaptive immune components and has become an attractive route for vaccine delivery. Langerhans cells are professional APC that are present in skin at a frequency of 500–1000 cells/mm^2. These cells take up epicutaneous antigens, emigrate into the regional draining lymph nodes, and present the processed antigens to the T and B cells to initiate an adaptive immune response which eliminates the invading pathogens (Partidos, 2003). However, the epithelial integrity has to be disrupted to initiate an immune response to vaccines. To achieve this investigators used a variety of approaches, such as hydration, electroporation, microneedles, epidermal powder immunization and particle-mediated (gene gun) and mechanical disruption to deliver viral vectors, peptide, proteins or DNA with or without adjuvant formulations. Antigen delivery through the skin route induces enhanced cellular and humoral immune responses in animal models. Transcutaneous delivery of influenza vaccines is presently at preclinical stage; the first human trials are also underway. Epidermal powder immunization (EPI) elicited both serum and mucosal antibodies to an inactivated influenza virus vaccine, which were further enhanced by codelivery of cholera toxin (CT), a synthetic oligodeoxynucleotide containing immunostimulatory CpG motifs (CpG DNA), or the combination of these two adjuvants. A single dose of adjuvanted vaccine conferred complete protection of mice against lethal challenge with an influenza virus. However, two immunizations were required for protection in the absence of an adjuvant (Chen et al., 2001). Subsequent studies in Rhesus macaques demonstrated that a QS-21 adjuvanted influenza vaccine induced significantly higher serum HAI titers than nonadjuvanted vaccine either by transcutaneous delivery or i.m. injection (Chen et al., 2003a). Another strategy that has been explored in mice is one in which the antigen is delivered by parenteral route and the adjuvant by skin patch delivery. Application of a patch containing LT adjuvant at the site of vaccination immediately after i.m., intradermal or subcutaneous injection of influenza vaccine enhanced antibody responses to trivalent influenza vaccine (Guebre-Xabier et al., 2003). This strategy also enhanced mucosal antibody responses and T cell responses to influenza virus in the lung. Patch delivery of adjuvant was shown to facilitate the migration of dendritic cells into draining lymph node, thus enhancing systemic responses. However, the precise mechanism which induced mucosal antibody responses remains unclear. This strategy was also used to overcome poor immune responses in the aged mice (Guebre-Xabier et al., 2004). Phase I/II clinical trials using this strategy are underway. It will be interesting to see if the transcutaneous patch delivery system can be extended to include both vaccine antigen as well adjuvant to induce enhanced protective humoral, cellular and mucosal immune responses. The data from a number of preclinical studies using influenza vaccine are very encouraging.

VACCINES AGAINST PANDEMIC INFLUENZA

There is no doubt that the next influenza pandemic will greatly burden health services, and cause widespread social disruption and economic losses (Meltzer et al., 1999). Vaccines will be the primary means for the prevention and control of pandemic influenza. Global susceptibility to a pandemic strain bearing a novel HA is expected and the demand for vaccine will be great. The population's immune status in a pandemic situation differs from that seen during the interpandemic period. In the initial pandemic period, a higher proportion of deaths occur in persons <65 years of age than in an interpandemic period (Simonsen et al., 1998). This may occur because younger adults are immunologically naïve to the new strains, whereas older populations may possess some level of pre-existing immunity due to previous infections with related strains that circulated in earlier times. The vaccine antigen dose required to elicit satisfactory immune responses in naïve individuals is unclear but it is likely that two doses will be needed to provide optimal protection.

Vaccines against avian influenza viruses

Since the late 1990s, multiple subtypes of avian influenza viruses have crossed the species barrier to infect and cause disease in humans. Most notable of these are the highly pathogenic (HP) H5N1 viruses that were transmitted from infected poultry to humans in Asia in 1997 and 2003 and most recently in 2004. HP H5N1 viruses have likely become endemic in the Asia, elevating the risk for additional human cases. Of even greater concern, is the potential for reassortment with human viruses and/or substantial adaptation to the human host, which would be likely to yield a pandemic strain capable of efficient human-to-human transmission. While H5N1 viruses clearly have the highest and most urgent priority for vaccine development, avian influenza H9 and H7 subtypes, which have also transmitted from avian species to humans and caused mild and in one case fatal human illness, are also considered a pandemic threat for which vaccines are needed. Since H2N2 viruses circulated in humans from 1957 until 1968, individuals born after this time lack immunity to the H2 subtype and are therefore a susceptible population for reintroduction of the H2 subtype.

Development of vaccines against HP H5 and H7 viruses pose several problems not previously encountered in the generation of influenza vaccine candidates. First, HP avian viruses are lethal to embryonated eggs, which limits growth in large quantities. Second, the multibasic amino acid motif at the HA cleavage site is believed to contribute to virulence of these viruses in humans as well as in domestic poultry. Thus, vaccine preparation from wild-type HP avian virus would require heightened biocontainment to protect workers and the environment which would be prohibitive for large scale vaccine production. Following the emergence of HP H5N1 viruses in humans in 1997, two vaccine strategies were developed to overcome these limitations.

One approach was to use a 'surrogate' apathogenic H5 vaccine strain that was antigenically related but not identical to the HP H5N1 strain, overcoming the need to grow and purify a vaccine under high containment conditions. Vaccine strategies based on apathogenic H5 strains which possessed HAs that were antigenically similar to A/Hong Kong/156/97 (H5N1) were evaluated in mice (Lu et al., 1999; Takada et al., 1999; Wood, 2001). Inactivated whole virus vaccine delivered i.n. or alum-adjuvanted subunit vaccine delivered i.m. were capable of protecting mice against lethal H5N1 challenge. A vaccine produced from A/duck/Singapore/97 (H5N3) surface-antigen vaccine administered with or without MF59 adjuvant was evaluated in a randomized Phase I clinical trial (Nicholson et al., 2001). Two doses of 7.5, 15 or 30 μg of H5 HA were given three weeks apart. Although both vaccines were well tolerated, the non-adjuvanted vaccine was poorly immunogenic, with only a 36% response rate after two 30μg doses of vaccine. Individuals who received the H5N3 vaccine formulated with MF59 achieved significantly higher antibody responses with a majority of individuals showing a seroconversion to the vaccine strain. Follow-up re vaccination 16 months later, using the same vaccine formulation, substantially boosted antibody titers in those receiving the adjuvanted, but not the non-adjuvanted vaccine (Stephenson et al., 2003). Another approach relied on production of a baculovirus-expressed purified H5 HA protein based on the HA gene cloned from the prototype H5N1 strain, thus overcoming the need to grow large quantities of virus in eggs. As mentioned above, previous clinical trials had determined that responses to doses of 15–45μg of recombinant H1 and H3 antigens in primed populations were similar to those induced by licensed inactivated vaccines. However, in an unprimed population, even the highest dose (two doses of 90μg) of the baculovirus-expressed recombinant H5 vaccine elicited seroconversion in only 52% subjects as measured by a microneutralization assay (Treanor et al., 2001). In animal studies, baculovirus recombinant H5 and H7 vaccines were efficacious in protecting chickens against lethal virus challenge, even when the vaccine strain and challenge strain showed substantial differences in sequence homology (Crawford et al., 1999; Halvorson, 2002). Taken together, the results of H5 vaccine human trials to date indicated a need for adjuvants and/or alternate strategies to enhance the immunogenicity of H5 vaccines.

Traditional inactivated vaccines against avian H9N2 or an early human H2N2 have also been evaluated in preclinical and/or human studies (Chen et al., 2003b; Crawford et al., 1999; Lu et al., 2001; Stephenson et al., 2003). Stephenson et al., (2003) compared subunit versus whole virus A/Hong Kong/1073/99 (H9N2) vaccines in subjects aged 18–60 years. Surprisingly, individuals aged >32 years were found to have reactivity in prevac-cination sera to the H9N2 virus. This age-related antibody response was attributed to cross-reactivity with human influenza viruses that had been encountered early in life and this 'priming' was sufficient to elicit a level of anti-H9 HA antibody response that was associated with protection following a single dose of the H9N2 vaccine. However, in individuals younger than 32 years of age, who were essentially unprimed, even two doses of vaccine was suboptimal since a significant number of volunteers failed to achieve antibody titers associated with protection. In this naïve population, the whole virus vaccine was more immunogenic than the subunit vaccine. In another study, alum adjuvant was shown to enhance responses to whole virus H9N2 and H2N2 vaccines, although once again, unprimed individuals required two doses of vaccine to achieve maximal mean antibody titers (Hehme et al., 2002). Vaccine doses <15 μg formulated with adjuvant induced antibody titers similar to those induced by unadjuvanted full dose vaccine. Taken together, these studies suggest that in unprimed populations, two doses of inactivated avian influenza vaccines were necessary to elicit a protective antibody response to avian influenza viruses. Furthermore, the studies demonstrated the feasibility of using adjuvants to either increase vaccine immunogenicity or the use of reduced doses of vaccine antigen which may be important to extend a limited vaccine supply in a pandemic situation.

Pandemic vaccines generated by reverse genetics

Reverse genetics technology was first used to generate reassortant vaccine candidates based on the HA and NA genes from H5N1 strains isolated from humans in 1997 and either the internal protein genes of the A/Ann Arbor/6/60 *ca* vaccine donor virus (Li et al., 1999) or PR8, the high growth donor for inactivated vaccines (Subbarao et al., 2003). In both cases the H5 HA gene was genetically modified to remove the multibasic amino acid motif associated with pathogenicity for chickens. Both candidates were avirulent for experimentally infected chickens, and showed an attenuated phenotype in mammalian species. A formalin-inactivated vaccine prepared from the H5N1/PR8 transfectant virus was immunogenic and protected mice from subsequent challenge with *wt* viruses from the homologous and heterologous antigenic subgroups. A similar approach was used to generate an H5N3 vaccine candidate with potential veterinary applications (Liu et al., 2003).

The 2003 and 2004 HP H5N1 strains isolated from humans were antigenically and genetically distinct from those isolated in Hong Kong in 1997 and apathogenic strains antigenically related to these newer H5N1 viruses have not been identified (Edwards et al., 2004). Fortunately, recent advances in reverse genetics technology now allow for the generation of vaccine candidates in mammalian (Vero or MRC-5) cell lines qualified for human vaccine production (Ozaki et al., 2004). In a matter of months, vaccine candidates based on the 2003, and most recently the 2004 H5N1 strains isolated from humans, were prepared, characterized and safety tested and now await clinical evaluation (Webby et al., 2004). The WHO has developed a risk assessment outlining appropriate safety testing for vaccine candidates derived from HP avian strains (WHO, 2003). National regulatory authorities are working to develop additional guidelines for the use of these 'genetically modified' organisms in vaccine production while intellectual property issues are being resolved so that vaccines may be commercially produced. While, reverse genetics technology may be essential for the generation of reassortant vaccines from some HP avian strains, the eight–plasmid system may also be used to rapidly and reproducibly produce vaccine seeds with PR8 internal protein genes and surface glycoproteins from a variety of human or avian influenza subtypes, thus confirming the utility of this approach for the generation of vaccine strains for both pandemic and interpandemic periods (Hoffmann et al., 2002a).

CONCLUSION

Several challenges now exist in out efforts toward the improving the control and prevention of influenza through vaccination. As the global population ages, the need for more immunogenic and effective influenza vaccines, as well as larger quantities of vaccine, to protect this growing high risk population will increase dramatically. Vaccination strategies that can, at least in part, overcome the limitations of the aging immune system will be needed to improve vaccine immunogenicity and efficacy for this population. It is now well recognized that healthy children aged <2 years also have increased rates of hospitalization due to influenza, prompting the new recommendations for annual vaccination of children aged 6–23 months. Although current inactivated influenza vaccines are safe and effective in this population, alternative strategies that allow for ease of administration and the induction of broader immune responses, may provide improved vaccine efficacy, as well as coverage.

In any case, novel vaccine strategies for any age groups will need to improve upon the existing licensed vaccines, and be evaluated extensively for safety and the ability to protect against influenza infection. Next generation influenza vaccines will likely include the use of novel adjuvants and delivery systems that target the innate immune system and trigger desirable and selective immune responses. Vaccine safety will be a primary concern as adjuvanted vaccines are developed. Finally, as the prospect of another pandemic grows ever greater, pandemic vaccines have become an urgent priority. Reverse genetics technology now offers a rapid method for the generation of pandemic vaccine candidates, but such vaccines are only now entering Phase I clinical evaluation and certain regulatory, manufacturing, and intellectual property issues remain to be resolved. Adjuvants may be useful to improve the immunogenicity of certain avian subtype vaccines and/or to extend limited vaccine supplies. Global production capacity for pandemic vaccines must be expanded, possibly through the use of new cell-based production technologies. Clearly, the challenges are many. Recent advances in vaccine technology, as well as our understanding of the immune system, should lead to new and improved influenza vaccine strategies in the 21st century.

References

Ada, G. and Ramshaw, I. (2003). DNA vaccination. Expert Opin. Emerg. Drugs *8*, 27–35.

Araneo, B., Dowell, T., Woods, M. L., Daynes, R., Judd, M., and Evans, T. (1995). DHEAS as an effective vaccine adjuvant in elderly humans. Proof-of-principle studies. Ann. N. Y. Acad. Sci. *774*, 232–48.

Araneo, B. A., Shelby, J., Li, G. Z., Ku, W., and Daynes, R. A. (1993). Administration of dehydroepiandrosterone to burned mice preserves normal immunologic competence. Arch. Surg *128*, 318–25.

Bacon, A., Makin, J., Sizer, P. J., Jabbal-Gill, I., Hinchcliffe, M., Illum, L., Chatfield, S., and Roberts, M. (2000). Carbohydrate biopolymers enhance antibody responses to mucosally delivered vaccine antigens. Infect. Immun. *68*, 5764–70.

Baldridge, J. R., Yorgensen, Y., Ward, J. R., and Ulrich, J. T. (2000). Monophosphoryl lipid A enhances mucosal and systemic immunity to vaccine antigens following intranasal administration. Vaccine *18*, 2416–25.

Banzhoff, A., Nacci, P., and Podda, A. (2003). A new MF59-adjuvanted influenza vaccine enhances the immune response in the elderly with chronic diseases: results from an immunogenicity meta-analysis. Gerontology *49*, 177–84.

Barackman, J. D., Ott, G., and O'Hagan, D. T. (1999). Intranasal immunization of mice with influenza vaccine in combination with the adjuvant LT-R72 induces potent mucosal and serum immunity which is stronger than that with traditional intramuscular immunization. Infect. Immun. *67*, 4276–9.

Behboudi, S., Morein, B., and Villacres-Eriksson, M. (1996). In vitro activation of antigen-presenting cells (APC) by defined composition of Quillaja saponaria Molina triterpenoids. Clin. Exp. Immunol. *105*, 26–30.

Belshe, R. B., Gruber, W. C., Mendelman, P. M., Cho, I., Reisinger, K., Block, S. L., Wittes, J., Iacuzio, D., Piedra, P., Treanor, J., King, J., Kotloff, K., Bernstein, D. I., Hayden, F. G., Zangwill, K., Yan, L., and Wolff, M. (2000). Efficacy of vaccination with live attenuated, cold-adapted, trivalent, intranasal influenza virus vaccine against a variant (A/Sydney) not contained in the vaccine. J. Pediatr. *136*, 168–75.

Ben-Yehuda, A., Danenberg, H. D., Zakay-Rones, Z., Gross, D. J., and Friedman, G. (1998). The influence of sequential annual vaccination and of DHEA administration on the efficacy of the immune response to influenza vaccine in the elderly. Mech Ageing Dev. *102*, 299–306.

Black, R. A., Rota, P. A., Gorodkova, N., Klenk, H. D., and Kendal, A. P. (1993). Antibody response to the M2 protein of influenza A virus expressed in insect cells. J. Gen. Virol. *74 (Pt 1)*, 143–6.

Boyce, T. G., Hsu, H. H., Sannella, E. C., Coleman-Dockery, S. D., Baylis, E., Zhu, Y., Barchfeld, G., DiFrancesco, A., Paranandi, M., Culley, B., Neuzil, K. M., and Wright, P. F. (2000). Safety and immunogenicity of adjuvanted and unadjuvanted subunit influenza vaccines administered intranasally to healthy adults. Vaccine *19*, 217–26.

Brett, S. J., Gao, X. M., Liew, F. Y., and Tite, J. P. (1991). Selection of the same major T cell determinants of influenza nucleoprotein after vaccination or exposure to infectious virus. J. Immunol. *147*, 1647–52.

Bridges, C. B., Thompson, W. W., Meltzer, M. I., Reeve, G. R., Talamonti, W. J., Cox, N. J., Lilac, H. A., Hall, H., Klimov, A., and Fukuda, K. (2000). Effectiveness and cost-benefit of influenza vaccination of healthy working adults: A randomized controlled trial. J. Amer. Med. Assoc. *284*, 1655–63.

Bruhl, P., Kerschbaum, A., Kistner, O., Barrett, N., Dorner, F., and Gerencer, M. (2000). Humoral and cell-mediated immunity to Vero cell-derived influenza vaccine. Vaccine *19*, 1149–58.

Bungener, L., Serre, K., Bijl, L., Leserman, L., Wilschut, J., Daemen, T., and Machy, P. (2002). Virosome-mediated delivery of protein antigens to dendritic cells. Vaccine *20*, 2287–95.

Byars, N. E., Allison, A. C., Harmon, M. W., and Kendal, A. P. (1990). Enhancement of antibody responses to influenza B virus haemagglutinin by use of a new adjuvant formulation. Vaccine *8*, 49–56.

Center for Biologics Evaluation and Research, Food and Drug Administration. Points to consider in the characterization of cell lines used to produce biologicals. 1987.

Centers for Disease Control and Prevention (2004). Update: Influenza Activity – United States and Worldwide, 2003–04 Season, and Composition of the 2004–05 Influenza Vaccine. Morb Mortal Wkly Rep. *53*, 547–552.

Chattaraj, S. C., Rathinavelu, A., and Das, S. K. (1999). Biodegradable microparticles of influenza viral vaccine: comparison of the effects of routes of administration on the in vivo immune response in mice. J. Control Release *58*, 223–32.

Chen, D., Endres, R., Maa, Y. F., Kensil, C. R., Whitaker-Dowling, P., Trichel, A., Youngner, J. S., and Payne, L. G. (2003a). Epidermal powder immunization of mice and monkeys with an influenza vaccine. Vaccine *21*, 2830–6.

Chen, D., Periwal, S. B., Larrivee, K., Zuleger, C., Erickson, C. A., Endres, R. L., and Payne, L. G. (2001). Serum and mucosal immune responses to an inactivated influenza virus vaccine induced by epidermal powder immunization. J. Virol. *75*, 7956–65.

Chen, H., Subbarao, K., Swayne, D., Chen, Q., Lu, X., Katz, J., Cox, N., and Matsuoka, Y. (2003b). Generation and evaluation of a high-growth reassortant H9N2 influenza A virus as a pandemic vaccine candidate. Vaccine *21*, 1974–9.

Chen, K. S. and Quinnan, G. V., Jr. (1989). Efficacy of inactivated influenza vaccine delivered by oral administration. Curr. Top Microbiol. Immunol. *146*, 101–6.

Chuang, T. H., Lee, J., Kline, L., Mathison, J. C., and Ulevitch, R. J. (2002). Toll-like receptor 9 mediates CpG-DNA signaling. J. Leukoc Biol. *71*, 538–44.

Cooper, C. L., Davis, H. L., Morris, M. L., Efler, S. M., Krieg, A. M., Li, Y., Laframboise, C., Al Adhami, M. J., Khaliq, Y., Seguin, I., and Cameron, D. W. (2004). Safety and immunogenicity of CPG 7909 injection as an adjuvant to Fluarix influenza vaccine. Vaccine *22*, 3136–43.

Couch, R. B., Kasel, J. A., Gerin, J. L., Schulman, J. L., and Kilbourne, E. D. (1974). Induction of partial immunity to influenza by a neuraminidase-specific vaccine. J. Infect. Dis. *129*, 411–20.

Coulter, A., Wong, T. Y., Drane, D., Bates, J., Macfarlan, R., and Cox, J. (1998). Studies on experimental adjuvanted influenza vaccines: comparison of immune stimulating complexes (Iscoms) and oil-in-water vaccines. Vaccine *16*, 1243–53.

Cox, N. J., Brammer, T. L., and Regnery, H. L. (1994). Influenza: global surveillance for epidemic and pandemic variants. Eur. J. Epidemiol. *10*, 467–70.

Crawford, J., Wilkinson, B., Vosnesensky, A., Smith, G., Garcia, M., Stone, H., and Perdue, M. L. (1999). Baculovirus-derived hemagglutinin vaccines protect against lethal influenza infections by avian H5 and H7 subtypes. Vaccine *17*, 2265–74.

Danenberg, H. D., Ben-Yehuda, A., Zakay-Rones, Z., and Friedman, G. (1995a). Dehydroepiandrosterone (DHEA) treatment reverses the impaired immune response of old mice to influenza vaccination and protects from influenza infection. Vaccine *13*, 1445–8.

Danenberg, H. D., Ben-Yehuda, A., Zakay-Rones, Z., and Friedman, G. (1995b). Dehydroepiandrosterone enhances influenza immunization in aged mice. Ann. N. Y Acad. Sci. *774*, 297–9.

de Bruijn, I. A., Nauta, J., Gerez, L., and Palache, A. M. (2004). Virosomal influenza vaccine: a safe and effective influenza vaccine with high efficacy in elderly and subjects with low pre-vaccination antibody titers. Virus Res. *103*, 139–45.

Deliyannis, G., Jackson, D. C., Dyer, W., Bates, J., Coulter, A., Harling-McNabb, L., and Brown, L. E. (1998). Immunopotentiation of humoral and cellular responses to inactivated influenza vaccines by two different adjuvants with potential for human use. Vaccine *16*, 2058–68.

Donnelly, J. J., Friedman, A., Martinez, D., Montgomery, D. L., Shiver, J. W., Motzel, S. L., Ulmer, J. B., and Liu, M. A. (1995). Preclinical efficacy of a prototype DNA vaccine: enhanced protection against antigenic drift in influenza virus. Nat. Med. *1*, 583–7.

Donnelly, J. J., Friedman, A., Ulmer, J. B., and Liu, M. A. (1997). Further protection against antigenic drift of influenza virus in a ferret model by DNA vaccination. Vaccine *15*, 865–8.

Durrer, P., Gluck, U., Spyr, C., Lang, A. B., Zurbriggen, R., Herzog, C., and Gluck, R. (2003). Mucosal antibody response induced with a nasal virosome-based influenza vaccine. Vaccine *21*, 4328–34.

Edwards, L., Williams, A. E., Krieg, A. M., Rae, A. J., Snelgrove, R. J., and Hussell, T. (2004). Stimulation via Toll-like receptor 9 reduces Cryptococcus neoformans-induced pulmonary inflammation in an IL-12-dependent manner. Eur. J. Immunol. *35*, 273–281.

Ennis, F. A., Cruz, J., Jameson, J., Klein, M., Burt, D., and Thipphawong, J. (1999). Augmentation of human influenza A virus-specific cytotoxic T lymphocyte memory by influenza vaccine and adjuvanted carriers (ISCOMS). Virology *259*, 256–61.

Fodor, E., Devenish, L., Engelhardt, O. G., Palese, P., Brownlee, G. G., and Garcia-Sastre, A. (1999). Rescue of influenza A virus from recombinant DNA. J. Virol. *73*, 9679–82.

Frace, A. M., Klimov, A. I., Rowe, T., Black, R. A., and Katz, J. M. (1999). Modified M2 proteins produce heterotypic immunity against influenza A virus. Vaccine *17*, 2237–44.

Frey, S., Poland, G., Percell, S., and Podda, A. (2003). Comparison of the safety, tolerability, and immunogenicity of a MF59-adjuvanted influenza vaccine and a non-adjuvanted influenza vaccine in non-elderly adults. Vaccine *21*, 4234–7.

Garcia-Sastre, A., Egorov, A., Matassov, D., Brandt, S., Levy, D. E., Durbin, J. E., Palese, P., and Muster, T. (1998). Influenza A virus lacking the NS1 gene replicates in interferon-deficient systems. Virology *252*, 324–30.

Gerdil, C. (2003). The annual production cycle for influenza vaccine. Vaccine *21*, 1776–9.

Gluck, R. and Metcalfe, I. C. (2002). New technology platforms in the development of vaccines for the future. Vaccine *20 Suppl 5*, 10–6.

Gluck, R., Mischler, R., Durrer, P., Furer, E., Lang, A. B., Herzog, C., and Cryz, S. J., Jr. (2000). Safety and immunogenicity of intranasally administered inactivated trivalent virosome-formulated influenza vaccine containing Escherichia coli heat-labile toxin as a mucosal adjuvant. J. Infect. Dis. *181*, 1129–32.

Govaert, T. M., Thijs, C. T., Masurel, N., Sprenger, M. J., Dinant, G. J., and Knottnerus, J. A. (1994). The efficacy of influenza vaccination in elderly individuals. A randomized double-blind placebo-controlled trial. J. Amer. Med. Assoc. *272*, 1661–5.

Griffiths, E. (1999). WHO requirements for the use of animal cells as in vitro substrates for the production of biologicals: application to influenza vaccine production. Dev. Biol. Stand *98*, 153–7.

Guebre-Xabier, M., Hammond, S. A., Ellingsworth, L. R., and Glenn, G. M. (2004). Immunostimulant patch enhances immune responses to influenza virus vaccine in aged mice. J. Virol. *78*, 7610–8.

Guebre-Xabier, M., Hammond, S. A., Epperson, D. E., Yu, J., Ellingsworth, L., and Glenn, G. M. (2003). Immunostimulant patch containing heat-labile enterotoxin from Escherichia coli enhances immune responses to injected influenza virus vaccine through activation of skin dendritic cells. J. Virol. *77*, 5218–25.

Guy, B., Pascal, N., Francon, A., Bonnin, A., Gimenez, S., Lafay-Vialon, E., Trannoy, E., and Haensler, J. (2001). Design, characterization and preclinical efficacy of a cationic lipid adjuvant for influenza split vaccine. Vaccine *19*, 1794–805.

Halperin, S. A., Smith, B., Mabrouk, T., Germain, M., Trepanier, P., Hassell, T., Treanor, J., Gauthier, R., and Mills, E. L. (2002). Safety and immunogenicity of a trivalent, inactivated, mammalian cell culture-derived influenza vaccine in healthy adults, seniors, and children. Vaccine *20*, 1240–7.

Halvorson, D. A. (2002). The control of H5 or H7 mildly pathogenic avian influenza: a role for inactivated vaccine. Avian Pathol. *31*, 5–12.

Hehme, N., Engelmann, H., Kunzel, W., Neumeier, E., and Sanger, R. (2002). Pandemic preparedness: lessons learnt from H2N2 and H9N2 candidate vaccines. Med. Microbiol. Immunol. (Berl) *191*, 203–8.

Heinen, P. P., Rijsewijk, F. A., de Boer-Luijtze, E. A., and Bianchi, A. T. (2002). Vaccination of pigs with a DNA construct expressing an influenza virus M2-nucleoprotein fusion protein exacerbates disease after challenge with influenza A virus. J. Gen. Virol. *83*, 1851–9.

Hemmi, H., Takeuchi, O., Kawai, T., Kaisho, T., Sato, S., Sanjo, H., Matsumoto, M., Hoshino, K., Wagner, H., Takeda, K., and Akira, S. (2000). A Toll-like receptor recognizes bacterial DNA. Nature *408*, 740–5.

Hjorth, R. N., Bonde, G. M., Piner, E. D., Goldberg, K. M., and Levner, M. H. (1997). The effect of Syntex adjuvant formulation (SAF-m) on humoral immunity to the influenza virus in the mouse. Vaccine *15*, 541–6.

Hoffmann, E., Krauss, S., Perez, D., Webby, R., and Webster, R. G. (2002a). Eight-plasmid system for rapid generation of influenza virus vaccines. Vaccine *20*, 3165–70.

Hoffmann, E., Mahmood, K., Yang, C. F., Webster, R. G., Greenberg, H. B., and Kemble, G. (2002b). Rescue of influenza B virus from eight plasmids. Proc. Natl. Acad. Sci. USA. *99*, 11411–6.

Hoffmann, E., Neumann, G., Hobom, G., Webster, R. G., and Kawaoka, Y. (2000). 'Ambisense' approach for the generation of influenza A virus: vRNA and mRNA synthesis from one template. Virology *267*, 310–7.

Horimoto, T., Takada, A., Iwatsuki-Horimoto, K., Hatta, M., Goto, H., and Kawaoka, Y. (2003). Generation of influenza A viruses with chimeric (type A/B) hemagglutinins. J. Virol. *77*, 8031–8.

Ikonomidis, G., Portnoy, D. A., Gerhard, W., and Paterson, Y. (1997). Influenza-specific immunity induced by recombinant Listeria monocytogenes vaccines. Vaccine *15*, 433–40.

Illum, L., Jabbal-Gill, I., Hinchcliffe, M., Fisher, A. N., and Davis, S. S. (2001). Chitosan as a novel nasal delivery system for vaccines. Adv. Drug Deliv Rev. *51*, 81–96.

Itamura, S., Iinuma, H., Shida, H., Morikawa, Y., Nerome, K., and Oya, A. (1990). Characterization of antibody and cytotoxic T lymphocyte responses to human influenza virus H3 haemagglutinin expressed from the haemagglutinin locus of vaccinia virus. J. Gen. Virol. *71 (Pt 12)*, 2859–65.

Jin, H., Lu, B., Zhou, H., Ma, C., Zhao, J., Yang, C. F., Kemble, G., and Greenberg, H. (2003). Multiple amino acid residues confer temperature sensitivity to human influenza virus vaccine strains (FluMist) derived from cold-adapted A/Ann Arbor/6/60. Virology *306*, 18–24.

Jin, H., Zhou, H., Lu, B., and Kemble, G. (2004). Imparting temperature sensitivity and attenuation in ferrets to A/Puerto Rico/8/34 influenza virus by transferring the genetic signature for temperature sensitivity from cold-adapted A/Ann Arbor/6/60. J. Virol. *78*, 995–8.

Johansson, B. E. (1999). Immunization with influenza A virus hemagglutinin and neuraminidase produced in recombinant baculovirus results in a balanced and broadened immune response superior to conventional vaccine. Vaccine *17*, 2073–80.

Johansson, M. and Lovgren-Bengtsson, K. (1999). Iscoms with different quillaja saponin components differ in their immunomodulating activities. Vaccine *17*, 2894–900.

Jones, T., Allard, F., Cyr, S. L., Tran, S. P., Plante, M., Gauthier, J., Bellerose, N., Lowell, G. H., and Burt, D. S. (2003). A nasal Proteosome influenza vaccine containing baculovirus-derived hemagglutinin induces protective mucosal and systemic immunity. Vaccine *21*, 3706–12.

Kaji, M., Kaji, Y., Ohkuma, K., Honda, T., Oka, T., Sakoh, M., Nakamura, S., Kurachi, K., and Sentoku, M. (1992). Phase 1 clinical tests of influenza MDP-virosome vaccine (KD-5382). Vaccine *10*, 663–7.

Karupiah, G., Ramsay, A. J., Ramshaw, I. A., and Blanden, R. V. (1992). Recombinant vaccine vector-induced protection of athymic, nude mice from influenza A virus infection. Analysis of protective mechanisms. Scand. J. Immunol. *36*, 99–105.

Katz, J. M., Wang, M., and Webster, R. G. (1990). Direct sequencing of the HA gene of influenza (H3N2) virus in original clinical samples reveals sequence identity with mammalian cell-grown virus. J. Virol. *64*, 1808–11.

Keitel, W. A., Cate, T. R., Atmar, R. L., Turner, C. S., Nino, D., Dukes, C. M., Six, H. R., and Couch, R. B. (1996). Increasing doses of purified influenza virus hemagglutinin and subvirion vaccines enhance antibody responses in the elderly. Clin. Diagn. Lab. Immunol. *3*, 507–10.

Keitel, W. A., Cate, T. R., Nino, D., Huggins, L. L., Six, H. R., Quarles, J. M., and Couch, R. B. (2001). Immunization against influenza: comparison of various topical and parenteral regimens containing inactivated and/or live attenuated vaccines in healthy adults. J. Infect. Dis. *183*, 329–332.

Keitel, W. A., Couch, R. B., Cate, T. R., Hess, K. R., Baxter, B., Quarles, J. M., Atmar, R. L., and Six, H. R. (1994). High doses of purified influenza A virus hemagglutinin significantly augment serum and nasal secretion antibody responses in healthy young adults. J. Clin. Microbiol. *32*, 2468–73.

Kendal, A. P. (1997). Cold-adapted live attenuated influenza vaccines developed in Russia: can they contribute to meeting the needs for influenza control in other countries? Eur. J. Epidemiol. *13*, 591–609.

Kilbourne, E. D., Pokorny, B. A., Johansson, B., Brett, I., Milev, Y., and Matthews, J. T. (2004). Protection of mice with recombinant influenza virus neuraminidase. J. Infect. Dis. *189*, 459–61.

Kistner, O., Barrett, P. N., Mundt, W., Reiter, M., Schober-Bendixen, S., Eder, G., and Dorner, F. (1999). Development of a Vero cell-derived influenza whole virus vaccine. Dev. Biol. Stand *98*, 101–10.

Klimov, A. I., Cox, N. J., Yotov, W. V., Rocha, E., Alexandrova, G. I., and Kendal, A. P. (1992). Sequence changes in the live attenuated, cold-adapted variants of influenza A/Leningrad/134/57 (H2N2) virus. Virology *186*, 795–7.

Lakey, D. L., Treanor, J. J., Betts, R. F., Smith, G. E., Thompson, J., Sannella, E., Reed, G., Wilkinson, B. E., and Wright, P. F. (1996). Recombinant baculovirus influenza A hemagglutinin vaccines are well tolerated and immunogenic in healthy adults. J. Infect. Dis. *174*, 838–41.

Lemoine, D., Deschuyteneer, M., Hogge, F., and Preat, V. (1999). Intranasal immunization against influenza virus using polymeric particles. J. Biomater Sci. Polym Ed *10*, 805–25.

Levi, R. and Arnon, R. (1996). Synthetic recombinant influenza vaccine induces efficient long-term immunity and cross-strain protection. Vaccine *14*, 85–92.

Li, S., Liu, C., Klimov, A., Subbarao, K., Perdue, M. L., Mo, D., Ji, Y., Woods, L., Hietala, S., and Bryant, M. (1999). Recombinant influenza A virus vaccines for the pathogenic human A/Hong Kong/97 (H5N1) viruses. J. Infect. Dis. *179*, 1132–8.

Liu, M., Wood, J. M., Ellis, T., Krauss, S., Seiler, P., Johnson, C., Hoffmann, E., Humberd, J., Hulse, D., Zhang, Y., Webster, R. G., and Perez, D. R. (2003). Preparation of a standardized, efficacious agricultural H5N3 vaccine by reverse genetics. Virology *314*, 580–90.

Lowell, G. H., Ballou, W. R., Smith, L. F., Wirtz, R. A., Zollinger, W. D., and Hockmeyer, W. T. (1988). Proteosome-lipopeptide vaccines: enhancement of immunogenicity for malaria CS peptides. Science *240*, 800–2.

Lu, X., Renshaw, M., Tumpey, T. M., Kelly, G. D., Hu-Primmer, J., and Katz, J. M. (2001). Immunity to influenza A H9N2 viruses induced by infection and vaccination. J. Virol. *75*, 4896–901.

Lu, X., Tumpey, T. M., Morken, T., Zaki, S. R., Cox, N. J., and Katz, J. M. (1999). A mouse model for the evaluation of pathogenesis and immunity to influenza A (H5N1) viruses isolated from humans. J. Virol. *73*, 5903–11.

Maassab, H. F., Kendal, A. P., Abrams, G. D., and Monto, A. S. (1982). Evaluation of a cold-recombinant influenza virus vaccine in ferrets. J. Infect. Dis. *146*, 780–90.

Massari, P., Henneke, P., Ho, Y., Latz, E., Golenbock, D. T., and Wetzler, L. M. (2002). Cutting edge: Immune stimulation by neisserial porins is toll-like receptor 2 and MyD88 dependent. J. Immunol. *168*, 1533–7.

Mbawuike, I. N., Acuna, C., Caballero, D., Pham-Nguyen, K., Gilbert, B., Petribon, P., and Harmon, M. (1996). Reversal of age-related deficient influenza virus-specific CTL responses and IFN-gamma production by monophosphoryl lipid A. Cell Immunol. *173*, 64–78.

Mbawuike, I. N. and Wyde, P. R. (1993). Induction of CD8+ cytotoxic T cells by immunization with killed influenza virus and effect of cholera toxin B subunit. Vaccine *11*, 1205–13.

Mbawuike, I. N., Wyde, P. R., and Anderson, P. M. (1990). Enhancement of the protective efficacy of inactivated influenza A virus vaccine in aged mice by IL-2 liposomes. Vaccine *8*, 347–52.

McEwen, J., Levi, R., Horwitz, R. J., and Arnon, R. (1992). Synthetic recombinant vaccine expressing influenza haemagglutinin epitope in Salmonella flagellin leads to partial protection in mice. Vaccine *10*, 405–11.

Meltzer, M. I., Cox, N. J., and Fukuda, K. (1999). The economic impact of pandemic influenza in the United States: priorities for intervention. Emerg. Infect. Dis. *5*, 659–671.

Moldoveanu, Z., Love-Homan, L., Huang, W. Q., and Krieg, A. M. (1998). CpG DNA, a novel immune enhancer for systemic and mucosal immunization with influenza virus. Vaccine *16*, 1216–24.

Morein, B., Sundquist, B., Hoglund, S., Dalsgaard, K., and Osterhaus, A. (1984). Iscom, a novel structure for antigenic presentation of membrane proteins from enveloped viruses. Nature *308*, 457–60.

Mowat, A. M., Maloy, K. J., and Donachie, A. M. (1993). Immune-stimulating complexes as adjuvants for inducing local and systemic immunity after oral immunization with protein antigens. Immunology *80*, 527–34.

Mozdzanowska, K., Feng, J., Eid, M., Kragol, G., Cudic, M., Otvos, L., Jr., and Gerhard, W. (2003). Induction of influenza type A virus-specific resistance by immunization of mice with a synthetic multiple antigenic peptide vaccine that contains ectodomains of matrix protein 2. Vaccine *21*, 2616–2626.

Murphy, B. R. and Coelingh, K. (2002). Principles underlying the development and use of live attenuated cold-adapted influenza A and B virus vaccines. Viral Immunol. *15*, 295–323.

Mustafaev, M. I. and Norimov, A. (1990). Polymer-metal complexes of protein antigens – new highly effective immunogens. Biomed Sci. *1*, 274–8.

Muszkat, M., Greenbaum, E., Ben-Yehuda, A., Oster, M., Yeu'l, E., Heimann, S., Levy, R., Friedman, G., and Zakay-Rones, Z. (2003). Local and systemic immune response in nursing-home elderly following intranasal or intramuscular immunization with inactivated influenza vaccine. Vaccine *21*, 1180–6.

Mutsch, M., Zhou, W., Rhodes, P., Bopp, M., Chen, R. T., Linder, T., Spyr, C., and Steffen, R. (2004). Use of the inactivated intranasal influenza vaccine and the risk of Bell's palsy in Switzerland. N. Engl. J. Med. *350*, 896–903.

Neirynck, S., Deroo, T., Saelens, X., Vanlandschoot, P., Jou, W. M., and Fiers, W. (1999). A universal influenza A vaccine based on the extracellular domain of the M2 protein. Nat. Med. *5*, 1157–63.

Neumann, G. and Kawaoka, Y. (2002). Generation of influenza A virus from cloned cDNAs – historical perspective and outlook for the new millenium. Rev. Med. Virol. *12*, 13–30.

Neumann, G., Watanabe, T., Ito, H., Watanabe, S., Goto, H., Gao, P., Hughes, M., Perez, D. R., Donis, R., Hoffmann, E., Hobom, G., and Kawaoka, Y. (1999). Generation of influenza A viruses entirely from cloned cDNAs. Proc. Natl. Acad. Sci. USA. *96*, 9345–50.

Nichol, K. L., Wuorenma, J., and von Sternberg, T. (1998). Benefits of influenza vaccination for low-, intermediate-, and high-risk senior citizens. Arch. Intern. Med. *158*, 1769–76.

Nicholson, K. G., Colegate, A. E., Podda, A., Stephenson, I., Wood, J., Ypma, E., and Zambon, M. C. (2001). Safety and antigenicity of non-adjuvanted and MF59-adjuvanted influenza A/Duck/Singapore/97 (H5N3) vaccine: a randomised trial of two potential vaccines against H5N1 influenza. Lancet *357*, 1937–43.

Nixon, D. F., Hioe, C., Chen, P. D., Bian, Z., Kuebler, P., Li, M. L., Qiu, H., Li, X. M.,

Singh, M., Richardson, J., McGee, P., Zamb, T., Koff, W., Wang, C. Y., and O'Hagan, D. (1996). Synthetic peptides entrapped in microparticles can elicit cytotoxic T cell activity. Vaccine *14*, 1523–30.

Ott, G., Barchfeld, G. L., Chernoff, D., Radhakrishnan, R., van Hoogevest, P., and Van Nest, G. (1995). MF59. Design and evaluation of a safe and potent adjuvant for human vaccines. Pharm Biotechnol. *6*, 277–96.

Ozaki, H., Govorkova, E. A., Li, C., Xiong, X., Webster, R. G., and Webby, R. J. (2004). Generation of high-yielding influenza A viruses in African green monkey kidney (Vero) cells by reverse genetics. J. Virol. *78*, 1851–7.

Palache, A. M. (1997). Influenza vaccines. A reappraisal of their use. Drugs *54*, 841–56.

Partidos, C. D. (2003). Antigens onto bare skin: a 'painless' paradigm shift in vaccine delivery. Expert Opin. Biol. Ther *3*, 895–902.

Patriarca, P. A., Weber, J. A., Parker, R. A., Hall, W. N., Kendal, A. P., Bregman, D. J., and Schonberger, L. B. (1985). Efficacy of influenza vaccine in nursing homes. Reduction in illness and complications during an influenza A (H3N2) epidemic. J. Amer. Med. Assoc. *253*, 1136–9.

Pau, M. G., Ophorst, C., Koldijk, M. H., Schouten, G., Mehtali, M., and Uytdehaag, F. (2001). The human cell line PER. C6 provides a new manufacturing system for the production of influenza vaccines. Vaccine *19*, 2716–21.

Pauksen, K., Linde, A., Hammarstrom, V., Sjolin, J., Carneskog, J., Jonsson, G., Oberga, G., Engelmann, H., and Ljungman, P. (2000). Granulocyte-macrophage colony-stimulating factor as immunomodulating factor together with influenza vaccination in stem cell Transplant. patients. Clin. Infect. Dis. *30*, 342–8.

Payne, L. G., Jenkins, S. A., Woods, A. L., Grund, E. M., Geribo, W. E., Loebelenz, J. R., Andrianov, A. K., and Roberts, B. E. (1998). Poly[di(carboxylatophenoxy) phosphazene] (PCPP) is a potent immunoadjuvant for an influenza vaccine. Vaccine *16*, 92–8.

Pertmer, T. M., Eisenbraun, M. D., McCabe, D., Prayaga, S. K., Fuller, D. H., and Haynes, J. R. (1995). Gene gun-based nucleic acid immunization: elicitation of humoral and cytotoxic T lymphocyte responses following epidermal delivery of nanogram quantities of DNA. Vaccine *13*, 1427–30.

Plante, M., Jones, T., Allard, F., Torossian, K., Gauthier, J., St-Felix, N., White, G. L., Lowell, G. H., and Burt, D. S. (2001). Nasal immunization with subunit proteosome influenza vaccines induces serum HAI, mucosal IgA and protection against influenza challenge. Vaccine *20*, 218–25.

Powers, D. C. (1997). Summary of a clinical trial with liposome-adjuvanted influenza A virus vaccine in elderly adults. Mech Ageing Dev. *93*, 179–88.

Powers, D. C., Smith, G. E., Anderson, E. L., Kennedy, D. J., Hackett, C. S., Wilkinson, B. E., Volvovitz, F., Belshe, R. B., and Treanor, J. J. (1995). Influenza A virus vaccines containing purified recombinant H3 hemagglutinin are well tolerated and induce protective immune responses in healthy adults. J. Infect. Dis. *171*, 1595–9.

Rimmelzwaan, G. F., Baars, M., van Beek, R., van Amerongen, G., Lovgren-Bengtsson, K., Claas, E. C., and Osterhaus, A. D. (1997). Induction of protective immunity against influenza virus in a macaque model: comparison of conventional and iscom vaccines. J. Gen. Virol. *78 (Pt 4)*, 757–65.

Rimmelzwaan, G. F., Nieuwkoop, N., Brandenburg, A., Sutter, G., Beyer, W. E., Maher, D., Bates, J., and Osterhaus, A. D. (2000). A randomized, double blind study in young healthy adults comparing cell mediated and humoral immune responses induced by influenza ISCOM vaccines and conventional vaccines. Vaccine *19*, 1180–7.

Robinson, H. L., Hunt, L. A., and Webster, R. G. (1993). Protection against a lethal influenza virus challenge by immunization with a haemagglutinin-expressing plasmid DNA. Vaccine *11*, 957–60.

Roman, M., Martin-Orozco, E., Goodman, J. S., Nguyen, M. D., Sato, Y., Ronaghy, A., Kornbluth, R. S., Richman, D. D., Carson, D. A., and Raz, E. (1997). Immunostimulatory DNA sequences function as T helper-1-promoting adjuvants. Nat. Med. *3*, 849–54.

Rudenko, L. G., Arden, N. H., Grigorieva, E., Naychin, A., Rekstin, A., Klimov, A. I., Donina, S., Desheva, J., Holman, R. C., DeGuzman, A., Cox, N. J., and Katz, J. M. (2000). Immunogenicity and efficacy of Russian live attenuated and US inactivated influenza vaccines used alone and in combination in nursing home residents. Vaccine *19*, 308–18.

Sambhara, S., Kurichh, A., Miranda, R., Tamane, A., Arpino, R., James, O., McGuinness, U., Kandil, A., Underdown, B., Klein, M., and Burt, D. (1998a). Enhanced immune responses and resistance against infection in aged mice conferred by Flu-ISCOMs vaccine correlate with up-regulation of costimulatory molecule CD86. Vaccine *16*, 1698–704.

Sambhara, S., Kurichh, A., Miranda, R., Tumpey, T., Rowe, T., Renshaw, M., Arpino, R., Tamane, A., Kandil, A., James, O., Underdown, B., Klein, M., Katz, J., and Burt, D. (2001). Heterosubtypic immunity against human influenza A viruses, including recently emerged avian H5 and H9 viruses, induced by FLU-ISCOM vaccine in mice requires both cytotoxic T-lymphocyte and macrophage function. Cell Immunol. *211*, 143–153.

Sambhara, S., Woods, S., Arpino, R., Kurichh, A., Tamane, A., Underdown, B., Klein, M., Lovgren, B. K., Morein, B., and Burt, D. (1998b). Heterotypic protection against influenza by immunostimulating complexes is associated with the induction of cross-reactive cytotoxic T lymphocytes. J. Infect. Dis. *177*, 1266–1274.

Schickli, J. H., Flandorfer, A., Nakaya, T., Martinez-Sobrido, L., Garcia-Sastre, A., and Palese, P. (2001). Plasmid-only rescue of influenza A virus vaccine candidates. Philos Trans R Soc. Lond B Biol. Sci. *356*, 1965–73.

Schild, G. C., Oxford, J. S., de Jong, J. C., and Webster, R. G. (1983). Evidence for host-cell selection of influenza virus antigenic variants. Nature *303*, 706–9.

Shinya, K., Hamm, S., Hatta, M., Ito, H., Ito, T., and Kawaoka, Y. (2004). PB2 amino acid at position 627 affects replicative efficiency, but not cell tropism, of Hong Kong H5N1 influenza A viruses in mice. Virology *320*, 258–66.

Simonsen, L., Clarke, M. J., Schonberger, L. B., Arden, N. H., Cox, N. J., and Fukuda, K. (1998). Pandemic versus epidemic influenza mortality: a pattern of changing age distribution. J. Infect. Dis. *178*, 53–60.

Singh, M., Briones, M., and O'Hagan, D. T. (2001). A novel bioadhesive intranasal delivery system for inactivated influenza vaccines. J. Control Release *70*, 267–76.

Skea, D. L., Douglas, A. R., Skehel, J. J., and Barber, B. H. (1993). The immunotargeting approach to adjuvant-independent immunization with influenza haemagglutinin. Vaccine *11*, 994–1002.

Slepushkin, V. A., Katz, J. M., Black, R. A., Gamble, W. C., Rota, P. A., and Cox, N. J. (1995). Protection of mice against influenza A virus challenge by vaccination with baculovirus-expressed M2 protein. Vaccine *13*, 1399–402.

Stepanova, L., Naykhin, A., Kolmskog, C., Jonson, G., Barantceva, I., Bichurina, M., Kubar, O., and Linde, A. (2002). The humoral response to live and inactivated influenza vaccines administered alone and in combination to young adults and elderly. J. Clin. Virol. *24*, 193–201.

Stephenson, I., Nicholson, K. G., Gluck, R., Mischler, R., Newman, R. W., Palache, A. M., Verlander, N. Q., Warburton, F., Wood, J. M., and Zambon, M. C. (2003). Safety and antigenicity of whole virus and subunit influenza A/Hong Kong/1073/99 (H9N2) vaccine in healthy adults: phase I randomised trial. Lancet *362*, 1959–66.

Subbarao, K., Chen, H., Swayne, D., Mingay, L., Fodor, E., Brownlee, G., Xu, X., Lu, X., Katz, J., Cox, N., and Matsuoka, Y. (2003). Evaluation of a genetically modified reassortant H5N1 influenza A virus vaccine candidate generated by plasmid-based reverse genetics. Virology *305*, 192–200.

Sundquist, B., Lovgren, K., and Morein, B. (1988). Influenza virus ISCOMs: antibody response in animals. Vaccine *6*, 49–53.

Taglietti, M. (1995). Vaccine adjuvancy: a new potential area of development for GM-CSF. Adv. Exp. Med. Biol. *378*, 565–9.

Takada, A., Kuboki, N., Okazaki, K., Ninomiya, A., Tanaka, H., Ozaki, H., Itamura, S., Nishimura, H., Enami, M., Tashiro, M., Shortridge, K. F., and Kida, H. (1999). Avirulent Avian influenza virus as a vaccine strain against a potential human pandemic. J. Virol. *73*, 8303–7.

Talon, J., Salvatore, M., O'Neill, R. E., Nakaya, Y., Zheng, H., Muster, T., Garcia-Sastre, A., and Palese, P. (2000). Influenza A and B viruses expressing altered NS1 proteins: A vaccine approach. Proc. Natl. Acad. Sci. USA. *97*, 4309–14.

Tamizifar, H., Robinson, A., Jennings, R., and Potter, C. W. (1995). Immune response and protection against influenza A infection in mice immunised with subunit influenza A vaccine in combination with whole cell or acellular DTP vaccine. Vaccine *13*, 1539–46.

Tamura, S., Ito, Y., Asanuma, H., Hirabayashi, Y., Suzuki, Y., Nagamine, T., Aizawa, C., and Kurata, T. (1992). Cross-protection against influenza virus infection afforded by trivalent inactivated vaccines inoculated intranasally with cholera toxin B subunit. J. Immunol. *149*, 981–8.

Tamura, S., Yamanaka, A., Shimohara, M., Tomita, T., Komase, K., Tsuda, Y., Suzuki, Y., Nagamine, T., Kawahara, K., Danbara, H., and et al. (1994). Synergistic action of cholera toxin B subunit (and Escherichia coli heat-labile toxin B subunit) and a trace amount of cholera whole toxin as an adjuvant for nasal influenza vaccine. Vaccine *12*, 419–26.

Thompson, W. W., Shay, D. K., Weintraub, E., Brammer, L., Cox, N., Anderson, L. J., and Fukuda, K. (2003). Mortality associated with influenza and respiratory syncytial virus in the United States. J. Amer. Med. Assoc. *289*, 179–86.

Tite, J. P., Gao, X. M., Hughes-Jenkins, C. M., Lipscombe, M., O'Callaghan, D., Dougan, G., and Liew, F. Y. (1990). Anti-viral immunity induced by recombinant nucleoprotein of influenza A virus. III. Delivery of recombinant nucleoprotein to the immune system using attenuated Salmonella typhimurium as a live carrier. Immunology *70*, 540–6.

Todd, C. W., Lee, E., Balusubramanian, M., Shah, H., Henk, W. G., Younger, L. E., and Newman, M. J. (1998). Systematic development of a block copolymer adjuvant for trivalent influenza virus vaccine. Dev. Biol. Stand *92*, 341–51.

Treanor, J. J., Betts, R. F., Smith, G. E., Anderson, E. L., Hackett, C. S., Wilkinson, B. E., Belshe, R. B., and Powers, D. C. (1996). Evaluation of a recombinant hemagglutinin expressed in insect cells as an influenza vaccine in young and elderly adults. J. Infect. Dis. *173*, 1467–70.

Treanor, J. J., Mattison, H. R., Dumyati, G., Yinnon, A., Erb, S., O'Brien, D., Dolin, R., and Betts, R. F. (1992). Protective efficacy of combined live intranasal and inactivated influenza A virus vaccines in the elderly. Ann. Intern. Med. *117*, 625–33.

Treanor, J. J., Tierney, E. L., Zebedee, S. L., Lamb, R. A., and Murphy, B. R. (1990). Passively transferred monoclonal antibody to the M2 protein inhibits influenza A virus replication in mice. J. Virol. *64*, 1375–7.

Treanor, J. J., Wilkinson, B. E., Masseoud, F., Hu-Primmer, J., Battaglia, R., O'Brien, D., Wolff, M., Rabinovich, G., Blackwelder, W., and Katz, J. M. (2001). Safety and immunogenicity of a recombinant hemagglutinin vaccine for H5 influenza in humans. Vaccine *19*, 1732–7.

Ulmer, J. B. (2002). Influenza DNA vaccines. Vaccine *20 Suppl 2*, 74–6.

Ulmer, J. B., Deck, R. R., DeWitt, C. M., Friedman, A., Donnelly, J. J., and Liu, M. A. (1994). Protective immunity by intramuscular injection of low doses of influenza virus DNA vaccines. Vaccine *12*, 1541–4.

Ulmer, J. B., Donnelly, J. J., Parker, S. E., Rhodes, G. H., Felgner, P. L., Dwarki, V. J., Gromkowski, S. H., Deck, R. R., DeWitt, C. M., Friedman, A., and et al. (1993). Heterologous protection against influenza by injection of DNA encoding a viral protein. Science *259*, 1745–9.

Ulmer, J. B., Fu, T. M., Deck, R. R., Friedman, A., Guan, L., DeWitt, C., Liu, X., Wang, S., Liu, M. A., Donnelly, J. J., and Caulfield, M.

J. (1998). Protective CD4+ and CD8+ T cells against influenza virus induced by vaccination with nucleoprotein DNA. J. Virol. *72*, 5648–53.

Watanabe, T., Watanabe, S., Kida, H., and Kawaoka, Y. (2002a). Influenza A virus with defective M2 ion channel activity as a live vaccine. Virology *299*, 266–70.

Watanabe, T., Watanabe, S., Neumann, G., Kida, H., and Kawaoka, Y. (2002b). Immunogenicity and protective efficacy of replication-incompetent influenza virus-like particles. J. Virol. *76*, 767–73.

Webby, R. J., Perez, D. R., Coleman, J. S., Guan, Y., Knight, J. H., Govorkova, E. A., McClain-Moss, L. R., Peiris, J. S., Rehg, J. E., Tuomanen, E. I., and Webster, R. G. (2004). Responsiveness to a pandemic alert: use of reverse genetics for rapid development of influenza vaccines. Lancet *363*, 1099–103.

Webster, R. G., Fynan, E. F., Santoro, J. C., and Robinson, H. (1994). Protection of ferrets against influenza challenge with a DNA vaccine to the haemagglutinin. Vaccine *12*, 1495–8.

Webster, R. G., Reay, P. A., and Laver, W. G. (1988). Protection against lethal influenza with neuraminidase. Virology *164*, 230–7.

WHO (2003). Production of pilot lots of inactivated influenza vaccine in response to a pandemic threat: an interim biosafety risk assessment. Weekly Epidemiological Record *78*, 405–408.

Wood, J. M. (2001). Developing vaccines against pandemic influenza. Philos Trans R Soc. Lond B Biol. Sci. *356*, 1953–60.

Yasui, H., Kiyoshima, J., Hori, T., and Shida, K. (1999). Protection against influenza virus infection of mice fed Bifidobacterium breve YIT4064. Clin. Diagn. Lab. Immunol. *6*, 186–92.

Yasui, H., Nagaoka, N., and Hayakawa, K. (1994). Augmentation of anti-influenza virus hemagglutinin antibody production by Peyer's patch cells with Bifidobacterium breve YIT4064. Clin. Diagn. Lab. Immunol. *1*, 244–6.

Zurbriggen, R. (2003). Immunostimulating reconstituted influenza virosomes. Vaccine *21*, 921–4.

CHAPTER 9

Epidemiology and control of human and animal influenza

Kanta Subbarao, David E. Swayne and Christopher W. Olsen

ABSTRACT

Influenza viruses are clinically and economically important agents of disease in people, horses, pigs, marine mammals and poultry. Human influenza results from infection with influenza A, B or C viruses and a wide variety of domestic and free-ranging wild animal species can be infected with influenza A viruses. Aquatic birds are the natural hosts of influenza A viruses and represent a vast, global reservoir of influenza genes. Because pandemic influenza is fundamentally a zoonotic disease involving interspecies transmission of viruses from animals, this chapter jointly reviews the epidemiology, ecology and evolution of influenza viruses among humans, birds, and pigs. The epidemiologic consequences of genetic reassortment and adaptation of influenza viruses in these species and interspecies transmission are discussed. The epidemiology of interpandemic human influenza is presented with an emphasis on events of the past decade.

Human influenza is an omnipresent disease with ancient roots (Potter 2001), and yet, because of continual virus evolution, also a perpetually re-emerging disease. Influenza viruses are negative sense, segmented, ribonucleic acid viruses of the family *Orthomyxoviridae*. Influenza viruses are classified into three genera or types (A, B and C) based on serological reactions of the conserved internal proteins, principally the nucleoprotein and matrix protein, in an agar gel immunodiffusion test (Beard 1970). The genus *Influenzavirus A* includes all avian and equine influenza viruses, and most influenza viruses of swine. Humans can have infections with influenza viruses A, B or C. Type A influenza viruses have eight gene segments that encode eleven different proteins (Chen, Calvo et al. 2001; Lamb and Krug 2001). *Influenzavirus A* can be further classified into subtypes based on the surface glycoproteins, i.e. the hemagglutinin (HA) and neuraminidase (NA). There are 16 HA (H1-16) and 9 NA (N1-9) recognized subtypes, based on hemagglutination-inhibition (HI) and neuraminidase-inhibition (NI) tests, respectively.

Human influenza A viruses evolve through two primary mechanisms (reviewed in (Webster, Bean et al. 1992; Wright and Webster 2001)). A slow, progressive process of point mutations produces subtle variants of the currently circulating viruses. These antigenic drift mutants account for the epidemics of influenza A that occur in most countries on an almost yearly basis. In contrast, influenza A viruses of an HA subtype different from those currently circulating have appeared periodically in history. These viruses constitute an 'antigenic shift' because the population is immunologically naive to the new subtype. As such, these viruses can replicate largely unabated and lead to massive pandemics of disease.

The impact of influenza in the human population, whether measured by morbidity, mortality, or economic costs, is clear and significant. Annual influenza epidemics in the

U.S. are associated with overall attack rates of 10–20%, an average of 114,000-142,000 hospitalizations, 20,000–36,000 deaths (~1% of all deaths in the U.S.), and up to $10 billion in medical expenses and lost income (Glezen 1996; Simonsen, Clarke et al. 1997; Klimov, Simonsen et al. 1999; Simonsen 1999; Cox and Subbarao 2000; Bridges, Fukuda et al. 2001). And the consequences of pandemic influenza are far greater. The 1957 'Asian flu' and 1968 'Hong Kong flu' pandemics caused 69,800 and 33,800 excess deaths, respectively, in the United States alone (Noble 1982; Klimov, Simonsen et al. 1999; Cox and Subbarao 2000), the 'Spanish flu' pandemic of 1918 is estimated to have made 25–30% of the human population ill and killed over 40 million people worldwide (Grove and Hetzel 1968; Crosby 1989; Taubenberger, Reid et al. 2000; Taubenberger, Reid et al. 2001) (See Chapter 11 for more information on the 1918 pandemic), and projections of the first year of the next influenza pandemic in the U.S. in the absence of intervention include 89,000–207,000 deaths, 314,000–734,000 hospitalizations, 38 to 89 million total illnesses and up to $166 billion in direct costs (Meltzer, Cox et al. 1999). The process of antigenic drift can be explained by the dynamics of influenza virus circulation and immunization within the human population. However, pandemic influenza is fundamentally a zoonotic disease involving interspecies transmission of viruses from animals. Therefore, this chapter will jointly review the epidemiology and evolution of influenza viruses among human beings, birds, and pigs.

INFLUENZA A VIRUS ECOLOGY, HOST RANGE AND SPECIES SPECIFICITY

Influenza A viruses infect a wide variety of domestic and free-ranging wild animal species. In horses, pigs, marine mammals and poultry, as in people, these viruses are clinically and/or economically important agents of disease (Geraci, St. Aubin et al. 1982; Acland, Silverman Bachin et al. 1984; Traub-Dargatz, Salman et al. 1991; Webster, Bean et al. 1992; Easterday, Hinshaw et al. 1997; Janke 1998; Van Reeth and Easterday 1999; Perkins and Swayne 2001; Wright and Webster 2001; Olsen 2002b; Olsen 2002a). In contrast, influenza A viruses in waterfowl and shorebird species are highly host-adapted. Infections in these birds are generally subclinical, and aquatic bird influenza viruses exhibit very low evolutionary rates. They are, therefore, considered to be in 'evolutionary stasis' (Gorman, Bean et al. 1990; Webster, Bean et al. 1992; Webster and Kawaoka 1994; Wright and Webster 2001). All 16 HA and 9 NA subtypes of influenza A viruses have been recovered from aquatic birds, and because influenza A viruses replicate primarily in epithelial cells of the gastrointestinal tract in waterfowl, they are shed in feces and readily contaminate the lakes and ponds the birds visit (Slemons, Johnson et al. 1974; Webster, Yakhno et al. 1978; Hinshaw, Webster et al. 1980; Halvorson, Karunakaran et al. 1983; Webster, Bean et al. 1992; Ito and Kawaoka 1998; Webster 1998; Laver, Bischofberger et al. 2000; Webby and Webster 2001; Wright and Webster 2001). Aquatic birds, thus, constitute an ever-present, vast, global reservoir of influenza viruses (Webster, Bean et al. 1992; Ito and Kawaoka 1998; Webby and Webster 2001). Reassortment among influenza viruses in aquatic birds results in the generation of viruses with novel genotypes that can co-circulate with or replace previously established genotypes (Chen, Deng et al. 2004; Li, Guan et al. 2004).

Beyond serving as a present day reservoir of viruses, phylogenetic data indicate that aquatic birds were also the evolutionary ancestral source of the current lineages of mammalian influenza A viruses (Gammelin, Altmuller et al. 1990; Gorman, Bean et al. 1990; Webster, Bean et al. 1992; Webster 1998). Historically, however, only viruses of H1,

H2, H3, N1 and N2 subtypes have circulated widely in the human population (Webster, Bean et al. 1992; Webster 1998; Wright and Webster 2001), and only H1, H3, N1 and N2 subtype viruses have been consistently isolated among pigs (Webby and Webster 2001; Olsen 2002b; Olsen 2002a). Since 1997, a limited number of human infections with avian influenza viruses of three subtypes (H5N1, H9N2 and H7N7) have occurred in association with poultry infections. Each of these examples of wholly avian virus infections raised concerns for a new pandemic, since the human population has not had immunologic experience with H5, H9 or H7 viruses. However, there was only minimal evidence for human-to-human transmission of these viruses (Claas, Osterhaus et al. 1998; Subbarao, Klimov et al. 1998; Yuen, Chan et al. 1998; Mounts, Kwong et al. 1999; Peiris, Yuen et al. 1999; Bridges, Katz et al. 2000; Lin, Shaw et al. 2000; Hatta, Gao et al. 2001; Webby and Webster 2001; Bridges, Lim et al. 2002; Hatta and Kawaoka 2002; http://www.who.int/csr/don/2003_04_24/en/), which is a prerequisite for generating a virus with true pandemic potential. (See Chapter 10 for more information on the H5 virus)

The lack of human-to-human spread of these viruses is consistent with the fact that there are host range barriers to transmission of influenza A viruses from one species to another. In particular, barriers to direct transmission of influenza viruses among human beings and birds have been recognized for many years. Studies have shown that avian influenza viruses generally replicate poorly in human beings and non-human primates, and vice versa, human influenza viruses do not typically replicate efficiently in birds (Hinshaw, Webster et al. 1983b; Snyder, Buckler-White et al. 1987; Beare and Webster 1991; Webster, Bean et al. 1992; Webby and Webster 2001; Wright and Webster 2001). Species-specificity of influenza viruses is most likely a multigenic trait and evidence has accumulated over many years of research for potential contributions by all of the viral gene products (Almond 1977; Scholtissek, Koennecke et al. 1978a; Mahy 1983; Scholtissek, Burger et al. 1985; Tian, Buckler-White et al. 1985; Buckler-White, Naeve et al. 1986; Snyder, Buckler-White et al. 1987; Klenk and Rott 1988; Snyder, Betts et al. 1988; Clements, Subbarao et al. 1992; Webster, Bean et al. 1992; Subbarao, London et al. 1993; Webby and Webster 2001; Wright and Webster 2001; Takahashi, Suzuki et al. 2003). More recently, avian- versus human-lineage signature sequences have been described within the genes encoding the nucleoprotein (NP), the M1 and M2 matrix proteins, and the PA, PB1 and PB2 polymerase proteins (Zhou, Shortridge et al. 1999b; Hiromoto, Yamazaki et al. 2000; Katz, Lu et al. 2000; Naffakh, Massin et al. 2000; Yao, Mingay et al. 2001; Reid, Fanning et al. 2002; Shaw, Cooper et al. 2002a). The HA, however, is considered a major factor in influenza virus species specificity because of its role as the receptor binding protein. Influenza viruses utilize sialic acid (SA) molecules linked to galactose (Gal) sugars on the surface of cells as receptors. Avian influenza viruses bind preferentially to sialyloligosaccharides with an $\alpha2,3$ linkage to galactose (SAα2,3Gal), whereas human influenza viruses prefer sialyloligosaccharides with an $\alpha2,6$ linkage (SAα2,6Gal) (Rogers and Paulson 1983; Rogers and D'Souza 1989; Connor, Kawaoka et al. 1994; Ito, Suzuki et al. 1997a; Matrosovich, Gambaryan et al. 1997; Ito, Suzuki et al. 2000; Wright and Webster 2001). These findings are in keeping with the fact that human tracheal epithelial cells express predominantly SAα2,6Gal molecules, while SAα2,3Gal is the major form of sialic acid in duck intestinal cells (Couceiro, Paulson et al. 1993; Ito and Kawaoka 1998; Ito 2000). Recent data from the study of cultures of differentiated human airway epithelial cells confirmed the predominance of $\alpha2,6$ linked sialic acids on the surface of non-ciliated cells in the human tracheal epithelium but also found that a substantial cellular subset of respiratory epithelium consisting of ciliated cells, express α2,3 linked sialic acid receptors and that the latter are present in sufficient density to allow entry and replication of avian viruses (Matrosovich, Matrosovich et al. 2004). As will be discussed below, pigs also express both avian and human virus receptors and, thus, may serve as intermediate hosts in transmission of influenza A viruses between birds and humans. (See Chapter 7 for more information on influenza virus receptor interactions).

HUMAN INFLUENZA
Introduction
Of the three types of influenza viruses, influenza A, B and C, influenza C virus infections usually cause mild infections in childhood while influenza A and B viruses co-circulate and cause epidemics of influenza that are associated with significant morbidity and mortality (Cox and Subbarao 2000). Human influenza A viruses were first isolated in 1933 (Smith, Andrewes et al. 1933) and influenza B viruses were first isolated in 1940 (Francis 1940). There are16 known HA subtypes and 9 known NA subtypes among influenza A viruses. However, as discussed above, only a limited number of HA and NA subtypes of influenza A have caused widespread disease in humans. In contrast to influenza A viruses, influenza B viruses are not divided into subtypes and humans are believed to be the only hosts for influenza B viruses.

Influenza viruses circulate in all parts of the world. In temperate climates, influenza is a winter illness while in tropical climates, the disease can occur year around (Cox and Subbarao 2000). Influenza epidemics in the United States can begin as early as October but analysis of 28 influenza seasons from 1976 until 2004 indicate that peak influenza activity occurs between December and March (Centers for Disease Control and Prevention 2004). The antigenicity of circulating influenza viruses is constantly changing through a process of antigenic drift that allows the virus to cause annual epidemics of illness. Antigenic drift occurs among influenza A and B viruses when mutations accumulate in the HA and NA genes that alter the antigenicity of these proteins and the drifted strains are no longer neutralized by antibodies that were directed against previously circulating strains. Thus, drift variant viruses can infect individuals and cause an epidemic in a population that was immune to previously circulating strains as a consequence of prior natural infection or immunization (Cox and Bender 1995; Cox and Subbarao 2000). The HA genes of influenza A and B viruses are under positive selection (Fitch, Bush et al. 1997; Zou, Prud'homme et al. 1997; Bush, Fitch et al. 1999). Antigenic drift among influenza A and B viruses is continuously monitored by the World Health Organization's global influenza surveillance program and the strains of influenza that are recommended for inclusion in the vaccine each year are updated to keep pace with antigenic drift (Cox and Subbarao 2000).

Far less commonly, antigenic shift occurs and a virus with a novel influenza A HA subtype (with or without an accompanying novel NA subtype) is introduced into the human population that lacks prior experience and immunity to the new subtype. These viruses can spread in the immunologically naive population, causing devastating morbidity and mortality (Cox and Subbarao 2000). Three pandemics occurred in the last century with the introduction of a new subtype of influenza A that replaced the previously circulating subtype- the H1N1 (Spanish influenza) pandemic of 1918, the H2N2 (Asian influenza) pandemic of 1957 and the H3N2 (Hong Kong influenza) pandemic of 1968 (Cox and Subbarao 2000). Antigenic shift does not occur in the case of influenza B viruses because different antigenic subtypes of influenza B viruses do not exist.

EPIDEMIC STRAINS OF HUMAN INFLUENZA

The clinical illness of influenza is caused by infection with influenza A or B viruses; both types of influenza viruses co-circulate and infection with the two types of influenza viruses cannot be distinguished based on clinical findings (Cox and Subbarao 1999). Since their appearance in 1968, influenza A H3N2 viruses have continued to circulate in the human population. In 1977, H1N1 viruses reappeared as a cause of epidemic influenza in humans after an absence of 20 years and have co-circulated with H3N2 viruses since then (Cox and Subbarao 1999; Cox and Subbarao 2000). The relative proportions of influenza A and B viruses and the subtypes of influenza A among the viruses characterized by the WHO Collaborating Center in London, UK for the period 1987–88 through 2001–2002 are presented in Figure 1 (Lin, Gregory et al. 2004).

The impact of influenza can be measured in terms of hospitalization rates and mortality. Severe illness requiring hospitalization and complications from influenza occur in persons over 65 years of age, young children and persons with certain underlying health conditions (Centers for Disease Control and Prevention 2004). Influenza-associated hospitalization rates from 1969 to 1995 ranged from approximately 16,000 to 220,000 per epidemic, with an average of 114,000 per year and 57% of the hospitalizations occurred among persons over 65 years of age (Centers for Disease Control and Prevention 2004). Since the appearance of the H3N2 viruses in 1968, epidemics associated with H3N2 viruses have caused the greatest number of influenza-associated hospitalizations, with an estimated average of 142,000 per year (Simonsen, Fukuda et al. 2000) and since 1990, epidemics caused by H3N2 viruses have been more severe than those caused by A/H1N1 or influenza B viruses with 36,000 excess deaths attributed to A/H3N2 epidemics (Thompson, Shay et al. 2003). In the United States, influenza A/H3N2 viruses were responsible for 9 of the past 11 influenza epidemics while A/H1N1 and influenza B viruses together were responsible for 2 epidemics (Table 1). Pneumonia and influenza deaths did not exceed the epidemic threshold in the two winters when A/H1N1 and B viruses were responsible for the epidemic, while pneumonia and influenza deaths exceeded the epidemic threshold for 5 to 22 weeks during A/H3N2 epidemics (Table 1) (Thompson, Shay et al. 2003).

Influenza surveillance in the United States is coordinated by the WHO Collaborating Center for Reference and Research on Influenza at the Centers for Disease Control and Prevention in Atlanta. Epidemiologic data regarding influenza in the United States is collected from several sources including sentinel physicians, state and territorial epidemiologists, and reports of pneumonia and influenza associated deaths from rapid reporting in 122 cities (Anonymous 2004d). Influenza and respiratory syncytial virus (RSV) epidemics occur each winter and it is now recognized that young children and the elderly are affected by both of these pathogens (Zambon, Stockton et al. 2001). With advances in diagnostic virology and virologic surveillance, the morbidity and mortality attributable to RSV and influenza can be calculated separately. Although the impact of influenza on mortality was traditionally reported in terms of deaths from pneumonia and influenza, it is now recognized that influenza contributes to total (all-cause) mortality and deaths from circulatory diseases as well. Therefore, new models have been developed that take these death rates and the timing and duration of RSV epidemics into account (Thompson, Shay et al. 2003).

Virologic surveillance includes antigenic and genetic analyses of influenza viruses; antigenic analysis identifies the type and subtype of influenza viruses and is used to determine whether circulating strains of influenza still resemble the corresponding vaccine strain and whether the vaccine strain needs to be updated to better represent epidemic strains. Antigenic

Table 9.1 Epidemic strains of human influenza in the United States, 1993–4 to 2003–4 seasons

Year	Epidemic virus type/subtype[1]			Number of weeks that P&I deaths exceeded threshold[2]	Predominant epidemic strain[3]	Estimated respiratory and circulatory deaths[4]
	A/H1N1	A/H3N2	B			
2003–04	New Caledonia/20/99	**Fujian/411/02**	Sichuan/379/99	9	H3 drift	
2002–03	**New Caledonia/20/99**	Panama/2007/99	**HK/330/01**	0	H1 + B	
2001–02	New Caledonia/20/99	**Panama/2007/99**	**Sichuan/379/99 + Shizuoka/15/01**	5	H3 + B	
2000–01	**New Caledonia/20/99+ Bayern/7/95**	Panama/2007/99	**Sichuan/379/99 + Beijing/184/93**	**0**	H1 + B	
1999–2000	Bayern/7/95 + Beijing/262/95 + New Caledonia/20/99	**Sydney/5/97**	Beijing/184/93	22	H3	
1998–99	Bayern/7/95	**Sydney/5/97**	**Beijing/184/93**	12	H3 + B	45793
1997–98	Johannesburg/82/96 + Beijing/262/95	**Sydney/5/97 (80%) +**	Harbin/7/94 + Beijing/184/93	10	H3 drift	51296
1996–97	Texas/36/91 + Taiwan/1/86	**Wuhan/359/95**	Harbin/7/94 + Beijing/184/93	10	H3	47934
1995–96	**Texas/36/91 +** Taiwan/1/86	Wuhan/359/95 +	Panama/45/90 + Beijing/184/93	6	H3 + H1	31614
1994–95	Texas/36/91 + Taiwan/1/86	**Johannesburg/33/94 Shangdong/9/93 +**	Beijing/184/93	6	H3 drift	29337
1993–94	Texas/36/91	**Beijing/36/92 +** Shangdong/9/93	Panama/45/90	9	H3	36134

1 Virus strains in bold type were the predominant circulating strains.

2 The epidemic threshold is 1.654 standard deviations above the seasonal baseline that is projected using a robust regression procedure in which a periodic regression model is applied to the observed percentage of deaths from pneumonia and influenza (Anonymous 1994b; Anonymous 1994c; Anonymous 1995c; Anonymous 1996b; Anonymous 1996c; Anonymous 1997b; Anonymous 1997c; Anonymous 1998b; Anonymous 1999c; Anonymous 2000c; Anonymous 2001c; Anonymous 2002c; Anonymous 2003c; Anonymous 2004c).

3 Indicates which type or subtype of virus was responsible for the epidemic.

4 Estimated annual influenza deaths extracted from the influenza and RSV model (Thompson, Shay et al. 2003). Data were not included for the years after 1999.

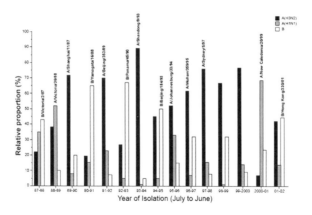

Figure 9.1 Relative proportions of influenza A and B viruses received by the WHO Collaborating Center for Reference and Research on Influenza, London from National Influenza Centers around the world for each year from 1987 to 2002 (Reprinted from Virus Research, Vol 103, Lin et al., Recent changes among human influenza viruses, page 48, Copyright 2004, with permission from Elsevier (Lin, Gregory et al. 2004)).

drift is an ongoing process and variant viruses that are antigenically distinguishable from previously circulating strains appear sporadically every few years. When viruses are antigenically distinct, it is often possible to identify genetic differences that correlate with the antigenic change. However, genetically distinct groups of viruses may not differ antigenically; viruses can belong to distinct phylogenetic clades and several such clades of viruses that co-circulate can be antigenically indistinguishable.

Viruses that belong to the same clade share one or more residues that can be targeted in the design of rapid genotyping assays. A fragment of the HA or NA gene that is amplified by reverse-transcriptase polymerase chain reaction can be digested with a selected restriction enzyme that targets the specific conserved nucleotide sequence that defines the clade. Thus, restriction fragment length polymorphisms can be applied to genotype influenza viruses more rapidly and easily than would be possible by sequence analysis. Genetic analysis of the HA and NA genes has greatly extended our understanding of the epidemiology of human influenza viruses. As a result of phylogenetic analysis of the HA and NA genes of circulating influenza viruses, it is now known that different lineages within a subtype co-circulate and that reassortment between the lineages is an important mechanism for continued evolution of influenza viruses (Lindstrom, Hiromoto et al. 1999; Xu, Lindstrom et al. 2004). By correlating phylogenetic data and antigenic analysis, it will be possible to determine the sequence of events that lead to the appearance of antigenic drift variants. Recent events in the epidemiology of influenza A/H1N1, A/H3N2 and B viruses are summarized below.

Influenza A/H1N1 viruses:

In the early 1990s phylogenetic analysis of the HA gene of H1N1 viruses identified two co-circulating lineages, the reference strains of which were A/Bayern/7/95 and A/Lanzhou/1/95. Viruses in the two lineages were antigenically indistinguishable until viruses in the Lanzhou/1/95-lineage sustained an amino acid deletion at residue 134; the reference strain for the group with the deletion was A/Beijing/262/95. A/Beijing/262/95-

like viruses were the predominant circulating H1N1 viruses in China for several years and there was limited co-circulation of A/Bayern/7/95- like viruses. In contrast to the pattern in China, from 1995 until 1999, A/Beijing/262/95-like viruses disappeared from circulation in the rest of the world, and A/Bayern/7/95-like viruses circulated alone (Daum, Canas et al. 2002). In 1999, A/New Caledonia/20/99-like viruses that evolved from A/Beijing/262/95-like viruses and shared the deletion at residue 134 appeared in the South Pacific, South Africa and Asia and were subsequently identified in South America (Daum, Canas et al. 2002). These viruses spread rapidly through North and South America and replaced A/Bayern/7/95-like viruses as the predominant H1N1 viruses in the 1999–2000 and 2000–2001 seasons (Daum, Canas et al. 2002). The HA and NA genes of H1N1 viruses continue to evolve with several genetically distinct clades of viruses that are not yet antigenically distinguishable. The reference strain A/Johannesburg/82/96 has evolved from A/Bayern/7/95 and A/New Caledonia/20/99 has evolved from A/Beijing/262/95.

Influenza A/H1N2 viruses:
In 2001, H1N2 reassortant viruses that derived the HA gene from H1N1 viruses and remaining gene segments from H3N2 viruses were isolated from people in several countries and became the predominant circulating strain for the season (Gregory, Bennett et al. 2002; Xu, Smith et al. 2002; Ellis, Alvarez-Aguero et al. 2003). Although previous instances of reassortment between H1N1 and H3N2 viruses were reported (Nishikawa and Sugiyama 1983; Guo, Xu et al. 1992), those reassortant viruses did not become established in the human population and there was no apparent epidemiological consequence observed in terms of altered virulence of the viruses. The recent establishment and circulation of the H1N2 viruses was widespread but was still of limited epidemiological consequence because the HA and NA were derived from viruses that were previously circulating and were already included in the trivalent vaccine; the HA of the H1N2 viruses was antigenically similar to that of the H1N1 viruses and the N2 NA was antigenically similar to the N2 NA of H3N2 viruses (Xu, Smith et al. 2002). From genetic analysis it is known that the H1N2 reassortant viruses of 2001 shared signature amino acids at residues Ala169, Thr193 and Asn196 in the HA gene (Xu, Smith et al. 2002) and Lys199 and Asn431 in the N2 NA gene (Gregory, Bennett et al. 2002) among themselves. It has been proposed that amino acid changes that had occurred in the H1 HA gene just before and following the reassortment with H3N2 viruses may have been important in accommodating functional compatibility with the N2 NA gene (Gregory, Bennett et al. 2002). Retrospective analysis suggests that the reassortment event occurred between 1999 and early 2001 (Ellis, Alvarez-Aguero et al. 2003). In the 2002–2003 season H1N1, H1N2 and H3N2 viruses co-circulated with influenza B viruses but H3N2 viruses predominated in the 2003–2004 season (Anonymous 2003c; Anonymous 2004d). It remains to be seen whether H1N1 and H1N2 viruses will continue to cocirculate with H3N2 viruses.

Influenza A/H3N2 viruses
Genetic and antigenic drift in the H3 HA and N2 NA genes has been an ongoing process since their introduction into the human population in 1968. Antibody pressure is the major force that drives the evolution of the HA and a number of codons in the H3 HA gene have been identified to be under positive selection (Fitch, Bush et al. 1997; Bush, Fitch et al. 1999). The location of deduced amino acid substitutions that have occurred among circulating influenza A/H3N2 viruses can be identified on the crystal structure of the H3 HA protein of an influenza virus

isolated in 1968 (Wilson, Skehel et al. 1981; Wilson and Cox 1990; Cox and Bender 1995). The residues that have changed are more likely to be on the surface of the molecule where they are likely to be exposed to antibody (Wilson and Cox 1990; Cox and Bender 1995). Smith et al. developed a new method of multidimensional scaling to quantify the antigenic evolution of influenza viruses and compared antigenic and genetic evolution of influenza A H3N2 viruses isolated from 1968 until 2003 (Smith, Lapedes et al. 2004). They observed remarkable correspondence between antigenic and genetic evolution, although there were important exceptions of epidemiological significance and the former was more punctuated than the latter (Smith, Lapedes et al. 2004). Antigenic drift variants appear more frequently among influenza A/H3N2 viruses than among influenza B viruses. This is consistent with the observation that the rate of evolution of the H3 HA exceeds that of A/H1N1 and B viruses (Rota, Wallis et al. 1990; Cox and Bender 1995; Lindstrom, Hiromoto et al. 1999). Over the past decade, phylogenetic analyses of the H3 HA gene have identified several co-circulating clades of viruses that share signature amino acid residues, but the viruses were initially antigenically indistinguishable. Eventually, viruses of one clade acquire sufficient genetic changes to become antigenically distinguishable from the circulating strain. In the past decade, antigenic drift variants of influenza A/H3N2 viruses appeared in 1994 (A/Johannesburg/33/94), 1995 (A/Wuhan/359/95), 1997 (A/Sydney/5/97) and 2003 (A/Fujian/411/2002). In each instance, the appearance of a drift variant resulted in a severe influenza epidemic associated with high rates of hospitalization and deaths (Thompson, Shay et al. 2003; Anonymous 2004d) (Table 1).

Influenza B viruses

Based on detailed antigenic and genetic analyses of circulating influenza B viruses, the following epidemiological scenario can be reconstructed. Two antigenically and genetically distinct lineages of influenza B viruses emerged in 1988 (Rota, Wallis et al. 1990). The prototype reference viruses of the two lineages were B/Victoria/2/87 and B/Yamagata/16/88. These viruses were antigenically divergent enough that they were not cross-reactive in HI assays and antiserum against one did not protect against infection with the other (Levandowski, Regnery et al. 1991; Rota, Hemphill et al. 1992). B/Victoria-like viruses were the predominant strain worldwide till the early 1990s, when B/Yamagata-like viruses replaced them and became dominant worldwide. Two phylogenetically distinct lineages of B/Yamagata-like viruses that were antigenically indistinguishable co-circulated in the 1990s; these clades were represented by the reference strains B/Harbin/7/94 and B/Beijing/184/93. During this period, B/Victoria-like viruses continued to evolve in East Asia where they co-circulated with B/Yamagata-like viruses, but they had disappeared from circulation in the rest of the world. Reassortant influenza B viruses like B/Johannesburg/15/94 that derived their HA gene from the Yamagata lineage and NA gene from the Victoria lineage were isolated between 1992 and 1996 in Africa, America and Europe (McCullers, Wang et al. 1999). Genetic analyses of the HA and NA genes of influenza B viruses have revealed a pattern of insertions and deletions of amino acids (Nerome, Hiromoto et al. 1998; McCullers, Wang et al. 1999). By the late 1990s, the reference strains of B/Victoria-like viruses were B/Shangdong/97 and B/Beijing/243/97 and that of the B/Yamagata lineage was B/Sichuan/379/99.

In 2001, B/Victoria-like viruses re-emerged in North America, Europe, the Middle East and the rest of Asia and caused significant outbreaks of disease (Shaw, Xu et al. 2002b). Over the course of the 2001–2002 season, B/Victoria-like viruses fell into two phylogenetic groups, represented by B/Hong Kong/330/2001 and B/Oman/16296/2001. Viruses of the B/Hong Kong/3301/01 clade shared Arg116, Asn121 and Glu 164 in the HA gene while viruses of the B/Oman/16296/2001 clade shared Thr121 compared with the sequence of the HA gene of B/Shangdong/7/97 (Shaw, Xu et al. 2002b). Similarly, the NA genes could be separated into two phylogenetic clades with conserved signature amino acid residues that define each clade (Shaw, Xu et al. 2002b). Over the course of the 2001–2002 season, reassortant viruses like B/New York/1/2002, B/Maryland/1/2002 and B/Hong Kong/1351/2002 were isolated that derived their HA gene from a Victoria-like virus and NA gene from B/Yamagata/16/88. Although influenza A/H3N2 viruses predominated in the United States in the 2003–2004 season, influenza B viruses from both lineages were also isolated.

Summary
Some of the recurring patterns that have been seen in the epidemiology of influenza A and B viruses in the past decade are as follows:

- genetically distinct clades of viruses co-circulate
- reassortment between the lineages and between subtypes is common
- amino acids can be deleted or inserted into the HA and NA genes
- periodically, a lineage of influenza viruses can disappear from large regions of the world but can persist in other regions of the world.
- these lineages of viruses can reappear later and become re-established as the predominant strain.

Some of the unanswered questions in the molecular epidemiology of influenza viruses are why lineages of viruses disappear from circulation in certain regions of the world, how (and why) they reappear years later, and whether reassortment between lineages confers any biological advantage.

PANDEMIC INFLUENZA VIRUSES
As mentioned above, three pandemics of influenza occurred in the last century. In each instance, an influenza A virus with a novel HA subtype was introduced into the human population. The virus caused widespread epidemics as it spread rapidly by efficient person-to-person transmission in the human population that lacked antibodies against the new subtype and ultimately replaced the previously circulating strain. The 1918 Spanish flu pandemic was caused by influenza A H1N1 viruses, the 1957 Asian flu pandemic by H2N2 viruses and the 1968 Hong Kong flu pandemic by H3N2 viruses.

Although the number of deaths attributed to the 1918 pandemic is a matter of discussion, it is estimated that worldwide, over 40 million deaths were caused by the influenza pandemic of 1918 (Taubenberger, Reid et al. 2000). In the United States alone, 675,000 deaths were attributed to the 1918 pandemic (Taubenberger, Reid et al. 2000), and 69,800 and 33,800 excess deaths were attributed to the to the 1957 and1968 pandemics, respectively (Cox and Subbarao 2000). The pattern of age-specific mortality during the first year of a pandemic differs from that seen in epidemic influenza; the pattern is U-shaped in epidemic influenza because of peaks at the extremes of age while the pattern in the 1918 pandemic has been described as a W-shape with the expected peaks at the extremes of age and a unique peak in young adults (Reid, Taubenberger et al. 2001). The 1918 pandemic occurred at a time of great upheaval and troop movements associated with World War I. Medical historians have reported the history of the 1918 pandemic (Crosby 1989). The temporal spread of the 1957 and 1968 pandemics was reviewed by Cox and Subbarao (Cox and Subbarao 2000). A recent analysis indicates that death rates among elderly above the age of 85 years did not increase in the 1968 pandemic (Simonsen, Reichert et al. 2004) because this age group had pre-existing immunity against H3 viruses from circulation of H3 subtype viruses in an earlier era.

Genetic characterization of the virus that caused the 1918 pandemic has been undertaken by Jeffery Taubenberger and his colleagues and is discussed in Chapter 11. Genetic characterization of the 1957 H2N2 and 1968 H3N2 pandemic strains of influenza established that they arose from reassortment between an avian influenza A virus and the circulating human influenza virus. The 1957 H2N2 virus derived the HA, NA and PB1 gene segments from an avian influenza virus and the remaining gene segments from the previously circulating human H1N1 virus (Kawaoka, Krauss et al. 1989) while the 1968 H3N2 virus derived the HA and PB1 gene segments from an avian influenza virus and remaining gene segments from the circulating human H2N2 virus (Laver and Webster 1973; Scholtissek, Rohde et al. 1978b; Kawaoka, Krauss et al. 1989). It is not entirely clear whether the NA gene of the 1968 pandemic H3N2 virus was derived from an avian influenza strain or the circulating human H2N2 virus.

AVIAN INFLUENZA
History
In 1878, a severe rapidly spreading disease in chickens with high mortality was described in Italy by Perroncito (Stubbs 1948). The disease, termed fowl plague, spread over all of Europe during the late 1800s and early 1900s via movement of exhibition and hobby poultry, and became endemic in domestic poultry until it was eradicated in the 1930s. Outbreaks of fowl plague were reported in the early 1900s in North and South America, Africa and Asia (Swayne and Halvorson 2003). The etiology of fowl plague was determined to be a virus in 1901, but the virus was not identified and classified until 1955 when Schafer determined it to be Type A influenza virus (orthomyxovirus) (Schafer 1955). Beginning in the 1960s, mild respiratory disease or drops in egg production caused by influenza viruses were reported in turkeys from the USA and Canada. These non-fowl plague syndromes were termed mildly pathogenic or low pathogenicity (LP) avian influenza (AI) (Swayne, Senne et al. 1998b). In 1968, antibodies to influenza A viruses were detected in migratory waterfowl (Easterday, Trainer et al. 1968). In 1972, influenza A viruses were isolated from cloacal samples of various species of migratory ducks in California and such ducks were determined to be the principal reservoir and a natural host of LPAI viruses (Slemons, Johnson et al. 1974). Additional studies have identified other wild bird species, especially from aquatic habitats, as reservoirs of AI viruses (Stallknecht 1998). Most AI viruses in wild birds cause asymptomatic infections. By 1979, it was established that differences in enzyme cleavability of the HA surface glycoprotein have a major impact on virulence in chickens and related galliforme birds, but a specific constellation of the other gene segments was necessary for maximal expression of virulence (Bosch, Orlich et al. 1979). At the First International Symposium on Avian Influenza in 1981 the term 'fowl plague' was abandoned for the more accurate term 'highly pathogenic avian influenza' (HPAI) (Bankowski 1981), while at the Fifth International Symposium on Avian Influenza, all other infections caused by AI viruses were lumped under the term 'low pathogenicity avian influenza' (LPAI) (International Symposium on Avian Influenza 2003). For more details concerning pathobiology, epidemiology and control of AI viruses, the readers are referred to recently published reviews (Swayne 2000; Swayne and Suarez 2000; Swayne and Halvorson 2003).

The original fowl plague viruses (pre-1955) were of the H7 hemagglutinin subtype, but in 1959, an outbreak of HPAI resulted from H5 hemagglutinin subtype (Table 2), and since then, several additional outbreaks of H5 HPAI have been reported. The remaining 14 hemagglutinin subtypes have not been reported to cause HPAI outbreaks.

AVIAN INFLUENZA VIRUS VIRULENCE

Avian influenza viruses are classified into two pathotypes based on the following laboratory criteria: a) experimental *in vivo* studies in chickens for H1-16 subtypes, and b) sequence data of the HA proteolytic cleavage site for H5 and H7 subtypes. High pathogenicity, or HPAI viruses cause 75% or greater mortality, i.e. high lethality, in intravenously inoculated chickens, or have intravenous pathogenicity indices (IVPI) reflecting combined morbidity and mortality scores of 1.2 or greater (Alexander 1998; OIE 2003). In addition, if the HA proteolytic cleavage sites of H5 or H7 AI viruses have the amino acid sequence motif compatible with HPAI viruses, the virus is considered HP irrespective of *in vivo* test results (OIE 2003). By contrast, AI viruses that do not meet *in vivo* or HA sequence requirements of a HPAI virus are termed AI virus of low pathogenicity; i.e. LPAI. All HPAI viruses have been of the H5 and H7 subtypes. By contrast, LPAI viruses comprise all AI viruses of the H1-4, H6, and H8-16 subtypes, and most of those of H5 and H7 subtypes.

The term 'highly pathogenic' indicates that the AI virus strain causes high lethality or severe disease in experimentally infected chickens and does not imply experimental high lethality or severe disease for other bird species, especially wild ducks or geese (Order: *Anseriformes*). However, if an AI virus causes high lethality in experimentally inoculated chickens, the virus will usually cause high lethality in other birds within the Order *Galliformes*, family *Phasianidae*, such as turkeys (*Meleagris gallopavo)* and Japanese quail (*Coturnix japonica*) (Alexander, Parsons et al. 1986). By contrast, most HPAI viruses usually cause asymptomatic infections or mild disease in experimentally inoculated domestic ducks and geese (Alexander, Parsons et al. 1986; Forman, Parsonson et al. 1986). The majority of AI viruses isolated from domestic ducks and geese have been associated with mild-to-moderate mortality and respiratory disease. However, these AI viruses do not cause high lethality in experimental studies with chickens, ducks or geese (Alexander 1993). Recently, an H5N1 AI virus isolated from a sick domestic goose in the People's Republic of China was HP for chickens based on the sequence of the viral HA proteolytic cleavage site, as listed in criterion b) above (Xu, Subbarao et al. 1999b). In addition, several viruses isolated from dead captive and wild waterfowl and other wild birds were highly lethal for chickens and ducks (Sturm-Ramirez, Ellis et al. 2004), but such high lethality of HPAI viruses for ducks has been rare.

Pathobiology

Infectivity of AI viruses requires extracellular or intracellular enzymatic cleavage of the whole hemagglutinin protein (HA0) into the HA1 and HA2 proteins (Bosch, Orlich et al. 1979; Bosch, Garten et al. 1981). For LPAI viruses, such HA0 cleavage occurs via trypsin-like enzymes either within respiratory and gastrointestinal epithelial cells, or extracellularly in the respiratory lumen by exoproteases secreted from respiratory epithelial cells or some bacteria (Slemons, Byrum et al. 1998; Swayne and Halvorson 2003). For HPAI viruses, cleavage occurs in a variety of cells throughout the body as a result of cleavage by a ubiquitous class of enzymes; i.e. subtilisin-like endoprotease furin-like enzymes (Stieneke Grober, Vey et al. 1992).

The H5 and H7 AI viruses are maintained in the natural reservoir as LPAI viruses while HPAI viruses originate from these LPAI viruses via mutation in the HA proteolytic cleavage site (Rohm, Horimoto et al. 1995). Most H5 LPAI viruses have a HA proteolytic cleavage site motif of B-X-X-R/G while most H7 LPAI viruses have a motif of B-X-R/G where B = basic amino acids arginine or lysine, X = nonbasic amino acid and R = arginine (Senne, Panigrahy et al. 1996). Shifts in the field from LP to HP have been clinically abrupt and have been best documented in the USA (H5N2 – 1983–84), Mexico (H5N2 – 1994–95), Italy (H7N1 – 1999–2000), Chile (H7N3 – 2002) and Canada (H7N3 – 2004) over the past two decades (Perdue, Garcia et al. 1997; Suarez, Senne et al. 2004) (http://www.avian-influenza.com/recent_outbreaks/canada_2004.asp). A similar abrupt change in virulence has been reproduced in the laboratory with some H5 and H7 LPAI viruses and such viruses had sequence changes in the HA proteolytic cleavage site similar to those of naturally occurring HPAI viruses (Horimoto and Kawaoka 1995; Swayne, Beck et al. 1998a). Shifts in virulence from LP to HP have resulted from insertions of multiple basic amino acids or substitution of non-basic with basic amino acids at the HA proteolytic cleavage site, loss of glycosylation sites such that proteolytic cleavage sites are uncovered, or insertions of large amounts of genetic information at the cleavage site. As mentioned above, the proposed minimal motif for HPAI is B-X-B-R/G (Kawaoka and Webster 1988; Vey, Orlich et al. 1992; Senne, Panigrahy et al. 1996). There are some exceptions to these criteria: some AI viruses that do not have this motif are HP *in vivo* for chickens while a few AI virus strains with this motif do not meet the *in vivo* definition for HP (Senne, Panigrahy et al. 1996; Swayne, Beck et al. 1998a). Although the HA is the major determinant of virulence, maximum expression of virulence requires an optimal constellation of internal genes (Bosch, Orlich et al. 1979).

ECOLOGY AND EPIDEMIOLOGY OF AVIAN INFLUENZA VIRUSES
Ecology
From an ecological perspective, LPAI viruses are maintained worldwide in wild bird reservoirs, particularly in aquatic birds of the orders Anseriformes (waterfowl) and Chadriiformes (shorebirds) (Stallknecht 1998). Infections by LPAI viruses in wild birds have usually been asymptomatic enteric infections. By contrast, HPAI viruses are not maintained in a wild bird reservoir, but arise by introduction and circulation of H5 and H7 LPAI viruses in domestic poultry with subsequent mutation to HPAI viruses (Rohm, Horimoto et al. 1995; Swayne and Halvorson 2003).

Epidemiology
Chickens and other gallinaceous poultry species (including turkeys, quails, partridges, etc.) are not natural reservoirs of AI viruses (Perdue, Suarez et al. 1999; Suarez and Schultz-Cherry 2000). Mankind has altered the natural ecosystems of wild birds through captivity, domestication, industrial agricultural practices, the need for national and international commerce, and non-traditional raising practices, and have thereby created new niches for AI influenza viruses including opportunities for infection in new host species (Swayne 2000). Variations in poultry rearing and husbandry practices around the world have created introduction points for AI viruses from the wild bird reservoir and have established permissive environments for the spread and maintenance of AI viruses in some domestic poultry populations. The man-made ecosystems can be divided into five major categories (Swayne 2000): 1) integrated indoor industrial poultry systems, 2) outdoor raised poultry (industrial

Table 9.2 Twenty-two documented outbreaks of highly pathogenic avian influenza since discovery of type A influenza virus as cause of fowl plague in 1955 (modified from (Swayne and Suarez 2000))

Prototype Avian Influenza Virus from Outbreak (Type/host species/ geographic location/strain number/ year of isolation)	Subtype	Number and type of birds affected with high mortality or culled (a)
A/chicken/Scotland/59	H5N1	2 flocks of chickens (*Gallus gallus domesticus*). Total number of birds affected not reported – See (Swayne and Suarez 2000)
A/tern/South Africa/61	H5N3	1,300 common terns (*Sterna hirundo*)
A/turkey/England/63	H7N3	29,000 breeder turkeys (*Meleagridis gallapavo*)
A/turkey/Ontario/7732/66	H5N9	8,100 breeder turkeys
A/chicken/Victoria/76	H7N7	25,000 laying chickens, 17,000 broilers and 16,000 ducks (*Anas platyrhyncos*)
A/turkey/England/199/79	H7N7	3 commercial farms of turkeys. Total number of birds affected not reported
A/chicken/Pennsylvania/1370/83	H5N2	17 million birds in 452 flocks; most were chickens or turkeys, a few chukar partridges (*Alectoris chukar*) and guinea-fowl (*Numida meleagris*)
A/turkey/Ireland/1378/83	H5N8	800 meat turkeys died on original farm; 8,640 turkeys, 28,020 chickens and 270,000 ducks were depopulated on original and 2 adjacent farms
A/chicken/Victoria/85	H7N7	24,000 broiler breeders, 27,000 laying chickens, 69,000 broilers and 118,518 chickens of unspecified type
A/turkey/England/50-92/91	H5N1	8,000 turkeys
A/chicken/Victoria/92	H7N3	12,700 broiler breeders and 5,700 ducks
A/chicken/Queensland/95	H7N3	22,000 laying chickens
A/chicken/Puebla/8623-607/94 A/Queretaro/14588-19/95	H5N2	Chickens [b]
A/chicken/Pakistan/447/95 A/chicken/Pakistan/1369-CR2/95	H7N3	3.2 million broilers and broiler breeder chickens in 1994–95 outbreak; 300,000 chickens in 2001; 3.5 million chickens in 2003 [c] Total number of birds affected is unknown.
A/goose/Guangdong/1/1996, A/chicken/Hong Kong/220/1997	H5N1	Culled or died during 1997 in Hong Kong, 1.4 million chickens and various lesser numbers of other domestic birds in contact with the chickens on farms and in the live-bird market system; during 2001 in Hong Kong, 1.6 million depopulated; during 2002 in Hong Kong, 950,000 depopulated; August 2003–March 2004, over 100 million birds in 8 Asian countries (Cambodia, China, Indonesia, Japan, Laos, South Korea, Taiwan,[d] Thailand, Vietnam)[e]

a Most outbreaks were controlled by 'stamping out' or culling policies for infected and/or exposed populations of birds. Chickens, turkeys and birds of the order *Galliformes* had clinical signs and mortality patterns consistent with highly pathogenic avian influenza while ducks, geese and other birds usually lacked or had low mortality rates or infrequent presence of clinical signs.

b A 'stamping-out' policy was not used for control. The outbreak of AI had concurrent circulation of mildly pathogenic avian influenza and highly pathogenic avian influenza (HPAI) virus strains. However, HPAI virus strains were present only from late 1994 to mid-1995. Estimates of the number of birds infected with HPAI strains

Prototype Avian Influenza Virus from Outbreak (Type/host species/ geographic location/strain number/ year of isolation)	Subtype	Number and type of birds affected with high mortality or culled (a)
A/chicken/New South Wales/1651/1997	H7N4	128,000 broiler breeders, 33,000 broilers and 261 emu (*Dromaius novaehollandiae*)
A/chicken/Italy/330/1997	H5N2	2,116 chickens, 1,501 turkeys, 731 guinea-fowl, 2,322 ducks, 204 quail (species unknown), 45 pigeons (*Columbia livia*), 45 geese (species unknown) and 1 pheasant (species unknown)
A/turkey/Italy/1999	H7N1	297 farms; 5.8 million laying chickens, 2.2 million meat and breeder turkeys, 1.6 million broiler breeders and broilers, 156,000 guinea-fowl, 168,000 quail, 577 backyard poultry and 200 ostriches; numbers include affected and culled
A/chicken/Chile/184240/2002	H7N3	617,800 broiler (meat-type chickens) breeders affected or depopulated. Two houses of turkey breeders affected – numbers unknown. (Rojas, Moreira et al. 2002; Jones and Swayne 2004; Suarez, Senne et al. 2004)
A/chicken/Netherlands/1/03	H7N7	30 million birds (mostly chickens), 255 affected flocks with 1,381 commercial flocks and 16,521 backyard/smallholders depopulated. (Elbers, Kamps et al. 2004)
A/chicken/Canada/AVFV2/04	H7N3	1,204,173 affected birds, mainly chickens and turkeys, on 42 commercial and 11 backyard premises in British Columbia. 19 million poultry pre-emptively slaughtered in the region. http://www.inspection.gc.ca/english/anima/heasan/disemala/avflu/situatione.shtml, http://www.oie.int/eng/info/hebdo/AIS_46.HTM#Sec4, http://www.oie.int/eng/info/hebdo/AIS_50.HTM#Sec2, accessed 23 June 2004.
A/chicken/Texas/ 298313-2/2004	H5N2	6608 chickens, one Texas farm and 2300 birds in five live-poultry markets in Houston. Virus had hemagglutinin proteolytic cleavage site consistent with HPAI virus, but *in vivo* tests in chickens failed to produce clinical signs of disease or death. http://www.oie.int/eng/info/hebdo/AIS_55.HTM#Sec8, accessed 23 June 2004

are unavailable, but 360 commercial chicken flocks were 'depopulated' for AI in 1995 through controlled marketing.
c A 'stamping-out' policy was not used for control. Surveillance, quarantine, vaccination and culling or controlled marketing were used as the control strategy. The numbers affected are crude estimates.
d Consisted of 6 red-faced ducks in sea adjacent to Quernoy Islands, no outbreaks on main island of Formosa
e Outbreak unusually complex. The H5 and N1 gene segments have recurred in multiple outbreaks from 1996–2004 in eastern Asia. However, the 6 internal gene segments have undergone reassortment with additional avian influenza viruses.

or organic systems), 3) poultry raised on small holdings and distributed through live and wet poultry market systems (LPM), 4) village (backyard), exhibition and hobby flocks (including fighting cocks), and 5) wild bird collection and trading systems.

Direct exposure to wild waterfowl and shorebirds are a risk factor for the introduction of AI viruses to any of the five man-made ecosystems, while the lack of biosecurity and frequent contacts between premises favor maintenance of AI viruses in these systems. For example, domestic Peking ducks (*Anas platyrhynchos*) raised outdoors on ponds have a high risk for introduction of AI viruses because there is the potential for direct exposure to a natural wild bird reservoir of AI viruses, especially mallards (Anas *platyrhynchos)*, and these AI viruses are already adapted for efficient replication in domestic duck species (Shortridge 1992; Alexander 2000). By comparison, chickens raised in indoor industrial production have negligible risk for introduction and maintenance of AI viruses because of low exposure risk to wild waterfowl and chickens are resistant to infection by waterfowl AI viruses (Lee, Senne et al. 2004). The raising of a single poultry species on a farm, practice of biosecurity principles to exclude disease causing agents and preventing direct access of poultry to wild birds serve to prevent the introduction, adaptation, and transmission of AI viruses in domestic poultry from wild birds.

Frequency of AI infections within each man-made eco-system has varied in different countries. Most integrated commercial chicken meat and egg production systems in developed countries rarely have AI infections, considering the 25–30 billion chickens raised each year (Agriculture 1999). For example, a 1997 serological survey by the National Poultry Improvement Plan of industrial raised poultry in the United States (Swayne 2004) reported no infections based on 265,960 and 201,321 serum samples tested from broilers (meat birds) and broiler breeders, which represented samples obtained from populations of 54 million (3972 flocks) and 7.6 billion (47,533 flocks) layer and broiler chickens, respectively. By contrast, sampling chickens that produced table eggs identified less than 1 million birds (12 farms in one state) that were infected with an H7N2 LPAI virus from a national sampling population of 3969 farms representing 259 million chickens. However, when AI infections have occurred in industrial poultry such as chickens and turkeys, they spread rapidly from farm-to-farm and throughout the integrated system resulting in epizootics of AI. Most recently, during the middle to late 1990s some developing countries of Asia and the Middle East had H9N2 LPAI become endemic in integrated, commercial broiler farms (Swayne and Halvorson 2003).

Historically, HPAI was endemic in poultry populations of Europe and some parts of Asia between 1900 and 1930 (Swayne and Halvorson 2003). During the 1940s and 50s, HPAI disappeared from poultry populations and there were infrequent reports of outbreaks. Since the identification of orthomyxoviruses as the cause of fowl plague in 1955, there have been 22 outbreaks of HPAI in the world (Table 2). Of these, 15 outbreaks involved less than a million birds, two outbreaks involved less than 10 million, and an additional four outbreaks involved less than 30 million birds. The H5N1 HPAI outbreak in Asia has been the largest with over 100 million birds dead or culled over more than eight years (1996–2004). Most HPAI outbreaks primarily involved chickens and other domestic poultry, but one outbreak has occurred in wild birds, common terns in South Africa, and mortality has been reported in wild and captive non-domesticated birds in Hong Kong during 2002 and most recently in various Asian countries with the H5N1 HPAI outbreak (Guan, Peiris et al. 2003; Sturm-Ramirez, Ellis et al. 2004). Most HPAI outbreaks have involved a single defined lineage of HPAI virus and were eliminated in less than a year through comprehensive eradication programs, but the Asian H5N1 has been unprecedented with the detection of multiple reassortants in domestic poultry over the eight-year outbreak period (Guan, Peiris et al. 2003). The Asian H5N1 HPAI viruses have maintained the original lineage of hemagglutinin (H5) gene contributed by the progenitor A/goose/Guangdong/1/1996 HPAI virus (Xu, Subbarao et al. 1999a) and the NA (N1) gene, but have reassorted to acquire different genes that encode the internal proteins (Guan, Peiris et al. 2003; Sturm-Ramirez, Ellis et al. 2004). By contrast, the Pakistan H7N3 HPAI outbreak involved a single virus lineage with distinctly different geographic episodes in 1994, 2001 and 2003. The Mexico 1994–95 H5 outbreak had two lineages of H5N2 HPAI virus, but these viruses originated from a single LPAI progenitor with two separate cleavage site mutations (Garcia, Crawford et al. 1996; Perdue, Suarez et al. 1999).

SWINE INFLUENZA
Introduction
Influenza A virus infection is a clinically and economically important cause of respiratory disease in pigs throughout large parts of the world (Janke 1998; Van Reeth and Easterday 1999). Swine influenza often presents as explosive outbreaks of fever, lethargy, inappetance, nasal discharge, coughing, and dyspnea. Additionally, influenza A virus infections can also contribute to the more insidious porcine respiratory disease complex, in conjunction with porcine reproductive and respiratory syndrome virus, *Mycoplasma hyopneumoniae*, and bacterial pathogens. Infection of pigs with influenza A viruses also poses important human public health concerns. As such, data from swine influenza virus surveillance studies and characterization of influenza virus isolates from pigs are critical to an overall understanding of long-term evolutionary and epidemiological patterns of human influenza and pandemic preparedness.

Classical H1N1 swine influenza viruses
Influenza was initially recognized as a disease of pigs in 1918 (Chun 1919; Koen 1919; Van Reeth and Easterday 1999), coincident both in time and clinical presentation with the devastating 'Spanish influenza' pandemic in the human population. In 1930, influenza viruses were first isolated from pigs (Shope 1931). These viruses were the initial examples of what has become known as the classical H1N1 lineage of swine influenza A viruses. Since 1930, these viruses have circulated within swine populations in North America (Hinshaw, Bean et al. 1978; Morin, Phaneuf et al. 1981; Chambers, Hinshaw et al. 1991; Arora, N'Diaye et al. 1997; Olsen, Carey et al. 2000), South America (Cunha, Varges Vinha et al. 1978), Europe (Blakemore and Gledhill 1941; Harnach, Hubik et al. 1950; Kaplan and Payne 1959; Nardelli, Pascucci et al. 1978; Roberts, Cartwright et al. 1987; Donatelli, Campitelli et al. 1991; Brown, Harris et al. 1995b; Brown 2000), and Asia (Yip 1976; Yamane, Arikawa et al. 1978; Shortridge and Webster 1979; Kupradinun, Peanpijit et al. 1991). Recent research has demonstrated that the early swine viruses and the human viruses of 1918 were closely related to one another antigenically and genetically, although it remains unclear as to whether a progenitor virus was transmitted from pigs to people or from people to pigs (Taubenberger, Reid et al. 1997; Reid, Fanning et al. 1999; Reid, Fanning et al. 2000; Taubenberger, Reid et al. 2000; Basler, Reid et al. 2001; Taubenberger, Reid et al. 2001; Reid, Fanning et al. 2002). (See Chapter 11 for more information on the 1918 influenza pandemic). The classical H1N1 swine influenza viruses in the U.S. remained antigenically and genetically highly conserved from 1965 through the 1980s (Sheerar, Easterday et al.

1989; Luoh, McGregor et al. 1992; Noble, McGregor et al. 1993), but antigenic and genetic variants of classical H1N1 viruses were isolated during the 1990s (Dea, Bilodeau et al. 1992; Olsen, McGregor et al. 1993; Rekik, Arora et al. 1994; Olsen, Carey et al. 2000).

Reassortant influenza A viruses within swine populations

It has been hypothesized that pigs can serve two very important roles in the development of pandemic influenza viruses: pigs may act as hosts for adaptation of avian influenza viruses to replication in mammals; and, they may serve as hosts for genetic reassortment between human and avian viruses. However, it is becoming increasingly clear that genetic reassortment of influenza A viruses in pigs is not only a concern for creation of pandemic human viruses, but also for creation of novel viruses of importance to the pigs themselves. Likewise, it is no longer appropriate to think about reassortment in pigs as occurring only in the Southeast Asia. Southeast Asia has historically been considered the 'epicenter' for pandemic virus generation. The 1957 and 1968 viruses first appeared in Asia, and agricultural practices such as fish farming in rural Asia bring people, pigs and waterfowl together in ways that enhance the potential for interspecies transmission of influenza viruses and reassortment in pigs (Shortridge and Stuart-Harris 1982; Scholtissek, Burger et al. 1985; Scholtissek and Naylor 1988; Webster, Bean et al. 1992; Webby and Webster 2001). However, reassortant viruses have been isolated from pigs in many areas of the world in recent years (Table 3).

Reassortant H3N2 swine influenza viruses

H3N2 viruses with both human and classical swine virus genes have been isolated from pigs in Asia and the United States (Shu, Lin et al. 1994; Nerome, Kanegae et al. 1995; Zhou, Senne et al. 1999a) and H3N2 viruses with human HA and NA genes and avian internal protein genes have been isolated from pigs in Europe (Castrucci, Donatelli et al. 1993; Campitelli, Donatelli et al. 1997; Lin, Bennett et al. 2003) and Asia (Peiris, Guan et al. 2000). However, since 1998, 'triple reassortant' H3N2 viruses have been isolated widely throughout the United States. These viruses contain HA, NA and PB1 polymerase genes of human influenza virus origin, NP, M and NS genes of classical swine H1N1 virus origin, and PB2 and PA polymerase genes of North American avian virus origin (Zhou, Senne et al. 1999a; Karasin, Schutten et al. 2000c; Webby, Swenson et al. 2000).

Above and beyond their triple reassortant genotype, several features of these viruses are of interest. First, infection with some of the triple reassortant viruses has been associated with not only typical respiratory disease, but also spontaneous abortion in sows, and death of even adult pigs (Zhou, Senne et al. 1999a; Karasin, Schutten et al. 2000c). These presentations are extraordinary for swine influenza. Infections of pigs with classical swine H1N1 viruses rarely cause death, and, with the exception of a few reports (Young and Underdahl 1949; Woods and Mansfield 1974; Madec, Kaiser et al. 1989), swine influenza viruses are not thought to directly target the reproductive tract of pigs. It remains to be determined whether the triple reassortant virus-associated abortions were due to direct viral effects or simply high fevers in the affected animals. Secondly, phylogenetic analyses of the HA genes of triple reassortant H3N2 viruses isolated from pigs in 1998–1999 suggest that these viruses either rapidly evolved into three distinct lineages (with associated antigenic differences) via genetic drift, or that they were derived from multiple reassortment events (Webby, Swenson et al. 2000). Thirdly, these viruses may provide important research tools for studying swine host-range determinants for influenza A viruses. For instance, differences in pathogenicity between viruses from the three different phylogenetic clades (Richt, Lager

Table 9.3 Examples of Reassortant Influenza Virus Genotypes Isolated from Pigs Around the World.

Virus subtype (Geographical region of isolation)					
	H3N2	H3N2	H3N2	H3N2	H1N2
RNA segment (Gene)	(Asia)[b]	(N. America)[c]	(Europe, Asia)[d]	(N. America)[e]	(N. America)[f]
1 (PB2)	S[a]	S	A	A	A
2 (PB1)	S	H	A	H	H
3 (PA)	S	S	A	A	A
4 (HA)	H	H	H	H	S
5 (NP)	S	S	A	S	S
6 (NA)	H	H	H	H	H
7 (M)	S	S	A	S	S
8 (NS)	S	S	A	S	S

Virus subtype (Geographical region of isolation)					
	H1N1	H1N2	H1N2	H1N7	H3N1
RNA segment (Gene)	(N. America)[g]	(Asia)[h]	(Europe)[i]	(United Kingdom)[j]	(Taiwan)[k]
1 (PB2)	A	S	A	H	?
2 (PB1)	H	S	A	H	?
3 (PA)	A	S	A	H	?
4 (HA)	S	S	H (A)	H	H
5 (NP)	S	S	A	H	?
6 (NA)	S	H	H	E	S
7 (M)	S	S	A	E	?
8 (NS)	S	S	A	H	?

a Phylogenetic lineages: H = human or human-like swine lineage; S = classical swine H1N1 lineage; A = avian or avian-like swine lineage; E = equine (A/Equine/Prague/1/56)-like lineage; ? = data not available to define the lineage

b (Shu, Lin et al. 1994; Nerome, Kanegae et al. 1995)

c (Zhou, Senne et al. 1999c)

d (Castrucci, Donatelli et al. 1993; Campitelli, Donatelli et al. 1997; Peiris, Guan et al. 2000; Lin, Bennett et al. 2003)

e (Karasin, Schutten et al. 2000c; Webby, Swenson et al. 2000; Zhou, Senne et al. 2000)

f (Karasin, Olsen et al. 2000b; Choi, Goyal et al. 2002a; Choi, Goyal et al. 2002b; Karasin, Landgraf et al. 2002)

g (Webby, Swenson et al. 2000); Olsen et al., unpublished

h (Nerome, Yoshioka et al. 1985; Ouchi, Nerome et al. 1996; Ito, Kawaoka et al. 1998b; Shimada, Ohtsuka et al. 2003)

i (Brown, Chakraverty et al. 1995a; Brown, Harris et al. 1998; Van Reeth, Brown et al. 2000; Marozin, Gregory et al. 2002; Schrader and Suss 2003; Van Reeth, Gregory et al. 2003) (The majority of these viruses contained human-lineage HA genes, but some isolates from Italy contained avian-like swine HA genes.)

j (Brown, Alexander et al. 1994; Brown, Hill et al. 1997a)

k (Tsai and Pan 2003)

et al. 2003) have been demonstrated. Furthermore, it has been shown that a representative triple reassortant H3N2 swine virus is more readily infectious and more highly pathogenic in pigs than a wholly human H3N2 virus that was isolated from a pig in 1997 (Landolt, Karasin et al. 2003). The later result is of particular interest because the initially described triple reassortant viruses contained at least 12 amino acid differences in their HA genes compared to the 1995 human-lineage viruses to which they were most closely related, including three amino acids residues within the receptor-binding pocket, and the loss of a potential N-linked glycosylation site (Zhou, Senne et al. 1999a; Karasin, Schutten et al. 2000c). These HA differences, differences between the triple reassortant viruses and wholly human viruses in their NA genes (Zhou, Senne et al. 1999a), and the potential impact of constellations of the other swine- and avian-lineage genes on infectivity and pathogenesis in pigs are currently being assessed through reverse genetics studies (Olsen et al., unpublished).

Reassortant H1N2 swine influenza viruses
Beginning in November 1999, approximately one year after the initial isolation of triple reassortant H3N2 viruses from pigs in the United States, influenza-like illness and, again, abortions in sows, were associated with infection of pigs in the United States with an H1N2 virus. Phylogenetic analyses demonstrated that this virus had been derived through reassortment between a triple reassortant H3N2 swine virus and a classical swine H1N1 virus (Karasin, Olsen et al. 2000b). This H1N2 virus retained the entire genetic backbone of the triple reassortant virus, but had acquired a classical swine H1 HA gene. This lineage of H1N2 viruses has subsequently spread throughout the swine population of the United States (Choi, Goyal et al. 2002a; Choi, Goyal et al. 2002b; Karasin, Landgraf et al. 2002), as well as into the domestic turkey population (Suarez, Woolcock et al. 2002) and, in at least one instance, into wild waterfowl (Olsen, Karasin et al. 2003).

These were the first H1N2 viruses isolated from pigs in North America, but H1N2 viruses had been recovered from pigs in Japan in 1978–1980, 1989–1992 (Sugimura, Yonemochi et al. 1980; Nerome, Yoshioka et al. 1985; Ouchi, Nerome et al. 1996; Ito, Kawaoka et al. 1998b), and 1999–2001 (Shimada, Ohtsuka et al. 2003), and in Taiwan in 2002 (Tsai and Pan 2003). All of these viruses were reassortants between human (or human-like swine) H3N2 and classical swine H1N1 viruses. H1N2 viruses have also been recovered from pigs in the United Kingdom since 1994 (Brown, Chakraverty et al. 1995a; Brown, Harris et al. 1998), and thereafter from pigs in Belgium since 1999 (Van Reeth, Brown et al. 2000; Van Reeth, Gregory et al. 2003), France and Italy since 1997 (Marozin, Gregory et al. 2002), and Germany since 2002 (Schrader and Suss 2003). The majority of these viruses contained human-lineage HA and NA genes and internal protein genes derived from the avian-like swine European H1N1 viruses, although some of the Italian isolates had also acquired an avian-like swine HA (Marozin, Gregory et al. 2002).

Reassortant H1N1 swine influenza viruses
Viruses of a third reassortant genotype have been isolated in the United States since 1998. These are H1N1 subtype viruses with the HA and NA genes derived from a classical swine H1N1 virus and the remaining genes derived from the American triple reassortant H3N2 or H1N2 swine viruses. This genotype was first recovered from a 57-year-old man with influenza-like illness in Wisconsin who had had direct contact with a pig (Cooper et al., unpublished). Since 2001, however, viruses of the same genotype have been isolated

frequently from pigs in the United States (Olsen et al., unpublished; Webby, Swenson et al. 2000). In fact, it appears from current surveillance activities that this is becoming the predominant genotype of H1N1 virus within the American swine population (Olsen and Webby, unpublished).

Reassortant H1N7 and H3N1 swine influenza viruses

These two unusual subtypes of influenza A viruses have been recovered on a limited basis from pigs. H1N7 viruses were isolated from pigs on a single farm in the United Kingdom in 1992. The NA and M genes were A/Equine/Prague/1/56-like, while the remaining genes were of human influenza virus origin. This virus was of low pathogenicity in experimentally infected pigs (Brown, Alexander et al. 1994; Brown, Hill et al. 1997a). H3N1 viruses that were reassortants between human H3N2 and classical swine H1N1 viruses have been recovered from pigs in Taiwan (Tsai and Pan 2003) and the U.K. (Brown, unpublished results).

Pigs as intermediate hosts in human influenza virus evolution and epidemiology

Pigs are susceptible to infection with both avian and human influenza A viruses because the cells of their respiratory tract express both the SAα2,3Gal receptors preferred by avian influenza viruses and the SAα2,6Gal receptors preferred by human influenza viruses (Ito, Couceiro et al. 1998a; Ito 2000). Furthermore, pigs express α2,3 receptors that contain, in addition to N-acetylneuraminic acid (NAc), sialic acid of the N-glycolneuraminic (NGc) form (Suzuki, Horiike et al. 1997). Receptors containing NGc are also present in the duck intestine (Ito, Suzuki et al. 2000). Thus, avian viruses may initially infect pigs via NAcα2,3Gal or NGcα2,3Gal receptors and then mutate during replication in pigs so as to acquire the ability to use NAcα2,6Gal receptors, and thereby adapt to replication in mammals (Ito 2000).

According to the 'mixing vessel' hypothesis (Scholtissek, Burger et al. 1985; Scholtissek and Naylor 1988; Webster, Bean et al. 1992), if avian and human influenza A viruses co-infect a single host, then during replication these viruses can exchange RNA segments and create new viruses containing mixtures of genes from each parental virus. Genetic evidence indicates that the 1957 H2N2 and the 1968 H3N2 pandemic strains arose through genetic reassortment between an avian influenza virus and the pre-existing human viruses (Laver and Webster 1973; Scholtissek, Rohde et al. 1978b; Kawaoka, Krauss et al. 1989; Webster, Bean et al. 1992). Given the limited susceptibility of birds to infection with human influenza viruses, and conversely, humans to infection with avian viruses (Hinshaw, Webster et al. 1983a; Snyder, Buckler-White et al. 1987; Beare and Webster 1991; Webster, Bean et al. 1992; Webby and Webster 2001; Wright and Webster 2001), it is unlikely that the origin of pandemic viruses rests solely in either birds or humans. Rather, such pandemic viruses may owe their creation to pigs. While there is no direct evidence that the 1957 and 1968 pandemic viruses arose in pigs, there is ample evidence that influenza viruses can move across species barriers from birds to pigs, from people to pigs, and from pigs to people, the three steps that would be necessary to create a pandemic virus in pigs and move it from pigs into the human population.

ADAPTATION AND TRANSMISSION

From an epidemiological perspective, type A influenza viruses have caused serious disease outbreaks in poultry, principally chickens and turkeys, as well as mammals such as pigs, horses, mink, seals and humans. Avian influenza viruses show various degrees of host adaptation. Avian influenza viruses exhibit maximal adaptation to a single host genus or even a species, and such adaptation results in easy transmission of the AI virus between birds of the same species and less easy transmission to birds that are more distantly related (Swayne 2000). For example, the H7N2 LPAI virus that caused an outbreak in commercial farms of the Shenandoah Valley (Virginia) during 2002 had a higher degree of adaptation to turkeys, as indicated by a low mean bird infective dose (BID_{50}) of $10^{0.7}$ mean embryo infectious doses (EID_{50}), than did the virus in chickens (10^3 BID_{50}), and this difference partially explained the preponderance of affected operations being turkey farms (Tumpey, Kapczynski et al. 2004b). Furthermore, because of lesser adaptation and poorer replication in chickens, this H7N2 LPAI virus infected meat-type chickens (broilers) without production of clinical disease (Akey 2003). Adaptation can be achieved by passage of the virus in the host species, resulting in greater replication of the virus. For example, a H7N2 LPAI virus isolated from Bobwhite quail (A/Bobwhite quail/Pennsylvania/20304/1998) replicated poorly in chickens (average peak titer in oropharynx, $10^{2.5}$ EID_{50}/ml) while 15 sequential intranasal passages in chickens greatly improved adaptation to the respiratory tract of chickens as evident by increased virus replication (average peak titer in oropharynx, $10^{4.3}$ EID_{50}/ml) (D. Swayne, unpublished data). One exception to the concept of maximal species adaptation has been the classic H1N1 swine influenza viruses. These viruses have produced natural infections of turkey breeder hens with serious drops in egg production and economic losses, and H1N1 influenza vaccination to protect turkey breeders is practiced in geographic regions where the pig production industry is also present (Swayne 2000; Swayne and Halvorson 2003; Swayne and Akey 2004).

A high degree of species adaptation is important for efficient transmission of AI viruses to other birds; i.e. the most highly adapted viruses for a host species transmit more easily between individuals of that species or closely related species (Swayne 2000). For adaptation to occur, an AI virus strain must have repeated exposure to a new population of host species, infrequent infection of individuals within the population and repeated passage in individuals of the new host species with selection of a strain adapted for the new species. For example, from 1978–1995, turkeys raised outdoors in Minnesota for the Thanksgiving and Christmas markets have experienced outbreaks of AI as a result of contact with wild ducks (Halvorson, Karunakaran et al. 1983; Halvorson, Kelleher et al. 1985). The first infected turkeys had asymptomatic infections detected in processing plants by serology, but over time, the virus adapts with passage in turkeys and transmits much more easily with resulting clinical respiratory disease in later flocks (Swayne 2000). Alternatively, the transmission of influenza from wild waterfowl to chickens and turkeys may occur through intermediates, in particular domestic ducks and geese that are reared or marketed in close association with chickens and turkeys. The mixing of poultry species in Village (backyard) or Live Poultry Marketing systems and access to wild birds creates a high-risk agricultural production environment for introduction of wild bird AI viruses, adaptation to gallinaceous birds and spread to commercial indoor poultry production operations. The latter usually occurs through fomites such as shoes, clothing or equipment contaminated with AI virus.

INTERSPECIES TRANSMISSION

In general, transmission of AI viruses from birds to mammals is a rare event, especially to humans. When examining the exposure of poultry workers to H5N1 AI viruses via live poultry markets in Hong Kong during 1997, the number of exposures to poultry were high for poultry workers, but the infection rate was low (10%) suggesting poor adaptation to human beings compared with human influenza type A strains which are easily transmitted and spread rapidly among health care workers (Bridges, Lim et al. 2002; Bridges, Kuehnert et al. 2003). However, reassortment of AI viruses with human influenza A viruses (as occurred in 1957 and 1968), has led to emergence of new subtypes of hemagglutinin in influenza A viruses that can be transmitted person-to-person and resulted in influenza pandemics in humans (Kawaoka, Krauss et al. 1989; Fanning, Slemons et al. 2002). The risks of AI virus transmission to humans varies with multiple factors including virus dose, exposure material and virus strain. For example, the risk of human infection, illness and death in the 1997 H5N1 outbreak in Hong Kong was associated with exposure to live poultry 1 week before the person's illness and was not associated with preparing or eating poultry meat (Mounts, Kwong et al. 1999). Direct exposure to infected birds poses the greatest exposure risk because high levels of virus are excreted in respiratory secretions and feces. In the recent H5N1 outbreaks, human infections in Thailand and Vietnam were associated with direct exposure to infected poultry at the village or smallholdings level (Anonymous 2004a; Tran, Nguyen et al. 2004).

Avian-to-human transmission of influenza viruses

Until 1997, it was believed that avian influenza viruses were limited in their ability to infect humans without adaptation or reassortment with human influenza virus genes. Reports have been published of human infections with wholly avian influenza viruses followed accidental exposure (Taylor and Turner 1977; Webster, Geraci et al. 1981; Kurtz, Manvell et al. 1996) or intentional exposure in which avian influenza viruses were administered to volunteers in an attempt to determine the susceptibility of humans to infection with avian influenza viruses (Beare and Webster 1991). The conclusion from these reports was that humans could be infected following direct inoculation onto a mucosal surface of certain subtypes of avian influenza viruses but that the infections tended to be self-limited without serious complications or spread to contacts. The reader is referred to a review by Subbarao and Katz for a detailed discussion of these reports (Subbarao and Katz 2000).

In 1997, 2003 and 2004, human infections with highly pathogenic avian influenza A/H5N1 viruses occurred in Hong Kong, Vietnam and Thailand during large-scale outbreaks of H5N1 infections in poultry. These human infections have been associated with severe disease and a high case fatality rate and are discussed in greater detail in Chapter 10. In 2003, human infections with a highly pathogenic avian influenza A/H7N7 virus occurred among persons involved in control of a large-scale poultry outbreak of HPAI in The Netherlands.

Additional human infections with avian influenza viruses that have been reported since the review by Subbarao and Katz (Subbarao and Katz 2000) and the H5N1 and H7N7 infections mentioned above include the following:

1 An H7N2 infection in the United States in 2002 that was identified in an individual who was involved in culling 4.7 million poultry infected with a low pathogenic avian influenza H7N2 virus in Virginia- the patient complained of upper respiratory symptoms from which he made a complete recovery and was subsequently found to have antibodies to the H7N2 virus (Anonymous 2004d).

2 An H9N2 infection in Hong Kong in December 2003 that was reported in a 5-year-old child who was hospitalized for two days following symptoms of fever, cough and a

runny nose and made a complete recovery - the virus isolated from this patient was genetically related to H9N2 viruses isolated from poultry in the live bird markets of Hong Kong. This case of human H9N2 infection was similar to the two cases of febrile pharyngitis that were reported in children in Hong Kong in 1999; the infections were self-limited and did not result in spread to contacts and the virus isolates were genetically closely related to H9N2 viruses isolated from poultry (Anonymous 2004d). Although direct epidemiologic links to infected poultry were not demonstrated among the three children with laboratory confirmed H9N2 infections in Hong Kong, the potential for exposure to infected poultry was present. H9N2 viruses have been widely prevalent in poultry in the live bird markets of Hong Kong since 1997 (Choi, Ozaki et al. 2004) and live poultry are sold in stalls and markets that are close to living quarters in Hong Kong.

3 Two laboratory-confirmed reports of H7N3 infection in Canada in 2004 in persons involved in culling poultry infected with an avian influenza A/H7N3 virus- these individuals developed conjunctivitis and upper respiratory symptoms or headache, were treated with oseltamivir and their symptoms resolved. While these two cases were the only laboratory confirmed cases of infection, there were reports of conjunctivitis and/or upper respiratory symptoms in 10 additional persons who were exposed to infected poultry. There were no reports of person-to-person transmission of H7N3 infection (Anonymous 2004d).

The following observations can be made from considering the reports of avian influenza virus infections in humans in the past decade along with the sporadic reports from accidental and experimental infections reported earlier:

- Wholly avian influenza viruses can infect humans and the consequences of the infections can range from mild febrile upper respiratory tract infections (H9) to conjunctivitis with or without upper respiratory symptoms (H7) to febrile respiratory illness that can progress to acute respiratory distress syndrome with multi-organ dysfunction (H5 and H7).
- The infections occur when there are contemporaneous infections in poultry and tend to occur in persons who are exposed to infected poultry.
- The virulence of avian influenza viruses for humans is likely to be determined by multiple gene segments of the virus (Hatta, Gao et al. 2001; Seo, Hoffmann et al. 2002) and host factors such as age may also play a role; the illnesses caused by the 1997 H5N1 and 1999 H9N2 viruses in Hong Kong were remarkably different in severity (Yuen, Chan et al. 1998; Peiris, Yuen et al. 1999) but the viruses shared significant homology in their non-surface glycoprotein genes (Lin, Shaw et al. 2000), suggesting that the HA and NA genes were major determinants of virulence. With the exception of the first case that occurred in a young child (Subbarao, Klimov et al. 1998), the serious illnesses and deaths from H5N1 infections in 1997 occurred in older individuals (Yuen, Chan et al. 1998). However, young children became severely ill during the 2004 H5N1 outbreak in Vietnam and Thailand (Anonymous 2004a; Tran, Nguyen et al. 2004). Additional genetic markers for virulence have been identified in the PB2 gene (Hatta, Gao et al. 2001; Fouchier, Schneeberger et al. 2004) and the NS1 gene segment (Seo, Hoffmann et al. 2002; Seo, Hoffmann et al. 2004) of the 1997 viruses. Careful analysis of the viruses from the 2004 outbreaks may explain the differences in the epidemiological patterns of disease.

- H7 viruses appear to have a propensity for causing conjunctivitis; the tropism of these viruses for conjunctival cells should be investigated (Taylor and Turner 1977; Webster, Geraci et al. 1981; Kurtz, Manvell et al. 1996; Koopmans, Wilbrink et al. 2004).

Swine-to-human transmission of influenza viruses (zoonotic swine influenza)

Zoonotic infections with swine influenza A viruses have been reported in North America, Europe and Asia (Smith, Burgert et al. 1976; Kendal, Goldfield et al. 1977; Top and Russell 1977; Hinshaw, Bean et al. 1978; Beare, Kendal et al. 1980; Kaplan 1982; Dacso, Couch et al. 1984; Patriarca, Kendal et al. 1984; de Jong, Paccaud et al. 1988; Kaufman, Hassan et al. 1988; Rota, Rocha et al. 1989; Claas, Kawaoka et al. 1994; Wentworth, Thompson et al. 1994; Wentworth, McGregor et al. 1997; Kimura, Adlakha et al. 1998; Alexander and Brown 2000; Gregory, Lim et al. 2001; Rimmelzwaan, de Jong et al. 2001; Gregory, Bennett et al. 2003). Many of these infections involved classical H1N1 swine influenza viruses (Smith, Burgert et al. 1976; Top and Russell 1977; Hinshaw, Bean et al. 1978; Dacso, Couch et al. 1984; Patriarca, Kendal et al. 1984; Rota, Rocha et al. 1989; Wentworth, Thompson et al. 1994; Wentworth, McGregor et al. 1997; Kimura, Adlakha et al. 1998). However, zoonotic infections from swine have also occurred with the wholly avian H1N1 swine viruses in Europe (de Jong, Paccaud et al. 1988; Rimmelzwaan, de Jong et al. 2001; Gregory, Bennett et al. 2003), as well as reassortant H3N2 viruses containing avian internal protein genes in Europe (Claas, Kawaoka et al. 1994) and Hong Kong (Gregory, Lim et al. 2001), and a reassortant H1N1 virus containing genes of human, swine and avian viruses in the United States (Cooper et al., unpublished). At present, there do not appear to be distinctive clinical features that distinguish zoonotic swine virus from conventional human influenza virus infections in human beings. However, it should be noted that in a number of the zoonotic cases, the infections proved fatal (Smith, Burgert et al. 1976; Top and Russell 1977; Patriarca, Kendal et al. 1984; Rota, Rocha et al. 1989; Wentworth, Thompson et al. 1994; Kimura, Adlakha et al. 1998).

Although neither influenza B or C are major disease concerns in pigs, there is some evidence for interspecies transmission of these viruses among people and pigs. For instance, some influenza C viruses isolated from people in Japan were more closely related to Chinese swine isolates of influenza C than other human isolates (Kimura, Abiko et al. 1997). In addition, there is serologic evidence of human influenza B virus infection in pigs in Europe (Brown, Harris et al. 1995b) and pigs have been shown to be susceptible to human influenza B virus infection following experimental infection (Takatsy, Farkas et al. 1969).

Not surprisingly, the vast majority of cases in which human beings have been infected with swine-origin influenza viruses have involved individuals in direct contact with pigs. Consistent with that fact, seroepidemiological studies in both the United States and Europe have suggested increased rates of swine influenza virus exposure among persons in contact with pigs (Kluska, Hanson et al. 1961; Woods, Hanson et al. 1968; Schnurrenberger, Woods et al. 1970; Woods, Schnurrenberger et al. 1981; Nowotny, Deutz et al. 1997). Most recently, a study of swine farmers, employees, and their family members in the United States compared to an urban control population from the same geographic region found that swine influenza virus seropositivity was statistically associated with being a farm owner or farm family member, living on a farm, or entering a swine barn 4 or more days per week (Olsen, Brammer et al. 2002). Nonetheless, there are also examples of infections without apparent animal contact, including children in Europe and Hong Kong, and those involved in the Fort Dix, NJ outbreak in 1976 (Goldfield, Noble et al. 1977; Hodder, C. et al. 1977; Kendal, Goldfield et al. 1977; Top and Russell 1977;

Claas, Kawaoka et al. 1994; Brown 2000; Gregory, Lim et al. 2001). These cases raise the possibility of human-to-human spread of a virus after initial interspecies transmission from a pig to a person. However, with the exception of the Ft. Dix incident in 1976, there is little evidence for spread of swine viruses from person-to-person. Although avian- versus human-lineage signature sequences have been described within the NP, M1, M2, PA, PB1 and PB2 influenza A virus genes (Zhou, Shortridge et al. 1999c; Hiromoto, Yamazaki et al. 2000; Katz, Lu et al. 2000; Naffakh, Massin et al. 2000; Yao, Mingay et al. 2001; Reid, Fanning et al. 2002; Shaw, Cooper et al. 2002a), descriptions of genetic differences that distinguish swine- versus human-lineage viruses are much more limited (Castrucci, Campitelli et al. 1994; Wentworth, Thompson et al. 1994; Bikour, Frost et al. 1995; Wentworth, McGregor et al. 1997; Gregory, Bennett et al. 2003). Therefore, it remains to be determined whether specific viral factors impact human-to-human transmission of swine influenza A viruses.

Avian-to-swine transmission of influenza viruses

Kida and colleagues have demonstrated experimentally that pigs can be infected with a wide range (H1-H13 subtypes) of avian influenza viruses (Kida, Ito et al. 1994), but naturally acquired infection of pigs with avian viruses has also been documented repeatedly. One of the most notable examples is that of the wholly avian H1N1 virus that crossed the species barrier *in toto* to infect pigs in Italy in the late 1970s. This virus ultimately adapted to use of SAα2,6Gal receptors (Rogers and D'Souza 1989; Ito, Couceiro et al. 1998a), and spread widely throughout much of the European continent and United Kingdom, becoming the dominant cause of swine influenza in these areas (Pensaert, Ottis et al. 1981; Scholtissek, Burger et al. 1983; Hinshaw, Alexander et al. 1984; Donatelli, Campitelli et al. 1991; Schultz, Fitch et al. 1991; Webster, Bean et al. 1992; Brown, Ludwig et al. 1997b; Webby and Webster 2001). Elsewhere in the world, avian H1N1, H3N2, and H9N2 viruses have been recovered from pigs in Asia (Kida, Shortridge et al. 1988; Peiris, Guan et al. 2000; Peiris, Guan et al. 2001), seropositivity to H4, H5, and H9 viruses has been documented among pigs in China (Ninomiya, Takada et al. 2002), and avian H4N6 (Karasin, Brown et al. 2000a) and H3N3 and H1N1 (Karasin, West et al. 2004) viruses have been isolated from pigs in Canada. Of these, the H9N2 and H4N6 viruses are of particular concern from a public health perspective because these are subtypes to which the human population is immunologically-naïve. In addition, it should be noted that the H4N6 viruses, though phylogenetically wholly avian, contained sequence signatures (Karasin, Brown et al. 2000a) previously associated with SAα2,6Gal receptor binding (Rogers, Paulson et al. 1983; Naeve, Hinshaw et al. 1984; Connor, Kawaoka et al. 1994; Ito, Suzuki et al. 1997b). Finally, it should be noted that interspecies transmission between avian species and pigs is bidirectional, with evidence for spread of wholly avian, classical swine, and reassortant viruses from pigs to turkeys and ducks (Hinshaw, Webster et al. 1983a; Wright, Kawaoka et al. 1992; Ludwig, Haustein et al. 1994; Wood, Banks et al. 1997; Suarez, Woolcock et al. 2002; Olsen, Karasin et al. 2003).

Human-to-swine transmission of influenza viruses

Infection of pigs with human H3N2 influenza viruses has occurred frequently in Europe (Tumova, Veznikova et al. 1980; Ottis, Sidoli et al. 1982; Castrucci, Campitelli et al. 1994; Brown, Harris et al. 1995b) and Asia (Kundin 1970; Shortridge, Webster et al. 1977; Shortridge, Cherry et al. 1979; Shortridge and Webster 1979; Nerome, Ishida et al. 1981; Nakajima, Nakajima et al. 1982; Mancini, Donatelli et al. 1985; Shu, Lin et al. 1994; Katsuda, Shirahata et al. 1995; Nerome, Kanegae et al. 1995), but less commonly in North

America (Hinshaw, Bean et al. 1978; Bikour, Frost et al. 1995; Karasin, Schutten et al. 2000c). In addition, it has been suggested that older lineages of human H3N2 viruses may be maintained by circulation in pigs beyond the time of their active circulation among human beings (Shortridge, Webster et al. 1977; Nakajima, Nakajima et al. 1982; Ottis, Sidoli et al. 1982; Haesebrouck, Biront et al. 1985; Mancini, Donatelli et al. 1985; Pritchard, Dick et al. 1987; Castrucci, Campitelli et al. 1994). This is a concern because of the potential for re-introduction of such viruses back into the human population, and particularly infection of young children who would be immunologically naive. Finally, beyond H3N2 viruses, there is also limited evidence for infection of pigs with human H1N1 viruses (Aymard, Brigaud et al. 1980; Nerome, Ishida et al. 1982; Brown, Harris et al. 1995b; Katsuda, Shirahata et al. 1995; Brown 2000).

Vaccination and Antiviral Control of Influenza in Humans and Swine

Vaccines and antiviral drugs are the two specific methods that are available for the control of influenza. The principle of the licensed vaccines is to elicit HA-specific protective humoral immunity. In humans, this is achieved using an inactivated virus administered parenterally or a live attenuated vaccine administered intranasally (Centers for Disease Control and Prevention 2004). There are two classes of antiviral drugs with activity against influenza A viruses: the adamantane compounds and the neuraminidase inhibitors. The latter are effective against influenza B viruses as well. Vaccines represent the mainstay of control, but antiviral drugs are a very useful adjunct in specific circumstances (Centers for Disease Control and Prevention 2004). (See chapter 7 for discussion of antiviral drugs and chapter 8 for a discussion of vaccines for control of human influenza). Control of influenza in poultry and pigs is discussed below.

VACCINATION TO CONTROL INFLUENZA VIRUS INFECTIONS
Vaccination against human influenza virus infection
Epidemic influenza

The World Health Organization's (WHO) global influenza surveillance program monitors antigenic and genetic drift among influenza A and B viruses and makes recommendations for the composition of the influenza vaccine so that vaccine strains match epidemic strains. The vaccines that are in use are trivalent vaccines with components that represent circulating influenza A (H1N1 and H3N2 subtypes) and B viruses. Two types of trivalent vaccines are licensed for use in the United States- an inactivated vaccine and a live attenuated vaccine. The efficacy of the inactivated vaccine is 70–90% in healthy adults and lower in the elderly (Cox and Subbarao 1999). The efficacy of the live attenuated vaccine was 91–96% in young children (Belshe, Mendelman et al. 1998). The current formulation of the live attenuated vaccine has not been directly compared with the inactivated vaccine.

The composition of the vaccine is updated when there is evidence that a circulating strain has drifted antigenically from the corresponding vaccine component. The primary factor in the decision to update a vaccine strain is evidence that an influenza virus that is an antigenic variant of the vaccine strain is associated with epidemics or outbreaks in distant regions. An antigenic variant is defined as a strain that can be distinguished from the vaccine strain using specific post-infection ferret antisera in an HI assay. Evidence of genetic drift is an additional consideration; ideally the vaccine strain is selected to be an antigenic and genetic match of the predicted epidemic strain in both the HA and NA genes. Since 1999 separate recommendations are made for the composition of influenza vaccine for use in the

Southern hemisphere. Table 4 lists the epidemic strains and vaccine strains recommended by the WHO for the past 10 years. The Advisory Committee on Immunization Practices of the Centers for Disease Control & Prevention and the Committee on Infectious Diseases of the American Academy of Pediatrics make recommendations for the use of influenza vaccine in the United States (American Academy of Pediatrics 2003; Centers for Disease Control and Prevention 2004).

Pandemic influenza

The experience with vaccines against potential pandemic strains of influenza is limited and is an area of increasing interest in view of the outbreaks of human H5N1 infections in 1997 (Subbarao, Klimov et al. 1998; Yuen, Chan et al. 1998) and 2004 (Anonymous 2004a; Tran, Nguyen et al. 2004) and reports of H9N2 infections in 1999 (Peiris, Yuen et al. 1999). The options that have been considered for H5N1 vaccines include the use of an inactivated vaccine made from a surrogate related avian influenza virus of the same subtype that is less pathogenic than the wild-type virus (Nicholson, Colegate et al. 2001), an inactivated vaccine generated from a reverse genetics-derived seed virus with a genetically modified HA gene that lacks the multi-basic acid cleavage site (Subbarao, Chen et al. 2003), a similar candidate live attenuated vaccine (Li, Liu et al. 1999) or purified HA protein (Treanor, Wilkinson et al. 2001). An inactivated vaccine generated from the wild-type H9N2 virus influenza A/Hong Kong/1073/99 has been evaluated in clinical trials; prior immunity to the H2 HA protein affected the immunogenicity of the H9N2 vaccine (Hehme, Engelmann et al. 2002; Stephenson, Nicholson et al. 2003). Approaches to enhance the immunogenicity of candidate inactivated and subunit vaccines must be explored and the infectivity and immunogenicity of candidate live attenuated pandemic vaccines must be investigated as part of preparation for the next influenza pandemic.

Vaccination against swine influenza virus infection

Vaccination is practiced widely for the control of influenza A virus infections in pigs. Although DNA-based (Eriksson, Yao et al. 1998; Macklin, McCabe et al. 1998; Olsen 2000; Larsen, Karasin et al. 2001; Heinen, Rijsewijk et al. 2002; Larsen and Olsen 2002) and recombinant viral vectored vaccines (Tang, Harp et al. 2002) have been studied experimentally, virtually all commercially-produced swine influenza virus vaccines are traditional, adjuvanted, inactivated virus preparations. These vaccines are typically produced regionally so as to correspond to the subtypes of viruses that predominate in different areas of the world. Vaccination regimes may be targeted toward breeder sows, young piglets, and/or older feeder pigs. When new subtypes of virus are introduced into and spread widely within a swine population, the vaccines are updated to include the new subtype, as occurred subsequent to the emergence of H3N2 viruses in North America in 1998. The swine vaccines are not, however, re-evaluated for matching to prevailing antigenic variants on a yearly basis as occurs for human influenza virus vaccines. This is due, in part, to the lack of large-

---

Note to Table 9.4 (*opposite*)

1 Recommendations made by the WHO, published in the Weekly Epidemiological Record in February and October for vaccines intended for use the following winter in the Northern and Southern hemisphere, respectively (Anonymous 1993; Anonymous 1994a; Anonymous 1995a; Anonymous 1996a; Anonymous 1997a; Anonymous 1998a; Anonymous 1999b; Anonymous 1999a; Anonymous 2000b; Anonymous 2000a; Anonymous 2001a; Anonymous 2001b; Anonymous 2002a; Anonymous 2002b; Anonymous 2003a; Anonymous 2003b; Anonymous 2004b).

Table 9.4 Epidemic Strains of Influenza and Recommended Composition of Influenza Virus Vaccines[1], by Type and Subtype

Virus type/ subtype	Year	Epidemic strain(s)	Vaccine strain	
			Northern Hemisphere	Southern Hemisphere (year)
A/H1N1	2004–05		New Caledonia/20/99	
	2003–04	New Caledonia/20/99	New Caledonia/20/99	New Caledonia/20/99 (2004)
	2002–03	New Caledonia/20/99	New Caledonia/20/99	New Caledonia/20/99 (2003)
	2001–02	New Caledonia/20/99	New Caledonia/20/99	New Caledonia/20/99 (2002)
	2000–01	New Caledonia/20/99 + Bayern/7/95	New Caledonia/20/99	New Caledonia/20/99 (2001)
	1999–2000	Bayern/7/95 + Beijing/262/95 + New Caledonia/20/99	Beijing/262/95	New Caledonia/20/99 (2000)
	1998–99	Bayern/7/95	Beijing/262/95	Beijing/262/95 (1999)
	1997–98	Johannesburg/82/96 + Beijing/262/95	Bayern/7/95	
	1996–97	Texas/36/91 + Taiwan/1/86	Singapore/6/86	
	1995–96	Texas/36/91 + Taiwan/1/86	Singapore/6/86	
	1994–95	Texas/36/91 + Taiwan/1/86	Singapore/6/86	
	1993–94	Texas/36/91	Singapore/6/86	
A/H3N2	2004–05		Fujian/411/2002	
	2003–04	Fujian/411/02	Moscow/10/99	Fujian/411/2002 (2004)
	2002–03	Panama/2007/99	Moscow/10/99	Moscow/10/99 (2003)
	2001–02	Panama/2007/99	Moscow/10/99	Moscow/10/99 (2002)
	2000–01	Panama/2007/99	Moscow/10/99	Moscow/10/99 (2001)
	1999–2000	Sydney/5/97	Sydney/5/97	Sydney/5/97 (2000)
	1998–99	Sydney/5/97	Sydney/5/97	Sydney/5/97 (1999)
	1997–98	Sydney/5/97 (80%) + Wuhan/359/95	Wuhan/359/95	
	1996–97	Wuhan/359/95	Wuhan/359/95	
	1995–96	Wuhan/359/95 + Johannesburg/33/94	Johannesburg/33/94	
	1994–95	Shangdong/9/93 + Johannesburg/33/94	Shangdong/9/93	
	1993–94	Beijing/32/92 + Shangdong/9/93	Beijing/32/92	
B	2004–05		Shanghai/361/2002	
	2003–04	Sichuan/379/99	HK/330/2001	HK/330/2001 (2004)
	2002–03	HK/330/01	HK/330/2001	HK/330/2001 (2003)
	2001–02	Sichuan/379/99 + Shizuoka/15/01	Sichuan/379/99	Sichuan/379/99 (2002)
	2000–01	Sichuan/379/99 + Beijing/184/93	Beijing/184/93	Sichuan/379/99 (2001)
	1999–2000	Beijing/184/93	Beijing/184/93 OR Shangdong/7/97	Beijing/184/93 OR Shangdong/7/97 (2000)
	1998–99	Beijing/184/93	Beijing/184/93	Beijing/184/93 (1999)
	1997–98	Harbin/7/94 + Beijing/184/93	Beijing/184/93	
	1996–97	Harbin/7/94 + Beijing/184/93	Beijing/184/93	
	1995–96	Panama/45/90 + Beijing/184/93	Beijing/184/93	
	1994–95	Beijing/184/93	Panama/45/90	
	1993–94	Panama/45/90	Panama/45/90	

257

scale, integrated swine influenza virus surveillance systems in many countries and because, historically, swine influenza viruses appeared to be antigenically conserved (Sheerar, Easterday et al. 1989). However, it is now clear that genetic/antigenic drift does occur among swine influenza viruses (Haesebrouck and Pensaert 1988; Dea, Bilodeau et al. 1992; Olsen, McGregor et al. 1993; Castrucci, Campitelli et al. 1994; Rekik, Arora et al. 1994; de Jong, van Nieuwstadt et al. 1999; Brown 2000; de Jong, Heinen et al. 2001; Marozin, Gregory et al. 2002). The mechanism of antigenic drift among swine viruses does not appear to be the same as that of the human influenza A viruses. Among pigs, different antigenic drift variants appear to develop locally within individual production units, and as such, single drift variants do not predominate at a population-wide level (de Jong, Heinen et al. 2001). Thus, some producers are moving from the use of commercially distributed vaccines to autogenous vaccines.

ANTIVIRAL DRUGS FOR THE CONTROL OF INFLUENZA VIRUS INFECTIONS
Antiviral drugs for human influenza virus infection
Epidemic influenza
There are two classes of licensed drugs with antiviral activity against influenza. The adamantane compounds amantadine and rimantadine have activity against influenza A viruses and the neuraminidase inhibitors have activity against both influenza A and B viruses. The Advisory Committee on Immunization Practices of the Centers for Disease Control & Prevention and the Committee on Infectious Diseases of the American Academy of Pediatrics provide recommendations on the use of antiviral compounds for the control of human influenza (American Academy of Pediatrics 2003; Centers for Disease Control and Prevention 2004). Amantadine and rimantadine are approved for the prevention of influenza in adults and children while oseltamivir is approved for this indication in children over the age of 13 years. Zanamivir, oseltamivir, amantadine and rimantadine are approved for treatment of influenza in adults. Amantadine is approved for treatment of influenza in children; zanamivir and oseltamivir are approved for this indication in children over the age of 7 years and 1 year of age, respectively (Centers for Disease Control and Prevention 2004). Further discussion of the two classes of antiviral drugs can be found in Chapter 7.

Pandemic influenza
Antiviral drugs can be very important for prophylaxis and treatment of potential pandemic strains of influenza. The susceptibility of avian influenza viruses to the two classes of antiviral drugs should be investigated and that of potential pandemic strains should be established early to ensure the judicious use of antiviral drugs for control and treatment of influenza. The WHO recommends the use of antiviral drugs in conjunction with personal protective equipment to protect agricultural workers involved in depopulating poultry infected with highly pathogenic avian influenza viruses as was done in the Netherlands (Fouchier, Schneeberger et al. 2004; Koopmans, Wilbrink et al. 2004).

Amantadine resistance among swine influenza viruses
Antiviral drugs are not used therapeutically to treat influenza virus infections in pigs. However, amantadine resistance has been noted among wholly avian-like H1N1 and reassortant H3N2 (with avian internal protein genes) influenza A viruses isolated from

pigs in Europe (Gregory, Lim et al. 2001; Lin, Bennett et al. 2003), and in a European-like reassortant H3N2 virus isolated from a child in Hong Kong (Gregory, Lim et al. 2001). It is not known whether some compound that pigs are fed or otherwise exposed to shares sufficient structural similarity to amantadine so as to drive the acquisition of resistance.

CONTROL OF AVIAN INFLUENZA IN POULTRY

Avian influenza control programs are designed to achieve one of three different goals or outcomes in poultry: 1) prevention of infection, 2) management of existing infection or disease to minimize economic losses, or 3) eradication (Swayne and Akey 2004). Strategies have been developed using various combinations of five different components to achieve these outcomes that include:

1 biosecurity procedures, including quarantine or movement restrictions, to prevent introduction or escape of AI virus from a farm or premise;
2 diagnostic and surveillance programs to detect premises with AI virus-infected birds or birds previously infected by an AI virus;
3 elimination of AI virus infected poultry through stamping-out of acutely infected birds, or controlled-marketing of recovered or vaccinated birds;
4 increasing host resistance to AI virus infection through vaccination or other preventatives, or improvements in host genetics, and
5 education of all persons in contact with poultry on AI biology and control strategies.

Multiple factors must be considered in selecting an acceptable outcome and developing a specific control strategy to achieve the desired outcome that may include virus strain (such as pathotype and HA subtype), the poultry species affected (chickens, turkeys, ducks, etc.), the production sector of poultry involved (industrial, commercial organic, live poultry market, village, recreational, etc.), the density of poultry in the region, export market requirements, federal versus state regulatory authority, availability of financial compensation and public health threat. Finally, the success of any control program will be dependent upon transparency and industry-government trust, cooperation and interaction. In the United States, the federal government has regulatory authority over eradication of HPAI viruses while the state governments have jurisdiction over LPAI viruses.

With regard to HPAI, prevention and eradication are the usually desired outcomes, although under some circumstances such as a large HPAI outbreak, a period of management may be necessary before eradication becomes achievable. Furthermore, H5 and H7 LPAI should be eradicated to prevent the opportunity for mutation with the emergence of HPAI virus strains. By contrast, prevention of LPAI and management of LPAI (H1-4, H6 and H8-16) when it occurs may be acceptable alternatives, especially for H1 and H3 influenza A virus infections in turkeys when located in geographic regions with a large swine population having endemic swine influenza which can infect turkeys.

Each of the components of the five control strategies is discussed below.

Biosecurity
Biosecurity is an important part of any AI control strategy and is structured to serve one of two purposes (Swayne and Suarez 2000): inclusion, to prevent escape of the AI virus from infected farms or premises, often termed biocontainment; or exclusion, to prevent

introduction of the AI virus onto naïve premises. However, biosecurity is not only the physical inclusion or exclusion practice, but also is a mind-set or attitude that acknowledges human behavior is the central risk for transmission, and looks for ways to reduce risk by eliminating risky behaviors. Furthermore, biosecurity must be implemented at all levels within the poultry sector from large industrial operations, including allied groups and all employees, to subsistence farmers and government workers who come into contact with birds. On infected premises as well as neighboring farms, movement of birds, equipment, supplies and personnel should be restricted through effective quarantine or movement restrictions to prevent the spread of AI virus from infected to uninfected areas.

Because feces and respiratory secretions of affected birds can contain the AI virus, any physical item from an infected premise can be a fomite with the potential to carry the virus off the farm (Swayne and Suarez 2000; Swayne and Halvorson 2003). Proper cleaning and disinfection are essential to kill the AI virus and prevent transmission. Therefore all equipment, shoes, boots, clothing and personal items should be properly cleaned and disinfected before leaving the premises. In unaffected commercial operations, employees should shower upon entry to the farm and the company should supply clothing and shoes to be worn only on the farm. Preferably, employees should be restricted to one farming operation, and they should not visit or work on other poultry farms and should not own or tend poultry at their homes. When workers must be shared between farms, such as vaccination, feed delivery and load-out crews, the highest level of biosecurity must be practiced. On AI-affected farms, the employees should shower upon leaving the farm and all work related clothing and shoes should be left on the farm. Finally, the success of biosecurity does not lie in the written policy, but in the practice of biosecurity, from the simplest detail, by every individual concerned, and on all farms.

Surveillance and diagnostics

A comprehensive, integrated and transparent surveillance and diagnostic program is essential to ascertain the national or regional prevalence of AI; such information is used to establish quarantine zones for management or eradication, or zones free of AI (Swayne and Suarez 2000; Swayne and Halvorson 2003; Swayne and Akey 2004). Surveillance can be active or passive. Active surveillance is based on statistical sampling of a population to determine the presence or absence of AI infection and typically has utilized detection of AI specific antibody. However, during an AI outbreak within the infected zone, virologic surveillance is necessary. By contrast, passive surveillance, usually in the form of diagnostic investigations of respiratory, reproductive or high mortality diseases, is based on clinical submissions and looks for the etiology of disease either by isolating the agent, or by detecting specific nucleic acids or proteins, in this case, of AI virus. Passive surveillance is used to detect the first HPAI cases in an AI-free area and for identifying additional cases within an infected zone but lack of clinical cases cannot be used as the criteria to demonstrate eradication of HPAI or freedom from AI infections.

To achieve a successful AI eradication program, rapid diagnosis is essential to initiate depopulation before the virus spreads to neighboring farms or premises (Akey 2003). Real-time reverse transcriptase polymerase chain reaction (RRT-PCR) diagnostic tests have greatly assisted in AI eradication efforts (Spackman, Senne et al. 2002; Akey 2003; Spackman, Senne et al. 2003). Furthermore, a national or regional veterinary medical diagnostic infrastructure is essential for rapid and accurate diagnosis of AI.

Elimination of infected poultry

All H5 and H7 LP or HPAI virus-infected poultry should be eliminated. Although individual birds are not persistently infected with the AI virus (Swayne and Akey 2004), collectively as a population, sufficient numbers of susceptible birds may be maintained on the farm to establish an endemic infection, especially in an AI virus-contaminated environment. Continuing AI infections will occur in susceptible birds following the addition of new naïve birds as a result of inefficient transmission and incomplete infection during the initial outbreak, or following the decline in immunity of previously infected birds. With HPAI, birds from affected farms should be humanely euthanized and disposed of within the quarantine zone, or should be transported in decontaminated, sealed trucks to suitable disposal locations. Disposal methods used can be incineration, composting, alkaline hydrolysis, fermentation, rendering, or burial in accordance with environmental standards and laws (Swayne and Suarez 2000). However, such stamping-out programs for HPAI must be initiated early in an outbreak to be effective and economical, and these programs tend to be expensive and subject to institutional inertia. HPAI virus-infected birds should not be used for food.

Stamping-out of H5 and H7 LPAI can be an acceptable strategy. Under specific conditions, LPAI virus-infected poultry (chickens and turkeys) can be eliminated through controlled marketing of convalescent (recovered) birds. The birds should be moved only after the infection has subsided and virus can no longer be detected, usually a minimum of three weeks after the first appearance of clinical signs of AI. Such poultry should be transported along routes that avoid poultry farms and should be processed at the end of the day.

Increasing resistance in poultry to AI virus

Various methods can be used to increase resistance of poultry to infections by AI viruses. There is some evidence that strains or breeds of chickens have different susceptibilities to LPAI viruses, but no such genetic resistance has been reported to HPAI viruses (Swayne, Beck et al. 1996). Properly administered, efficacious vaccines can be used to increase resistance of poultry to AI viruses and have been reviewed elsewhere (Swayne 2003). However, vaccination should be used only as one component of a comprehensive control strategy (Swayne and Suarez 2000). Unrestricted use of AI vaccines will be counterproductive to eradication efforts (Swayne and Suarez 2000).

Historically, AI vaccines have been used mostly for prevention or management of LPAI viruses, but in the last 10 years, AI vaccines have been used in some HPAI prevention and management programs, specifically in Mexico, Pakistan, Hong Kong, China and Indonesia. Vaccines can be helpful in decreasing the number of AI virus-infected birds, reducing environmental contamination with the AI virus, preventing spread of AI viruses between farms, and minimizing economic losses. With HPAI, vaccination may help bring an uncontrolled outbreak into a manageable situation, but eradication can only be accomplished if vaccination is accompanied by the other components of a control strategy. Furthermore, only through active surveillance can prevention or eradication be demonstrated and such surveillance strategies must differentiate vaccinated from infected animals (DIVA). The DIVA approach can be achieved through serological monitoring:

- antibodies against influenza A virus in unvaccinated sentinel birds (Swayne 2003);
- antibodies to the NA subtype of the field AI virus in birds vaccinated with an inactivated homologous-HA-heterologous-NA AI virus vaccine (Capua, Terregino et al. 2003);
- antibodies against influenza A nucleoprotein or matrix protein in birds vaccinated with

a recombinant vaccine containing only the HA subtype of the field AI virus (Swayne, Beck et al. 1997), or antibodies against influenza A NS1 protein in birds vaccinated with inactivated AI virus or recombinant vaccines (Tumpey, Alvarez et al. 2004a).

Antibodies against the HA protein are the major protective immune response in poultry to AI viruses, but antibodies to the NA can contribute to protection (McNulty, Allan et al. 1986; Swayne and Halvorson 2003). Such protection is HA or NA subtype-specific. For example, an H5 vaccine protects against an H5 field virus, but not against an H7 field virus, and vice versa (Brugh, Beard et al. 1979; Wood, Kawaoka et al. 1985; Stone 1987; Stone 1988). For the HA, protection is most consistent when vaccine and field viruses have at least 87% similarity in HA amino acid sequence (Swayne, Garcia et al. 2000), and indicates that influenza vaccine strains will not need to be changed as often as those used for human influenza in order to maintain efficacy.

Avian influenza vaccines provide protection through prevention of clinical signs and death, decreasing AI virus infection rates and reducing the titer of field AI virus replication in respiratory and intestinal tracts, which translates into decreased AI virus into the environment and increases the AI virus dose needed to produce infection (Swayne 2003; Capua, Terregino et al. 2004a). Vaccines can provide protection for over twenty weeks following a single vaccine dose (Swayne, Beck et al. 1997; Swayne, Beck et al. 1999), but under some field conditions, multiple (up to four) vaccinations may be necessary to obtain consistent, long-term protection (Capua, Marangon et al. 2004b). Efficacious vaccines can provide protection even against high doses of AI viruses (Swayne 2003).

Many experimental studies have been published using a myriad of different vaccine technologies, but only two have been licensed and used in the field: inactivated whole virus vaccines of various HA subtypes, and a recombinant fowl pox vaccine with an H5 AI HA gene insert. In poultry, vaccination against AI is not routine. Vaccination is used only in areas at high risk for AI infections. For example, during 2001 in the USA, only 3.5 million doses of inactivated vaccine were used, mostly H1N1 or H1N2 vaccine to protect turkey breeders from swine influenza in states where swine and turkey production occur in the same geographic regions (Swayne 2001). No vaccine was used in meat chickens (broilers). By contrast, H9N2 AI is endemic in broilers and broiler breeders in the Middle East and most chickens are vaccinated with inactivated H9N2 AI virus vaccine. In addition, since 1995, 850 million doses of recombinant fowlpox-AI-H5 and 1.3 billion doses of inactivated H5N2 AI vaccine have been used in Mexico.

Antivirals are not approved for use in food producing animals and, experimental usage of amantadine resulted in rapid emergence of amantadine-resistant strains (Beard, Brugh et al. 1987; Swayne and Suarez 2000).

Education

All persons with poultry contact should be educated in the clinical features of AI and their individual responsibility to use biosecurity principles and practices to prevent introduction of the virus (Swayne and Suarez 2000). Deficiencies in communicating critical information will encourage complacency and will result in a failure to comply or participate in the AI control program.

References

Acland, H. M., Silverman Bachin, L. A. and Eckroade, R. J. (1984). Lesions in broiler and layer chickens in an outbreak of highly pathogenic avian influenza virus infection. Vet. Pathol. *21*, 564–9.

Agriculture, U. S. Department of. (1999). Agricultural Statistics (1999). Washington D. C., USDA.

Akey, B. L. (2003). Low pathogenicity H7N2 avian influenza outbreak in Virginia during (2002). Avian Dis. *47*, 1099–1103.

Alexander, D. J. (1993). Orthomyxovirus infections. Virus infections of birds. J. B. McFerran and M. S. McNulty. London, Elsevier Science, 287–316.

Alexander, D. J. (1998). Control strategies of the International Office of Epizooties, the European Union and the harmonization of international standards for the diagnosis of avian influenza. Proceedings of the Fourth International Symposium on Avian Influenza. D. E. Swayne and R. D. Slemons. Richmond, Virginia, United States Animal Health Association, 353–357.

Alexander, D. J. (2000). A review of avian influenza in different bird species. Vet. Microbiol. *74*, 3–13.

Alexander, D. J. and Brown, I. H. (2000). Recent zoonoses caused by influenza A viruses. Rev. Sci. Tech. *19*, 197–225.

Alexander, D. J., Parsons, G. and Manvell, R. J. (1986). Experimental assessment of the pathogenicity of eight influenza A viruses of H5 subtype for chickens, turkeys, ducks and quail. Avian Pathology. *15*, 647–662.

Almond, J. W. 1977. A single gene determines the host range of influenza virus. Nature *270*, 617–8.

American Academy of Pediatrics (2003). Influenza. In Red Book, 2003. Report of the committee on infectious diseases. L. K. Pickering. Elk Grove Village, IL, American Academy of Pediatrics, 382–391.

Anonymous. (1993). Recommended composition of influenza virus vaccines for use in the 1993–1994. season. Wkly. Epidemiol. Rec. *68*, 57–60.

Anonymous. (1994a). Recommended composition of influenza virus vaccines for use in the 1994–1995. season. Influenza activity, October 1993-February 1994. Wkly. Epidemiol. Rec. *69*, 53–6.

Anonymous. (1994b). Update, influenza activity–United States and worldwide, 1993–94 season, and composition of the 1994–95 influenza vaccine. MMWR – Morbidity & Mortality Weekly Report *43*, 179–83.

Anonymous. (1994c). Update, influenza activity–worldwide, 1994. MMWR – Morbidity & Mortality Weekly Report *43*, 691–3.

Anonymous. (1995a). Recommended composition of influenza virus vaccines for use in the 1995–1996. season. Wkly. Epidemiol. Rec. *70*, 53–6.

Anonymous. (1995b). Update, influenza activity–United States and worldwide, 1994–95 season, and composition of the 1995–96 influenza vaccine. MMWR – Morbidity & Mortality Weekly Report *44*, 292–5.

Anonymous. (1995c). Update, influenza activity–worldwide, 1995. MMWR – Morbidity & Mortality Weekly Report *44*, 644–5, 651–2.

Anonymous. (1996a). Influenza vaccine formula for 1996–1997. Wkly. Epidemiol. Rec. *71*, 57–61.

Anonymous. (1996b). Update, influenza activity–United States and Worldwide, 1995–96 season, and composition of the 1996–97 influenza vaccine. MMWR – Morbidity & Mortality Weekly Report *45*, 326–9.

Anonymous. (1996c). Update, influenza activity–worldwide, 1996. MMWR – Morbidity & Mortality Weekly Report *45*, 816–9.

Anonymous. (1997a). Recommended composition of influenza virus vaccines for use in the 1997–1998. season. Wkly. Epidemiol. Rec. *72*, 57–61.

Anonymous. (1997b). Update, influenza activity–United States and worldwide, 1996–97 season, and composition of the 1997–98 influenza vaccine. MMWR – Morbidity & Mortality Weekly Report *46*, 325–30.

Anonymous. (1997c). Update, influenza activity–worldwide, March-August 1997. MMWR – Morbidity & Mortality Weekly Report *46*, 815–8.

Anonymous. (1998a). Recommended composition of influenza virus vaccines for use in the 1998–1999. season. Wkly. Epidemiol. Rec. *73*, 56–61.

Anonymous. (1998b). Update, influenza activity–United States and worldwide, 1997–98 season, and composition of the 1998–99 influenza vaccine. MMWR – Morbidity & Mortality Weekly Report *47*, 280–4.

Anonymous. (1999a). Recommended composition of influenza virus vaccines for use in 2000. Wkly. Epidemiol. Rec. *74*, 321–5.

Anonymous. (1999b). Recommended composition of influenza virus vaccines for use in the 1999–2000. season. Wkly. Epidemiol. Rec. *74*, 57–61.

Anonymous. (1999c). Update, influenza activity–United States and worldwide, 1998–99 season, and composition of the 1999–2000. influenza vaccine. MMWR – Morbidity & Mortality Weekly Report *48*, 374–8.

Anonymous. (2000a). Recommended composition of influenza virus vaccines for use in the 2000–2001. season. Wkly. Epidemiol. Rec. *75*, 61–68.

Anonymous. (2000b). Recommended composition of influenza virus vaccines for use in the 2001. influenza season. Wkly. Epidemiol. Rec. *75*, 330–3.

Anonymous. (2000c). Update, influenza activity–United States and worldwide, 1999–2000. season, and composition of the 2000-01 influenza vaccine. MMWR – Morbidity & Mortality Weekly Report *49*, 375–81.

Anonymous. (2001a). Recommended composition of influenza virus vaccines for use in the 2001–2002. season. Wkly. Epidemiol. Rec. *76*, 58–61.

Anonymous. (2001b). Recommended composition of influenza virus vaccines for use in the 2002. influenza season. Wkly. Epidemiol. Rec. *76*, 311–4.

Anonymous. (2001c). Update, influenza activity–United States and worldwide, 2000-01 season, and composition of the 2001–02 influenza vaccine. MMWR – Morbidity & Mortality Weekly Report *50*, 466–79.

Anonymous. (2002a). Recommended composition of influenza virus vaccines for use in the 2002–2003. season. Wkly. Epidemiol. Rec. *77*, 62–6.

Anonymous. (2002b). Recommended composition of influenza virus vaccines for use in the 2003. influenza season. Wkly. Epidemiol. Rec. *77*, 344–8.

Anonymous. (2002c). Update, Influenza activity–United States and worldwide, 2001-02 season, and composition of the 2002-03 influenza vaccine. MMWR – Morbidity & Mortality Weekly Report *51*, 503–6.

Anonymous. (2003a). Recommended composition of influenza virus vaccines for use in the 2003–2004. influenza season. Wkly. Epidemiol. Rec. *78*, 58–62.

Anonymous. (2003b). Recommended composition of influenza virus vaccines for use in the 2004. influenza season. Wkly. Epidemiol. Rec. *78*, 375–9.

Anonymous. (2003c). Update, Influenza activity- United States and worldwide, 2002-03 season, and composition of the 2003-04 influenza vaccine. MMWR Morb Mortal Wkly Rep. *52*, 516–521.

Anonymous. (2004a). Avian influenza A(H5N1). Wkly. Epidemiol. Rec. *79*, 65–76.

Anonymous. (2004b). Recommended composition of influenza virus vaccines for use in the 2004–2005. influenza season. Wkly. Epidemiol. Rec. *79*, 88–92.

Anonymous. (2004c). Update, influenza activity–United States and worldwide, 2003-04 season, and composition of the 2004-05 influenza vaccine. MMWR – Morbidity & Mortality Weekly Report *53*, 547–52.

Anonymous. (2004d). Update, influenza activity–United States, 2003-04 season. MMWR – Morbidity & Mortality Weekly Report *53*, 284–7.

Arora, D. J., N'Diaye, M. and Dea, S. (1997). Genomic study of hemagglutinins of swine influenza (H1N1) viruses associated with acute and chronic respiratory diseases in pigs. Arch. Virol. *142*, 401–12.

Aymard, M., Brigaud, M., Fontaine, M., Tillon, J. P. and Vannier, P. (1980). Relations between swine and human influenza, data obtained from serological studies in France. Annales de Virologie *131*, 527–528.

Bankowski, R. A. (1981). Introduction and objectives of the symposium. Proceedings of the First International Symposium on Avian Influenza. R. A. Bankowski. Richmond, Virginia, U. S. Animal Health Association, vii-xiv.

Basler, C. F., Reid, A. H., Dybing, J. K., Janczewski, T. A., Fanning, T. G., et al. (2001). Sequence of the 1918 pandemic influenza virus nonstructural gene (NS) segment and characterization of recombinant viruses bearing the 1918 NS genes. Proc. Natl. Acad. Sci. USA *98*, 2746–51.

Beard, C. W. 1970. Demonstration of type-specific influenza antibody in mammalian and avian sera by immunodiffusion. Bulletin of the World Health Organization. *42*, 779–785.

Beard, C. W., Brugh, M. and Webster, R. G. (1987). Emergence of amantadine-resistant H5N2 avian influenza virus during a simulated layer flock treatment program. Avian Dis. *31*, 533–7.

Beare, A. S., Kendal, A. P. and Craig, J. W. (1980). Further studies in man of Hsw1N1 influenza viruses. J. Med. Virol. *5*, 33–8.

Beare, A. S. and Webster, R. G. (1991). Replication of avian influenza viruses in humans. Archives of Virol. *119*, 37–42.

Belshe, R. B., Mendelman, P. M., Treanor, J., King, J., Gruber, W. C., et al. (1998). The efficacy of live attenuated, cold-adapted, trivalent, intranasal influenzavirus vaccine in children. N. Engl. J. Med. *338*, 1405–12.

Bikour, M. H., Frost, E. H., Deslandes, S., Talbot, B., Weber, J. M., et al. (1995). Recent H3N2 swine influenza virus with haemagglutinin and nucleoprotein genes similar to 1975 human strains. J. Gen. Virol. 76 (Pt 3), 697–703.

Blakemore, F. and Gledhill, A. W. 1941. Discusion on swine influenza in the British Isles. Proc. Royal Soc. Med. *34*, 611–615.

Bosch, F. X., Garten, W., Klenk, H. D. and Rott, R. (1981). Proteolytic cleavage of influenza virus hemagglutinins, Primary structure of the connecting peptide between HA1 and HA2 determines proteolytic cleavability and pathogenicity of avian influenza viruses. Virol. *113*, 725–735.

Bosch, F. X., Orlich, M., Klenk, H. D. and Rott, R. (1979). The structure of the hemagglutinin, a determinant for the pathogenicity of influenza viruses. Virol. *95*, 197–207.

Bridges, C. B., Fukuda, K., Cox, N. J. and Singleton, J. A. (2001). Prevention and control of influenza. Recommendations of the Advisory Committee on Immunization Practices (ACIP). MMWR – Morbidity & Mortality Weekly Report *50*, 1–44.

Bridges, C. B., Katz, J. M., Seto, W. H., Chan, P. K. S., Tsang, D., et al. (2000). Risk of influenza A (H5N1) infection among health care workers exposed to patients with influenza A (H5N1), Hong Kong. J. Infect. Dis. *181*, 344–348.

Bridges, C. B., Kuehnert, M. J. and Hall, C. B. (2003). Transmission of influenza, Implications for control in health care settings. Clinical Infectious Diseases. *37*, 1094–1101.

Bridges, C. B., Lim, W., Hu-Primmer, J., Sims, L., Fukuda, K., et al. (2002). Risk of influenza A (H5N1) infection among poultry workers, Hong Kong, 1997–1998. J. Infect. Dis. *185*, 1005–1010.

Brown, I. H. (2000). The epidemiology and evolution of influenza viruses in pigs. Vet. Microbiol. *74*, 29–46.

Brown, I. H., Alexander, D. J., Chakraverty, P., Harris, P. A. and Manvell, R. J. (1994). Isolation of an influenza A virus of unusual subtype (H1N7) from pigs in England, and the subsequent experimental transmission from pig to pig. Vet. Microbiol. *39*, 125–34.

Brown, I. H., Chakraverty, P., Harris, P. A. and Alexander, D. J. (1995a). Disease outbreaks in pigs in Great Britain due to an influenza A virus of H1N2 subtype. Vet. Rec. *136*, 328–9.

Brown, I. H., Harris, P. A. and Alexander, D. J. (1995b). Serological studies of influenza viruses in pigs in Great Britain 1991–2. Epidemiol. Infect. *114*, 511–20.

Brown, I. H., Harris, P. A., McCauley, J. W. and Alexander, D. J. (1998). Multiple genetic reassortment of avian and human influenza A viruses in European pigs, resulting in the emergence of novel H1N2 genotype. Journal of General Virol. *79*, 2947–2955.

Brown, I. H., Hill, M. L., Harris, P. A., Alexander, D. J. and McCauley, J. W. (1997a). Genetic characterisation of an influenza A virus of unusual subtype (H1N7) isolated from pigs in England. Arch. Virol. *142*, 1045–50.

Brown, I. H., Ludwig, S., Olsen, C. W., Hannoun, C., Scholtissek, C., et al. (1997b). Antigenic and genetic analyses of H1N1 influenza A viruses from European pigs. J. Gen. Virol. *78*, 553–62.

Brugh, M., Beard, C. W. and Stone, H. D. (1979). Immunization of chickens and turkeys against avian influenza with monovalent and polyvalent oil emulsion vaccines. American Journal of Veterinary Research. *40*, 165–9.

Buckler-White, A. J., Naeve, C. W. and Murphy, B. R. (1986). Characterization of a gene coding for M proteins which is involved in host range restriction of an avian influenza A virus in monkeys. J. Virol. *57*, 697–700.

Bush, R. M., Fitch, W. M., Bender, C. A. and Cox, N. J. (1999). Positive selection on the H3 hemagglutinin gene of human influenza virus A. Mol. Biol. Evol. *16*, 1457–65.

Campitelli, L., Donatelli, I., Foni, E., Castrucci, M. R., Fabiani, C., et al. (1997). Continued evolution of H1N1 and H3N2 influenza viruses in pigs in Italy. Virol. *232*, 310–8.

Capua, I., Terregino, C., Cattoli, G., Mutinelli, F. and Rodriguez, J. F. (2003). Development of a DIVA (Differentiating Infected from Vaccinated Animals) strategy using a vaccine containing a heterologous neuraminidase for the control of avian influenza. Avian Pathol. *32*, 47–55.

Capua, I., Terregino, C., Cattoli, G. and Toffan, A. (2004a). Increased resistance of vaccinated turkeys to experimental infection with an H7N3 low-pathogenicity avian influenza virus. Avian Pathol. *33*, 158–163.

Capua, L., Marangon, S. and Bonfanti, L. (2004b). Eradication of low pathogenicity avian influenza of the H7N3 subtype from Italy. Vet. Rec. *154*, 639–640.

Castrucci, M. R., Campitelli, L., Ruggieri, A., Barigazzi, G., Sidoli, L., et al. (1994). Antigenic and sequence analysis of H3 influenza virus haemagglutinins from pigs in Italy. J. Gen. Virol. *75*, 371–9.

Castrucci, M. R., Donatelli, I., Sidoli, L., Barigazzi, G., Kawaoka, Y., et al. (1993). Genetic reassortment between avian and human influenza A viruses in Italian pigs. Virol. *193*, 503–6.

Centers for Disease Control and Prevention. (2004). Prevention and control of influenza, recommendations of the Advisory Committee on Immunization Practices (ACIP). MMWR – Morbidity & Mortality Weekly Report *53*, 1–40.

Chambers, T. M., Hinshaw, V. S., Kawaoka, Y., Easterday, B. C. and Webster, R. G. (1991). Influenza viral infection of swine in the United States 1988–1989. Arch. Virol. *116*, 261–5.

Chen, H., Deng, G., Li, Z., Tian, G., Li, Y., et al. (2004). The evolution of H5N1 influenza viruses in ducks in southern China. Proc. Natl. Acad. Sci. U S A. *101*, 10452–7.

Chen, W., Calvo, P. A., Malide, D., Gibbs, J., Schubert, U., et al. (2001). A novel influenza A virus mitochondrial protein that induces cell death. Nat. Med. *7*, 1306–12.

Choi, Y. K., Goyal, S. M., Farnham, M. W. and Joo, H. S. (2002a). Phylogenetic analysis of H1N2 isolates of influenza A virus from pigs in the United States. Virus Res. *87*, 173–9.

Choi, Y. K., Goyal, S. M. and Joo, H. S. (2002b). Prevalence of swine influenza virus subtypes on swine farms in the United States. Arch. Virol. *147*, 1209–20.

Choi, Y. K., Ozaki, H., Webby, R. J., Webster, R. G., Peiris, J. S., et al. (2004). Continuing evolution of H9N2 influenza viruses in Southeastern China. J. Virol. *78*, 8609–14.

Chun, J. 1919. Influenza including its infection among pigs. Natl. Med. Jo. China *5*, 34–44.

Claas, E. C., Kawaoka, Y., de Jong, J. C., Masurel, N. and Webster, R. G. (1994). Infection of children with avian-human reassortant influenza virus from pigs in Europe. Virol. *204*, 453–7.

Claas, E. C., Osterhaus, A. D., van Beek, R., De Jong, J. C., Rimmelzwaan, G. F., et al. (1998). Human influenza A H5N1 virus related to a highly pathogenic avian influenza virus. Lancet. *351*, 472–7.

Clements, M. L., Subbarao, E. K., Fries, L. F., Karron, R. A., London, W. T., et al. (1992). Use of single-gene reassortant viruses to study the role of avian influenza A virus genes in attenuation of wild-type human influenza A virus for squirrel monkeys and adult human volunteers. J. Clin. Microbiol. *30*, 655–662.

Connor, R. J., Kawaoka, Y., Webster, R. G. and Paulson, J. C. (1994). Receptor specificity in human, avian, and equine H2 and H3 influenza virus isolates. Virol. *205*, 17–23.

Couceiro, J. N., Paulson, J. C. and Baum, L. G. (1993). Influenza virus strains selectively recognize sialyloligosaccharides on human respiratory epithelium; the role of the host cell in selection of hemagglutinin receptor specificity. Virus Res. *29*, 155–65.

Cox, N. J. and Bender, C. A. (1995). The molecular epidemiology of influenza viruses. Seminars in Virol. *6*, 359–370.

Cox, N. J. and Subbarao, K. (1999). Influenza. Lancet. *354*, 1277–82.

Cox, N. J. and Subbarao, K. (2000). Global epidemiology of influenza, past and present. Ann. Rev. Med. *51*, 407–421.

Crosby, A. W. (1989). America's Forgotten Pandemic. Cambridge, Cambridge University Press.

Cunha, R. G., Varges Vinha, V. R. and Passos, W. D. (1978). Isolation of a strain of Myxovirus influenzae-A suis from swine slaughtered in Rio de Janeiro. Rev. Bras. Biol. *38*, 13–7.

Dacso, C. C., Couch, R. B., Six, H. R., Young, J. F., Quarles, J. M., et al. (1984). Sporadic occurrence of zoonotic swine influenza virus infections. J. Clin. Microbiol. *20*, 833–5.

Daum, L. T., Canas, L. C., Smith, C. B., Klimov, A., Huff, W., et al. (2002). Genetic and antigenic analysis of the first A/New Caledonia/20/99-like H1N1 influenza isolates reported in the Americas. Emerg. Infect. Dis. *8*, 408–12.

de Jong, J. C., Heinen, P. P., Loeffen, W. L., van Nieuwstadt, A. P., Claas, E. C., et al. (2001). Antigenic and molecular heterogeneity in recent swine influenza A(H1N1) virus isolates with possible implications for vaccination policy. Vaccine. *19*, 4452–64.

de Jong, J. C., Paccaud, M. F., de Ronde-Verloop, F. M., Huffels, N. H., Verwei, C., et al. (1988). Isolation of swine-like influenza A(H1N1) viruses from man in Switzerland and The Netherlands. Ann. Inst Pasteur Virol. *139*, 429–37.

de Jong, J. C., van Nieuwstadt, A. P., Kimman, T. G., Loeffen, W. L., Bestebroer, T. M., et al. (1999). Antigenic drift in swine influenza H3 haemagglutinins with implications for vaccination policy. Vaccine. *17*, 1321–8.

Dea, S., Bilodeau, R., Sauvageau, R., Montpetit, C. and Martineau, G. P. (1992). Antigenic variant of swine influenza virus causing proliferative and necrotizing pneumonia in pigs. J. Vet. Diagn. Invest. *4*, 380–92.

Donatelli, I., Campitelli, L., Castrucci, M. R., Ruggieri, A., Sidoli, L., et al. (1991). Detection of two antigenic subpopulations of A(H1N1) influenza viruses from pigs, antigenic drift or interspecies transmission? J. Med. Virol. *34*, 248–57.

Easterday, B. C., Hinshaw, V. S. and Halvorson, D. A. (1997). Influenza. Diseases of Poultry. B. W. Calnek. Ames, Iowa State University Press, 583–605.

Easterday, B. C., Trainer, D. O., Tumova, B. and Pereira, H. G. 1968. Evidence of infection with influenza viruses in migratory waterfowl. Nature *219*, 523–524.

Elbers, A. R., Kamps, B. and Koch, G. (2004). Performance of gross lesions at post-mortem for the detection of outbreaks during the avian influenza A virus (H7N7) epidemic in the Netherlands in 2003. Avian Pathol. *33*, 418–22.

Ellis, J. S., Alvarez-Aguero, A., Gregory, V., Lin, Y. P., Hay, A., et al. (2003). Influenza AH1N2 viruses, United Kingdom, 2001-02 influenza season. Emerg. Infect. Dis. J. *9*, 304–10.

Eriksson, E., Yao, F., Svensjo, T., Winkler, T., Slama, J., et al. (1998). In vivo gene transfer to skin and wound by microseeding. J. Surgical Res. *78*, 85–91.

Fanning, T. G., Slemons, R. D., Reid, A. H., Janczewski, T. A., Dean, J., et al. (2002). 1917 avian influenza virus sequences suggest that the 1918 pandemic virus did not acquire its hemagglutinin directly from birds. J. Virol. *76*, 7860–7862.

Fitch, W. M., Bush, R. M., Bender, C. A. and Cox, N. J. (1997). Long term trends in the evolution of H(3) HA1 human influenza type A. Proc. Natl. Acad. Sci. USA. *94*, 7712–8.

Forman, A. J., Parsonson, I. M. and Doughty, W. J. (1986). The pathogenicity of an avian influenza virus isolated in Victoria. Aust. Vet. J. *63*, 294–296.

Fouchier, R. A., Schneeberger, P. M., Rozendaal, F. W., Broekman, J. M., Kemink, S. A., et al. (2004). Avian influenza A virus (H7N7) associated with human conjunctivitis and a fatal case of acute respiratory distress syndrome. Proc. Natl. Acad. Sci. U S A. *101*, 1356–61.

Francis, T. J. (1940). New type of virus from epidemic influenza. Science. *91*, 405–408.

Gammelin, M., Altmuller, A., Reinhardt, U., Mandler, J., Harley, V. R., et al. (1990). Phylogenetic analysis of nucleoproteins suggests that human influenza A viruses emerged from a 19th-century avian ancestor. Mol. Biol. Evol., *7*, 194–200.

Garcia, M., Crawford, J. M., Latimer, J. W., Rivera-Cruz, M. V. Z. E. and Perdue, M. L. (1996). Heterogeneity in the hemagglutinin gene and emergence of the highly pathogenic phenotype among recent H5N2 avian influenza viruses from Mexico. J. Gen. Virol. *77*, 1493–1504.

Geraci, J. R., St. Aubin, D. J., Barker, I. K., Webster, R. G., Hinshaw, V. S., et al. (1982). Mass mortality of harbor seals, pneumonia associated with influenza A virus. Science. *215*, 1129–31.

Glezen, W. P. (1996). Emerging infections, pandemic influenza. Epidemiologic Rev. *18*, 64–76.

Goldfield, M., Noble, G. R. and Dowdle, W. R. (1977). Identification and preliminary antigenic analyses of swine influenza-like viruses isolated during an influenza outbreak at Fort Dix, New Jersey. J. Infect. Dis. *136*, 381–385.

Gorman, O. T., Bean, W. J., Kawaoka, Y. and Webster, R. G. (1990). Evolution of the nucleoprotein gene of influenza A virus. J. Virol. *64*, 1487–97.

Gregory, V., Bennett, M., Orkhan, M. H., Al Hajjar, S., Varsano, N., et al. (2002). Emergence of influenza A H1N2 reassortant viruses in the human population during (2001). Virol. *300*, 1–7.

Gregory, V., Bennett, M., Thomas, Y., Kaiser, L., Wunderli, W., et al. (2003). Human infection by a swine influenza A (H1N1) virus in Switzerland. Arch. Virol. *148*, 793–802.

Gregory, V., Lim, W., Cameron, K., Bennett, M., Marozin, S., et al. (2001). Infection of a child in Hong Kong by an influenza A H3N2 virus closely related to viruses circulating in European pigs. J. Gen. Virol. *82*, 1397–406.

Grove, R. D. and Hetzel, A. M. 1968. Vital statistics rates in the United States, 1940–1960. Washington DC, US Government Printing Office.

Guan, Y., Peiris, J. S. M., Poon, L. L. M., Dyrting, K. C., Ellis, T. M., et al. (2003). Reassortants of H5N1 influenza viruses recently isolated from aquatic poultry in Hong Kong SAR. Avian Dis. *47*, 911–913.

Guo, Y. J., Xu, X. Y. and Cox, N. J. (1992). Human influenza A (H1N2) viruses isolated from China. J. Gen. Virol. 73 (Pt 2), 383–7.

Haesebrouck, F., Biront, P., Pensaert, M. B. and Leunen, J. (1985). Epizootics of respiratory tract disease in swine in Belgium due to H3N2 influenza virus and experimental reproduction of disease. Am. J. Vet. Res. *46*, 1926–8.

Haesebrouck, F. and Pensaert, M. (1988). Influenza in swine in Belgium 1969–1986, epizootiologic aspects. Comp. Immunol. Microbiol. Infect. Dis. *11*, 215–22.

Halvorson, D., Karunakaran, D., Senne, D., Kelleher, C., Bailey, C., et al. (1983). Epizootiology of avian influenza–simultaneous monitoring of sentinel ducks and turkeys in Minnesota. Avian Dis. *27*, 77–85.

Halvorson, D. A., Kelleher, C. J. and Senne, D. A. (1985). Epizootiology of avian influenza, Effect of season on incidence in sentinel ducks and domestic turkeys in Minnesota. Appl. Environ. Microbiol. *49*, 914–919.

Harnach, R., Hubik, R. and Chvatal, O. 1950. Isolation of the virus of swine influenza in

Czechoslovakia. Casopis Ceskoslov. Vet. *5*, 289.

Hatta, M., Gao, P., Halfmann, P. and Kawaoka, Y. (2001). Molecular basis for high virulence of Hong Kong H5N1 influenza A viruses. Science. *293*, 1840–2.

Hatta, M. and Kawaoka, Y. (2002). The continued pandemic threat posed by avian influenza viruses in Hong Kong. Trends in Microbiol. *10*, 340–4.

Hehme, N., Engelmann, H., Kunzel, W., Neumeier, E. and Sanger, R. (2002). Pandemic preparedness, lessons learnt from H2N2 and H9N2 candidate vaccines. Med. Microbiol. Immunol. (Berl). *191*, 203–8.

Heinen, P. P., Rijsewijk, F. A., de Boer-Luijtze, E. A. and Bianchi, A. T. (2002). Vaccination of pigs with a DNA construct expressing an influenza virus M2-nucleoprotein fusion protein exacerbates disease after challenge with influenza A virus. J. Gen. Virol. *83*, 1851–9.

Hinshaw, V. S., Alexander, D. J., Aymard, M., Bachmann, P. A., Easterday, B. C., et al. (1984). Antigenic comparisons of swine-influenza-like H1N1 isolates from pigs, birds and humans, an international collaborative study. Bull. World Health Organization. *62*, 871–8.

Hinshaw, V. S., Bean, W. J., Jr., Webster, R. G. and Easterday, B. C. (1978). The prevalence of influenza viruses in swine and the antigenic and genetic relatedness of influenza viruses from man and swine. Virol. *84*, 51–62.

Hinshaw, V. S., Webster, R. G., Bean, W. J., Downie, J. and Senne, D. A. (1983a). Swine influenza-like viruses in turkeys, potential source of virus for humans? Science. *220*, 206–8.

Hinshaw, V. S., Webster, R. G., Naeve, C. W. and Murphy, B. R. (1983b). Altered tissue tropism of human-avian reassortant influenza viruses. Virol. *128*, 260–3.

Hinshaw, V. S., Webster, R. G. and Turner, B. (1980). The perpetuation of orthomyxoviruses and paramyxoviruses in Canadian waterfowl. Can. J. Microbiol. *26*, 622–9.

Hiromoto, Y., Yamazaki, Y., Fukushima, T., Saito, T., Lindstrom, S. E., et al. (2000). Evolutionary characterization of the six internal genes of H5N1 human influenza A virus. J. Gen. Virol. *81*, 1293–1303.

Hodder, R. A., C., G. J., Allen, J. C., Top, F. H., Nowosiwsky, T., et al. 1977. Swine influenza A viruses with pandemic potential. J. Virol. *72*, 7367–7373.

Horimoto, T. and Kawaoka, Y. (1995). Molecular changes in virulent mutants arising from avirulent avian influenza viruses during replication in 14-day-old embryonated eggs. Virol. *206*, 755–9.

International Symposium on Avian Influenza. (2003). Recommendations of the 5th International Symposium on avian influenza. Avian Dis. *47*, 1260–1261.

Ito, T. (2000). Interspecies transmission and receptor recognition of influenza A viruses. Microbiology & Immunology. *44*, 423–430.

Ito, T., Couceiro, J. N., Kelm, S., Baum, L. G., Krauss, S., et al. (1998a). Molecular basis for the generation in pigs of influenza A viruses with pandemic potential. J. Virol. *72*, 7367–73.

Ito, T. and Kawaoka, Y. (1998). Avian influenza. Textbook of influenza. K. G. Nicholson, R. G. Webster and A. J. Hay. Oxford, Blackwell Science Ltd, 126–136.

Ito, T., Kawaoka, Y., Vines, A., Ishikawa, H., Asai, T., et al. (1998b). Continued circulation of reassortant H1N2 influenza viruses in pigs in Japan. Archives of Virol. *143*, 1773–82.

Ito, T., Suzuki, Y., Mitnaul, L., Vines, A., Kida, H., et al. (1997a). Receptor specificity of influenza A viruses correlates with the agglutination of erythrocytes from different animal species. Virol. *227*, 493–9.

Ito, T., Suzuki, Y., Suzuki, T., Takada, A., Horimoto, T., et al. (2000). Recognition of N-glycolylneuraminic acid linked to galactose by the alpha2,3 linkage is associated with intestinal replication of influenza A virus in ducks. J. Virol. *74*, 9300–5.

Ito, T., Suzuki, Y., Takada, A., Kawamoto, A., Otsuki, K., et al. (1997b). Differences in sialic acid-galactose linkages in the chicken egg amnion and allantois influence human influenza virus receptor specificity and variant selection. J. Virol. *71*, 3357–62.

Janke, B. H. (1998). Classic swine influenza. Large Animal Practice. *19*, 24–29.

Jones, Y. L. and Swayne, D. E. (2004). Comparative pathobiology of low and high pathogenicity H7N3 Chilean avian influenza viruses in chickens. Avian Dis. *48*, 119–128.

Kaplan, M. M. (1982). The epidemiology of influenza as a zoonosis. Vet. Rec. *110*, 395–9.

Kaplan, M. M. and Payne, A. M. (1959). Serological survey in animals for type A influenza in relation to the 1957 pandemic. Bull. World Health Organ. *20*, 465–88.

Karasin, A. I., Brown, I. H., Carman, S. and Olsen, C. W. (2000a). Isolation and characterization of H4N6 avian influenza viruses from pigs with pneumonia in Canada. J. Virol. *74*, 9322–9327.

Karasin, A. I., Landgraf, J., Swenson, S., Erickson, G., Goyal, S., et al. (2002). Genetic characterization of H1N2 influenza A viruses isolated from pigs throughout the United States. J. Clin. Microbiol. *40*, 1073–9.

Karasin, A. I., Olsen, C. W. and Anderson, G. A. (2000b). Genetic characterization of an H1N2 influenza virus isolated from a pig in Indiana. J. Clin. Microbiol. *38*, 2453–2456.

Karasin, A. I., Schutten, M. M., Cooper, L. A., Smith, C. B., Subbarao, K., et al. (2000c). Genetic characterization of H3N2 influenza viruses isolated from pigs in North America, 1977–1999, evidence for wholly human and reassortant virus genotypes. Virus Res. *68*, 71–85.

Karasin, A. I., West, K., Carman, S. and Olsen, C. W. (2004). Characterization of avian H3N3 and H1N1 influenza A viruses isolated from pigs in Canada. J. Clin. Microbiol. *42*, 4349–54.

Katsuda, K., Shirahata, T., Kida, H. and Goto, H. (1995). Antigenic and genetic analyses of the hemagglutinin of influenza viruses isolated from pigs in (1993). J. Vet. Med. Sci. *57*, 1023–7.

Katz, J. M., Lu, X., Tumpey, T. M., Smith, C. B., Shaw, M. W., et al. (2000). Molecular correlates of influenza A H5N1 virus pathogenesis in mice. J. Virol. *74*, 10807–10.

Kaufman, J., Hassan, M., Rytel, M., Chayer, R., Volkert, P., et al. (1988). Human infection with swine influenza virus-Wisconsin.

MMWR – Morbidity & Mortality Weekly Report *37*, 661–663.

Kawaoka, Y., Krauss, S. and Webster, R. G. (1989). Avian-to-human transmission of the PB1 gene of influenza A viruses in the 1957 and 1968 pandemics. J. Virol. *63*, 4603–8.

Kawaoka, Y. and Webster, R. G. (1988). Sequence requirements for cleavage activation of influenza virus hemagglutinin expressed in mammalian cells. Proc. Natl., Acad. Sci. USA. *85*, 324–8.

Kendal, A. P., Goldfield, M., Noble, G. R. and Dowdle, W. R. (1977). Identification and preliminary antigenic analysis of swine influenza-like viruses isolated during an influenza outbreak at Fort Dix, New Jersey. J. Infect. Dis. *136*, S381–5.

Kida, H., Ito, T., Yasuda, J., Shimizu, Y., Itakura, C., et al. (1994). Potential for transmission of avian influenza viruses to pigs. J. Gen. Virol. *75*, 2183–8.

Kida, H., Shortridge, K. F. and Webster, R. G. (1988). Origin of the hemagglutinin gene of H3N2 influenza viruses from pigs in China. Virol. *162*, 160–6.

Kimura, H., Abiko, C., Peng, G., Muraki, Y., Sugawara, K., et al. (1997). Interspecies transmission of influenza C virus between humans and pigs. Virus Res. *48*, 71–9.

Kimura, K., Adlakha, A. and Simon, P. M. (1998). Fatal case of swine influenza virus in an immunocompetent host. Mayo Clinic Proceedings. *73*, 243–5.

Klenk, H. D. and Rott, R. (1988). The molecular biology of influenza virus pathogenicity. Adv. Virus Res. *34*, 247–81.

Klimov, A., Simonsen, L., Fukuda, K. and Cox, N. (1999). Surveillance and impact of influenza in the United States. Vaccine. *17*, S42–6.

Kluska, V., Hanson, L. E. and Hatch, R. D. (1961). Evdence for swine influenza antibodies in human. Cesk. Pediatr. *116*, 408–414.

Koen, J. S. (1919). A practical method for field diagnosis of swine diseases. Am. J. Vet. Med. *14*, 468–470.

Koopmans, M., Wilbrink, B., Conyn, M., Natrop, G., van der Nat, H., et al. (2004). Transmission of H7N7 avian influenza A virus to human beings during a large outbreak in commercial poultry farms in the Netherlands. Lancet. *363*, 587–93.

Kundin, W. D. (1970). Hong Kong A-2 influenza virus infection among swine during a human epidemic in Taiwan. Nature *228*, 857.

Kupradinun, S., Peanpijit, P., Bhodhikosoom, C., Yoshioka, Y., Endo, A., et al. (1991). The first isolation of swine H1N1 influenza viruses from pigs in Thailand. Arch. Virol. *118*, 289–97.

Kurtz, J., Manvell, R. J. and Banks, J. (1996). Avian influenza virus isolated from a woman with conjunctivitis. Lancet. *348*, 901–2.

Lamb, R. A. and Krug, R. M. (2001). Orthomyxoviridae, The viruses and their replication. Fields Virol. D. M. Knipe and P. M. Howley. Philadeiphia, PA, Lippincott Williams and Wilkins. p. 1487–1532.

Landolt, G. A., Karasin, A. I., Phillips, L. and Olsen, C. W. (2003). Comparison of the pathogenesis of two genetically different H3N2 influenza A viruses in pigs. J. Clin. Microbiol. *41*, 1936–41.

Larsen, D. L., Karasin, A. and Olsen, C. W. (2001). Immunization of pigs against influenza virus infection by DNA vaccine priming followed by killed-virus vaccine boosting. Vaccine. *19*, 2842-2853.

Larsen, D. L. and Olsen, C. W. (2002). Effects of DNA dose, route of vaccination, and coadministration of porcine interleukin-6 DNA on results of DNA vaccination against influenza virus infection in pigs. Am. J. Vet. Res. *63*, 653–9.

Laver, W. G., Bischofberger, N. and Webster, R. G. (2000). The origin and control of pandemic influenza. Perspectives in Biology & Medicine. *43*, 173–192.

Laver, W. G. and Webster, R. G. (1973). Studies on the origin of pandemic influenza. 3. Evidence implicating duck and equine influenza viruses as possible progenitors of the Hong Kong strain of human influenza. Virol. *51*, 383–91.

Lee, C. W., Senne, D. A., Linares, J. A., Woolcock, P. R., Stallknecht, D. E., et al. (2004). Characterization of recent H5 subtype avian influenza viruses from US poultry. Avian Pathol. *33*, 288–97.

Levandowski, R. A., Regnery, H. L., Staton, E., Burgess, B. G., Williams, M. S., et al. (1991). Antibody responses to influenza B viruses in immunologically unprimed children. Pediatrics. *88*, 1031–6.

Li, K. S., Guan, Y., Wang, J., Smith, G. J., Xu, K. M., et al. (2004). Genesis of a highly pathogenic and potentially pandemic H5N1 influenza virus in eastern Asia. Nature *430*, 209–13.

Li, S., Liu, C., Klimov, A., Subbarao, K., Perdue, M. L., et al. (1999). Recombinant influenza A virus vaccines for the pathogenic human A/Hong Kong/97 (H5N1) viruses. J. Infect. Dis. *179*, 1132–8.

Lin, Y. P., Bennett, M., Gregory, V., Grambas, S., Ragazzoli, V., et al. (2003). Emergence of distinct avian-like influenza A H1N1 viruses in pigs in Ireland and their reassortment with cocirculating H3N2 viruses. International Conference on Options for the Control of Influenza V, Okinawa, Japan, Elsevier.

Lin, Y. P., Gregory, V., Bennett, M. and Hay, A. (2004). Recent changes among human influenza viruses. Virus Res. *103*, 47–52.

Lin, Y. P., Shaw, M., Gregory, V., Cameron, K., Lim, W., et al. (2000). Avian-to-human transmission of H9N2 subtype influenza A viruses, relationship between H9N2 and H5N1 human isolates. Proc. Natl. Acad. Sci. USA. *97*, 9654–8.

Lindstrom, S. E., Hiromoto, Y., Nishimura, H., Saito, T., Nerome, R., et al. (1999). Comparative analysis of evolutionary mechanisms of the hemagglutinin and three internal protein genes of influenza B virus, Multiple cocirculating lineages and frequent reassortment of the NP, M, and NS genes. J. Virol. *73*, 4413-26.

Ludwig, S., Haustein, A., Kaleta, E. F. and Scholtissek, C. (1994). Recent influenza A (H1N1) infections of pigs and turkeys in northern Europe. Virol. *202*, 281–6.

Luoh, S. M., McGregor, M. W. and Hinshaw, V. S. (1992). Hemagglutinin mutations related to antigenic variation in H1 swine influenza viruses. J. Virol. *66*, 1066–73.

Macklin, M. D., McCabe, D., McGregor, M. W., Neumann, V., Meyer, T., et al. (1998). Immunization of pigs with a particle-mediated DNA vaccine to influenza A virus protects against challenge with homologous virus. J. Virol. *72*, 1491–6.

Madec, F., Kaiser, C., Gourreau, J. M. and Martinat-Botte, F. (1989). Pathologic consequences of a severe influenza outbreak (swine virus A/H1N1) under natural conditions in the non-immune sow at the beginning of pregnancy. Comp. Immunol. Microbiol. Infect. Dis. *12*, 17-27.

Mahy, B. W. J. (1983). Mutants of influenza virus. Genetics of Influenza Viruses. P. Palese and D. W. Kingsbury. New York, Springer-Verlag, 192-254.

Mancini, G., Donatelli, I., Rozera, C., Arangio Ruiz, G. and Butto, S. (1985). Antigenic and biochemical analysis of influenza 'A' H3N2 viruses isolated from pigs. Arch. Virol. *83*, 157–67.

Marozin, S., Gregory, V., Cameron, K., Bennett, M., Valette, M., et al. (2002). Antigenic and genetic diversity among swine influenza A H1N1 and H1N2 viruses in Europe. J. Gen. Virol. *83*, 735–45.

Matrosovich, M. N., Gambaryan, A. S., Teneberg, S., Piskarev, V. E., Yamnikova, S. S., et al. (1997). Avian influenza A viruses differ from human viruses by recognition of sialyloligosaccharides and gangliosides and by a higher conservation of the HA receptor-binding site. Virol. *233*, 224–34.

Matrosovich, M. N., Matrosovich, T. Y., Gray, T., Roberts, N. A. and Klenk, H. D. (2004). Human and avian influenza viruses target different cell types in cultures of human airway epithelium. Proc. Natl. Acad. Sci. U S A. *101*, 4620–4.

McCullers, J. A., Wang, G. C., He, S. and Webster, R. G. (1999). Reassortment and insertion-deletion are strategies for the evolution of influenza B viruses in Nature J. Virol. *73*, 7343–8.

McNulty, M. S., Allan, G. M. and Adair, B. M. (1986). Efficacy of avian influenza neuraminidase-specific vaccines in chickens. Avian Pathol. *15*, 107–115.

Meltzer, M. I., Cox, N. J. and Fukuda, K. (1999). The economic impact of pandemic influenza in the United States, priorities for intervention. Emerg. Infect. Dis. *5*, 659–71.

Morin, M., Phaneuf, J. B., Sauvageau, R., DiFranco, E., Marsolais, G., et al. (1981). An epizootic of swine influenza in Quebec. Can. Vet. J. *22*, 204–5.

Mounts, A. W., Kwong, H., Izurieta, H. S., Ho, Y., Au, T., et al. (1999). Case-control study of risk factors for avian influenza A (H5N1) disease, Hong Kong, 1997. J. Infect. Dis. *180*, 505–8.

Naeve, C. W., Hinshaw, V. S. and Webster, R. G. (1984). Mutations in the hemagglutinin receptor-binding site can change the biological properties of an influenza virus. J. Virol. *51*, 567–9.

Naffakh, N., Massin, P., Escriou, N., Crescenzo-Chaigne, B. and van der Werf, S. (2000). Genetic analysis of the compatibility between polymerase proteins from human and avian strains of influenza A viruses. J. Gen. Virol. *81*, 1283–1291.

Nakajima, K., Nakajima, S., Shortridge, K. F. and Kendal, A. P. (1982). Further genetic evidence for maintenance of early Hong Kong-like influenza A(H3N2) strains in swine until 1976. Virol. *116*, 562–72.

Nardelli, L., Pascucci, S., Gualandi, G. L. and Loda, P. (1978). Outbreaks of classical swine influenza in Italy in 1976. Zentralbl. Veterinarmed. B. *25*, 853–7.

Nerome, K., Ishida, M., Nakayama, M., Oya, A., Kanai, C., et al. (1981). Antigenic and genetic analysis of A/Hong Kong (H3N2) influenza viruses isolated from swine and man. J. Gen. Virol. *56*, 441–5.

Nerome, K., Ishida, M., Oya, A., Kanai, C. and Suwicha, K. (1982). Isolation of an influenza H1N1 virus from a pig. Virol. *117*, 485–9.

Nerome, K., Kanegae, Y., Shortridge, K. F., Sugita, S. and Ishida, M. (1995). Genetic analysis of porcine H3N2 viruses originating in southern China. J. Gen. Virol. *76*, 613-24.

Nerome, K., Yoshioka, Y., Sakamoto, S., Yasuhara, H. and Oya, A. (1985). Characterization of a 1980-swine recombinant influenza virus possessing H1 hemagglutinin and N2 neuraminidase similar to that of the earliest Hong Kong (H3N2) virus. Arch. Virol. *86*, 197-211.

Nerome, R., Hiromoto, Y., Sugita, S., Tanabe, N., Ishida, M., et al. (1998). Evolutionary characteristics of influenza B virus since its first isolation in *1940*, dynamic circulation of deletion and insertion mechanism. Archives of Virol. *143*, 1569–83.

Nicholson, K. G., Colegate, A. E., Podda, A., Stephenson, I., Wood, J., et al. (2001). Safety and antigenicity of non-adjuvanted and MF59-adjuvanted influenza A/Duck/Singapore/97 (H5N3) vaccine, a randomised trial of two potential vaccines against H5N1 influenza. Lancet. *357*, 1937–43.

Ninomiya, A., Takada, A., Okazaki, K., Shortridge, K. F. and Kida, H. (2002). Seroepidemiological evidence of avian H4, H5, and H9 influenza A virus transmission to pigs in southeastern China. Vet. Microbiol. *88*, 107–14.

Nishikawa, F. and Sugiyama, T. (1983). Direct isolation of H1N2 recombinant virus from a throat swab of a patient simultaneously infected with H1N1 and H3N2 influenza A viruses. J. Clin. Microbiol. *18*, 425–7.

Noble, G. R. (1982). Epidemiological and clinical aspects of influenza. Basic and Applied Influenza Research. A. S. Beare. Boca Raton, CRC Press, 11–50.

Noble, S., McGregor, M. S., Wentworth, D. E. and Hinshaw, V. S. (1993). Antigenic and genetic conservation of the haemagglutinin in H1N1 swine influenza viruses. J. Gen. Virol. *74*, 1197–200.

Nowotny, N., Deutz, A., Fuchs, K., Schuller, W., Hinterdorfer, F., et al. (1997). Prevalence of swine influenza and other viral, bacterial, and parasitic zoonoses in veterinarians. J. Infect. Dis. *176*, 1414–5.

OIE. (2003). Manual of standards for diagnostic tests and vaccines, 2000. Paris, Office International des Epizooties.

Olsen, C. W. (2000). DNA vaccination against influenza viruses, a review with emphasis on equine and swine influenza. Vet. Microbiol. *74*, 149–164.

Olsen, C. W. (2002a). Emergence of novel strains of swine influenza virus in North America. Trends in Emerging Viral Infections of Swine. A. Morilla, K.-J. Yoon and J. J. Zimmerman. Ames, Iowa State University Press, 37–43.

Olsen, C. W. (2002b). The emergence of novel swine influenza viruses in North America. Virus Res. *85*, 199-210.

Olsen, C. W., Brammer, L., Easterday, B. C., Arden, N., Belay, E., et al. (2002). Serologic evidence of H1 swine Influenza virus infection in swine farm residents and employees. Emerg. Infect. Dis. *8*, 814–9.

Olsen, C. W., Carey, S., Hinshaw, L. and Karasin, A. I. (2000). Virologic and serologic surveillance for human, swine and avian influenza virus infections among pigs in the north-central United States. Archives of Virol. *145*, 1399–1419.

Olsen, C. W., Karasin, A. and Erickson, G. (2003). Characterization of a swine-like reassortant H1N2 influenza virus isolated from a wild duck in the United States. Virus Res. *93*, 115-21.

Olsen, C. W., McGregor, M. W., Cooley, A. J., Schantz, B., Hotze, B., et al. (1993). Antigenic and genetic analysis of a recently isolated H1N1 swine influenza virus. American Journal of Veterinary Research. *54*, 1630–6.

Ottis, K., Sidoli, L., Bachmann, P. A., Webster, R. G. and Kaplan, M. M. (1982). Human influenza A viruses in pigs, isolation of a H3N2 strain antigenically related to A/England/42/72 and evidence for continuous circulation of human viruses in the pig population. Arch. Virol. *73*, 103–8.

Ouchi, A., Nerome, K., Kanegae, Y., Ishida, M., Nerome, R., et al. (1996). Large outbreak of swine influenza in southern Japan caused by reassortant (H1N2) influenza viruses, its epizootic background and characterization of the causative viruses. J. Gen. Virol. *77*, 1751–9.

Patriarca, P. A., Kendal, A. P., Zakowski, P. C., Cox, N. J., Trautman, M. S., et al. (1984). Lack of significant person-to-person spread of swine influenza-like virus following fatal infection in an immunocompromised child. American Journal of Epidemiology. *119*, 152–8.

Peiris, J. S. M., Guan, Y., Ghose, P., Markwell, D., Krauss, S., et al. (2000). Co-circulation of avian H9N2 and human H3N2 viruses in pigs in southern China. Options for the Control of Influenza IV, Crete, Greece, Excerpta Medica.

Peiris, J. S. M., Guan, Y., Markwell, D., Ghose, P., Webster, R. G., et al. (2001).

Cocirculation of avian H9N2 and contemporary 'human' H3N2 influenza A viruses in pigs in southeastern China, Potential for genetic reassortment? J. Virol. *75*, 9679–9686.

Peiris, M., Yuen, K. Y., Leung, C. W., Chan, K. H., Ip, P. L., et al. (1999). Human infection with influenza H9N2. Lancet. *354*, 916–7.

Pensaert, M., Ottis, K., Vandeputte, J., Kaplan, M. M. and Bachmann, P. A. (1981). Evidence for the natural transmission of influenza A virus from wild ducts to swine and its potential importance for man. Bull World Health Organ. *59*, 75–8.

Perdue, M. L., Garcia, M., Senne, D. A. and Fraire, M. (1997). Virulence-associated sequence duplication at the hemagglutinin cleavage site of avian influenza viruses. Virus Res. *49*, 173–186.

Perdue, M. L., Suarez, D. L. and Swayne, D. E. (1999). Avian Influenza in the 1990s. Poultry and Avian Biology Reviews. *11*, 1–20.

Perkins, L. E. L. and Swayne, D. E. (2001). Pathobiology of A/Chicken/Hong Kong/220/97 (H5N1) avian influenza virus in seven Gallinaceous species. Veterinary Pathology. *38*, 149–164.

Potter, C. W. (2001). A history of influenza. Journal of Applied Microbiology. *91*, 572–579.

Pritchard, G. C., Dick, I. G., Roberts, D. H. and Wibberley, G. (1987). Porcine influenza outbreak in East Anglia due to influenza A virus (H3N2). Vet. Rec. *121*, 548.

Reid, A. H., Fanning, T. G., Hultin, J. V. and Taubenberger, J. K. (1999). Origin and evolution of the 1918 'Spanish' influenza virus hemagglutinin gene. Proc. Natl., Acad. Sci. U S A. *96*, 1651–56.

Reid, A. H., Fanning, T. G., Janczewski, T. A., McCall, S. and Taubenberger, J. K. (2002). Characterization of the 1918 'Spanish' influenza virus matrix gene segment. J. Virol. *76*, 10717–23.

Reid, A. H., Fanning, T. G., Janczewski, T. A. and Taubenberger, J. K. (2000). Characterization of the 1918 'Spanish' influenza virus neuraminidase gene. Proc. Natl. Acad. Sci. USA. *97*, 6785–6790.

Reid, A. H., Taubenberger, J. K. and Fanning, T. G. (2001). The 1918 Spanish influenza,

integrating history and biology. Microbes & Infection. *3*, 81–87.

Rekik, M. R., Arora, D. J. and Dea, S. (1994). Genetic variation in swine influenza virus A isolate associated with proliferative and necrotizing pneumonia in pigs. J. Clin. Microbiol. *32*, 515–8.

Richt, J. A., Lager, K. M., Janke, B. H., Woods, R. D., Webster, R. G., et al. (2003). Pathogenic and antigenic properties of phylogenetically distinct reassortant H3N2 swine influenza viruses cocirculating in the United States. J. Clin. Microbiol. *41*, 3198–205.

Rimmelzwaan, G. F., de Jong, J. C., Bestebroer, T. M., van Loon, A. M., Claas, E. C., et al. (2001). Antigenic and genetic characterization of swine influenza A (H1N1) viruses isolated from pneumonia patients in The Netherlands. Virol. *282*, 301–6.

Roberts, D. H., Cartwright, S. F. and Wibberley, G. (1987). Outbreaks of classical swine influenza in pigs in England in 1986. Vet. Rec. *121*, 53–5.

Rogers, G. N. and D'Souza, BL, (1989). Receptor binding properties of human and animal H1 influenza virus isolates. Virol. *173*, 317-22.

Rogers, G. N. and Paulson, J. C. (1983). Receptor determinants of human and animal influenza virus isolates, differences in receptor specificity of the H3 hemagglutinin based on species of origin. Virol. *127*, 361–73.

Rogers, G. N., Paulson, J. C., Daniels, R. S., Skehel, J. J., Wilson, I. A., et al. (1983). Single amino acid substitutions in influenza haemagglutinin change receptor binding specificity. Nature *304*, 76–8.

Rohm, C., Horimoto, T., Kawaoka, Y., Suss, J. and Webster, R. G. (1995). Do hemagglutinin genes of highly pathogenic avian influenza viruses constitute unique phylogenetic lineages? Virol. *209*, 664–70.

Rojas, H., Moreira, R., Avalos, P., Capua, I. and Marangon, S. (2002). Avian influenza in poultry in Chile. Vet. Rec. *151*, 188.

Rota, P. A., Hemphill, M. L., Whistler, T., Regnery, H. L. and Kendal, A. P. (1992). Antigenic and genetic characterization of the haemagglutinins of recent cocirculating strains of influenza B virus. J. Gen. Virol. 73 (Pt 10), 2737–42.

Rota, P. A., Rocha, E. P., Harmon, M. W., Hinshaw, V. S., Sheerar, M. G., et al. (1989). Laboratory characterization of a swine influenza virus isolated from a fatal case of human influenza. J. Clin. Microbiol. 27, 1413–6.

Rota, P. A., Wallis, T. R., Harmon, M. W., Rota, J. S., Kendal, A. P., et al. (1990). Cocirculation of two distinct evolutionary lineages of influenza type B virus since 1983. Virol. 175, 59–68.

Schafer, W. 1955. Vergleichende sero-immunologische untersuchungen uber die viren der influenza unf klassichen geflugelpest. Z. 10B, 81–91.

Schnurrenberger, P. R., Woods, G. T. and Martin, R. J. 1970. Serologic evidence of human infection with swine influenza virus. Am. Rev. Respir. Dis. 102, 356–61.

Scholtissek, C., Burger, H., Bachmann, P. A. and Hannoun, C. (1983). Genetic relatedness of hemagglutinins of the H1 subtype of influenza A viruses isolated from swine and birds. Virol. 129, 521–3.

Scholtissek, C., Burger, H., Kistner, O. and Shortridge, K. F. (1985). The nucleoprotein as a possible major factor in determining host specificity of influenza H3N2 viruses. Virol. 147, 287–94.

Scholtissek, C., Koennecke, I. and Rott, R. (1978a). Host range recombinants of fowl plague (influenza A) virus. Virol. 91, 79–85.

Scholtissek, C. and Naylor, E. (1988). Fish farming and influenza pandemics. Nature 331, 215.

Scholtissek, C., Rohde, W., Von Hoyningen, V. and Rott, R. (1978b). On the origin of the human influenza virus subtypes H2N2 and H3N2. Virol. 87, 13–20.

Schrader, C. and Suss, J. (2003). Genetic characterization of a porcine H1N2 influenza virus strain isolated in Germany. InterVirol. 46, 66–70.

Schultz, U., Fitch, W. M., Ludwig, S., Mandler, J. and Scholtissek, C. (1991). Evolution of pig influenza viruses. Virol. 183, 61–73.

Senne, D. A., Panigrahy, B., Kawaoka, Y., Pearson, J. E., Suss, J., et al. (1996). Survey of the hemagglutinin (HA) cleavage site sequence of H5 and H7 avian influenza viruses, amino acid sequence at the HA cleavage site as a marker of pathogenicity potential. Avian Dis. 40, 425–37.

Seo, S. H., Hoffmann, E. and Webster, R. G. (2002). Lethal H5N1 influenza viruses escape host anti-viral cytokine responses. Nat. Med. 8, 950–4.

Seo, S. H., Hoffmann, E. and Webster, R. G. (2004). The NS1 gene of H5N1 influenza viruses circumvents the host anti-viral cytokine responses. Virus Res. 103, 107–13.

Shaw, M. W., Cooper, L., Xu, X., Thompson, W. K., Krauss, S., et al. (2002a). Molecular changes associated with the transmission of avian influenza A H5N1 and H9N2 viruses to humans. Journal of Medical Virol. 66, 107–114.

Shaw, M. W., Xu, X., Li, Y., Normand, S., Ueki, R. T., et al. (2002b). Reappearance and global spread of variants of influenza B/Victoria/2/87 lineage viruses in the 2000–2001 and 2001–2002. seasons. Virol. 303, 1–8.

Sheerar, M. G., Easterday, B. C. and Hinshaw, V. S. (1989). Antigenic conservation of H1N1 swine influenza viruses. J. Gen. Virol. 70, 3297–303.

Shimada, S., Ohtsuka, T., Tanaka, S., Mimura, M., Shinohara, M., et al. (2003). Existence of reassortant A (H1N2) swine influenza viruses in Saitama Prefecture, Japan. Options for the Control of Influenza V, Okinawa, Japan.

Shope, R. E. 1931. Swine influenza. J. Exp. Med. 54, 373–385.

Shortridge, K. F. (1992). Pandemic influenza, a zoonosis? Seminars in Respiratory Infections. 7, 11–25.

Shortridge, K. F., Cherry, A. and Kendal, A. P. (1979). Further studies of the antigenic properties of H3N2 strains of influenza A isolated from swine in South East Asia. J. Gen. Virol. 44, 251–4.

Shortridge, K. F. and Stuart-Harris, C. H. (1982). An influenza epicentre? Lancet. 2, 812–3.

Shortridge, K. F. and Webster, R. G. (1979). Geographical distribution of swine (Hsw1N1) and Hong Kong (H3N2) influenza virus variants in pigs in Southeast Asia. InterVirol. 11, 9–15.

Shortridge, K. F., Webster, R. G., Butterfield, W. K. and Campbell, C. H. (1977). Persistence of Hong Kong influenza virus variants in pigs. Science. *196*, 1454–5.

Shu, L. L., Lin, Y. P., Wright, S. M., Shortridge, K. F. and Webster, R. G. (1994). Evidence for interspecies transmission and reassortment of influenza A viruses in pigs in southern China. Virol. *202*, 825–33.

Simonsen, L. (1999). The global impact of influenza on morbidity and mortality. Vaccine. *17*, S3–10.

Simonsen, L., Clarke, M. J., Williamson, G. D., Stroup, D. F., Arden, N. H., et al. (1997). The impact of influenza epidemics on mortality, introducing a severity index. American J. Public Health *87*, 1944–50.

Simonsen, L., Fukuda, K., Schonberger, L. B. and Cox, N. J. (2000). The impact of influenza epidemics on hospitalizations. J. Infect. Dis. *181*, 831–837.

Simonsen, L., Reichert, T. A. and Miller, M. A. (2004). The virtues of antigenic sin, consequences of pandemic recycling on influenza-associated mortality. Options for the Control of Influenza V, Okinawa, Japan, Elsevier, 791–794.

Slemons, R. D., Byrum, B. and Swayne, D. E. (1998). Bacterial proteases and co-infections as enhancers of virulence. Proceedings of the Fourth International Symposium on Avian Influenza. D. E. Swayne and R. D. Slemons. Richmond, Virginia, U. S. Animal Health Association, 203–208.

Slemons, R. D., Johnson, D. C., Osborn, J. S. and Hayes, F. 1974. Type-A influenza viruses isolated from wild free-flying ducks in California. Avian Dis. *18*, 119–124.

Smith, D. J., Lapedes, A. S., de Jong, J. C., Bestebroer, T. M., Rimmelzwaan, G. F., et al. (2004). Mapping the antigenic and genetic evolution of influenza virus. Science. *305*, 371–6.

Smith, T. F., Burgert, E. O., Dowdle, W. R., Noble, G. R., Campbell, R. J., et al. 1976. Isolation of swine influenza virus from autopsy lung tissue of man. New Engl. J. Med. *294*, 708–10.

Smith, W., Andrewes, C. H. and Laidlaw, P. P. (1933). A virus obtained from influenza patients. Lancet. 66–8.

Snyder, M. H., Betts, R. F., DeBorde, D., Tierney, E. L., Clements, M. L., et al. (1988). Four viral genes independently contribute to attenuation of live influenza A/Ann Arbor/6/60 (H2N2) cold-adapted reassortant virus vaccines. J. Virol. *62*, 488–495.

Snyder, M. H., Buckler-White, A. J., London, W. T., Tierney, E. L. and Murphy, B. R. (1987). The avian influenza virus nucleoprotein gene and a specific constellation of avian and human virus polymerase genes each specify attenuation of avian-human influenza A/Pintail/79 reassortant viruses for monkeys. J. Virol. *61*, 2857–63.

Spackman, E., Senne, D. A., Bulaga, L. L., Myers, T. J., Perdue, M. L., et al. (2003). Development of real-time RT-PCR for the detection of avian influenza virus. Avian Dis. *47*, 1079–1082.

Spackman, E., Senne, D. A., Myers, T. J., Bulaga, L. L., Garber, L. P., et al. (2002). Development of a real-time reverse transcriptase PCR assay for type A influenza virus and the avian H5 and H7 hemagglutinin subtypes. J. Clin. Microbiol. *40*, 3256–3260.

Stallknecht, D. E. (1998). Ecology and epidemiology of avian influenza viruses in wild bird populations, waterfowl, shorebirds, pelicans, cormorants, etc. Proceedings of the Fourth International Symposium on Avian Influenza. D. E. Swayne and R. D. Slemons. Richmond, Virginia, U. S. Animal Health Association, 61–69.

Stephenson, I., Nicholson, K. G., Gluck, R., Mischler, R., Newman, R. W., et al. (2003). Safety and antigenicity of whole virus and subunit influenza A/Hong Kong/1073/99 (H9N2) vaccine in healthy adults, phase I randomised trial. Lancet. *362*, 1959–66.

Stieneke Grober, A., Vey, M., Angliker, H., Shaw, E., Thomas, G., et al. (1992). Influenza virus hemagglutinin with multibasic cleavage site is activated by furin, a subtilisin-like endoprotease. EMBO J. *11*, 2407–2414.

Stone, H. D. (1987). Efficacy of avian influenza oil-emulsion vaccines in chickens of various ages. Avian Dis. *31*, 483–490.

Stone, H. D. (1988). Optimization of hydrophile-lipophile balance for improved efficacy of Newcastle disease and avian influenza oil-emulsion vaccines. Avian Dis. *32*, 68–73.

Stubbs, E. L. (1948). Fowl pest. Diseases of Poultry. H. E. Biester and L. H. Schwarte. Ames, Iowa, Iowa State University Press, 603–614.

Sturm-Ramirez, K. M., Ellis, T., Bousfield, B., Bissett, L., Dyrting, K., et al. (2004). Reemerging H5N1 influenza viruses in Hong Kong in 2002. are highly pathogenic to ducks. J. Virol. *78*, 4892–4901.

Suarez, D. L. and Schultz Cherry, S. (2000). Immunology of avian influenza, a review. Developmental and Comparative Immunology. in press.

Suarez, D. L., Senne, D. A., Banks, J., Brown, I. H., Essen, S. C., et al. (2004). Recombination resulting in virulence shift in avian influenza outbreak, Chile. Emerg. Infect. Dis. J. *10*, 693–699.

Suarez, D. L., Woolcock, P. R., Bermudez, A. J. and Senne, D. A. (2002). Isolation from turkey breeder hens of a reassortant H1N2 influenza virus with swine, human, and avian lineage genes. Avian Dis. *46*, 111–21.

Subbarao, E. K., London, W. and Murphy, B. R. (1993). A single amino acid in the PB2 gene of influenza A virus is a determinant of host range. J. Virol. *67*, 1761–4.

Subbarao, K., Chen, H., Swayne, D., Mingay, L., Fodor, E., et al. (2003). Evaluation of a genetically modified reassortant H5N1 influenza A virus vaccine candidate generated by plasmid-based reverse genetics. Virol. *305*, 192–200.

Subbarao, K. and Katz, J. (2000). Avian influenza viruses infecting humans. Cell. Mol. Life Sci. *57*, 1770–84.

Subbarao, K., Klimov, A., Katz, J., Regnery, H., Lim, W., et al. (1998). Characterization of an avian influenza A (H5N1) virus isolated from a child with a fatal respiratory illness. Science. *279*, 393–6.

Sugimura, T., Yonemochi, H., Ogawa, T., Tanaka, Y. and Kumagai, T. (1980). Isolation of a recombinant influenza virus (Hsw 1 N2) from swine in Japan. Arch. Virol. *66*, 271–4.

Suzuki, T., Horiike, G., Yamazaki, Y., Kawabe, K., Masuda, H., et al. (1997). Swine influenza virus strains recognize sialylsugar

chains containing the molecular species of sialic acid predominantly present in the swine tracheal epithelium. FEBS Lett. *404*, 192–6.

Swayne, D. E. (2000). Understanding the ecology and epidemiology of avian influenza viruses, implications for zoonotic potential. Emerging diseases of animals. C. C. Brown and C. A. Bolin. Washington, D. C., ASM Press, 101–130.

Swayne, D. E. (2001). Avian influenza vaccine use during 2001, Richmond, Virginia, USAHA, 469–471.

Swayne, D. E. (2003). Vaccines for list A poultry diseases, emphasis on avian influenza. Developments in Biologics. *114*, 201–212.

Swayne, D. E. and Suarez, D. L. (2005). Strategies for controlling influenza in birds and animals, The U. S. programs. The Threat of Pandemic Influenza, Are we ready? Washington D. C., Institute of Medicine, National Academies of Science, 233–243

Swayne, D. E. and Akey, B. (2005). Avian influenza control strategies in the United States of America. Proceedings of the Wageningen Frontis International Workshop on Avian Influenza Prevention and Control. G. Koch. Dordrecht, Kluwer Academic Publishers, 113–130.

Swayne, D. E., Beck, J. R., Garcia, M., Perdue, M. L. and Brugh, M. (1998a). Pathogenicity shifts in experimental avian influenza virus infections in chickens. Proceedings of the Fourth International Symposium on Avian Influenza. D. E. Swayne and R. D. Slemons. Richmond, Virginia, U. S. Animal Health Association, 171–181.

Swayne, D. E., Beck, J. R., Garcia, M. and Stone, H. D. (1999). Influence of virus strain and antigen mass on efficacy of H5 avian influenza inactivated vaccines. Avian Pathol. *28*, 245–255.

Swayne, D. E., Beck, J. R. and Mickle, T. R. (1997). Efficacy of recombinant fowl pox vaccine in protecting chickens against highly pathogenic Mexican-origin H5N2 avian influenza virus. Avian Dis. *41*, 910–922.

Swayne, D. E., Beck, J. R., Perdue, M. L., Brugh, M. and Slemons, R. D. (1996). Assessment of the ability of ratite-origin influenza viruses to Infect. and produce disease in rheas and chickens. Avian Dis. *40*, 438–447.

Swayne, D. E., Garcia, M., Beck, J. R., Kinney, N. and Suarez, D. L. (2000). Protection against diverse highly pathogenic avian influenza viruses in chickens immunized with a recombinant fowl pox vaccine containing an H5 avian influenza hemagglutinin gene insert. Vaccine. *18*, 1088–1095.

Swayne, D. E. and Halvorson, D. A. (2003). Influenza. Diseases of Poultry. Y. M. Saif, H. J. Barnes, A. M. Fadlyet al. Ames, IA, Iowa State University Press, 135–160.

Swayne, D. E., Senne, D. A. and Beard, C. W. (1998b). Influenza. Isolation and identification of avian pathogens. D. E. Swayne, J. R. Glisson, M. W. Jackwood, J. E. Pearson and W. M. Reed. Pennsylvania, Kennett Square, 150–155.

Swayne, D. E. and Suarez, D. L. (2000). Highly pathogenic avian influenza. Revue Scientifique et Technique Office International des Epizooties. *19*, 463–482.

Takahashi, T., Suzuki, T., Hidari, K. I., Miyamoto, D. and Suzuki, Y. (2003). A molecular mechanism for the low-pH stability of sialidase activity of influenza A virus N2 neuraminidases. FEBS Lett. *543*, 71–5.

Takatsy, G., Farkas, E. and Romvary, J. (1969). Susceptibility of the domestic pig to influenza B virus. Nature *222*, 184–5.

Tang, M., Harp, J. A. and Wesley, R. D. (2002). Recombinant adenovirus encoding the HA gene from swine H3N2 influenza virus partially protects mice from challenge with heterologous virus, A/HK/1/68 (H3N2). Arch. Virol. *147*, 2125–41.

Taubenberger, J. K., Reid, A. H. and Fanning, T. G. (2000). The 1918 influenza virus, A killer comes into view. Virol. *274*, 241–245.

Taubenberger, J. K., Reid, A. H., Janczewski, T. A. and Fanning, T. G. (2001). Integrating historical, clinical and molecular genetic data in order to explain the origin and virulence of the 1918 Spanish influenza virus. Philos. Trans. R. Soc. Lond. B Biol. Sci. *356*, 1829–39.

Taubenberger, J. K., Reid, A. H., Krafft, A. E., Bijwaard, K. E. and Fanning, T. G. (1997). Initial genetic characterization of the 1918 'Spanish' influenza virus. Science. *275*, 1793–1796.

Taylor, H. R. and Turner, A. J. (1977). A case report of fowl plague keratoconjunctivitis. British Journal of Ophthalmology. *61*, 86–88.

Thompson, W. W., Shay, D. K., Weintraub, E., Brammer, L., Cox, N., et al. (2003). Mortality associated with influenza and respiratory syncytial virus in the United States. JAMA. *289*, 179–86.

Tian, S. F., Buckler-White, A. J., London, W. T., Reck, L. J., Chanock, R. M., et al. (1985). Nucleoprotein and membrane protein genes are associated with restriction of replication of influenza A/Mallard/NY/78 virus and its reassortants in squirrel monkey respiratory tract. J. Virol. *53*, 771–5.

Top, F. H., Jr. and Russell, P. K. (1977). Swine influenza A at Fort Dix, New Jersey (January-February 1976). IV. Summary and speculation. J. Infect. Dis. 136 Suppl, S376–80.

Tran, T. H., Nguyen, T. L., Nguyen, T. D., Luong, T. S., Pham, P. M., et al. (2004). Avian influenza A (H5N1) in 10 patients in Vietnam. N. Engl. J. Med. *350*, 1179–88.

Traub-Dargatz, J. L., Salman, M. D. and Voss, J. L. (1991). Medical problems of adult horses, as ranked by equine practitioners. J. Am. Vet. Med. Assoc. *198*, 1745–7.

Treanor, J. J., Wilkinson, B. E., Masseoud, F., Hu-Primmer, J., Battaglia, R., et al. (2001). Safety and immunogenicity of a recombinant hemagglutinin vaccine for H5 influenza in humans. Vaccine. *19*, 1732–1737.

Tsai, C. P. and Pan, M. J. (2003). New H1N2 and H3N1 influenza viruses in Taiwanese pig herds. Vet. Rec. *153*, 408.

Tumova, B., Veznikova, D., Mensik, J. and Stumpa, A. (1980). Surveillance of influenza in pig herds in Czechoslovakia in 1974–(1979). 1. Introduction of influenza epidemic A (H3N2) viruses into pig herds. Zentralbl. Veterinarmed. B. *27*, 517–23.

Tumpey, T. M., Alvarez, R., Swayne, D. E. and Suarez, D. L. (2004a). Diagnostic approach for differentiating infected from vaccinated poultry on the basis of antibodies to NS1, the nonstructural protein of influenza A virus. J. Clin. Microbiol. *43*, 676–83.

Tumpey, T. M., Kapczynski, D. R. and Swayne, D. E. (2004b). Comparative susceptibility of chickens and turkeys to avian influenza A H7N2 virus infection and protective efficacy of a commercial avian influenza H7N2 virus vaccine. Avian Dis. *48*, 167–176.

Van Reeth, K., Brown, I. H. and Pensaert, M. (2000). Isolations of H1N2 influenza A virus from pigs in Belgium. Vet. Rec. *146*, 588–9.

Van Reeth, K. and Easterday, B. C. (1999). Swine influenza. Diseases of Swine. B. E. Straw, S. D'Allaire, W. L. Mengeling and D. J. Taylor. Ames, Iowa State University Press, 277–290.

Van Reeth, K., Gregory, V., Hay, A. and Pensaert, M. (2003). Protection against a European H1N2 swine influenza virus in pigs previously infected with H1N1 and/or H3N2 subtypes. Vaccine. *21*, 1375–81.

Vey, M., Orlich, M., Adler, S., Klenk, H. D., Rott, R., et al. (1992). Hemagglutinin activation of pathogenic avian influenza viruses of serotype H7 requires the protease recognition motif R-X-K/R-R. Virology. *188*, 408–413.

Webby, R. J., Swenson, S. L., Krauss, S. L., Gerrish, P. J., Goyal, S. M., et al. (2000). Evolution of swine H3N2 influenza viruses in the United States. J. Virol. *74*, 8243–8251.

Webby, R. J. and Webster, R. G. (2001). Emergence of influenza A viruses. Phil. Trans. Roy. Soc. London – Series B, Biol. Sci. *356*, 1817–1828.

Webster, R. G. (1998). Influenza, an emerging disease. Emerg. Infect. Dis. J. *4*, 436–41.

Webster, R. G., Bean, W. J., Gorman, O. T., Chambers, T. M. and Kawaoka, Y. (1992). Evolution and ecology of influenza A viruses. Microbiol. Rev. *56*, 152–79.

Webster, R. G., Geraci, J., Petursson, G. and Skirnisson, K. (1981). Conjunctivitis in human beings caused by influenza A virus of seals. New Engl. J. Med. *304*, 911.

Webster, R. G. and Kawaoka, Y. (1994). Influenza- an emerging and re-emerging disease. Seminars in Virol. *5*, 103–111.

Webster, R. G., Yakhno, M., Hinshaw, V. S., Bean, W. J. and Murti, K. G. (1978). Intestinal influenza, replication and characterization of influenza viruses in ducks. Virol. *84*, 268–78.

Wentworth, D. E., McGregor, M. W., Macklin, M. D., Neumann, V. and Hinshaw, V. S. (1997). Transmission of swine influenza virus to humans after exposure to experimentally infected pigs. J. Infect. Dis. *175*, 7–15.

Wentworth, D. E., Thompson, B. L., Xu, X., Regnery, H. L., Cooley, A. J., et al. (1994). An influenza A (H1N1) virus, closely related to swine influenza virus, responsible for a fatal case of human influenza. J. Virol. *68*, 2051–8.

Wilson, I. A. and Cox, N. J. (1990). Structural basis of immune recognition of influenza virus hemagglutinin. Annu. Rev. Immunol. *8*, 737–71.

Wilson, I. A., Skehel, J. J. and Wiley, D. C. (1981). Structure of the haemagglutinin membrane glycoprotein of influenza virus at 3 A resolution. Nature *289*, 366–73.

Wood, G. W., Banks, J., Brown, I. H., Strong, I. and Alexander, D. J. (1997). The nucleotide sequence of the HA1 of the hemagglutinin of an H1 avian influenza virus isolate from turkeys in Germany provides additional evidence suggesting recent transmission from pigs. Avian Pathol. *26*, 347–355.

Wood, J. M., Kawaoka, Y., Newberry, L. A., Bordwell, E. and Webster, R. G. (1985). Standardization of inactivated H5N2 influenza vaccine and efficacy against lethal A/Chicken/Pennsylvania/1370/83 infection. Avian Dis. *29*, 867–72.

Woods, G. T., Hanson, L. E. and Hatch, R. D. (1968). Investigation of four outbreaks of acute respiratory disease in swine and isolation of swine influenza virus. Health Lab. Sci. *5*, 218–24.

Woods, G. T. and Mansfield, M. E. (1974). Transplacental migration of swine influenza virus in gilts exposed experimentally. Res. Commun. Chem. Pathol. Pharm. *7*, 629–32.

Woods, G. T., Schnurrenberger, P. R., Martin, R. J. and Tompkins, W. A. (1981). Swine influenza virus in swine and man in Illinois. J. Occup. Med. *23*, 263–7.

Wright, P. F. and Webster, R. G. (2001). Orthomyxoviruses. Fields Virol. D. M. Knipe, P. M. Howley, D. E. Griffen et al.

Philadelphia, Lippincott Williams and Wilkens. p. 1533–1580.

Wright, S. M., Kawaoka, Y., Sharp, G. B., Senne, D. A. and Webster, R. G. (1992). Interspecies transmission and reassortment of influenza A viruses in pigs and turkeys in the United States. Am. J. Epidemiol. *136*, 488–97.

Xu, X., Lindstrom, S. E., Shaw, M. W., Smith, C. B., Hall, H. E., et al. (2004). Reassortment and evolution of current human influenza A and B viruses. Virus Res. *103*, 55–60.

Xu, X., Smith, C. B., Mungall, B. A., Lindstrom, S. E., Hall, H. E., et al. (2002). Intercontinental circulation of human influenza A(H1N2) reassortant viruses during the 2001–2002. influenza season. J. Infect. Dis. *186*, 1490–3.

Xu, X., Subbarao, Cox, N. J. and Guo, Y. (1999). Genetic characterization of the pathogenic influenza A/Goose/Guangdong/1/96 (H5N1) virus, similarity of its hemagglutinin gene to those of H5N1 viruses from the 1997 outbreaks in Hong Kong. Virol. *261*, 15–9.

Yamane, N., Arikawa, J., Odagiri, T., Kumasaka, M. and Ishida, N. (1978). Distribution of antibodies against swine and Hong Kong influenza viruses among pigs in 1977. Tohoku J. Exp. Med. *126*, 199–200.

Yao, Y., Mingay, L. J., McCauley, J. W. and Barclay, W. S. (2001). Sequences in influenza A virus PB2 protein that determine productive infection for an avian influenza virus in mouse and human cell lines. J. Virol. *75*, 5410–5.

Yip, T. K. S. (1976). Serological survey on the influenza antibody status in pigs of the Takwuling pig breeding centre. Agric. Hong Kong. *1*, 446–458.

Young, G. A. and Underdahl, N. A. (1949). Swine influenza as a possible factor in suckling pig mortalities. I. Seasonal occurrence in adult swine as indicated by hemagglutinin inhibitors in serum. Cornell Veterinarian. *39*, 105–119.

Yuen, K. Y., Chan, P. K., Peiris, M., Tsang, D. N., Que, T. L., et al. (1998). Clinical features and rapid viral diagnosis of human disease associated with avian influenza A H5N1 virus. Lancet. *351*, 467–71.

Zambon, M. C., Stockton, J. D., Clewley, J. P. and Fleming, D. M. (2001). Contribution of influenza and respiratory syncytial virus to community cases of influenza-like illness, an observational study. Lancet. *358*, 1410–6.

Zhou, N. N., Senne, D. A., Landgraf, J. S., Swenson, S. L., Erickson, G., et al. (1999a). Genetic reassortment of avian, swine, and human influenza A viruses in American pigs. J. Virol. *73*, 8851–8856.

Zhou, N. N., Senne, D. A., Landgraf, J. S., Swenson, S. L., Erickson, G., et al. (2000). Emergence of H3N2 reassortant influenza A viruses in North American pigs. Vet. Microbiol. *74*, 47–58.

Zhou, N. N., Shortridge, K. F., Claas, E. C., Krauss, S. L. and Webster, R. G. (1999b). Rapid evolution of H5N1 influenza viruses in chickens in Hong Kong. J. Virol. *73*, 3366–74.

Zou, S., Prud'homme, I. and Weber, J. M. (1997). Evolution of the hemagglutinin gene of influenza B virus was driven by both positive and negative selection pressures. Virus Genes. *14*, 181–5.

CHAPTER 10

H5 influenza viruses

Robert G. Webster

Abstract

Influenza viruses are classified into 16 subtypes on the basis of the hemagglutinin (HA) that they carry. Two subtypes found in aquatic birds worldwide – the H5 and H7 subtypes – are unique in having the ability to become highly pathogenic for domestic poultry and, occasionally, for humans after interspecies transmission. This chapter will primarily describe H5 influenza viruses. After transfer from the reservoir in aquatic birds to other avian species, H5 viruses can undergo rapid evolution: they may acquire multiple basic amino acids in the connecting peptide of the HA cleavage site, lose carbohydrate-bearing residues from the HA, undergo shortening of the neuraminidase (NA) stalk length, and acquire mutations in multiple genes encoding internal proteins including the polymerase PB2 and the nonstructural (NS) protein. Each of these events and the ecological conditions promoting their occurrence are considered in this chapter.

NATURAL RESERVOIRS AND THE EVOLUTION OF PATHOGENIC STRAINS

H5 influenza viruses are more frequently isolated from shorebirds and gulls (order Charadriiformes) than from wild ducks (order Anseriformes) (Alexander, 2000; Krauss et al., 2004). In wild waterfowl, H5 viruses are usually nonpathogenic, but highly pathogenic strains possessing polybasic amino acids at the cleavage site of the HA have, very rarely, been isolated from terns (Becker, 1966), from ducks associated with infected turkeys (Kawaoka et al., 1987), and from a variety of dead wild waterfowl linked to the H5N1 infections in Hong Kong in 2002 (Sturm-Ramirez et al., 2004; Ellis et al., 2004).

Influenza A viruses circulate widely in aquatic birds throughout the world.

Phylogenetic and genetic evidence indicates that:

- there are a limited number of host-specific lineages of influenza A viruses, including those in aquatic birds, humans, pigs, horses, sea mammals, and domestic avian species
- the viruses can be separated on the basis of geography into Eurasian and American lineage
- influenza A viruses in their natural avian reservoirs are in evolutionary stasis
- rapid evolution of the virus occurs after it is transferred to a new host species
- influenza viruses in their wild-bird reservoirs are usually not pathogenic and exist in equilibrium with their host

The following principles have been proposed concerning the ecology of influenza viruses in their natural hosts (Webster et al., 1992):

- wild aquatic birds provide the natural reservoir for all influenza A viruses that infect other species, including humans
- in wild aquatic birds, influenza viruses replicate predominantly in the intestinal tract and are spread by fecal-oral transmission, often through water
- most interspecies transmissions of virus are transitory and do not result in a stable lineage
- intermediate hosts involved in the interspecies transmission of avian influenza viruses include pigs, chickens, and quail.

Although highly pathogenic H5 influenza viruses have been isolated from wild waterfowl, there is currently no convincing evidence that these highly pathogenic viruses are maintained by wild waterfowl (see below). The available evidence indicates that each highly pathogenic lineage of H5 influenza viruses originates from a nonpathogenic precursor (Röhm et al., 1995; Banks et al., 2001). Highly pathogenic H5 influenza viruses from the American clade of influenza viruses include A/turkey/Ontario/7732/66 (H5N9) (Lang et al., 1968), A/chicken/Pennsylvania/1370/83 (H5N2) (Bean et al., 1985), and A/chicken/Mexico/31381-1/95 (H5N2) (Garcia et al., 1997). The Pennsylvanian and Mexican H5N2 viruses emerged from nonpathogenic precursors in the American clade (Saito et al., 1994, Horimoto et al 1995). Far less information is available about the reservoirs of H5 influenza viruses in European wild aquatic birds, but the highly pathogenic H7N7 virus found in domestic poultry in The Netherlands in 2003 and H7N3 virus found in poultry in Italy in 2002 probably came from wild aquatic birds (Fouchier et al., 2003; Campitelli et al., 2004). The precursors of the highly pathogenic H5N1 viruses that infected poultry and humans in Hong Kong in 1997 came from domestic geese (A/goose/Guangdong/1/96 [H5N1]; Tang et al., 1998; Xu et al., 1999); the wild-bird precursor of this virus is unknown. Because there is a large population of domestic ducks and geese in southeast Asia (including China), the natural reservoir of influenza viruses in Asia may include domestic as well as wild aquatic birds (Shortridge, 1992).

After transfer to an alternative avian host or to a mammalian host, influenza viruses undergo rapid evolution (Ludwig et al., 1995; Zhou et al., 1999). Avian influenza viruses including those of H5 subtype can be transferred throughout poultry in backyard flocks or live markets, where ducks, geese, quail, pheasants, chickens, etc., are raised or housed together (Webster, 2004). Because H5 influenza virus is transmitted primarily in feces, insufficient biosecurity can result in the spread of influenza virus by people or trucks traveling between live markets and industrial poultry farms. Once an influenza virus reaches a commercial poultry farm, which contains a large number of susceptible poultry, the conditions are optimal for rapid viral evolution.

PATHOGENESIS
The role of hemagglutinin and neuraminidase
Although the pathogenicity of influenza viruses is a polygenic trait, HA plays a pivotal role in determining whether the pathogenicity is high or low (Webster and Rott, 1987). The cleavage site of the HA of nonpathogenic avian H5 influenza viruses in the aquatic bird reservoirs usually contains few basic amino acids, but during rapid evolution of H5 and H7 influenza viruses, it acquires multiple basic amino acids. These basic amino acids are thought to be acquired through a 'stuttering' action by the polymerase during replication (Perdue et al., 1996). The HAs of the highly pathogenic influenza viruses, all of which possess

multiple basic amino acids at their cleavage site, are cleaved by the ubiquitous subtilisin-like proprotein convertases furin and PC6 (Stieneke-Grober et al., 1992; Horimoto and Kawaoka 1994). The ease of cleavage conferred by the presence of multiple basic amino acids results, after infection, results in systemic spread of the virus, its replication in all organs, and death of chickens, turkeys, and most gallinaceous birds. The cleavability of HA is also affected by carbohydrate-bearing residues in the vicinity of the HA cleavage site, the presence of which blocks access by activating enzymes (Kawaoka et al., 1984), and by mutations that result in the loss of carbohydrate-bearing residues near the cleavage site or in the lengthening of the cleavage-site sequence, either of which events increases the cleavability of the HA (Horimoto and Kawaoka, 1994). The HA of nonpathogenic influenza viruses in domestic poultry, in contrast, is typically less easily cleavable or less accessible to the activating enzymes. These nonpathogenic viruses are therefore unable to enter cells and effect systemic infection, and they replicate predominantly in the respiratory and intestinal tracts. Signs of disease resulting from infection with a virus of low pathogenicity can vary from imperceptible to severe, depending on the presence of co-infecting bacteria or viruses (or both).

An alternative strategy by which a nonpathogenic avian influenza virus can become highly pathogenic is recombination. Recombination, which involves the insertion of genetic material from another virus or host gene has been detected in H7 but not in H5 influenza viruses. High pathogenicity was recently acquired by recombination in, for example, A/chicken/Chile/4322/02 (H7N3), in which 30 nucleotides were inserted into the region of the HA gene encoding the cleavage site. The inserted sequence was similar to a portion (positions 1268–1297) of the nucleoprotein (NP) gene of A/gull/Maryland/704/77 (H13N6) (Suarez et al., 2004). The resulting highly pathogenic H7N3 Chilean variant was derived from a virus of low pathogenicity and was remarkable in having only three basic amino acids at the HA cleavage site.

High pathogenicity has also been acquired in the laboratory by recombination occurring during in vitro passaging of A/turkey/Oregon/71 (H7N3) and A/seal/Massachusetts/1/80 (H7N7) (Khatchikian et al., 1989; Orlich et al., 1994). Preliminary reports suggest that the highly pathogenic Canadian H7N3/04 virus that originated in British Columbia became highly pathogenic as a result of recombination acquiring an insertion of 21 nucleotides from its own matrix gene segment.

There is increasing evidence of a balance between HA and NA activity in influenza virus replication and pathogenesis. A deletion in the region of the NA gene encoding the protein's stalk has frequently been found in influenza viruses from the aquatic bird reservoir that have recently adapted to terrestrial domestic poultry (Matrosovich et al., 1999). The shortening of the NA stalk that results from such deletion is associated with reduced enzymatic activity of the NA and consequent inefficient release of progeny virus from infected cells. Compensation for low NA activity is achieved by increasing the glycosylation of the globular head of the HA: this change reduces the affinity of the virus for cellular receptors and thereby facilitates the release of the progeny virus by preventing its aggregation at the cell surface (Mitnaul et al., 2000). Another functional feature of NA, which is unresolved but is conserved among some avian virus subtypes, is its ability to bind to sialic acid at a second active site (the HA site) (Laver et al., 1984; Kobasa et al., 1997). Mutations at this site occur after transmission of the virus to pigs and humans; this observation raises the possibility that the HA site in NA plays a role in host adaptation.

Receptor specificity and cell tropism

Avian influenza viruses, including H5 strains, bind to cell surface glycoproteins containing terminal sialyl-galactosyl residues with 2-3 linkage [Neu5Ac(α2-3)Gal], whereas human influenza viruses bind to residues with 2-6 linkages (Paulson, 1985). The avian H5N1 viruses that infected humans in 1997, however, retained their avian specificity (α2-3 linkage) (Matrosovich et al., 1999). How did they bind to human cells and cause infection? The available evidence indicates that avian influenza viruses and egg-adapted human influenza viruses infect mainly ciliated cells in the human respiratory tract, but human influenza viruses infect mainly nonciliated cells (Matrosovich et al., 2004). This finding correlates with the presence of predominantly α2-3 linked (avian type) receptors on ciliated and α2-6 linked (human type) receptors on nonciliated cells. Although avian influenza viruses clearly can infect human airway cells, their further replication is usually restricted. However, why the avian H5N1/97 influenza viruses failed to transmit from human to human (Mounts et al., 1999) is unresolved, as is the reason for the infection of very few humans by these viruses.

The role of the PB2 and NS genes

The pathogenicity of the H5N1/97 viruses has been studied in a number of mammalian models, including mice (Gao et al., 1999; Lu et al., 1999; Tumpey et al., 2000), ferrets (Zitzow et al., 2002), pigs (Shortridge et al., 1998), and cynomolgus macaques (Rimmelzwaan et al., 2001), but the results are conflicting. Unlike other human and avian influenza A viruses, the human and avian H5N1/97 isolates do not require adaptation to be pathogenic in mice and are categorized as viruses that either are of high pathogenicity and replicate systemically (including in the brain) or are of low pathogenicity and replicate only in the lungs and upper respiratory tract of mice (Gao et al., 1999; Lu et al., 1999). In general, pathogenicity of H5N1/97 isolates in mice corresponds to severity of disease in humans (Katz et al., 2000). High virulence and systemic replication of A/Hong Kong/483/97 (H5N1) virus in mice is dependent on the presence of a lysine at position 627 in PB2 (Hatta et al., 2001). The amino acid at this position in PB2 determines the efficiency of viral replication in mouse (not avian) cells, but this amino acid does not determine viral tropism toward different organs in the mouse (Shinya et al., 2004). Although structural aspects of PB2 and HA are associated with the pathogenicity of H5N1/97 viruses in mice, other genotypes of H5N1 viruses that emerged in 2001 and are neurotropic in mice do not posses a lysine at residue 627 of PB2. These highly pathogenic H5N1 variants have mutations in all gene segments except those encoding the PB1, NP, and NS1 proteins but have no common set of mutations (Lipatov et al., 2003). Therefore multiple gene constellations and residues determine the pathogenicity of influenza viruses in mice.

Despite differential pathogenicity of H5N1/97 influenza viruses in mice, all of these viruses cause systemic infection in ferrets (Zitzow et al., 2002). In studies with cynomolgus macaques they caused severe respiratory disease but did not spread systemically (Rimmelzwaan et al., 2001), and in studies with pigs the H5N1/97 viruses replicated (to modest titers) only in the respiratory tract and caused no disease signs (Shortridge et al., 1998). All of these findings indicate that multiple gene constellations are involved in influenza virus pathogenicity and that the outcome of infection is host dependent.

Studies on the NS gene of Hong Kong H5N1/97 viruses have shown that this gene has a role in the determination of high pathogenicity of the viruses in mammals. Seo et al. (2002) demonstrated that the NS gene of H5N1/97 virus dramatically increases the pathogenicity of A/PR/8/34 (H1N1) virus in pigs. These authors hypothesized that the NS gene of H5N1/97 viruses confers resistance to the antiviral effects of interferons (IFNs) and tumor necrosis factor alpha (TNF-α). The NS gene segment of influenza A viruses encodes two proteins: NS1 and nuclear export protein. NS1 contributes to viral pathogenesis by allowing the virus to disarm the host's IFN defense system in multiple ways (reviewed by Garcia-Sastre, 2001, 2002; Krug et al., 2003). In mice, A/WSN/33 (H1N1) reassortants with the complete NS gene or with only the NS1 segment of the NS gene from the 1918 pandemic influenza virus are less pathogenic than the original A/WSN/33 virus (Basler et al., 2001). On the other hand, a virus containing the NS gene of the 1918 pandemic strain blocks the expression of IFN-regulated genes in human lung cells more efficiently than does its parental A/WSN/33 virus (Geiss et al., 2002).

In contrast, a study in which primary human monocyte-derived macrophages served as an in vitro model showed that transcription of TNF-α and IFN-β genes is induced by viruses containing the NS gene of H5N1/97 viruses as well as particular constellations of their internal genes rather than the genes encoding the surface proteins (Cheung et al., 2002).

Studies of reassortant viruses containing the NS gene of the highly pathogenic H5N1/97 virus in two mammalian models support the theory that the NS gene of H5N1/97 viruses can confer and support high pathogenicity when it is inserted into a virus (e.g., A/PR/8/34) that is pathogenic in one model (in this case, mice) and nonpathogenic in another (i.e., pigs) (Seo et al., 2002). The high pathogenicity of viruses containing the NS gene of the H5N1/97 virus may result from induction of a cytokine imbalance; this view is supported by findings from the detailed pathological examination of two persons who died of H5N1 pneumonia in Hong Kong in 1997 (Yuen et al., 1998; To et al., 2001). A cytokine imbalance could explain, at least partially, the unusual severity of illness caused by infection with H5N1/97 influenza virus.

It is unlikely that the products of the NS gene of the H5N1/97 virus are unique in causing a cytokine imbalance: other gene products will certainly play a role, especially in different hosts. Studies of human H5N1/03 isolates, which possess a different NS gene from that of H5N1/97 viruses, show that these human influenza viruses also induce the expression of high levels of TNF-α and IFN-induced protein 10 (IP-10) in infected patients and in a cell model in vitro (Guan et al., 2004; Peiris et al., 2004). Therefore, the ability to induce a cytokine imbalance is probably an important factor in influenza virus pathogenicity and a polygenic property that is substantially influenced by host factors.

H5 INFLUENZA VIRUSES IN THE AMERICAS

Multiple H5N2 influenza viruses have transmitted from wild aquatic birds to domestic poultry in North America in the past 70 years and caused outbreaks of influenza. The most extensively studied of these highly pathogenic viruses include the strains causing the outbreaks in Pennsylvania from 1980 to 1989 (Ck/PA/83) and the outbreak in Mexico from 1993 to the present (Ck/Mex/93). Both were different introductions and originated from American-lineage viruses (Suarez, 2000). The Ck/PA/83 virus spread locally and was eradicated by culling, whereas the Ck/Mex/83 virus was widespread when first detected and began as a nonpathogenic strain before evolving into a highly pathogenic strain (Horimoto et al., 1995; Garcia et al., 1997). The Mexican outbreak was controlled by selective culling and vaccination of domestic poultry, but the virus has not been eradicated from the wider region, and descendants of the Ck/Mex/83 H5N2 virus continue to infect domestic poultry in Central America.

Analysis of the Ck/PA/83 (H5N2) phylogeny indicates that the virus may have circulated in domestic poultry for some time before it was detected (Suarez, 2000). Two features of this virus may partly explain why its high pathogenicity was masked. The initial H5N2 isolate possessed an additional carbohydrate-bearing residue that masked the multiple basic amino acids at the cleavage site of the HA (Kawaoka et al., 1984; Deshpande et al., 1987). In addition, the initial isolates of highly pathogenic Ck/PA/83 had defective interfering (di) viruses cocirculating with them (Bean et al., 1985). The presence of di viruses in chickens can mask high pathogenicity by inhibiting the multiplication of the highly pathogenic virus (Chambers and Webster, 1987). Ck/PA/83 H5N1 was spread in Pennsylvania by humans; no evidence was found for spread by wildlife (Nettles et al., 1985). Throat and nasal swabs from people involved in the culling of the infected flocks of poultry were positive for H5N2 virus when these persons left the infected poultry houses, but negative the next morning. This finding indicates that this virus lacked replicative capacity in humans (Bean et al., 1985). An important epidemiological finding, however, was that live-poultry markets are involved in the transmission and maintenance of highly pathogenic avian influenza viruses (Senne et al., 1992).

The Ck/Mex/93 virus had spread, in a nonpathogenic form, throughout most Mexican commercial poultry before it was recognized and before it evolved into a highly pathogenic strain. The use of vaccine to control this highly pathogenic Ck/Mex/93 H5N2 strain reduced its spread. However, because some of the H5N2 vaccines used in Mexico were not standardized, they were not uniformly efficacious (Garcia et al., 1998). This lack of efficacy probably contributed to the selection of variants and to the endemicity of highly pathogenic H5N2 virus in the Central American region, including Guatemala in 2000 and El Salvador in 2001. The available evidence suggests that a single strain of H5N2 evolved into several distinct lineages in the region (Lee et al., 2004).

H5N1 IN ASIA
The A/goose/Guangdong/1/96 (H5N1) precursor viruses
The precursor of the H5N1 influenza viruses that emerged in humans in Hong Kong in 1997 was A/goose/Guangdong/1/96 (H5N1). This highly pathogenic avian influenza virus killed 40% of geese on a farm in Guangdong, China in the summer of 1996 (Tang et al., 1998) and was the source of the HA gene in the A/Hong Kong/156/97 (H5N1) virus that transmitted to a child in Hong Kong in 1997 (de Jong et al., 1997; Xu et al., 1999). It is unusual for influenza viruses to kill aquatic birds such as ducks (Alexander et al., 1986), but little is known about their lethality to geese. Experimental inoculation of geese with A/goose/Guangdong/1/96-like H5N1 viruses isolated in Hong Kong in 1999 (A/goose/Hong Kong/437-4/99) resulted in the death of the inoculated birds, systemic spread of the virus in the inoculated birds, and transmission of the virus to contact birds (geese, ducks, quails, and chickens) (Webster et al., 2002). The ducks showed no disease signs, but all contact quails and chickens died with systemic infection (Webster et al., 2002).

The importance of the A/goose/Guangdong/1/96-like H5N1 viruses as a continuing source of virus in southeastern China has been very much underestimated in the re-emergence of H5N1 viruses in Hong Kong poultry markets in 2001, 2002, and 2003 (Figure 1). The A/goose/Guangdong/1/96-like H5N1 viruses continued to circulate in southern China after the eradication, in 1998, of the human and poultry H5N1 viruses (Cauthen et al., 2000; Webster et al., 2002). Additionally, this H5N1 goose virus reassorted with influenza viruses in domestic ducks and transmitted to ducks in the region (Chen et al., 2004). These goose H5N1 viruses were the source of both the HA and the NA genes of the H5N1 viruses that became dominant throughout eastern Asia in 2004 (see below).

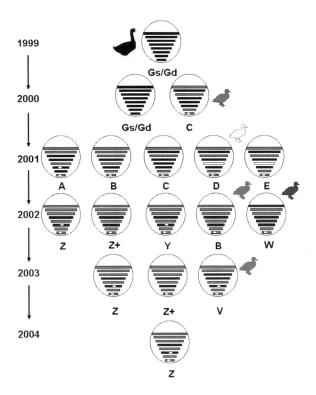

Figure 10.1 The genesis of different genotypes of H5N1 influenza viruses. Reassortment between influenza viruses is proposed to generate viruses with different gene constellations. A/goose/ Guangdong/1/96 (Gs/Gd/96)-like viruses reassorted with viruses from wild aquatic birds, and multiple H5N1 genotypes appeared in Hong Kong poultry markets in 2001. Some 2001 H5N1 genotypes may have been transmitted from domestic poultry back to the wild aquatic bird reservoir, where the next reassortant events may have occurred. As a result, multiple H5N1 genotypes, mostly different from those of 2001, were isolated from domestic poultry and wild birds in Hong Kong in 2002 and 2003. The eight gene segments schematically shown in each virus particle encode (top to bottom) polymerase complex proteins (PB2, PB1, and PA), HA, nucleo-protein (NP), NA, matrix protein (M), and nonstructural (NS) protein. Shading indicates virus lineages. A gap in the NA or NS gene segment denotes a deletion. Deletions within the NA gene appear to result from adaptation to poultry, although the exact role of this phenomenon is unclear. NS gene deletions in H5N1 viruses have recently been observed, but the biological significance of these deletions is unknown. The mechanism of selection of the Z genotype is unresolved. (Modified from Li et al., 2004.). To view this figure in colour, go to http://www.horizonpress.com/ hsp/supplementary/flu.

The human H5N1 1997 viruses

The HA on the H5N1 virus that emerged in a child in May 1997 was first isolated in April 1997 when some 2,000 chickens on a farm in the New Territories of Hong Kong died of H5N1 influenza (Simms et al., 2003). The virus was almost 100% lethal on the first farm and about 70% lethal on the second and third farms to which it spread. At that time there were reports of a high mortality rate among poultry in Guangzhou, but the causative agent was never identified. The index case of H5N1 infection in humans occurred in a 3-year-old boy in May 1997 and was temporarily related to the outbreak of H5N1 infection on the third farm (Simms et al., 2003). The child died of acute respiratory distress with pneumonia after being treated with aspirin and developing Reye syndrome and multiorgan failure (Yuen et al., 1998). No further cases of human infection occurred until November 1997, and by the end of December that year, 17 additional cases and a total of 8 deaths had occurred (Yuen et al., 1998).

Most of those who died were elderly persons; the only child younger than 5 years who died was the child with the index case of infection (Yuen et al., 1998). Symptoms in the severe cases included respiratory distress syndrome affecting the upper and lower respiratory tracts, gastrointestinal manifestations, elevated liver enzymes, and renal failure. Post-mortem examination of two patients who died of pneumonia revealed elevated cytokine levels (To et al., 2001).

The transmission of H5N1 virus to humans in Hong Kong in 1997 was the first confirmed avian-to-human transmission of influenza virus. Virologic and epidemiologic studies revealed that the live-poultry markets in Hong Kong were the source of the virus (Shortridge et al., 1998; Bridges et al., 2000). The detection of highly pathogenic avian H5N1 virus in each of the live-poultry markets tested in Hong Kong in 1997 was surprising, because no poultry were dying in any of the markets. The probable explanation for this lack of deaths is that H9N2 viruses were cocirculating in the poultry and providing cross-reactive cell-mediated immunity. In-vivo and in-vitro studies (Seo and Webster, 2001) demonstrated that cell-mediated immunity induced by H9N2 influenza viruses that share the six internal genes of H5N1/97 provided protection from overt H5N1-mediated disease but permitted shedding of H5N1 virus by these birds. Thus cell-mediated immunity can prevent clinical disease and permit the shedding of a sufficient amount of H5N1 virus to infect humans in the markets.

The ongoing epidemic ceased when the entire poultry population of Hong Kong was culled in December 1997. The epidemic curve fell to zero after culling of the poultry. This event dramatically illustrates the role of poultry and poultry markets in the spread of H5N1 influenza viruses to humans. These viruses had not acquired the capacity for human-to-human transmission. No additional cases of human H5N1 infections were detected, and this genotype of H5N1 has not been isolated since.

Emergence of H5N1 viruses with novel properties

Although from 1997 through early 2002 the H5N1 influenza viruses in Asia acquired different constellations of internal genes, their HA and NA remained antigenically conserved (Guan et al., 2002; Chen et al., 2004). However, in 2002 the H5N1 virus underwent marked antigenic drift (Guan et al., 2004; Sturm-Ramirez et al., 2004). The resulting viruses were highly pathogenic in chickens but nonpathogenic in ducks. This situation changed dramatically in November 2002 when H5N1 viruses were isolated from dead wild birds in Hong Kong. The most remarkable property of the H5N1 genotype from late 2002 was its high pathogenicity for ducks and other aquatic birds (Figure 2): a property rarely found in nature. The previous event of significance to aquatic birds had occurred in 1961, when A/tern/South Africa/61 (H5N3) killed terns (Becker, 1966).

In early February 2003, H5N1 virus genetically similar to the virus killing aquatic birds re-emerged in a family in Fujian, China. The daughter died of a respiratory infection of undiagnosed cause while visiting Fujian; the father and son developed severe respiratory illness after their return to Hong Kong. The father died and the son recovered. Infection with H5N1 influenza virus was confirmed in father and son.

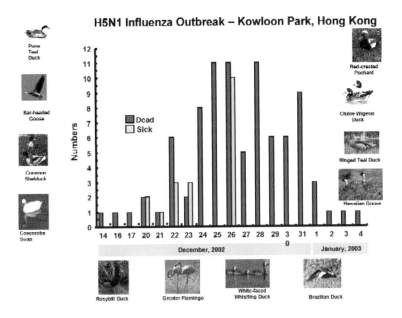

Figure 10.2 H5N1 avian influenza virus outbreak in Kowloon Park, Hong Kong SAR, China. Shown are the numbers of birds identified as sick or dead on each calendar day. H5N1 influenza virus infection was confirmed by virus isolation from the affected birds. The species killed included Rosybill Pochard *(Netta peposaca)*, Coscoroba Swan *(Coscoroba coscoroba)*, Chestnut-breasted Teal *(Anas castanea)* Red-crested Pochard *(Netta rufina)*, Chiloe Wigeon *(Anas sibilatrix)*, Brazilian Teal *(Amazonetta basiliensis)*, Greater Flamingo *(Phoenicopterus rubber)*, Falcated Teal *(Anas falcata)*, White-faced Whistling Duck *(Dendrocygna viduata)*, Ringed Teal *(Callonetta leucophrys)*, Common Shelduck *(Tadorna tadorna)*, Hawaiian Goose *(Branta sandvicensis)*, Bar-headed Goose *(Anser indicus)*. To view this figure in colour, go to http://www.horizonpress.com/hsp/supplementary/flu.

In 2001 and 2002 there was a multiplicity of different H5N1 genotypes cocirculating in poultry in southeastern China (Figure 1). At least nine genotypes were circulating in southern China in 2002 (Li et al., 2004). The H5N1 virus that infected humans in 2003 (A/HK/212/03 [H5N1]) was the forerunner of the H5N1 virus that would become dominant throughout Asia in 2004. These human isolates were of the Z genotype but without the shortened stalk of the NA and were designated genotype Z^+ (Figure 1).

The evolution of H5N1 viruses in Asia

The highly pathogenic avian H5N1 influenza viruses that were widespread among ducks in the coastal provinces of southern China from 1999 to 2002 (Chen et al., 2004) were probably the source of influenza viruses in gallinaceous poultry in southern China. Virologic surveillance in live-poultry markets in Guangdong province of southern China and in Hong Kong in 2000, with expansion to Hunan and Yunnan provinces in 2002, established the presence of highly pathogenic avian H5N1 viruses in aquatic birds in the poultry markets in 2000 but not in terrestrial poultry (Li et al., 2004). By 2001, highly pathogenic avian H5N2 viruses were isolated from both aquatic and terrestrial poultry in southern China and in Hong Kong. The rate of isolation of highly pathogenic avian H5N1 influenza viruses from terrestrial poultry in Hong Kong and southern China increased in 2002 and 2003.

However, in 2004, no H5N1 viruses were isolated from terrestrial poultry and only one isolate was obtained from an aquatic bird in Hong Kong. This situation contrasts with the continuing isolation of highly pathogenic avian influenza viruses from both aquatic and terrestrial poultry in southern China in 2004.

Prospective virologic surveillance in aquatic birds and in live-poultry markets during this period (2000–2004) revealed a seasonal pattern of appearance of the H5N1 viruses that corresponded with the onset of cooler weather (Figure 3). The H5N1 viruses were isolated only from aquatic birds in the live-poultry markets in 2000, but by 2002 they were isolated from both aquatic and terrestrial poultry. The frequency of isolation of H5N1 from terrestrial poultry increased in 2002 and continued to increase each autumn through 2003 and 2004.

The avian and human H5N1 influenza epidemic of 2004

During 2003, the Z genotype became dominant in southern China: 60 of 62 isolates from poultry were of the Z genotype (Li et al., 2004). The unprecedented magnitude of the bird flu epidemic in Asian countries in 2004 – when H5N1 virus was infecting birds in China, Japan, South Korea, Thailand, Vietnam, Indonesia, Cambodia, and Laos – resulted in the destruction of hundreds of millions of poultry, mainly chickens. In most of these countries, outbreaks of highly lethal H5N1 avian influenza were confined to poultry, but in at least two countries the virus transmitted to humans, and most of the persons infected died (20 deaths in Vietnam and 11 deaths in Thailand). (See Postscript p. 294.)

The time course of the spread of the dominant Z genotype across Asia is not known. There were unofficial reports of highly pathogenic H5N1 virus in poultry in Indonesia, Vietnam, and Thailand in June 2003, but these cases were not reported officially until the disease got out of control and humans were infected in December 2003 and January 2004. Analysis of the clinical features of the first 10 human cases (Hein et al., 2004) established that children (average age 13.7 years) were at the highest risk of lethal infection resulting in acute respiratory distress syndrome, high fever, lymphopenia, and diarrhea. The human viruses in Vietnam and Thailand were most closely related phylogenetically to H5N1 viruses isolated from poultry in Vietnam and Thailand and to H5N1 viruses isolated from wild birds in Hong Kong in 2002 (Li et al., 2004). In contrast, the H5N1 isolates from Indonesia were shown to belong to a separate group of viruses more closely related to H5N1 isolates from Yunnan, China (Li et al., 2004). It is noteworthy that the H5N1 viruses in Indonesia did not infect humans. (See Postscript p. 294.)

The human isolates from Vietnam, A/Vietnam/1203/04 (H5N1), and Thailand were antigenic drift variants of the 2003 human isolates from Hong Kong (A/Hong Kong/213/03 [H5N1]). Most had acquired a potential N-linked glycosylation site at amino acid positions 154 to 156 of HA. Each of the Z-genotype viruses from humans and poultry in Vietnam and Thailand that was tested for pathogenicity in chickens and ducks was highly pathogenic. In ferrets, all of the human isolates and some of the avian isolates were remarkably pathogenic: they spread systemically and caused weight loss and hind-leg paralysis (Govorkova et al., 2005). A subset of the avian H5N1 isolates replicated in ferrets but were not neurotropic and did not spread systemically. Virologic and serologic studies in pigs in Vietnam have revealed that the H5N1 virus can infect pigs but has not caused widespread infection in the regions tested (Y.K. Choi unpublished).

By June 2004, each of the countries that had had H5N1 influenza virus infections in poultry considered that their domestic poultry were free of H5N1 virus. However, the H5N1 virus resurged in July 2004 in poultry in Vietnam, Thailand, and China, and the first cases

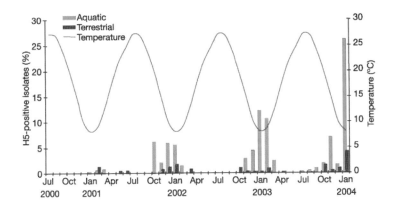

Figure 10.3 Seasonality of the isolation of avian H5N1 viruses from domestic poultry from July 2000 through January 2004 in mainland China. The mean monthly temperature in southern China (approximated from the monthly average temperatures of the cities of Changsha, Kunming, and Xiamen) is shown for reference. To view this figure in colour, go to http://www.horizonpress.com/hsp/supplementary/flu.

of H5N1 infection in poultry appeared in Malaysia in August 2004. Three people died of H5N1 infection in Vietnam in August 2004. The resurgence of H5N1 virus in the summer months was of particular concern, and the possibility was considered that the clean-up and disinfection done in January 2004 had been incomplete or that the H5N1 virus had become established in birds (perhaps ducks) in the region. Additional studies of the ecology of H5N1 viruses in Asia are yet to be done.

The spread of H5N1 influenza across Asia in 2003 and 2004

The cause of the rapid spread of H5N1 influenza across nine countries in eastern Asia in 2003 and 2004 is unresolved – humans or migrating birds could have been the culprit. Let us first consider the more likely culprit – humans. The poultry industry in Asia is huge and integrated: a number of firms have branches in China, Vietnam, Thailand, and Indonesia. Do people, poultry, or fomites move viruses between these countries? In previous outbreaks of infection with highly pathogenic H5 or H7 viruses in different countries, the spread of virus was directly or indirectly attributable to humans. If humans were the spreading agent in Asia, we might expect a single genotype to dominate. However, the above considerations indicate that although the Z genotype is dominant, the viruses in Indonesia are different from those in Vietnam and Thailand (Li et al., 2004), and this difference argues against the spread of the virus from a single source by humans.

The seasonal pattern of appearance of H5N1 virus in poultry in southeastern China corresponds to the autumn migration of wild birds. In late November 2002, H5N1 virus was isolated from a dead little egret (*Egretta garetta*) and from a Canada goose (*Branta canadensis*) in a nature park in Hong Kong (Ellis et al., 2004). In mid December 2002 there was a second outbreak of H5N1 infection at another nature park in Hong Kong (Kowloon Park). The virus spread to all of the aquatic birds in Kowloon Park and killed most species there (Ellis et al., 2004; Sturm-Ramirez et al., 2004), including a wide range of exotic birds such as greater flamingos (*Phoenicopterus aruber*). The virus causing this outbreak was first isolated from a grey heron (*Ardea cinerea*). Notably, each lethal outbreak was caused by H5N1 viruses of different genotypes that had not previously been isolated from domestic birds, and this finding implicates wild aquatic birds in the introduction of these viruses.

In addition to being isolated from the wild birds described above, H5N1 virus was also isolated, during 2003, from a black-headed gull (*Larus ridibundus*), from a feral pigeon (*Columba livia*), and from a tree sparrow (*Passer montanus*). Prospective virologic surveillance of other aquatic and passerine birds in Hong Kong during 2003 failed to detect H5N1 virus in other species, but in early 2004, H5N1 virus was isolated from two dead peregrine falcons (*Falco peregrinus*) in Hong Kong. Each of the H5N1 influenza viruses isolated intermittently from wild birds in Hong Kong since 2003 has been of the Z genotype.

The presence of H5N1 virus in some of the wild birds in Hong Kong may be explained by scavenging or feeding by these birds on virus-infected farms. Such might be the case, for example, for the black-headed gull, tree sparrow, feral pigeon, and egret, but this explanation fails for grey herons, which feed on live fish and are very shy of humans. The dead herons in Hong Kong were found on fish farms a considerable distance from poultry farms. However, one could argue that they became infected through contaminated water. It is difficult to explain how peregrine falcon(s) became infected – these birds feed on live small animals caught on the wing. Their diet and feeding habit make them unlikely to have become infected through contact with domestic poultry but they could have been infected from infected small birds (e.g. an infected tree sparrow).

The first outbreak of H5N1 infection in chickens in Japan occurred on January 11, 2004, in Yamaguchi; the second, on February 16, 2004, in Onita; and the third, on February 26, 2004, in Kyoto. These places are widely separated and operate strict quarantine and eradication procedures, but it is difficult to document human movement between them. The occurrence of these outbreaks in mid-winter means it is unlikely that wild migrating birds were responsible for spreading the virus. However, H5N1 viruses were isolated from dead crows (*Corvus brachyrhynchos*) near infected chicken houses, and this fact implicates the local population of wild birds in local and regional spread of the virus.

The presence of H5N1 virus over much of eastern Asia is consistent with a role for migrating birds in the spreading of the virus over large areas; local spread may be mediated by humans or resident birds. However, there is no definitive evidence supporting the role of migrating birds in the spread of H5N1/04 virus. Much of the available evidence is circumstantial; more definitive information is urgently needed from serologic and virologic studies on wild birds in Asia and throughout the world.

The possibility that migrating birds are involved in the spread of the H5N1/04 virus in Asia has led to consideration of culling of migrating birds. This strategy must be strongly discouraged, because it has unknown ecological consequences. Influenza is a noneradicable zoonotic disease that the poultry industry is learning to live with. Wild birds of all kinds must be kept out of poultry farms, and chicken farms must be located separately from duck and pig farms.

Is H5N1 influenza endemic in Asia?

The resurgence of highly pathogenic H5N1 influenza viruses on poultry farms in Thailand, Vietnam, and China in July 2004 and the subsequent infection and death of humans in August 2004 raises the possibility that highly pathogenic avian H5N1 influenza viruses are now endemic in poultry in Asia. The available evidence indicates that H5N1 is endemic in domestic ducks in southern China (Chen et al., 2004), and the above information supports the likelihood that H5N1 virus is now endemic in domestic ducks throughout southern Asia. Although the Z genotype of H5N1 influenza virus has been dominant in Asia, the virus is heterogeneous in its pathogenicity in ducks. Examination of multiple H5N1 isolates of the Z genotype for pathogenicity in ducks reveals that these isolates can be divided into three broad groups: those that kill all inoculated ducks that have neurologic signs, those that kill a percentage of ducks that do not have neurologic signs, and those that replicate but cause no overt disease (K.M. Sturm-Ramirez, unpublished). All of these H5N1 viruses are highly pathogenic in gallinaceous poultry.

SUMMARY AND CONCLUDING REMARKS

H5 influenza is a continuing threat in both the veterinary and human public health settings. The precursor of the highly pathogenic H5 influenza viruses continues to circulate worldwide in a nonpathogenic form in wild aquatic birds. After transmission to land-based poultry, nonpathogenic H5 viruses evolve rapidly and become highly pathogenic. Since 1997, the highly pathogenic Asian H5N1 strains have transferred from geese to domestic ducks and have become increasingly pathogenic for mammals, including humans. There is concern that these highly pathogenic viruses may have transmitted back to migrating birds. Therefore there is a need for increased biosecurity and improvement in sanitation at poultry farms, for the ending of the age-old practice of mixed poultry and pig farming, and for the modification of live-poultry markets, which are the recognized breeding ground of influenza viruses.

Continuing evolution by reassortment and mutation of the Asian H5N1 viruses has resulted in the emergence of a dominant Z genotype that is highly transmissible in poultry and is becoming increasingly transmittable to humans. These Z-genotype viruses have killed 31 humans, and in poultry, ferrets, and mice they can spread systemically and cause neurologic disease. Although the time-honored strategy for the control of highly pathogenic avian influenza viruses is their eradication through culling, this strategy may not be economically feasible in some countries, and the adoption of a vaccination and culling program as a control measure remains controversial. The advent of reverse genetics has resulted in significant progress in the understanding of pathogenesis and in vaccine development for humans. However, agricultural vaccines are not standardized on the basis of antigen content, and the use of substandard vaccines raises concern about the possibility of enhanced selection of variants having an increased host range.

Pandemic preparedness for influenza is necessary – there will eventually be another human influenza pandemic – and the rapid evolution displayed by H5N1 influenza viruses in Asia makes these viruses prime candidates to cause that pandemic.

Acknowledgements

We thank Janet R. Davies for editorial assistance and Julie Groff for illustrations. We also thank Drs. Richard Webby, Elena Govorkova, and Erich Hoffmann and the Influenza Group at St. Jude Children's Research Hospital; Drs. Malik Peiris and Yi Guan at the University of Hong Kong; Dr. Amin Soebandrio, Assistant Deputy Minister for Medical & Health Services, Ministry of Research & Technology, Jakarta, Indonesia; Dr. Nguyen Tien Dung, Ministry of Agriculture & Rural Health Development, National Institute of Veterinary Research, Hanoi, Vietnam; and Dr. Chantanee Buranathai, Division of Veterinary

Epidemiology, Bureau of Disease Control and Veterinary Services, Department of Livestock Development, Bangkok, Thailand. Influenza research at St. Jude Children's Research hospital is supported by NIAID contract AI95357 and Cancer Center Support (CORE) grant CA-21765 from the National Institutes of Health and by the American Lebanese Syrian Associated Charities (ALSAC). At the University of Hong Kong, influenza studies were supported by contract AI95357 from the U.S. National Institutes of Health, by Wellcome Trust grant 067072/D/02/Z, and by the Ellison Medical Foundation. We thank Carol Walsh and Francis Wong for manuscript preparation and administrative assistance.

Postscript

By the time this article went to press the H5N1 influenza virus was confirmed in 117 humans with 41 deaths in Vietnam, 12 deaths in Thailand, 4 deaths in Cambodia, and at least 3 deaths in Indonesia. The highly pathogenic H5N1 virus has been isolated from a large die off of bar-headed geese (*Anser indicus*) in Qinghai Lake, China and the virus has continued to spread westwards to Southern Russia, Kazakhstan, and Mongolia with possible isolations from poultry farms in Turkey in Mid October 2005. The highly pathogenic H5N1 avian influenza virus is now endemic in wild migratory birds; wild ducks are of major concern for they can act as 'Trojan horses' and show no disease signs. It is probable that global spread of this virus will occur and that pandemic preparedness is fully merited.

References

Alexander, D. J. (2000). A review of avian influenza in different bird species. Vet. Microbiol. *77*, 3–13.

Alexander, D. J., Parsons, G., and Manvell, R. J. (1986). Experimental assessment of the pathogenicity of eight avian influenza A viruses of H5 subtypes for chickens, turkeys, ducks and quail. Avian Pathol. *15*, 647–662.

Banks, J., Speidel, E. S., Moore, E., Plowright, L., Piccirilo, A., Capua, I., Cordioli, P., Fioretti, A., and Alexander, D. J. (2001). Changes in the hemagglutinin and the neuraminidase genes prior to the emergence of highly pathogenic H7N1 avian influenza viruses in Italy. Arch. Virol. *146*, 963–973.

Basler, C., Reid, A. H., Dybing, J. K., Janczewski, T. A., Fanning, T. G., Zheng, H., Salvatore, M., Perdue, M. L., Swayne, D. E., Garcia-Sastre, A., Palese, P., and Taubenberger, J. K. (2001). Sequence of the 1918 pandemic influenza virus nonstructural gene (NS) segment and characterization of recombinant viruses bearing the 1918 NS genes. Proc. Natl. Acad. Sci. USA *98*, 2746–2751.

Bean, W. J., Kawaoka, Y., Wood, J. M., Pearson, J. E., and Webster, R. G. (1985). Characterization of virulent and avirulent A/Chicken/Pennsylvania/83 influenza A viruses: potential role of defective interfering RNAs in nature. J. Virol. *54*, 151–160.

Becker, W. B. (1966). The isolation and classification of tern virus: influenza virus A/Tern/South Africa/61. J. Hygiene *64*, 309–320.

Bridges, C. F., Katz, J. M., Seto, W. H., Chan, P. K., Tsang, D., Ho, W., Mak, K. H., Lim, W., Tam, J. S., Clarke, M., Williams, S. G., Mounts, A. W., Bresee, J. S., Conn, L. A., Rowe, T., Hu-Primmer, J., Abernathy, R. A., Lu, X., Cox, N. J., and Fukuda, K. (2000). Risk of influenza A (H5N1) infection among health care workers exposed to patients with influenza A (H5N1), Hong Kong. J. Infect. Dis. *181*, 344–348.

Campitelli, L., Mogavero, E., De Marco, M. A., Delogu, M., Puzelli, S., Frezza, F., Facchini, M., Chiapponi, C., Foni, E., Cordioli, P., Webby, R., Barigazzi, G., Webster, R. G., and Donatelli, I. (2004). Interspecies transmission of an H7N3 influenza virus from wild birds to intensively reared domestic poultry in Italy. Virology *323*, 24–36.

Cauthen, A. N., Swayne, D. E., Schultz-Cherry, S., Perdue, M. L., and Suarez, D. L. (2000).

Continued circulation in China of highly pathogenic avian influenza viruses encoding the hemagglutinin gene associated with the 1997 H5N1 outbreak in poultry and humans. J. Virol. *74*, 6592–6599.

Chambers, T. M. and Webster, R. G. (1987). Defective interfering virus associated with A/Chicken/Pennsylvania/83 influenza virus. J. Virol. *61*, 1517–1523.

Chen, H., Deng, G., Li, Z., Tian, G., Li, Y., Jiao, P., Zhang, L., Liu, Z., Webster, R. G., and Yu, K. (2004). The evolution of H5N1 influenza viruses in ducks in southern China. Proc. Natl. Acad. Sci. USA *101*, 10452–10457.

Cheung, C. Y., Poon, L. L. M., Lau, A. S., Luk, W., Lau, Y. L., Shortridge, K. F., Gordon, S., Guan, Y., and Peiris, J. S. M. (2002). Induction of proinflammatory cytokines in human macrophages by influenza A (H5N1) viruses: a mechanism for the unusual severity of human disease? Lancet *360*, 1831–1837.

de Jong, J. C., Claas, E. C., Osterhaus, D., Webster, R., and Lim, W. (1997). A pandemic warning? Nature *389*, 554.

Deshpande, K., Fried, V. A., Ando, M., and Webster, R. G. (1987). Glycosylation affects cleavage of an H5N2 influenza virus hemagglutinin and regulates virulence. Proc. Natl. Acad. Sci. USA *84*, 36–40.

Ellis, T. M., Bousefield, R. B., Bissett, L. A., Dyrting, K. C., Luk, G. S. M., Tsim, S. T., Sturm-Ramirez, K., Webster, R. G., and Guan, Y. (2004). Investigation of outbreaks of highly pathogenic H5N1 avian influenza in waterfowl and wild birds in Hong Kong in late 2002. Avian Pathology *33*, 1–14.

Fouchier, R. A. M., Olsen, B., Bestebroer, T. M., Herfst, S., van der Kemp, L., Rimmelzwaan, G. F., and Osterhaus, A. D. M. E. (2003). Influenza A virus surveillance in birds in Northern Europe in 1999 and 2000. Avian Dis. *47*, 857–860.

Gao, P., Watanabe, S., Ito, T., Goto, H., Wells, K., McGregor, M., Cooley, A. J., and Kawaoka, Y. (1999). Biological heterogeneity, including systemic replication in mice, of H5N1 influenza A virus isolates from humans in Hong Kong. J. Virol. *73*, 3184–3189.

Garcia, A., Johnson, H., Srivastava, D. K., Jayawardene, D. A., Wehr, D. R., Webster, R. G. (1998). Efficacy of inactivated H5N2 influenza vaccines against lethal A/Chicken/Queretaro/19/95 infection. Avian Dis *42*, 248–256.

Garcia, M., Suarez, D. L., Crawford, J. M., Latimer, J. W., Slemons, R. D., Swayne, D. E., and Perdue, M. L. (1997). Evolution of H5 subtype avian influenza A viruses in North America. Virus Res. *51*, 115–124.

Garcia-Sastre, A. (2001). Inhibition of interferon-mediated antiviral responses by influenza A viruses and other negative-strand RNA viruses. Virology *279*, 375–384.

Garcia-Sastre, A. (2002). Mechanisms of inhibition of the host interferon ?/?-mediated antiviral response by viruses. Microbes Infect. *4*, 647–655.

Geiss, G. K., Salvatore, M., Tumpey, T. M., Carter, V. S., Wang, X., Basler, C. F., Taubenberger, J. K., Bumgarner, R. E., Palese, P., Katze, M. G., and Garcia-Sastre, A. (2002). Cellular transcriptional profiling in influenza A virus-infected lung epithelial cells: The role of the nonstructural NS1 protein in the evasion of the host innate defense and its potential contribution for pandemic influenza. Proc. Natl. Acad. Sci. USA *99*, 10736–10741.

Govorkova, E. A., Rehg, J. E., Krauss, S., Yen, H. L., Guan, Y., Peiris, J. S. M., Dung, N. T., Hahn, N. T. H., Puthavathana, P., Long, H. T., Buranthai, C., Lim, W., Webster, R. G., and Hoffmann, E. (2004). Lethality to ferrets of H5N1 influenza viruses isolated from humans and poultry in 2005. J. Virol. *79*, 2191–2198.

Guan, Y., Peiris, J. S. M., Lipatov, A. S., Ellis, T. M., Kyrting, K. C., Krauss, S., Zhang, L. J., Webster, R. G., and Shortridge, K. F. (2002). Emergence of multiple genotypes of H5N1 avian influenza viruses in Hong Kong SAR. Proc. Natl. Acad. Sci. USA *99*, 8950–8955.

Guan., Y., Poon, L. L. M., Cheung, C. Y., Ellis, T. M., Lim, W., Lipatov, A. S., Chan, K. H., Sturm-Ramirez, K. M., Cheung, C. L., Leung, Y. H. C., Yuen, K. Y., Webster, R. G., and Peiris, J. S. M. (2004). H5N1 influenza:

a protean pandemic threat. Proc. Natl. Acad. Sci. USA *101*, 8156–8161.

Hatta, M. P., Gao, P., Halfmann, P., and Kawaoka, Y. (2001). Molecular basis for high virulence of Hong Kong H5N1 influenza A viruses. Science *293*, 1840–1842.

Hein, T. T., Liem, N. R., Dung, N. T., San, L. T., Mai, P. P., Chau, N. V., Suu, P. T., Dong, V. C., Mai, L. T. Q., Thi, N. T., Khoa, D. B., Phat, L. P., Truong, N. T., Long, H. T., Tung, C. V., Giang, L. T., Tho, N. D., Nga, L. H., Tien, N. T. K., San, L. H., Tuan, L. V., Dolecek, C., Thanh, T. T., de Jong, M., Schultsz, C., Cheng, P., Lim, W., Horby, P., Farrar, J., and The World Health Organization International Avian Influenza Investigative Team (2004). Avian influenza A (H5N1) in 10 patients in Vietnam. N. Engl. J. Med. *350*, 1179–1188.

Horimoto, T. and Kawaoka, Y. (1994). Reverse genetics provides direct evidence for a correlation of hemagglutinin cleavability and virulence of an avian influenza A virus. J. Virol. *68*, 3120–3128.

Horimoto, T., Rivera, E., Pearson, J. Senne, D., Krauss, S., Kawaoka, Y., and Webster, R. G. (1995). Origin and molecular changes associated with emergence of a highly pathogenic H5N2 influenza virus in Mexico. Virology *213*, 223–230.

Katz, J. M., Lu X., Tumpey, T. M., Smith, C. B., Shaw, M. W., and Subbarao, K. (2000). Molecular correlates of influenza A H5N1 virus pathogenesis in mice. J. Virol. *74*, 10807–10810.

Kawaoka, Y., Naeve, C. W., and Webster, R. G. (1984). Is virulence of H5N2 influenza viruses in chickens associated with loss of carbohydrate from hemagglutinin? Virology *139*, 303–316.

Kawaoka, Y., Nestorowicz, A., Alexander, D. J., and Webster, R. G. (1987). Molecular analyses of the hemagglutinin genes of H5 influenza viruses: origin of a virulent turkey strain. Virology *158*, 218–227.

Khatchikian, D., Orlich, M., and Rott, R. (1989). Increased viral pathogenicity after insertion of a 28S ribosomal RNA sequence into the hemagglutinin gene of an influenza virus. Nature *340*, 156–157.

Kobasa, D., Rodgers, M. E., Wells, K., and Kawaoka, Y. (1997). Neuraminidase hemadsorption activity, conserved in avian influenza A viruses, does not influence viral replication in ducks. J. Virol. *71*, 6706–6713.

Krauss, S., Walker, D., Pryor, S. P., Niles, L., Chenghong, L., Hinshaw, V. S., and Webster, R. G. (2004). Influenza A viruses of migrating wild aquatic birds in North America. Vector Borne Zoonotic Dis. *4*, 177–189.

Krug, R. M., Yuan, W., Noah, D. L., and Latham, A. G. (2003). Intracellular warfare between human influenza viruses and human cells: the roles of the viral NS1 protein. Virology *309*, 181–189.

Lang, G., Narayan, O., Rouse, B. T., Ferguson, A. E., Connell, M. C. 1968. A new influenza virus A infection in turkeys. II. A highly pathogenic variant, A/turkey/Ontario/7732/66. Can. Vet. J. *9*, 151–160.

Laver, W. G., Colman, P. M., Webster, R. G., Hinshaw, V. S., and Air, G. M. (1984). Influenza virus neuraminidase with hemagglutinin activity. Virology *137*, 314–323.

Lee, C. W., Senne, D. A., and Suarez, D. L. (2004). Effect of vaccine use in the evolution of Mexican lineage H5N2 avian influenza virus. J. Virol. *78*, 8372–8381.

Li, K. S., Guan, Y., Wang, J., Smith, G. J., Xu, K. M., Duan, L., Rahardjo, A. P., Puthavathana, P., Buranathai, C., Nguyen, T. D., Estoepangestie, A. T., Chaisingh, A., Auewarakul, P., Long, H. T., Hanh, N. T., Webby, R. J, Poon, L. L., Chen, H., Shortridge, K. F., Yuen, K. Y., Webster, R. G., and Peiris, J. S. (2004). Genesis of a highly pathogenic and potentially pandemic H5N1 influenza virus in eastern Asia. Nature *430*, 209–13.

Lipatov, A. S., Krauss, S., Guan, Y., Peiris, M., Rehg, R. E., Perez, D. R., and Webster, R. G. (2003). Neurovirulence in mice of H5N1 influenza virus genotypes isolated from Hong Kong poultry in 2001. J. Virol. *77*, 3816–3823.

Lu, X., Tumpey, T. M., Morken, T., Zaki, S. R., Cox, N. J., and Katz, J. M. (1999). A mouse model for the evaluation of pathogenesis and immunity to influenza A (H5N1) viruses isolated from humans. J. Virol. *73*, 5903–5911.

Ludwig, S., Stitz, L., Planz, O., Van, H., Fitch, W. M., and Scholtissek, C. (1995). European swine virus as a possible source for the next influenza pandemic? Virology *212*, 555–561.

Matrosovich, M., Zhou, N., Kawaoka, Y., and Webster, R. (1999). The surface glycoproteins of H5 influenza viruses isolated from humans, chickens, and wild aquatic birds have distinguishable properties. J. Virol. *73*, 1146–1155.

Matrosovich, M. N., Matrosovich, T. Y., Gray, T., Roberts, N. A., and Klenk, H. D. (2004). Human and avian influenza viruses target different cell types in cultures of human airway epithelium. Proc. Natl. Acad. Sci. USA *101*, 4620–4624.

Mitnaul, L. J., Matrosovich, M. N., Castrucci, M. R., Tuzikov, A. B., Bovin, N. V., Kobasa, D., and Kawaoka, Y. (2000). Balanced hemagglutinin and neuraminidase activities are critical for efficient replication of influenza A virus. J. Virol. *74*, 6015–6020.

Mounts, A. W., Kwong, H., Izurieta, H. S., Ho, Y., Au, T., Lee, M., Buxton-Bridges, C., Williams, S. W., Mak, K. H., Katz, J. M., Thompson, W. W., Cox, N. J., and Fukuda, K. (1999). Case-control study of risk factors for avian influenza A (H5N1) disease, Hong Kong, 1997. J. Infect. Dis. *180*, 505–508.

Nettles, V. F., Wood, J. M., and Webster, R. G. (1985). Wildlife surveillance associated with an outbreak of lethal H5N2 avian influenza in domestic poultry. Avian Dis. *29*, 733–741.

Orlich, M., Gottwald, H., and Rott, R. (1994). Nonhomologous recombination between the hemagglutinin gene and the nucleoprotein gene of an influenza virus. Virology *204*, 462–465.

Paulson, J. C. (1985). In The Receptors, P. M. Conn, ed. (Orlando, FL: Academic Press), Vol. 2, pp. 131–219.

Peiris, J. S. M., Yu, W. C., Leung, C. W., Cheung, C. Y., Ng, W. F., Nicholls, J. M., Ng, T. K., Chan, K. H., Lai, S. T., Lim, W. L., Yuen, K. Y., and Guan, Y. (2004). Re-emergence of fatal human influenza A subtype H5N1 disease. Lancet *363*, 617–619.

Perdue, M. L., Garcia, M., Beck, J., Brugh, M., and Swayne, D. E. (1996). An Arg-Lys insertion at the hemagglutinin cleavage site of an H5N2 avian influenza isolate. Virus Genes *12*, 77–84.

Rimmelzwaan, G. F., Kuiken, T., van Amerongen, G., Bestebroer, T. M., Fouchier, R. A., and Osterhaus, A. D. (2001). Pathogenesis of influenza A (H5N1) virus infection in a primate model. J. Virol. *75*, 6687–6691.

Röhm, C., Horimoto, T., Kawaoka, Y., Süss, J., and Webster, R. G. (1995). Do hemagglutinin genes of highly pathogenic avian influenza viruses constitute unique phylogenetic lineages? Virology *209*, 664–670.

Saito, T., Horimoto, T., Kawaoka, Y., Senne, D. A., and Webster, R. G. (1994). Emergence of a potentially pathogenic H5N2 influenza virus in chickens. Virology *201*, 277–284.

Senne, D. A., Pearson, J. E., and Panigrahy, B. (1992). Live poultry markets: a missing link in the epidemiology of avian influenza. In Proceedings of the Third International Symposium on Avian Influenza, May 27–29, 1992, Richmond, VA, USA (Madison, WI: Animal Health Association), pp. 7–9.

Seo, S. H. and Webster, R. G. (2001). Cross-reactive, cell-mediated immunity and protection of chickens from lethal H5N1 influenza virus infection in Hong Kong poultry markets. J. Virol. *75*, 2516–2525.

Seo, S. H., Hoffmann, E., and Webster, R. G. (2002). Lethal H5N1 influenza viruses escape host anti-viral cytokine responses. Nat. Med. *8*, 950–954.

Shinya, K., Hamm, S., Hatta, M., Ito, H., Ito, T., and Kawaoka, Y. (2004). PB2 amino acid at position 627 affects replicative efficiency, but not cell tropism, of Hong Kong H5N1 influenza A viruses in mice. Virology *320*, 258–266.

Shortridge, K. F. (1992). Pandemic influenza: a zoonosis? Semin. Respir. Infect. *7*, 11–25.

Shortridge, K. F., Zhou, N. N., Guan, Y., Gao, P., Ito, T., Kawaoka, Y., Kodihalli, S., Krauss, S., Markwell, D., Murti, K. G., Norwood, M., Senne, D., Sims, L., Takada, A., and Webster, R. G. (1998). Characterization of avian H5N1 influenza viruses from poultry in Hong Kong. Virology *252*, 331–342.

Simms, L. D., Ellis, T. M., Liu, K. K., Dyrting, K., Wong, H., Peiris, M., Guan, Y., and Shortridge, K. F. (2003). Avian influenza in Hong Kong 1997–2002. Avian Dis. *47*, 832–838.

Stieneke-Grober, A., Vey, M., Angliker, H., Shaw, E., Thomas, G., Roberts, C., Klenk, H. D., and Garten, W. (1992). Influenza virus hemagglutinin with multibasic cleavage site is activated by furin, a subtilisin-like endoprotease. EMBO J. *11*, 2401–2414.

Sturm-Ramirez, K. M., Ellis, T., Bousefield, B., Bissett, L., Kyrting, K., Rehg, J. E., Poon, L., Guan, Y., Peiris, M., and Webster, R. G. (2004). Re-emerging H5N1 influenza viruses in Hong Kong in 2002 are highly pathogenic to ducks. J. Virol. *78*, 4892–4901.

Suarez, D. L. (2000). Evolution of avian influenza viruses. Vet. Microbiol. *74*, 15–27.

Suarez, D. L., Senne, D. A., Banks, J., Brown, I. H., Essen, S. C., Lee, C. W., Manvell, R. J., Mathieu-Benson, C., Moreno, V., Pedersen, J. C., Panigrahy, B., Rojas, H., Spackman, E., and Alexander, D. J. (2004). Recombination resulting in virulence shift in avian influenza outbreak, Chile. Emerg. Infect. Dis. *10*, 693–699.

Tang, X., Tian, G., Zhao, J., and Zhou, K. Y. (1998). Isolation and characterization of prevalent strains of avian influenza viruses in China. Chin. J. Anim. Poult. Infect. Dis. *20*, 105 (in Chinese).

To, K. F., Chan, P. K., Chan, K. F., Lee, W. K., Lam, W. Y., Wong, K. F., Tang, N. L., Tsang, D. N., Sung, R. Y., Buckley, T. A., Tam, J. S., and Cheng, A. F. (2001). Pathology of fatal human infection associated with avian influenza A H5N1 virus. J. Med. Virol. *63*, 242–246.

Tumpey, T. M., Lu, X., Morken, T., Zaki, S. R., and Katz, J. M. (2000). Depletion of lymphocytes and diminished cytokine production in mice infected with a highly virulent influenza A (H5N1) virus isolated from humans. J. Virol. *74*, 6105–6116.

Webster, R. G. (2004). Wet markets – a continuing source of severe acute respiratory syndrome and influenza? Lancet *363*, 234–236.

Webster, R. G. and Rott, R. (1987). Influenza virus A pathogenicity: the pivotal role of hemagglutinin. Cell *50*, 665–666.

Webster, R. G., Bean, W. J., Gorman, O. T., Chambers, T. M., and Kawaoka, Y. (1992). Evolution and ecology of influenza A viruses. Microbiol. Rev. *56*, 152–179.

Webster, R. G., Guan, Y., Peiris, M., Walker, D., Krauss, S., Zhou, N. N., Govorkova, E. A., Ellis, T. M., Dyrting, K. C., Sit, T., Perez, D. R., and Shortridge, K. F. (2002). Characterization of H5N1 influenza viruses that continue to circulate in geese in southeastern China J. Virol. *76*, 118–126.

Xu, S. K., Subbarao, K., Cox, N. J., and Guo, Y. (1999). Genetic characterization of the pathogenic influenza A/goose/Guangdong/1/96 (H5N1) virus: similarity of its hemagglutinin gene to those of H5N1 viruses from the 1997 outbreaks in Hong Kong. Virology *261*, 15–19.

Yuen, K. Y., Chan, P. K. S., Peiris, M., Tsang, D. N. C., Que, T. L., Shortridge, K. F., Cheung, P. T., To, W. K., Ho, E. T. F., Sung, R., Cheng, A. F. B., and members of the H5N1 study group. Clinical features and rapid viral diagnosis of human disease associated with avian influenza A H5N1 virus. Lancet *351*, 467–471, 1998.

Zhou, N. N., Shortridge, K. F., Claas, E. C., Krauss, S. L., and Webster, R. G. (1999). Rapid evolution of H5N1 influenza viruses in chickens in Hong Kong. J. Virol. *73*, 3366–3374.

Zitzow, L. A., Rowe, T., Morken, T., Shieh, W. J., Zaki, S., and Katz, J. M. (2002). Pathogenesis of avian influenza A (H5N1) viruses in ferrets. J. Virol. *76*, 4420–4429.

CHAPTER 11

The origin and virulence of the 1918 'Spanish' influenza virus

Jeffery K. Taubenberger and Peter Palese

Abstract

The 'Spanish' influenza pandemic of 1918–1919 caused acute illness in 25–30% of the world's population and resulted in the death of an estimated 40 million people. Using fixed and frozen lung tissue of 1918 influenza victims, the complete genomic sequence of the 1918 influenza virus is being deduced. Sequence and phylogenetic analysis of the completed 1918 influenza virus genes shows them to be the most avian-like among the mammalian-adapted viruses. This finding supports the hypothesis that (1) the pandemic virus contains genes derived from avian-like influenza virus strains and that (2) the 1918 virus is the common ancestor of human and classical swine H1N1 influenza viruses. The relationship of the 1918 virus with avian and swine influenza viruses is further supported by recent work in which the 1918 hemagglutinin (HA) protein crystal structure was resolved. Neither the 1918 hemagglutinin (HA) nor the neuraminidase (NA) genes possess mutations known to increase tissue tropicity that account for virulence of other influenza virus strains like A/WSN/33 or the highly pathogenic avian influenza H5 or H7 viruses. Using reverse genetics approaches, influenza virus constructs containing the 1918 HA and NA on an A/WSN/33 virus background were lethal in mice. The genotypic basis of this virulence has not yet been elucidated. The complete sequence of the non-structural (NS) gene segment of the 1918 virus was deduced and also tested for the hypothesis that enhanced virulence in 1918 could have been due to type I interferon inhibition by the NS1 protein. Results from these experiments suggest that in human cells the 1918 NS1 is a very effective interferon antagonist, but the 1918 NS1 gene does not have the amino acid change that correlates with virulence in the H5N1 virus strains identified in 1997 in Hong Kong. Sequence analysis of the 1918 pandemic influenza virus is allowing us to test hypotheses as to the origin and virulence of this strain. This information should help elucidate how pandemic influenza virus strains emerge and what genetic features contribute to virulence in humans.

INTRODUCTION

Influenza A viruses are negative strand RNA viruses of the genus *Orthomyxoviridae*. They continually circulate in humans in yearly epidemics (mainly in the winter in temperate climates) and antigenically novel virus strains emerge sporadically as pandemic viruses (Cox and Subbarao, 2000). In the United States, influenza is estimated to kill 30,000 people in an average year (Simonsen et al., 2000; Thompson et al., 2003). Every few years, influenza epidemics boost the annual number of deaths past the average, causing 10–15,000 additional deaths. Occasionally, and unpredictably, influenza sweeps the world, infecting 20% to 40% of the population in a single year. In these pandemic years, the numbers of deaths can be dramatically above average. In 1957–1958, a pandemic was estimated to cause 66,000 excess deaths in the United States (Simonsen et al., 1998). In 1918, the worst pandemic in recorded history was associated with approximately 675,000 total deaths in the United States (United States Department of Commerce, 1976), and killed an estimated 40 million people worldwide (Crosby, 1989; Johnson and Mueller, 2002; Patterson and Pyle, 1991).

Influenza A viruses constantly evolve by the mechanisms of antigenic drift and shift (Webster et al., 1992). Consequently they should be considered emerging infectious disease agents, perhaps 'continually' emerging pathogens. The importance of predicting the emergence of new circulating influenza virus strains for subsequent annual vaccine development cannot be underestimated (Gensheimer et al., 1999). Pandemic influenza viruses have emerged three times in this century: in 1918 ('Spanish' influenza, H1N1), in 1957 ('Asian' influenza, H2N2), and in 1968 ('Hong Kong' influenza, H3N2) (Cox and Subbarao, 2000; Webby and Webster, 2003). Recent circulation of highly pathogenic avian H5N1 viruses in Asia from 1997–2004 has caused a small number of human deaths (Claas et al., 1998; Peiris et al., 2004; Subbarao et al., 1998; Tran et al., 2004). How and when novel influenza viruses emerge as pandemic virus strains and how they cause disease is still not understood.

Studying the extent to which the 1918 influenza was like other pandemics may help us to understand how pandemic influenzas emerge and cause disease in general. On the other hand, if we determine what made the 1918 influenza different from other pandemics, we may use the lessons of 1918 to predict the magnitude of public health risks a new pandemic virus might pose.

ORIGIN OF PANDEMIC INFLUENZA VIRUSES

The predominant natural reservoir of influenza viruses is thought to be wild waterfowl (Webster et al., 1992). Periodically, genetic material from avian virus strains is transferred to virus strains infectious to humans by a process called reassortment. Human influenza virus strains with recently acquired avian surface and internal protein-encoding RNA segments were responsible for the pandemic influenza outbreaks in 1957 and 1968 (Kawaoka et al., 1989; Scholtissek et al., 1978a). The change in the hemagglutinin subtype or the hemagglutinin and the neuraminidase subtype is referred to as antigenic shift. Since pigs can be infected with both avian and human virus strains, and various reassortants have been isolated from pigs, they have been proposed as an intermediary in this process (Ludwig et al., 1995; Scholtissek, 1994). Until recently there was only limited evidence that a wholly avian influenza virus could directly infect humans, but in 1997 eighteen people were infected with avian H5N1 influenza viruses in Hong Kong and six died of complications after infection (Claas et al., 1998; Ludwig et al., 1995; Scholtissek, 1994; Subbarao et al., 1998). Although these viruses were very poorly or non-transmissible (Claas et al., 1998; Katz et al., 1999; Ludwig et al., 1995; Scholtissek, 1994; Subbarao et al., 1998), their isolation from infected patients indicates that humans can be infected with wholly avian influenza virus strains. In 2003–2004, H5N1 outbreaks in poultry have become widespread in Asia (Tran et al., 2004), and at least 23 people have died of complications of infection in Vietnam and Thailand (W.H.O., 2004). In 2003, a highly pathogenic H7N7 outbreak occurred in poultry farms in the Netherlands. This virus caused infections (predominantly conjunctivitis) in 86 poultry handlers and in 3 secondary contacts. One of the infected individuals died of pneumonia (Fouchier et al., 2004; Koopmans et al., 2004; W.H.O., 2004). In 2004 an H7N3 influenza outbreak in poultry in Canada also resulted in the infection of a single individual

(W.H.O., 2004) and a patient in New York was reported to be sick following infection with an H7N2 virus (Lipsman, 2004). Therefore, it may not be necessary to invoke swine as the intermediary in the formation of a pandemic virus strain since reassortment between an avian and a human influenza virus could take place directly in humans.

While reassortment involving genes encoding surface proteins appears to be a critical event for the production of a pandemic virus, a significant amount of data exists to suggest that influenza viruses must also acquire specific adaptations to spread and replicate efficiently in a new host. Among other features, there must be functional HA receptor binding and interaction between viral and host proteins (Weis et al., 1988). Defining the minimal adaptive changes needed to allow a reassortant virus to function in humans is essential to understanding how pandemic viruses emerge.

Once a new virus strain has acquired the changes that allow it to spread in humans, virulence is affected by the presence of novel surface protein(s) which allow the virus to infect an immunologically naïve population (Kilbourne, 1977). This was the case in 1957 and 1968 and was almost certainly the case in 1918. While immunological novelty may explain much of the virulence of the 1918 influenza, it is likely that additional genetic features contributed to its exceptional lethality. Unfortunately not enough is known about how genetic features of influenza viruses affect virulence. The degree of illness caused by a particular virus strain, or virulence, is complex and involves host factors like immune status, and viral factors like host adaptation, transmissibility, tissue tropism, or viral replication efficiency. The genetic basis for each of these features is not yet fully characterized, but is most likely polygenic in nature (Kilbourne, 1977).

Prior to the analyses on the 1918 virus described in this review, only two pandemic influenza virus strains were available for molecular analysis: the H2N2 virus strain from 1957 and the H3N2 virus strain from 1968. The 1957 pandemic resulted from the emergence of a reassortant influenza virus in which both HA and NA had been replaced by gene segment closely related to those in avian virus strains (Schafer et al., 1993; Scholtissek et al., 1978b; Webster et al., 1995). The 1968 pandemic followed with the emergence of a virus strain in which the H2 subtype HA gene was exchanged with an avian-derived H3 HA RNA segment (Scholtissek et al., 1978b; Webster et al., 1995), while retaining the N2 gene derived in 1957. More recently it has been shown that the PB1 gene was replaced in both the 1957 and the 1968 pandemic virus strains, also with a likely avian derivation in both cases (Kawaoka et al., 1989). The remaining five RNA segments encoding the PA, PB2, nucleoprotein, matrix and non-structural proteins, all were preserved from the H1N1 virus strains circulating before 1957. These segments were likely the direct descendants of the genes present in the 1918 virus. Since only the 1957 and 1968 influenza pandemic virus strains have been available for sequence analysis, it is not clear what changes are necessary for the emergence of a virus strain with pandemic potential. Sequence analysis of the 1918 influenza virus allows us potentially to address the genetic basis of virulence and human adaptation.

HISTORICAL BACKGROUND

The influenza pandemic of 1918 was exceptional in both breadth and depth. Outbreaks of the disease swept not only North America and Europe but also spread as far as the Alaskan wilderness and the most remote islands of the Pacific. It has been estimated that one-third of the world's population (500 million people) may have been clinically infected during the pandemic (Burnet and Clark, 1942; Frost, 1920). The disease was also exceptionally severe, with mortality rates among the infected of over 2.5%, compared to less than 0.1% in other influenza epidemics (Marks and Beatty, 1976; Rosenau and Last, 1980). Total mortality attributable to the 1918 pandemic was probably around 40 million (Crosby, 1989; Johnson and Mueller, 2002; Patterson and Pyle, 1991).

Unlike most subsequent influenza virus strains that have developed in Asia, the 'first wave' or 'spring wave' of the 1918 pandemic seemingly arose in the United States in March, 1918 (Barry, 2004; Crosby, 1989; Jordan, 1927). However the near simultaneous appearance of influenza in March–April, 1918 in North America, Europe, and Asia makes definitive assignment of a geographic point of origin difficult (Jordan, 1927). It is possible that a mutation or reassortment occurred in the late summer of 1918, resulting in significantly enhanced virulence. The main wave of the global pandemic, the 'fall wave' or 'second wave,' occurred in September–November, 1918. In many places, there was yet another severe wave of influenza in early 1919 (Jordan, 1927).

Three extensive outbreaks of influenza within one year is unusual, and may point to unique features of the 1918 virus that could be revealed in its sequence. Interpandemic influenza outbreaks generally occur in a single annual wave in the late winter. The severity of annual outbreaks is affected by antigenic drift, with an antigenically modified virus strain emerging every two to three years. Even in pandemic influenza, while the normal late winter seasonality may be violated, the successive occurrence of distinct waves within a year is unusual. The 1890 pandemic began in the late spring of 1889 and took several months to spread throughout the world, peaking in northern Europe and the United States late in 1889 or early 1890. The second wave peaked in spring 1891 (over a year after the first wave) and the third wave in early 1892 (Jordan, 1927). As in 1918, subsequent waves seemed to produce more severe illness so that the peak mortality was reached in the third wave of the pandemic. The three waves, however, were spread over more than three years, in contrast to less than one year in 1918. It is unclear what gave the 1918 virus this unusual ability to generate repeated waves of illness. Perhaps the surface proteins of the virus drifted more rapidly than other influenza virus strains, or perhaps the virus had an unusually effective mechanism for evading the human immune system.

It has been estimated that the influenza epidemic of 1918 killed 675,000 Americans, including 43,000 servicemen mobilized for World War I (Crosby, 1989). The impact was so profound as to depress average life expectancy in the U.S. by over 10 years, (Figure 1) (Grove and Hetzel, 1968), and may have played a significant role in ending the World War I conflict (Crosby, 1989; Ludendorff, 1919).

The majority of individuals who died during the pandemic succumbed to secondary bacterial pneumonia (Jordan, 1927; LeCount, 1919; Wolbach, 1919), since no antibiotics were available in 1918. However, a subset died rapidly after the onset of symptoms often with either massive acute pulmonary hemorrhage or pulmonary edema, often in less than 5 days (LeCount, 1919; Winternitz et al., 1920; Wolbach, 1919). In the hundreds of autopsies performed in 1918, the primary pathologic findings were confined to the respiratory tree and death was due to pneumonia and respiratory failure (Winternitz et al., 1920). These findings are consistent with infection by a well-adapted influenza virus capable of rapid replication throughout the entire respiratory tree (Reid and Taubenberger, 1999; Taubenberger et al., 2001). There was no clinical or pathological evidence for systemic circulation of the virus (Winternitz et al., 1920).

Furthermore, in the 1918 pandemic most deaths occurred among young adults, a group which usually has a very low death rate from influenza. Influenza and pneumonia death

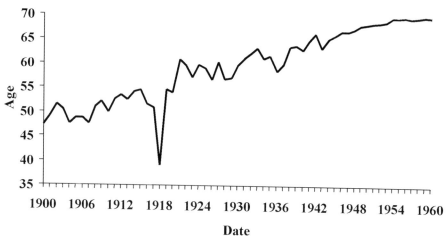

Figure 11.1 Life expectancy in the United States, 1900–1960 showing the impact of the 1918 influenza pandemic (Grove and Hetzel, 1968; Linder and Grove, 1943; United States Department of Commerce, 1976).

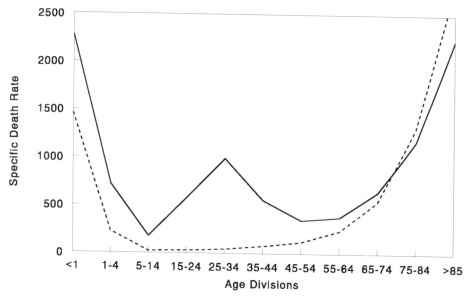

Figure 11.2 Influenza and pneumonia mortality by age, United States. Influenza and pneumonia specific mortality by age, including an average of the inter-pandemic years 1911–1915 (dashed line), and the pandemic year 1918 (solid line). Specific death rate is per 100,000 of the population in each age division (Grove and Hetzel, 1968; Linder and Grove, 1943; United States Department of Commerce, 1976).

rates for 15–34 year olds were more than 20 times higher in 1918 than in previous years, (Figure 2) (Linder and Grove, 1943; Simonsen et al., 1998). The 1918 pandemic is also unique among influenza pandemics in that absolute risk of influenza mortality was higher in those less than 65 years of age than in those greater than 65. Strikingly, persons less than 65 years old accounted for greater than 99% of all excess influenza-related deaths in 1918–19 (Simonsen et al., 1998). In contrast, the less-than-65 age group accounted for only 36% of all excess influenza-related mortality in the 1957 H2N2 pandemic and 48% in the 1968 H3N2 pandemic. Overall, nearly half of the influenza-related deaths in the 1918 influenza pandemic were young adults, age 20–40, (Figure 2) (Simonsen et al., 1998). Why this particular age group suffered such extreme mortality is not fully understood (see below).

The 1918 influenza had as another unique feature: the simultaneous infection of both humans and swine. Interestingly, swine influenza was first recognized as a clinical entity in that species in the fall of 1918 (Koen, 1919) concurrently with the spread of the second wave of the pandemic in humans (Dorset et al., 1922–23). Investigators were impressed by clinical and pathological similarities of human and swine influenza in 1918 (Koen, 1919; Murray and Biester, 1930). An extensive review by the veterinarian W.W. Dimoch of the diseases of swine published in August 1918 makes no mention of any swine disease resembling influenza (Dimoch, 1918–19). Thus, contemporary investigators were convinced that influenza virus had not circulated as an epizootic disease in swine before 1918 and that the virus spread from humans to pigs because of the appearance of illness in pigs after the first wave of the 1918 influenza in humans (Shope, 1936).

Thereafter the disease became widespread among swine herds in the U.S. Midwest. The epizootic of 1919–1920 was as extensive as in 1918–1919. The disease then appeared among swine in the Midwest every year, leading to Shope's isolation of the first influenza virus in 1930, A/swine/Iowa/30 (Shope and Lewis, 1931), three years before the isolation of the first human influenza virus, A/WS/33 by Smith, Andrewes, and Laidlaw (Smith et al., 1933). Classical swine viruses have continued to circulate not only in North American pigs, but also in swine populations in Europe and Asia (Brown et al., 1995; Kupradinun et al., 1991; Nerome et al., 1982).

During the fall and winter of 1918–19, severe influenza-like outbreaks were noted not only in swine in the United States, but also in Europe, and China (Beveridge, 1977; Chun, 1919; Koen, 1919). Since 1918 there have been many examples of both H1N1 and H3N2 human influenza A virus strains becoming established in swine (Brown et al., 1998; Castrucci et al., 1993; Zhou et al., 2000) while swine influenza A virus strains have been isolated only sporadically from humans (Gaydos et al., 1977; Woods et al., 1981).

The unusual severity of the 1918 pandemic and the exceptionally high mortality it caused among young adults have stimulated great interest in the influenza virus strain responsible for the 1918 outbreak (Crosby, 1989; Kolata, 1999; Monto et al., 1997). Since the first human and swine influenza A viruses were not isolated until the early 1930s (Shope and Lewis, 1931; Smith et al., 1933), characterization of the 1918 virus strain has had previously to rely on indirect evidence (Kanegae et al., 1994; Shope, 1958).

SEROLOGY AND EPIDEMIOLOGY OF THE 1918 INFLUENZA VIRUS

Analyses of antibody titers of 1918 influenza survivors from the late 1930s suggested correctly that the 1918 virus strain was an H1N1-subtype influenza A virus, closely related to what is now known as 'classic swine' influenza virus (Dowdle, 1999; Philip and Lackman, 1962; Shope, 1936). The relationship to swine influenza is also reflected in the simultaneous influenza outbreaks in humans and pigs around the world (Beveridge, 1977; Chun, 1919; Koen, 1919). While historical accounts described above suggest that the virus spread from humans to pigs in the fall of 1918, the relationship of these two species in the development of the 1918 influenza has not been resolved.

It is not known for certain what influenza A subtype(s) circulated before the 1918 pandemic. In a recent review of the existing archaeoserologic and epidemiologic data, Walter Dowdle concluded that an H3 subtype influenza A virus strain circulated from the 1889–1891 pandemic to 1918 when it was replaced by the novel H1N1 virus strain of the 1918 pandemic (Dowdle, 1999).

It is reasonable to conclude that the 1918 virus strain must have contained a hemagglutinin gene encoding a novel subtype such that large portions of the population did not have protective immunity (Kilbourne, 1977; Reid and Taubenberger, 1999). In fact, epidemiological data on influenza prevalence by age in the population collected between 1900 and 1918 provide good evidence for the emergence of an antigenically novel influenza virus in 1918 (Jordan, 1927). Jordan showed that from 1900–1917, the 5–15 age group accounted for 11% of total influenza cases in this series while the >65 age group similarly accounted for 6% of influenza cases. In 1918 the 5–15 year old group jumped to 25% of influenza cases, compatible with exposure to an antigenically novel virus strain. The >65 age group only accounted for 0.6% of the influenza cases in 1918. It is likely that this age group accounted for a significantly lower percentage of influenza cases because younger people were so susceptible to the novel virus strain (as seen in the 1957 pandemic (Ministry of Health, 1960; Simonsen et al., 1998)) but it is also possible that this age group had pre-existing H1 antibodies. Further evidence for pre-existing H1 immunity can be derived from the age adjusted mortality data in Figure 2. Those individuals >75 years had a lower influenza and pneumonia case mortality rate in 1918 than they had for the pre-pandemic period of 1911–1917.

When 1918 influenza case rates by age (Jordan, 1927) are superimposed on the familiar 'W' shaped mortality curve (seen in Figure 2), a different perspective emerges (Figure 3). As shown, those <35 years of age in 1918 accounted for a disproportionately high influenza incidence by age. Interestingly, the 5–14 age group accounted for a large fraction of 1918 influenza cases, but had an extremely low case mortality rate compared to other age groups (Figure 3). Why this age group had such a low case fatality rate cannot yet be fully explained. Conversely, why the 25–34 age group had such a high influenza and pneumonia mortality rate in 1918 remains enigmatic but it is one of the truly unique features of the 1918 influenza pandemic.

One theory that may explain these data concerns the possibility that the virus had an intrinsically high virulence that was only tempered in those patients who had been born before 1889. It can be speculated that the virus circulating prior to 1889 was an H1-like virus strain that provided partial protection against the 1918 virus strain (Ministry of Health, 1960; Simonsen et al., 1998; Taubenberger et al., 2001). Short of this cross-protection in patients older than 29 years of age, the pandemic of 1918 might have been even more devastating (Zamarin and Palese, 2004). In this context it is interesting to note that remote regions of Alaska showed a different epidemiological pattern. A possible explanation is that older people living in these far-away villages may not have been exposed to the pre-1889 virus and when they were exposed to the 1918 virus they had no cross-protective immunity at all and thus succumbed in even higher numbers to the disease. In some of these villages children and younger adults were the survivors rather than the older age groups (Figure 4).

Thus, it seems clear that the H1N1 virus of the 1918 pandemic contained an antigenically novel hemagglutinin to which most humans and swine were susceptible in 1918. Given the

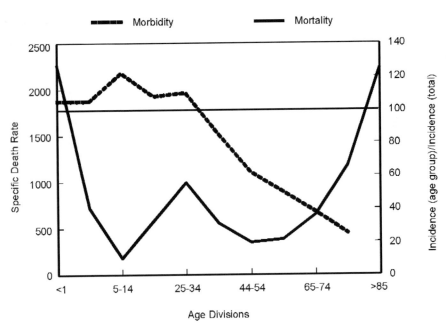

Figure 11.3 Influenza and pneumonia mortality by age (solid line), with influenza morbidity by age (dashed line) superimposed. Influenza and pneumonia mortality by age as in figure 2. Specific death rate per age group, left ordinal axis. Influenza morbidity presented as ratio of incidence in persons of each group to incidence in persons of all ages (=100), right ordinal axis. Horizontal line at 100 (right ordinal axis) represents average influenza incidence in the total population (Taubenberger et al., 2001) (Adapted from (Jordan, 1927)).

severity of the pandemic, it is also reasonable to suggest that the other dominant surface protein, NA, would also have been replaced by antigenic shift before the start of the pandemic (Reid and Taubenberger, 1999; Taubenberger et al., 2001). In fact, sequence and phylogenetic analyses suggest that the genes encoding these two surface proteins were derived from an avian-like influenza virus shortly before the start of the 1918 pandemic and that the precursor virus did not circulate widely in either humans or swine before 1918 (Fanning et al., 2002; Reid et al., 1999; Reid et al., 2000) (Figure 5). It is currently unclear what other influenza gene segments were novel in the 1918 pandemic virus in comparison to the previously circulating virus strain. It is possible that sequence and phylogenetic analyses of the gene segments of the 1918 virus may help elucidate this question.

GENETIC CHARACTERIZATION OF THE 1918 VIRUS
Sequence and functional analysis of the hemagglutinin and neuraminidase gene segments

Frozen and fixed lung tissue from five second-wave cases influenza victims (dating from September 1918 to February 1919) has been used to examine directly the genetic structure of the 1918 influenza virus. Two of the cases analyzed were U.S. Army soldiers who died in September, 1918, one in Camp Upton, New York and the other in Fort Jackson, South Carolina. The available material consists of formalin-fixed, paraffin-embedded (FFPE) autopsy tissue, hematoxylin and eosin-stained microscopic sections, and the clinical histories

Figure 11.4 During the 1918/1919 influenza pandemic the older age groups were hard hit in remote villages of Alaska. More than 40 influenza Inuit orphans from the village of Nushagak wait for the government to make arrangements for their permanent care. Printed with permission from the Alaska State Library, Core: Nushagak-People-4. Alaskan Packers Association. PCA 01-2432.

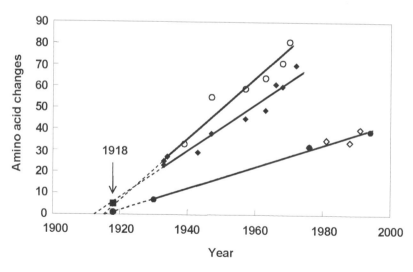

Figure 11.5 Change in hemagglutinin (HA) and neuraminidase (NA) proteins over time (Reid et al., 1999; Reid et al., 2000; Taubenberger et al., 2000). The number of amino acid changes from a hypothetical ancestor was plotted versus the date of viral isolation for viruses isolated from 1930–1993. Open circles, human HA; closed diamonds, human NA; closed circles, swine HA; open diamonds, swine NA. Regression lines were drawn, extrapolated to the x-intercept and then the 1918 data points, closed square, 1918 HA, closed circle, 1918 NA were added to the graph (arrow).

of these patients. A third sample was obtained from an Alaskan Inuit woman who had been interred in permafrost in Brevig Mission, Alaska, since her death from influenza in November 1918. The influenza virus sequences derived from these three cases have been called A/South Carolina/1/18 (H1N1), A/New York/1/18 (H1N1), and A/Brevig Mission/1/18 (H1N1), respectively. To date, five RNA segment sequences have been published (Basler et al., 2001; Reid et al., 1999; Reid et al., 2004; Reid et al., 2002; Reid et al., 2000). More recently, the HA sequence of two additional fixed autopsy cases of 1918 influenza victims from the Royal London Hospital were determined (Reid et al., 2003). The HA sequences from these five cases show >99% sequence identity, but differ at amino acid residue 225 (see below).

The sequence of the 1918 HA is most closely related to that of the A/swine/Iowa/30 virus. However, despite this similarity the sequence has many avian features. Of the 41 amino acids that have been shown to be targets of the immune system and subject to antigenic drift pressure in humans, 37 match the avian sequence consensus, suggesting that there was little immunologic pressure on the HA protein before the fall of 1918 (Reid et al., 1999). Another mechanism by which influenza viruses evade the human immune system is the acquisition of glycosylation sites to mask antigenic epitopes. The HAs from modern H1N1 viruses have up to five glycosylation sites in addition to the four found in all avian HAs. The HA of the 1918 virus has only the four conserved avian sites (Reid et al., 1999).

Influenza virus infection requires binding of the HA protein to sialic acid receptors on the host cell surface. The HA receptor binding site consists of a subset of amino acids that are invariant in all avian HAs but vary in mammalian-adapted HAs. Human-adapted influenza viruses preferentially bind sialic acid receptors with $\alpha(2-6)$ linkages. Those viral strains adapted to birds preferentially bind $\alpha(2-3)$ linked sugars (Gambaryan et al., 1997; Matrosovich et al., 1997; Weis et al., 1988). To shift from the proposed avian-adapted receptor binding site configuration (with a preference for α (2-3) sialic acids) to that of swine H1s (which can bind both α (2-3) and α (2-6)) requires only one amino acid change, E190D. The HA sequences of all five 1918 cases have the E190D change (Reid et al., 2003). In fact, the critical amino acids in the receptor-binding site of two of the 1918 cases are identical to that of the A/swine/Iowa/30 HA. The other three 1918 cases have an additional change from the avian consensus, G225D. Since swine viruses with the same receptor site as A/swine/Iowa/30 bind both avian and mammalian-type receptors (Gambaryan et al., 1997), A/New York/1/18 virus probably also had the capacity to bind both. The change at residue 190 may represent the minimal change necessary to allow an avian H1-subtype HA to bind mammalian-type receptors (Gamblin et al., 2004; Glaser et al., 2005; Reid et al., 1999; Reid et al., 2003; Stevens et al., 2004), a critical step in host adaptation.

The crystal structure analysis of the 1918 HA (Gamblin et al., 2004; Stevens et al., 2004) suggests that the overall structure of the receptor binding site is akin to that of an avian H5 HA in terms of its having a narrower pocket than that identified for the human H3 HA (Wilson et al., 1981). This provides an additional clue for the avian derivation of the 1918 HA. The four antigenic sites that have been identified for another H1 HA, the A/PR/8/34 virus HA (Caton et al., 1982), also appear to be the major antigenic determinants on the 1918 HA. The X-ray analyses suggest that these sites are exposed on the 1918 HA and thus they could be readily recognized by the human immune system.

The principal biological role of NA is the cleavage of the terminal sialic acid residues that are receptors for the virus' HA protein (Palese and Compans, 1976). The active site of the enzyme consists of 15 invariant amino acids that are conserved in the 1918 NA. The functional NA protein is configured as a homotetramer in which the active sites are found on a terminal knob carried on a thin stalk (Colman et al., 1983). Some early human virus strains have short (11–16 amino acids) deletions in the stalk region, as do many virus strains isolated from chickens. The 1918 NA has a full-length stalk and has only the glycosylation sites shared by avian N1 virus strains (Schulze, 1997). Although the antigenic sites on human-adapted N1 neuraminidases have not been definitively mapped, it is possible to align the N1 sequences with N2 subtype NAs and examine the N2 antigenic sites for evidence of drift in N1. There are 22 amino acids on the N2 protein that may function in antigenic epitopes (Colman et al., 1983). The 1918 NA matches the avian consensus at 21 of these sites (Reid et al., 2000). This finding suggests that the 1918 NA, like the 1918 HA, had not circulated long in humans before the pandemic and very possibly had an avian origin (Reid and Taubenberger, 2003).

Neither the 1918 HA nor NA genes have obvious genetic features that can be related directly to virulence. Two known mutations that can dramatically affect the virulence of influenza virus strains have been described. For viral activation HA must be cleaved into two pieces, HA1 and HA2 by a host protease (Lazarowitz and Choppin, 1975; Rott et al., 1995). Some avian H5 and H7 subtype viruses acquire a mutation that involves the addition of one or more basic amino acids to the cleavage site, allowing HA activation by ubiquitous proteases (Kawaoka and Webster, 1988; Webster and Rott, 1987). Infection with such a pantropic virus strain can cause systemic disease in birds with high mortality. This mutation was not observed in the 1918 virus (Reid et al., 1999; Taubenberger et al., 1997).

The second mutation with a significant effect on virulence through pantropism has been identified in the NA gene of two mouse-adapted influenza virus strains, A/WSN/33 and A/NWS/33. Mutations at a single codon (N146R or N146Y, leading to the loss of a glycosylation site) appear, like the HA cleavage site mutation, to allow the virus to replicate in many tissues outside the respiratory tract (Li et al., 1993). This mutation was also not observed in the NA of the 1918 virus (Reid et al., 2000).

Therefore, neither surface protein-encoding gene has known mutations that would allow the 1918 virus to become pantropic. Since clinical and pathological findings in 1918 showed no evidence of replication outside the respiratory system (Winternitz et al., 1920; Wolbach, 1919), mutations allowing the 1918 virus to replicate systemically would not have been expected. However, the relationship of other structural features of these proteins (aside from their presumed antigenic novelty) to virulence remains unknown. In their overall structural and functional characteristics, the 1918 HA and NA are avian-like but they also have mammalian-adapted characteristics.

Interestingly, recombinant influenza viruses containing the 1918 HA and NA and up to three additional genes derived from the 1918 virus (the other genes being derived from the A/WSN/33 virus) were all highly virulent in mice (Tumpey et al., 2004). Furthermore, expression microarray analysis performed on whole lung tissue of mice infected with the 1918 HA/NA recombinant showed increased upregulation of genes involved in apoptosis, tissue injury and oxidative damage (Kash et al., 2004). These findings were unusual because the viruses with the 1918 genes had not been adapted to mice. On the other hand replacement of only a single gene, the NS gene, in the A/WSN/33 virus background led to a dramatic decrease in the LD_{50} value for mice, suggesting that the lack of mouse adaptation by the interferon antagonist (coded for by the 1918 NS gene) is associated with a decrease in virulence (Basler et al., 2001). Again, the 1918 NS gene in the context of additional 1918 genes (the HA, NA, NP and M genes) appears to result in recombinant viruses with high virulence in mice (see above). One explanation is that the combination of the genes/proteins of the 1918 virus was 'optimal,' and that this virus – unfortunately for humans and pigs – was a 'lucky' winner. As a consequence, viruses with

Table 11.1 Hemagglutination Inhibition (HI) reactions of H1N1 virus variants with ferret antisera.

Virus	HI titer with ferret antisera*							
	1918	Sw/Ia/30	WS/33	PR/8/34	USSR/77	Chili/83	Tx/91	N.Cal/99
1918 HA	**2560**	1280	320	40	<10	10	80	20
Sw/Ia/30	1280	**2560**	20	320	80	10	80	20
WS/33	<10	<10	**640**	40	<10	<10	<10	40
PR/8/34	20	<10	160	**2560**	10	<10	10	10
USSR/77	<10	<10	10	<10	**1280**	20	<10	<10
Chili/83	<10	<10	10	<10	40	**320**	20	10
Tx/91	<10	<10	20	<10	<10	<10	**2560**	40
N.Cal/99	10	<10	10	20	<10	<10	40	**1280**

* Serum samples from ferrets infected with indicated H1N1 viruses. HI titers represent reciprocal of highest dilution of sera inhibiting agglutination of 0.5% chicken erythrocytes by 4 hemagglutination units of virus (Tumpey et al., 2004). Virus strain abbreviations used: 1918, A/South Carolina/1/1918; Sw/Ia/30, A/swine/Iowa/30; WS/33, A/WS/33; PR/8/34, A/Puerto Rico/8/34; USSR/77, A/USSR/90/77; Chile/83, A/Chile/1/83; Tx/91, A/Texas/36/91; N. Cal/99, A/New Caledonia/20/99.

many (or most) genes derived from the 1918 virus are highly pathogenic in different species because the 1918 genes possibly work synergistically in terms of virulence. The completion of the sequence of the entire genome of the 1918 virus and the reconstruction and characterization of viruses with 1918 genes under appropriate biosafety conditions will shed more light on this hypothesis and should allow a definitive examination of this explanation.

Antigenic analysis of recombinant viruses possessing the 1918 HA and NA by hemagglutination inhibition tests using ferret and chicken antisera suggested a close relationship with the A/swine/Iowa/30 virus and H1N1 viruses isolated in the 1930s (Tumpey et al., 2004), further supporting data of Shope from the 1930s (Shope, 1936) (Table 1). Interestingly, when mice were immunized with different H1N1 virus strains, challenge studies using the 1918-like viruses revealed partial protection by this treatment suggesting that current vaccination strategies are adequate against a 1918-like virus (Tumpey et al., 2004). In fact, the data may even allow us to suggest that the human population, having experienced a long period of exposure to H1N1 viruses, may be partially protected against a 1918-like virus (Tumpey et al., 2004).

Since virulence (in the immunologically naïve person) has not yet been mapped to particular sequence motifs of the 1918 HA and NA genes, what can gene sequencing tell us about the origin of the 1918 virus? The best approach to analyzing the relationships among influenza viruses is phylogenetics, whereby hypothetical family trees are constructed which take available sequence data and use them to make assumptions about the ancestral relationships between current and historical influenza virus strains (Fitch et al., 1991; Gammelin et al., 1990; Scholtissek et al., 1993) (Figure 6). Since influenza viruses possess eight discrete RNA segments that can move independently between virus strains by the process of reassortment, these evolutionary studies must be performed independently for each gene segment.

A comparison of the complete 1918 HA (Figure 6) and NA genes with those of numerous human, swine, and avian sequences demonstrates the following. Phylogenetic analyses based upon HA nucleotide changes (either total or synonymous) or HA amino acid changes always

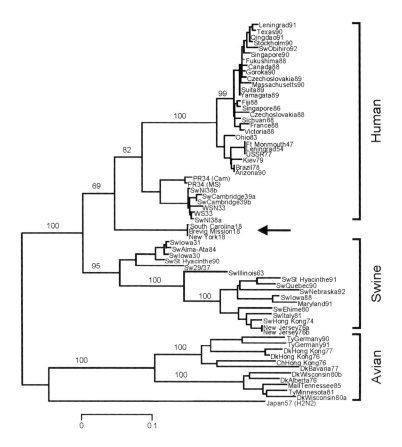

Figure 11.6 Phylogenetic tree of the influenza virus hemagglutinin gene segment. Amino acid changes in three lineages of the influenza virus hemagglutinin protein segment, HA1. Bootstrap values for key nodes are shown. Data derived from (Reid et al., 1999).

place the 1918 HA with the mammalian viruses, not with the avian viruses (Reid et al., 1999). In fact, both synonymous and nonsynonymous changes place the 1918 HA in the human clade. Phylogenetic analyses of total or synonymous NA nucleotide changes also place the 1918 NA sequence with the mammalian viruses, but analysis of nonsynonymous changes or amino acid changes places the 1918 NA with the avian viruses (Reid et al., 2000). Since the 1918 HA and NA have avian features and most analyses place HA and NA near the root of the mammalian clade (close to an ancestor of the avian genes), it is likely that both genes emerged from an avian-like influenza reservoir just prior to 1918 (Fanning et al., 2000; Fanning and Taubenberger, 1999; Reid et al., 1999; Reid et al., 2000; Reid and Taubenberger, 2003) (Figure 5). Clearly, by 1918 the virus had acquired enough mammalian-adaptive changes to function as a human pandemic virus and to form a stable lineage in swine.

Sequence and functional analysis of the non-structural gene segment

The complete coding sequence of the 1918 non-structural (NS) segment was completed (Basler et al., 2001). The functions of the two proteins, NS1 and NS2 (NEP), encoded by overlapping reading frames (Lamb and Lai, 1980) of the NS segment are still being elucidated (Garcia-Sastre, 2002; Garcia-Sastre et al., 1998; Krug et al., 2003; Li et al., 1998; O'Neill et al., 1998). The NS1 protein has been shown to prevent type I interferon (IFN) production, by preventing activation of the latent transcription factors IRF-3 (Talon et al., 2000) and NF-κB (Wang et al., 2000). One of the distinctive clinical characteristics of the 1918 influenza was its ability to produce rapid and extensive damage to both the upper and lower respiratory epithelium (Winternitz et al., 1920). Such a clinical course suggests a virus that replicated to a high titer and spread quickly from cell to cell. Thus, an NS1 protein that was especially effective at blocking the type I IFN system might have contributed to the exceptional virulence of the 1918 virus strain (Garcia-Sastre et al., 1998; Talon et al., 2000; Wang et al., 2000). To address this possibility, transfectant A/WSN/33 influenza viruses were constructed with the 1918 NS1 gene or with the entire 1918 NS segment (coding for both NS1 and NS2 (NEP) proteins) (Basler et al., 2001). In both cases, viruses containing 1918 NS genes were attenuated in mice compared to wild-type A/WSN/33 controls. The attenuation demonstrates that NS1 is critical for the virulence of A/WSN/33 in mice. On the other hand transcriptional profiling (microarray analysis) of infected human lung epithelial cells showed that a virus with the 1918 NS1 gene was more effective at blocking the expression of IFN-regulated genes than the isogenic parental mouse-adapted A/WSN/33 virus (Geiss et al., 2002) suggesting that the 1918 NS1 contributes virulence characteristics in human cells but not murine ones. The 1918 NS1 protein varies from that of the WSN virus at 10 amino acid positions. The amino acid differences between the 1918 and A/WSN/33 NS segments may be important in the adaptation of the latter virus strain to mice and likely account for the observed differences in virulence in these experiments. Recently, a single amino acid change (D92E) in the NS1 protein was associated with increased virulence of the 1997 Hong Kong H5N1 viruses in a swine model (Seo et al., 2002). This amino acid change was not found in the 1918 NS1 protein.

Sequence and functional analysis of the matrix gene segment

The coding region of influenza A RNA segment 7 from the 1918 pandemic virus, consisting of the open reading frames of the two matrix genes, M1 and M2, has been sequenced (Reid et al., 2002). While this segment is highly conserved among influenza virus strains, the 1918 sequence does not match any previously sequenced influenza virus strains. The 1918 sequence matches the consensus over the M1 RNA-binding domains and nuclear localization signal and the highly conserved transmembrane domain of M2. Amino acid changes that correlate with high yield and pathogenicity in animal models were not found in the 1918 virus strain.

Influenza A virus RNA segment 7 encodes two proteins, the matrix proteins M1 and M2. The M1 mRNA is colinear with the viral RNA, while the M2 mRNA is encoded by a spliced transcript (Lamb and Krug, 2001). The proteins encoded by these mRNAs share their initial nine amino acids and also have a stretch of 14 amino acids in overlapping reading frames. The M1 protein is a highly conserved 252 amino acid protein. It is the most abundant protein in the viral particle, lining the inner layer of the viral membrane and contacting the ribonucleoprotein core. M1 has been shown to have several functions (Lamb and Krug, 2001) including regulation of nuclear export of vRNPs, both permitting the transport of vRNP particles into the nucleus upon infection and preventing newly exported vRNP particles from re-entering the nucleus. The 97 amino acid M2 protein is a homo-tetrameric integral membrane protein that exhibits ion channel activity and is the target of the drug amantadine (Hay et al., 1985). The ion channel activity of M2 is important both during virion uncoating and during viral budding (Lamb and Krug, 2001).

Five amino acid sites have been identified in the transmembrane region of the M2 protein that are involved in resistance to the antiviral drug, amantadine: sites 26, 27, 30, 31 and 34 (Holsinger et al., 1994). The 1918 influenza M2 sequence is identical at these positions to that of the amantadine sensitive influenza virus strains. Thus, it was predicted that the M2 protein of the 1918 influenza virus would be sensitive to amantadine. This was recently demonstrated experimentally. A recombinant virus possessing the 1918 matrix segment was inhibited effectively both in tissue culture and *in vivo* by the M2 ion-channel inhibitors amantadine and rimantadine (Tumpey et al., 2002).

The phylogenetic analyses suggest that the 1918 matrix genes, while more avian-like than those of other mammalian influenza viruses, were mammalian adapted (Reid et al., 2002). For example, the extracellular domain of the M2 protein contains 4 amino acids that differ consistently between the avian and mammalian clades (M2 residues #14, 16, 18, and 20). The 1918 sequence matches the mammalian sequence at all four of these residues (Reid et al., 2002), suggesting that the matrix segment may have been circulating in humans virus strains for at least several years before 1918.

Sequence and functional analysis of the nucleoprotein gene segment

The nucleoprotein gene (NP) of the 1918 pandemic influenza A virus has been amplified and sequenced from archival material (Reid et al., 2004). The NP gene is known to be involved in many aspects of viral function and to interact with host proteins, thereby playing a role in host specificity (Portela and Digard, 2002). NP is highly conserved, with a maximum amino acid difference of 11% among virus strains probably because it must bind to multiple proteins, both viral and cellular. Numerous studies suggest that NP is a major determinant of host specificity (Scholtissek et al., 1985; Scholtissek et al., 1978a). The 1918 NP amino acid sequence differs at only 6 amino acids from avian consensus sequences, consistent with reassortment from an avian source shortly before 1918. However, the 1918 NP nucleotide sequence has more than 170 differences from avian consensus sequences, suggesting substantial evolutionary distance from known avian sequences. Both the 1918 NP gene and protein sequences fall within the mammalian clade upon phylogenetic analysis.

Phylogenetic analyses of NP sequences from many virus strains result in trees with two main branches, one consisting of mammalian-adapted virus strains and one of avian-adapted virus strains (Gammelin et al., 1990; Gorman et al., 1991; Shu et al., 1993). The NP gene segment was not replaced in the pandemics of 1957 and 1968, so it is likely that the sequences in the mammalian clade are descended from the 1918 NP segment. The mammalian branches, unlike the avian branch, show a slow but steady accumulation of changes over time. Extrapolation of the rate of change along the human branch back to a putative common ancestor suggests that this NP entered the mammalian lineage sometime after 1900 (Gammelin et al., 1990; Gorman et al., 1991; Shu et al., 1993). Separate analyses of synonymous and nonsynonymous substitutions also placed the 1918 virus NP gene in the mammalian clade (Reid et al., 2004). When synonymous substitutions were analyzed the 1918 virus gene was placed within and near the root of swine viruses. When nonsynonymous viruses were analyzed the 1918 virus gene was placed within and near the root of the human viruses.

The evolutionary distance of the 1918 NP from avian and mammalian sequences was examined using several different parameters. There are at least three possibilities for the origin of the 1918 NP gene segment (Reid et al., 2004). First, it could have been retained from the previously circulating human virus, as was the case with the 1957 and 1968 pandemic virus strains, whose NP segments are descendants of the 1918 NP. The large number of nucleotide changes from the avian consensus and the placement of the 1918 sequence in the mammalian clade are consistent with this hypothesis. Neighbor-joining (NJ) analyses of nonsynonymous nucleotide sequences or of amino acid sequences place the 1918 sequence within and near the root of the human clade. The 1918 NP has only a few amino acid differences from most bird virus strains but this consistent group of amino acid changes is shared by the 1918 NP and its subsequent mammalian descendants and is not found in any birds, resulting in the 1918 sequence being placed outside the avian clade (Reid et al., 2004). One or more of these amino acid substitutions may be important for adaptation of the protein to humans. However, the very small number of amino acid differences from the avian consensus argues for recent introduction from birds – 80 years after 1918, the NP genes of human influenza virus strains have accumulated over 30 additional amino acid differences from the avian consensus (a rate of 2.3 amino acid changes per year). Thus it seems unlikely that the 1918 NP, with only 6 amino acid differences from the avian consensus, could have been in humans for many years before 1918. This conclusion is supported by the regression analysis that suggests that the progenitor of the 1918 virus probably entered the human population around 1915 (Reid et al., 2004).

A second possible origin for the 1918 NP segment is direct reassortment from an avian virus. The small number of amino acid differences between 1918 and the avian consensus supports this hypothesis. While 1918 varies at many nucleotides from the nearest avian virus strain, avian virus strains are quite diverse at the nucleotide level. Synonymous/non-synonymous ratios between 1918 and avian virus strains are similar to the ratios between avian virus strains, opening the possibility that avian virus strains may exist that are more closely related to 1918. The great evolutionary distance between the 1918 sequence and the avian consensus suggests that no avian virus strain similar to those in the currently identified clades could have provided the 1918 virus strain with its NP segment.

A final possibility is that the 1918 gene segment was acquired shortly before 1918 from a source not currently represented in the database of influenza sequences. There may be a currently unknown influenza host that, while similar to currently characterized avian virus strains at the amino acid level, is quite different at the nucleotide level. It is possible that such a host was the source of the 1918 NP segment (Reid et al., 2004).

FUTURE WORK

Five of the eight RNA segments of the 1918 influenza virus have been sequenced and analyzed. Their characterization has shed light on the origin of the virus and strongly supports the hypothesis that the 1918 virus was the common ancestor of both subsequent human and swine H1N1 lineages. Sequence analysis of the genes to date offers no definitive clue as to the exceptional virulence of the 1918 virus strain. Thus, experiments testing models of virulence using reverse genetics approaches with 1918 influenza genes have begun.

In future work it is hoped that the 1918 pandemic virus strain can be placed in the context of influenza virus strains that preceded it and followed it. The direct precursor of the pandemic virus, the first or 'spring' wave virus strain, lacked the exceptional virulence of the fall wave virus strain. Identification of an influenza RNA-positive case from the first wave would have tremendous value in deciphering the genetic basis for virulence by allowing differences in the sequences to be highlighted. Identification of pre-1918 human influenza RNA samples would clarify which gene segments were novel in the 1918 virus.

In many respects, the 1918 influenza pandemic was similar to other influenza pandemics. In its epidemiology, disease course and pathology, the pandemic generally was different in degree but not in kind from previous and subsequent pandemics. Furthermore, laboratory experiments using recombinant influenza viruses containing genes from the 1918 virus suggest that the 1918 and 1918-like viruses would be as sensitive to the FDA-approved anti-influenza drugs rimantadine and oseltamivir as other virus strains (Tumpey et al., 2002). However, there are some characteristics of the pandemic that appear to be unique: Mortality was exceptionally high, ranging from five to twenty times higher than normal. Clinically and pathologically, the high mortality appears to be the result of a higher proportion of severe and complicated infections of the respiratory tract, not with systemic infection or involvement of organ systems outside the influenza virus' normal targets. The mortality was concentrated in an unusually young age group. Finally, the waves of influenza activity followed on each other unusually rapidly, resulting in three major outbreaks within a year's time. Each of these unique characteristics may find their explanation in genetic features of the 1918 virus. The challenge will be in determining the links between the biological capabilities of the virus and the known history of the pandemic.

Acknowledgements
This work has been partially supported by NIH grants to JKT and PP, and by grants from the Veterans Administration and the American Registry of Pathology (JKT). PP is a Senior Fellow of the Ellison Medical Foundation. We also thank the following colleagues and their groups for wonderful and stimulating collaborations during the last several years: Chris Basler, Roger Bumgarner, Thomas Fanning, Adolfo García-Sastre, Michael Katze, David Swayne, Terry Tumpey, Ian Wilson, and especially Ann Reid.

References

Barry, J. M. (2004). The Great Influenza: The Epic Story of the Deadliest Plague In History (New York, NY: Viking Press).

Basler, C. F., Reid, A. H., Dybing, J. K., Janczewski, T. A., Fanning, T. G., Zheng, H., Salvatore, M., Perdue, M. L., Swayne, D. E., Garcia-Sastre, A., et al. (2001). Sequence of the 1918 pandemic influenza virus nonstructural gene (NS) segment and characterization of recombinant viruses bearing the 1918 NS genes. Proc. Natl. Acad. Sci. USA. 98, 2746–2751.

Beveridge, W. (1977). Influenza: The Last Great Plague, an unfinished story of discovery (New York: Prodist).

Brown, I. H., Chakraverty, P., Harris, P. A., and Alexander, D. J. (1995). Disease outbreaks in pigs in Great Britain due to an influenza A virus of H1N2 subtype. Vet. Rec 136, 328–329.

Brown, I. H., Harris, P. A., McCauley, J. W., and Alexander, D. J. (1998). Multiple genetic reassortment of avian and human influenza A viruses in European pigs, resulting in the emergence of an H1N2 virus of novel genotype. J. Gen. Virol. 79, 2947–2955.

Burnet, F., and Clark, E. (1942). Influenza: A Survey of the Last 50 Years in the Light of Modern Work on the virus of Epidemic Influenza (Melbourne: MacMillan).

Castrucci, M. R., Donatelli, I., Sidoli, L., Barigazzi, G., Kawaoka, Y., and Webster, R. G. (1993). Genetic reassortment between avian and human influenza A viruses in Italian pigs. Virology 193, 503–506.

Caton, A. J., Brownlee, G. G., Yewdell, J. W., and Gerhard, W. (1982). The antigenic structure of the influenza virus A/PR/8/34 hemagglutinin (H1 subtype). Cell 31, 417–427.

Chun, J. (1919). Influenza Including its Infection Among Pigs. National Medical Journal (of China) 5, 34–44.

Claas, E. C., Osterhaus, A. D., van Beek, R., De Jong, J. C., Rimmelzwaan, G. F., Senne, D. A., Krauss, S., Shortridge, K. F., and Webster, R. G. (1998). Human influenza A H5N1 virus related to a highly pathogenic avian influenza virus. Lancet 351, 472–477.

Colman, P. M., Varghese, J. N., and Laver, W. G. (1983). Structure of the catalytic and antigenic sites in influenza virus neuraminidase. Nature 303, 41–44.

Cox, N. J., and Subbarao, K. (2000). Global epidemiology of influenza: past and present. Annu. Rev. Med. 51, 407–421.

Crosby, A. (1989). America's Forgotten Pandemic (Cambridge: Cambridge University Press).

Dimoch, W. W. (1918–19). Diseases of Swine. J. Amer. Vet. Med. Assn. 54, 321–340.

Dorset, M., McBryde, C. N., and Niles, W. B. (1922–23). Remarks on 'hog' flu. J. Amer. Vet. Med. Assn. 62, 162–171.

Dowdle, W. R. (1999). Influenza A virus recycling revisited. Bull. World Health Organ. 77, 820–828.

Fanning, T. G., Reid, A. H., and Taubenberger, J. K. (2000). Influenza A virus neuraminidase: regions of the protein potentially involved in virus-host interactions. Virology 276, 417–423.

Fanning, T. G., Slemons, R. D., Reid, A. H., Janczewski, T. A., Dean, J., and Taubenberger, J. K. (2002). 1917 avian influenza virus sequences suggest that the 1918 pandemic virus did not acquire its hemagglutinin directly from birds. J. Virol. 76, 7860–7862.

Fanning, T. G., and Taubenberger, J. K. (1999). Phylogenetically important regions of the influenza A H1 hemagglutinin protein. Virus Res. 65, 33–42.

Fitch, W., Leiter, J., Li, X., and Palese, P. (1991). Positive Darwinian evolution in human influenza A viruses. Proc. Natl. Acad. Sci. USA. 88, 4270–4274.

Fouchier, R. A., Schneeberger, P. M., Rozendaal, F. W., Broekman, J. M., Kemink, S. A., Munster, V., Kuiken, T., Rimmelzwaan, G. F., Schutten, M., Van

Doornum, G. J., et al. (2004). Avian influenza A virus (H7N7) associated with human conjunctivitis and a fatal case of acute respiratory distress syndrome. Proc. Natl. Acad. Sci. USA. 101, 1356–1361.

Frost, W. (1920). Statistics of Influenza Morbidity. Public Health Reports 35, 584–597.

Gambaryan, A., Tuzikov, A., Piskarev, V., Yamnikova, S., Lvov, D., Robertson, J., Bovin, N., and Matrosovich, M. (1997). Specification of Receptor-Binding Phenotypes of Influenza Virus Isolates from Different Hosts Using Synthetic Sialylglycopolymers: Non-Egg-Adapted Human H1 and H3 Influenza A and Influenza B Viruses Share a Common High Binding Affinity for 6'-Sialyl(N-acetyllactosamine). Virology 232, 345–350.

Gamblin, S. J., Haire, L. F., Russell, R. J., Stevens, D. J., Xiao, B., Ha, Y., Vasisht, N., Steinhauer, D. A., Daniels, R. S., Elliot, A., et al. (2004). The structure and receptor binding properties of the 1918 influenza hemagglutinin. Science 303, 1838–1842.

Gammelin, M., Altmuller, A., Reinhardt, U., Mandler, J., Harley, V., Hudson, P., Fitch, W., and Scholtissek, C. (1990). Phylogenetic analysis of nucleoproteins suggests that human influenza A viruses emerged from a 19th-century avian ancestor. Mol. Biol. Evol. 7, 194–200.

Garcia-Sastre, A. (2002). Mechanisms of inhibition of the host interferon alpha/beta-mediated antiviral responses by viruses. Microbes Infect. 4, 647–655.

Garcia-Sastre, A., Egorov, A., Matassov, D., Brandt, S., Levy, D. E., Durbin, J. E., Palese, P., and Muster, T. (1998). Influenza A virus lacking the NS1 gene replicates in interferon-deficient systems. Virology 252, 324–330.

Gaydos, J., Hodder, R., Top, F. J., Soden, V., Allen, R., Bartley, J., Zabkar, J., Nowosiwsky, T., and Russell, P. (1977). Swine influenza A at Fort Dix, New Jersey (January–February 1976). I. Case finding and clinical study of cases. J. Infect. Dis. 136, S356–S362.

Geiss, G. K., Salvatore, M., Tumpey, T. M., Carter, V. S., Wang, X., Basler, C. F., Taubenberger, J. K., Bumgarner, R. E.,

Palese, P., Katze, M. G., and Garcia-Sastre, A. (2002). Cellular transcriptional profiling in influenza A virus-infected lung epithelial cells: the role of the nonstructural NS1 protein in the evasion of the host innate defense and its potential contribution to pandemic influenza. Proc. Natl. Acad. Sci. USA. 99, 10736–10741.

Gensheimer, K. F., Fukuda, K., Brammer, L., Cox, N., Patriarca, P. A., and Strikas, R. A. (1999). Preparing for pandemic influenza: the need for enhanced surveillance. Emerg. Infect. Dis. 5, 297–299.

Glaser, L., Stevens, J., Zamarin, D., Wilson, I. A., García-Sastre, A., Tumpey, T. M., Basler, C. F., Taubenberger, J. K., and Palese, P. (2005). A Single Amino Acid Substitution in the 1918 Influenza Virus Hemagglutinin Changes the Receptor Binding Specificity. J. Virol. 79, 11533–11536.

Gorman, O., Bean, W., Kawaoka, Y., Donatelli, I., Guo, Y., and Webster, R. (1991). Evolution of influenza A virus nucleoprotein genes: implications for the origins of H1N1 human and classical swine viruses. J. Virol. 65, 3704–3714.

Grove, R. D., and Hetzel, A. M. (1968). Vital Statistics Rates in the United States: 1940–1960 (Washington, DC: US Government Printing Office).

Hay, A., Wolstenholme, A., Skehel, J., and Smith, M. (1985). The molecular basis of the specific anti-influenza action of amantadine. EMBO 4, 3021–3024.

Holsinger, L. J., Nichani, D., Pinto, L. H., and Lamb, R. A. (1994). Influenza A virus M2 ion channel protein: a structure-function analysis. J. Virol. 68, 1551–1563.

Johnson, N. P., and Mueller, J. (2002). Updating the accounts: global mortality of the 1918–1920 'Spanish' influenza pandemic. Bull. Hist. Med. 76, 105–115.

Jordan, E. (1927). Epidemic Influenza: A Survey (Chicago: American Medical Association).

Kanegae, Y., Sugita, S., Sortridge, K., Yoshioka, Y., and Nerome, K. (1994). Origin and Evolutionary Pathways of the H1 Hemagglutinin gene of Avian, Swine and Human Influenza Viruses: Cocirculation of two distinct lineages of swine viruses. Arch. Virol. 134, 17–28.

Kash, J. C., Basler, C. F., Garcia-Sastre, A., Carter, V., Billharz, R., Swayne, D. E., Przygodzki, R. M., Taubenberger, J. K., Palese, P., Katze, M. G., and Tumpey, T. M. (2004). The Global Host Immune Response: Contribution of HA and NA Genes from the 1918 Spanish Influenza to Viral Pathogenesis. J. Virol. 78, 9499–9511.

Katz, J., Lim, W., Bridges, C., Rowe, T., Hu-Primmer, J., Lu, X., Abernathy, R., Clarke, M., Conn, L., Kwong, H., et al. (1999). Antibody response in individuals infected with avian influenza A (H5N1) viruses and detection of anti-H5 antibody among household and social contacts. J. Infect. Dis. 180, 1763–1770.

Kawaoka, Y., Krauss, S., and Webster, R. G. (1989). Avian-to-human transmission of the PB1 gene of influenza A viruses in the 1957 and 1968 pandemics. J. Virol. 63, 4603–4608.

Kawaoka, Y., and Webster, R. G. (1988). Molecular mechanism of acquisition of virulence in influenza virus in nature. Microb Pathog 5, 311–318.

Kilbourne, E. (1977). Influenza pandemics in perspective. J. Amer. Med. Assoc. 237, 1225–1228.

Koen, J. S. (1919). A practical method for field diagnoses of swine diseases. Am. J. Vet. Med. 14, 468–470.

Kolata, G. B. (1999). Flu: The Story of the Great Influenza Pandemic of 1918 and the Search for the Virus That Caused It. (New York City: Farrar Straus & Giroux).

Koopmans, M., Wilbrink, B., Conyn, M., Natrop, G., van der Nat, H., Vennema, H., Meijer, A., van Steenbergen, J., Fouchier, R., Osterhaus, A., and Bosman, A. (2004). Transmission of H7N7 avian influenza A virus to human beings during a large outbreak in commercial poultry farms in the Netherlands. Lancet 363, 587–593.

Krug, R. M., Yuan, W., Noah, D. L., and Latham, A. G. (2003). Intracellular warfare between human influenza viruses and human cells: the roles of the viral NS1 protein. Virology 309, 181–189.

Kupradinun, S., Peanpijit, P., Bhodhikosoom, C., Yoshioka, Y., Endo, A., and Nerome, K. (1991). The first isolation of swine H1N1 influenza viruses from pigs in Thailand. Arch. Virol. *118*, 289–297.

Lamb, R., and Krug, R. (2001). Orthomyxoviridae: The Viruses and Their Replication, In Fields Virology, D. Knipe, and P. Howley, eds. (Philadelphia, PA: Lippincott Williams & Wilkins), pp. 1487–1531.

Lamb, R. A., and Lai, C. J. (1980). Sequence of interrupted and uninterrupted mRNAs and cloned DNA coding for the two overlapping nonstructural proteins of influenza virus. Cell *21*, 475–485.

Lazarowitz, S. G., and Choppin, P. W. (1975). Enhancement of the infectivity of influenza A and B viruses by proteolytic cleavage of the hemagglutinin polypeptide. Virology *68*, 440–454.

LeCount, E. R. (1919). The pathologic anatomy of influenzal bronchopneumonia. J. Am. Med. Assoc. *72*, 650–652.

Li, S., Schulman, J., Itamura, S., and Palese, P. (1993). Glycosylation of neuraminidase determines the neurovirulence of influenza A/WSN/33 virus. J. Virol. *67*, 6667–6673.

Li, Y., Yamakita, Y., and Krug, R. (1998). Regulation of a nuclear export signal by an adjacent inhibitory sequence: the effector domain of the influenza virus NS1 protein. Proc. Natl. Acad. Sci. USA. *95*, 4864–4869.

Linder, F. E., and Grove, R. D. (1943). Vital Statistics Rates in the United States: 1900–1940 (Washington, D.C.: Government Printing Office).

Lipsman, J. (2004). H7N2 Avian Influenza Identified in Westchester Resident (Westchester County Department of Health).

Ludendorff, E. (1919). Meine Kriegserinnerungen 1914–1918 (Berlin: Ernst Siegfried Mittler und Sohn Verlagsbuchhandlung).

Ludwig, S., Stitz, L., Planz, O., Van, H., Fitch, W., and Scholtissek, C. (1995). European Swine Virus as a Possible Source for the Next Influenza Pandemic? Virology *212*, 551–561.

Marks, G., and Beatty, W. K. (1976). Epidemics (New York City: Scribner).

Matrosovich, M., Gambaryan, A., Teneberg, S., Piskarev, V., Yamnikova, S., Lvov, D., Robertson, J., and Karlsson, K. (1997). Avian Influenza A Viruses Differ from Human Viruses by Recognition of Sialyloigosaccharides and Gangliosides and by a Higher Conservation of the HA Receptor-Binding Site. Virology *233*, 224–234.

Ministry of Health, U. K. (1960). The influenza epidemic in England and Wales 1957–1958, In Reports on Public Health and Medical Subjects (London: Ministry of Health).

Monto, A. S., Iacuzio, D. A., and La Montaigne, J. R. (1997). Pandemic Influenza: Confronting a Re-emergent Threat. J. Infect. Dis. *176*, S1-S3.

Murray, C., and Biester, H. E. (1930). Swine influenza. J. Amer. Vet. Med. Assn. *76*, 349–355.

Nerome, K., Ishida, M., Oya, A., and Oda, K. (1982). The possible origin H1N1 (Hsw1N1) virus in the swine population of Japan and antigenic analysis of the isolates. J. Gen. Virol. *62*, 171–175.

O'Neill, R. E., Talon, J., and Palese, P. (1998). The influenza virus NEP (NS2 protein) mediates the nuclear export of viral ribonucleoproteins. Embo J. *17*, 288–296.

Palese, P., and Compans, R. W. (1976). Inhibition of influenza virus replication in tissue culture by 2-deoxy- 2,3-dehydro-N-trifluoroacetylneuraminic acid (FANA): mechanism of action. J. Gen. Virol. *33*, 159–163.

Patterson, K. D., and Pyle, G. F. (1991). The Geography and Mortality of the 1918 Influenza Pandemic. Bull. Hist. Med. *65*, 4–21.

Peiris, J. S., Yu, W. C., Leung, C. W., Cheung, C. Y., Ng, W. F., Nicholls, J. M., Ng, T. K., Chan, K. H., Lai, S. T., Lim, W. L., et al. (2004). Re-emergence of fatal human influenza A subtype H5N1 disease. Lancet *363*, 617–619.

Philip, R. N., and Lackman, D. B. (1962). Observations on the present distribution of inlfuenza A/swine antibodies among Alaskan natives relative to the occurrence of influenza in 1918–1919. Amer. J. Hygiene *75*, 322–334.

Portela, A., and Digard, P. (2002). The influenza virus nucleoprotein: a multifunctional RNA-binding protein pivotal to virus replication. J. Gen. Virol. *83*, 723–734.

Reid, A. H., Fanning, T. G., Hultin, J. V., and Taubenberger, J. K. (1999). Origin and evolution of the 1918 'Spanish' influenza virus hemagglutinin gene. Proc. Natl. Acad. Sci. USA. *96*, 1651–1656.

Reid, A. H., Fanning, T. G., Janczewski, T. A., Lourens, R., and Taubenberger, J. K. (2004). Novel origin of the 1918 pandemic influenza virus nucleoprotein gene segment. J. Virol. *78*, 12462–12470.

Reid, A. H., Fanning, T. G., Janczewski, T. A., McCall, S., and Taubenberger, J. K. (2002). Characterization of the 1918 'Spanish' influenza virus matrix gene segment. J. Virol. *76*, 10717–10723.

Reid, A. H., Fanning, T. G., Janczewski, T. A., and Taubenberger, J. K. (2000). Characterization of the 1918 'Spanish' influenza virus neuraminidase gene. Proc. Natl. Acad. Sci. USA. *97*, 6785–6790.

Reid, A. H., Janczewski, T. A., Lourens, R. M., Elliot, A. J., Daniels, R. S., Berry, C. L., Oxford, J. S., and Taubenberger, J. K. (2003). 1918 Influenza Pandemic Caused By Highly Conserved Viruses With Two Receptor-Binding Variants. Emerging Infectious Diseases 9. 1249–53.

Reid, A. H., and Taubenberger, J. K. (1999). The 1918 flu and other influenza pandemics: 'over there' and back again. Lab. Invest. *79*, 95–101.

Reid, A. H., and Taubenberger, J. K. (2003). The origin of the 1918 pandemic influenza virus: a continuing enigma. J. Gen. Virol. *84*, 2285–2292.

Rosenau, M. J., and Last, J. M. (1980). Maxcy-Rosenau Preventative Medicine and Public Health (New York City: Appleton-Century-Crofts).

Rott, R., Klenk, H. D., Nagai, Y., and Tashiro, M. (1995). Influenza viruses, cell enzymes, and pathogenicity. Amer. J. Respir Crit. Care Med. *152*, S16-19.

Schafer, J. R., Kawaoka, Y., Bean, W. J., Suss, J., Senne, D., and Webster, R. G. (1993). Origin of the pandemic 1957 H2 influenza A virus and the persistence of its possible progenitors in the avian reservoir. Virology *194*, 781–788.

Scholtissek, C. (1994). Source for influenza pandemics. Eur. J. Epidemiol. *10*, 455–458.

Scholtissek, C., Burger, H., Kistner, O., and Shortridge, K. F. (1985). The nucleoprotein as a possible major factor in determining host specificity of influenza H3N2 viruses. Virology *147*, 287–294.

Scholtissek, C., Koennecke, I., and Rott, R. (1978a). Host range recombinants of fowl plague (influenza A) virus. Virology *91*, 79–85.

Scholtissek, C., Ludwig, S., and Fitch, W. (1993). Analysis of Influenza A Virus Nucleoproteins for the Assessment of Molecular Genetic Mechanisms Leading to new Phylogenetic Virus Lineages. Arch. Virol. *131*, 237–250.

Scholtissek, C., Rohde, W., Von Hoyningen, V., and Rott, R. (1978b). On the origin of the human influenza virus subtypes H2N2 and H3N2. Virology *87*, 13–20.

Schulze, I. T. (1997). Effects of glycosylation on the properties and functions of influenza virus hemagglutinin. Journal of Infectious Diseases *176(Suppl 1)*, S24-S28.

Seo, S. H., Hoffmann, E., and Webster, R. G. (2002). Lethal H5N1 influenza viruses escape host anti-viral cytokine responses. Nat. Med. *8*, 950–954.

Shope, R. (1958). Influenza: History, Epidemiology, and Speculation. Public Health Reports *73*, 165–178.

Shope, R. E. (1936). The incidence of neutralizing antibodies for swine influenza virus in the sera of human beings of different ages. J. Exp. Med. *63*, 669–684.

Shope, R. E., and Lewis, P. A. (1931). Swine influenza. J. Exp. Med. *54*.

Shu, L., Bean, W., and Webster, R. (1993). Analysis of the evolution and variation of the human influenza A virus nucleoprotein gene from 1933 to 1900. J. Virol. *67*, 2723–2729.

Simonsen, L., Clarke, M. J., Schonberger, L. B., Arden, N. H., Cox, N. J., and Fukuda, K. (1998). Pandemic versus Epidemic Influenza Mortality: A Pattern of Changing Age Distribution. Journal of Infectious Diseases *178*, 53–60.

Simonsen, L., Fukuda, K., Schonberger, L. B., and Cox, N. J. (2000). The Impact of Influenza Epidemics on Hospitalizations. J. Infect. Dis. *181*, 831–837.

Smith, W., Andrewes, C., and Laidlaw, P. (1933). A Virus Obtained from Influenza Patients. Lancet *225*, 66–68.

Stevens, J., Corper, A. L., Basler, C. F., Taubenberger, J. K., Palese, P., and Wilson, I. A. (2004). Structure of the uncleaved human H1 hemagglutinin from the extinct 1918 influenza virus. Science *303*, 1866–1870.

Subbarao, K., Klimov, A., Katz, J., Regnery, H., Lim, W., Hall, H., Perdue, M., Swayne, D., Bender, C., Huang, J., *et al.* (1998). Characterization of an avian influenza A (H5N1) virus isolated from a child with a fatal respiratory illness. Science *279*, 393–396.

Talon, J., Horvath, C. M., Polley, R., Basler, C. F., Muster, T., Palese, P., and Garcia-Sastre, A. (2000). Activation of interferon regulatory factor 3 is inhibited by the influenza A virus NS1 protein. J. Virol. *74*, 7989–7996.

Taubenberger, J. K., Reid, A. H., and Fanning, T. G. (2000). The 1918 influenza virus: A killer comes into view. Virology *274*, 241–245.

Taubenberger, J. K., Reid, A. H., Janczewski, T. A., and Fanning, T. G. (2001). Integrating historical, clinical and molecular genetic data in order to explain the origin and virulence of the 1918 Spanish influenza virus. Philos. Trans. R. Soc. Lond. B Biol. Sci. *356*, 1829–1839.

Taubenberger, J. K., Reid, A. H., Krafft, A. E., Bijwaard, K. E., and Fanning, T. G. (1997). Initial genetic characterization of the 1918 'Spanish' influenza virus. Science *275*, 1793–1796.

Thompson, W. W., Shay, D. K., Weintraub, E., Brammer, L., Cox, N., Anderson, L. J., and Fukuda, K. (2003). Mortality associated with influenza and respiratory syncytial virus in the United States. J. Amer. Med. Assoc. *289*, 179–186.

Tran, T. H., Nguyen, T. L., Nguyen, T. D., Luong, T. S., Pham, P. M., Nguyen, V. C., Pham, T. S., Vo, C. D., Le, T. Q., Ngo, T. T., *et al.* (2004). Avian influenza A (H5N1) in 10 patients in Vietnam. N. Engl. J. Med. *350*, 1179–1188.

Tumpey, T. M., Garcia-Sastre, A., Mikulasova, A., Taubenberger, J. K., Swayne, D. E., Palese, P., and Basler, C. F. (2002). Existing antivirals are effective against influenza viruses with genes from the 1918 pandemic virus. Proc. Natl. Acad. Sci. USA. *99*, 13849–13854.

Tumpey, T. M., Garcia-Sastre, A., Taubenberger, J. K., Palese, P., Swayne, D. E., and Basler, C. F. (2004). Pathogenicity and immunogenicity of influenza viruses with genes from the 1918 pandemic virus. Proc. Natl. Acad. Sci. USA. *101*, 3166–71.

United States Department of Commerce (1976). Historical Statistics of the United States: Colonial Times to 1970 (Washington, D.C.: Government Printing Office).

W.H.O. (2004). Avian influenza A(H7) human infections in Canada (WHO).

Wang, X., Li, M., Zheng, H., Muster, T., Palese, P., Beg, A. A., and Garcia–Sastre, A. (2000). Influenza A virus NS1 protein prevents activation of NF-kappaB and induction of alpha/beta interferon. J. Virol. *74*, 11566–11573.

Webby, R. J., and Webster, R. G. (2003). Are we ready for pandemic influenza? Science *302*, 1519–1522.

Webster, R., and Rott, R. (1987). Influenza Virus A Pathogenicity: The Pivotal Role of Hemagglutinin. Cell *50*, 665–666.

Webster, R. G., Bean, W. J., Gorman, O. T., Chambers, T. M., and Kawaoka, Y. (1992). Evolution and ecology of influenza A viruses. Microbiol. Rev. *56*, 152–179.

Webster, R. G., Sharp, G. B., and Claas, E. C. (1995). Interspecies transmission of influenza viruses. Amer. J. Respir. Crit. Care Med. *152*, S25-30.

Weis, W., Brown, J. H., Cusack, S., Paulson, J. C., Skehel, J. J., and Wiley, D. C. (1988). Structure of the influenza virus haemagglutinin complexed with its receptor, sialic acid. Nature *333*, 426–431.

Wilson, I. A., Skehel, J. J., and Wiley, D. C. (1981). Structure of the haemagglutinin membrane glycoprotein of influenza virus at 3 A resolution. Nature *289*, 366–373.

Winternitz, M. C., Wason, I. M., and McNamara, F. P. (1920). The Pathology of Influenza (New Haven: Yale University Press).

Wolbach, S. B. (1919). Comments on the pathology and bacteriology of fatal influenza cases, as observed at Camp Devens, Mass. Johns Hopkins Hospital Bulletin *30*, 104.

Woods, G. T., Schnurrenberger, P. R., Martin, R. J., and Tompkins, W. A. (1981). Swine influenza virus in swine and man in Illinois. J. Occup. Med. *23*, 263–267.

Zamarin, D., and Palese, P. (2004). Influenza virus: Lessons learned., In International Kilmer Conference Proceedings, J. B. Kowalski, and J. B. Morissey, eds. (Champlain, NY: Polyscience Publications, Inc.).

Zhou, N. N., Senne, D. A., Landgraf, J. S., Swenson, S. L., Erickson, G., Rossow, K., Liu, L., Yoon, K. J., Krauss, S., and Webster, R. G. (2000). Emergence of H3N2 reassortant influenza A viruses in North American pigs. Vet. Microbiol. *74*, 47–58.

CHAPTER 12

Signaling and apoptosis in influenza virus-infected cells

Stephan Ludwig

Abstract

Infection of cells with viruses commonly leads to the activation of a variety of different signaling pathways within the infected cells. It is quite obvious that DNA viruses or retroviruses with DNA intermediates in their replication cycle need these signaling processes to activate the cellular DNA and protein synthesis machinery. However, it is not so clear how RNA viruses, such as influenza viruses cope with intracellular signaling. Since intracellular signaling pathways represent key switches in the determination of cell fate, the knowledge about their activation and function in infected cells may help to unravel some of the secrets of virus-host cell interactions. In this overview we will focus on recent advances on the function of intracellular signaling pathways and the induction of the apoptotic program in influenza virus infected cells. The role of these signaling events in the constant struggle between efficient virus propagation and the innate antiviral defense will be discussed.

INTRACELLULAR SIGNALING CASCADES – MAP KINASES AND THE IKK/NF-κB MODULE

Cell fate decisions in response to extracellular agents, including pathogenic invaders are commonly mediated by phosphorylation-regulated signaling cascades that transduce signals into stimulus specific actions, e. g. changes in gene expression patterns or alterations in the metabolic state of the cell. Within this field MAP kinase cascades have gained much attention as being critical transducers to convert a variety of extracellular signals into a multitude of responses (English et al., 1999; Hazzalin and Mahadevan, 2002; Widmann et al., 1999). Thereby, these pathways regulate numerous cellular decision processes, such as proliferation and differentiation, but also cell activation and immune responses (Dong et al., 2002). Four different members of the MAPK family that are organized in separate cascades have been identified so far: ERK (extracellular signal regulated kinase), JNK (Jun-N-terminal kinase), p38 and BMK-1/ERK5 (Big MAP kinase) (Garrington and Johnson, 1999; Widmann et al., 1999). These MAP kinases are activated by a dual phosphorylation event on threonine and tyrosine mediated by MAP kinase kinases (MEKs or MKKs). The MAP kinase ERK is activated by the dual-specific kinase MEK that itself is activated by the serine threonine kinase Raf. Raf, MEK and ERK form the prototype module of a MAP kinase pathway and are also known as the classical mitogenic cascade. The MAP kinases p38 and JNK are activated by MKK3/6 and MKK4/7, respectively, and are predominantly activated by proinflammatory cytokines and certain environmental stress conditions. The MEK5/ERK5 module is both activated by mitogens and certain stress inducers. There is evidence that the different MAPK cascades are also activated upon infection with RNA viruses, including influenza viruses. Thus, these signaling cascades may serve different functions in viral replication and host cell response.

Another important signaling pathway which is commonly activated upon virus infection is the IκB kinase (IKK)/NF-κB signaling module (Hiscott et al., 2001). The NF-κB/IκB family of transcription factors promote the expression of well over 150 different genes, such as cytokine or chemokine genes, or genes encoding for adhesion molecules or anti- and pro-apoptotic proteins (Pahl, 1999). The canonical mechanism of NF-κB activation includes activation of IκB kinase (IKK) that phosphorylates the inhibitor of NF-κB, IκB and targets the protein for subsequent degradation (Delhase and Karin, 1999; Karin, 1999b). This leads to the release and migration of the transcriptionally active NF-κB factors, such as p65 or p50 to the nucleus (Ghosh, 1999; Karin and Ben-Neriah, 2000). The IKK complex consists of at least three isozymes of IKK, IKK1/IKKα, IKK2/IKKß and NEMO/IKKγ. The most important isozyme for NF-κB activation via the degradation of IκB is IKK2 (Karin, 1999a), while IKK1 seems to primarily phosphorylate other factors of the NF-κB/IKK family namely p100/p52 (Senftleben et al., 2001; Xiao et al., 2001). NEMO acts as a scaffolding protein for the large IKK complex (Courtois et al., 2001) that contains still other kinases such as MEKK1 (MAPK kinase kinase 1) (Lee et al., 1998), NIK (NF-κB inducing kinase) (Nemoto et al., 1998; Woronicz et al., 1997) and the dsRNA activated protein kinase PKR (Gil et al., 2000; Zamanian-Daryoush et al., 2000).

Both NF-κB and the JNK MAP kinase pathway regulate one of the most important antiviral gene expression events, the transcriptional induction of interferon beta (IFNß) (Maniatis et al., 1998). IFNß is one of the first antiviral cytokines to be expressed upon virus infection, initiating an auto-amplification loop to cause an efficient and strong type I IFN response (Taniguchi and Takaoka, 2002). The IFNß enhanceosome, which mediates the inducible expression of IFNß, carries binding sites for transcription factors of three families, namely the AP-1 family members and JNK targets c-Jun and ATF-2, the NF-κB factors p50 and p65, and the interferon-regulatory factors (IRFs) (Hiscott et al., 1999; Thanos and Maniatis, 1995). In the initial phase of a virus infection this promoter element has been demonstrated to specifically bind to constitutively expressed IRF-3 as a dimer, while other inducible factors, such as IRF-1 or IRF-7 are only bound during the late amplification phase (Taniguchi and Takaoka, 2002). AP-1 and NF-κB transcription factors are activated by a variety of stimuli, however, a strong IRF-3 activation is selectively induced upon RNA virus infection, specifically by the dsRNA which accumulates during replication (Lin et al., 1998; Yoneyama et al., 1998). Thus, IRF-3 is a main determinant of a strong virus- and dsRNA-induced IFNß response. While the signaling pathways leading to the activation of c-Jun and ATF-2 via JNK and NF-κB via IKK are well established there is only very limited information with regard to the intracellular signaling chains that mediate virus-induced IRF-3 activation, especially those which lead to the virus specific C-terminal phosphorylation (Servant et al., 2002; Servant et al., 2003). For a long time most attempts have failed to identify the crucial mediators but rather helped to exclude a variety of signaling components and pathways (Iwamura et al., 2001; Servant et al., 2001; Smith et al., 2001). Only recently a major breakthrough towards identification of the IRF-3 kinase was achieved (Fitzgerald et al., 2003; Sharma et al., 2003) which will be further discussed below.

With regard to DNA viruses and retroviruses with DNA intermediates in their replication cycle it is pretty obvious that these pathogens manipulate host-cell signaling to support their replication. DNA viruses require cells in S-phase to take advantage of the cellular DNA polymerases and therefore it is not surprising that they interfere with proliferative signaling and cell cycle regulators. The function of signaling pathways in cells infected by RNA viruses is not as clear. This chapter aims to collect and discuss recent findings how influenza viruses, interfere with intracellular signaling cascades and the apoptotic response, with a special focus on signaling through MAP kinase pathways and the IKK/NF-κB module.

INFLUENZA A VIRUS-INDUCED GENE EXPRESSION

Signaling in influenza virus infected cells has not been a focus of research for a long time, however, it has been noted quite early on, that infection of cells with these viruses leads to the induction of a variety of cytokine and chemokine genes, such as IFNα/β, TNF-α, IL-1ß, IL-6, IL-8, MCP-1, RANTES and many others in different cell types (reviewed in (Julkunen et al., 2000; Ludwig et al., 1999; Mogensen and Paludan, 2001). These findings have later been extended in transcriptional profiling approaches. In the first of these array assays the expression of 6.400 genes has been analyzed 4h and 8h post influenza virus infection with a multiplicity of infection of 50 (Geiss et al., 2001). All together 342 genes were deregulated in this study. This also included transcription factor genes, such as c-jun or the NF-κB factor p65 as well as signaling kinase genes such as the extracellular regulated kinase 3 (ERK3) gene (Geiss et al., 2001). Unfortunately for an unknown reason the study was performed in HeLa cells which do not support virus replication. Thus the relevance of the results to draw conclusion towards the molecular basis of viral pathogenicity is limited. In a later attempt the authors used the A549 lung epithelial cell line that is derived from the genuine target tissue of an influenza virus infection (Geiss et al., 2002). In this study 84 out of an array of 13.000 genes were deregulated in 8h post infection with the human influenza virus strain A/PR8/8/34 at a MOI below 10 (Geiss et al., 2002).

Unfortunately, in none of the studies the functional relevance of the deregulated genes for the outcome of influenza virus propagation has been directly addressed yet. Nevertheless, deregulation of such a variety of genes implicates a heavy engagement of intracellular signaling mediators during an influenza virus infection.

MAP KINASE CASCADES AND INFLUENZA VIRUS INFECTION: OPPOSITE ROLES OF THE JNK AND ERK PATHWAYS

Interestingly all four so far defined MAPK family members are activated upon an influenza virus infection (Kujime et al., 2000; Ludwig et al., 2001; Pleschka et al., 2001) (Korte et al., 2005, personal communication). While the function of BMK-1/ERK5 in virus-infected cells is still elusive, recent work has helped to get a clearer picture of the function of the p38, JNK and ERK signaling pathways.

By the use of specific kinase inhibitors p38 and JNK but not ERK have been linked to virus-induced expression of RANTES, a chemokine involved in the attraction of eosinophils during an inflammatory response (Kujime et al., 2000). In a more recent study using the same set of inhibitors the ERK and JNK, but not p38 pathway were shown to be involved in the expression of the inflammatory mediator cyclooxygenase (COX) and phosphorylation of cytosolic phospholipase A2 (cPLA2) in bronchial epithelial cells (Mizumura et al., 2003). Further, the inhibitors of all three MAPK pathways were effective to dose dependently-block prostaglandin E2 release by various extents (Mizumura et al., 2003), indicating that viral MAPK activation contributes to the onset of anti-inflammatory response. JNK and p38 activation has also been demonstrated *in vivo* in mice infected with a neurovirulent influenza A virus that caused lethal acute encephalitis, although it is not clear whether this activity is directly virus-induced or mediated by immunological or inflammatory responses

(Mori et al., 2003). In this study JNK but not p38 activity had been linked to the onset of apoptosis in the infected brain (Mori et al., 2003). In embryonic fibroblasts from mice genetically deficient for apoptosis-signal-regulated kinase (ASK-1) virus-induced p38 and JNK activation was blunted concomitant with an inhibition of caspase 3 activation and virus-induced apoptosis (Maruoka et al., 2003). ASK-1 is an ubiquitously expressed MAPK kinase kinase that activates the MAPK kinase 4/JNK and the MAPK kinase 6/p38 cascade. These findings not only identified a novel upstream component of virus induced-MAPK signaling but also linked activity of certain MAPKs to apoptosis induction (Maruoka et al., 2003).

The JNK subgroup of MAPKs further came into focus in the context of an influenza virus infection since a very early activation of activator-protein 1 (AP-1) transcription factors (Karin et al., 1997) was observed in productively infected cells (Ludwig et al., 2001). AP-1 factors include c-Jun and ATF-2 that are phosphorylated by JNKs to potentiate their transcriptional activity (Karin et al., 1997). Both factors are phosphorylated upon influenza virus infection (Ludwig et al., 2001; Ludwig et al., 2002). Accordingly, activation of JNK was observed with different virus strains in a variety of permissive cell lines (Kujime et al., 2000; Ludwig et al., 2001). JNK activation required productive replication and was induced by the accumulating RNA produced by the viral polymerase. As upstream activators in the viral context the MAPK kinases MKK4 and MKK7 have been identified. The AP-1 factors c-Jun and ATF-2 are critical for the expression of IFNß, a most potent antiviral cytokine (Stark et al., 1998). Accordingly, inhibition of the cascade by dominant-negative mutants of MKK7, JNK or c-Jun during a virus infection resulted in impaired transcription from the IFNß promoter and an enhanced virus production. Thus, the JNK pathway appears to be a crucial mediator of the antiviral response to an influenza virus infection by co-regulating IFNß expression (Ludwig et al., 2001).

The MAP kinase ERK is also activated upon productive influenza virus infection (Kujime et al., 2000), however, it appears to serve a mechanism that is beneficial for the virus (Pleschka et al., 2001). Strikingly, blockade of the pathway by specific inhibitors of the upstream kinase MEK, or dominant-negative mutants of ERK or the MEK activator Raf resulted in a strongly impaired growth of both, influenza A and B type viruses (Pleschka et al., 2001; Ludwig et al. 2004). Conversely, virus titers are enhanced in cells expressing active mutants of Raf or MEK (Ludwig et al., 2004). This indicates that activation of the Raf/MEK/ERK pathway, in contrast to the JNK cascade, is required for efficient virus growth. Strikingly, inhibition of the pathway did not affect viral RNA or protein synthesis (Pleschka et al., 2001). The pathway rather appears to control the active nuclear export of the viral RNP complexes. RNPs are readily retained in the nucleus upon blockade of the signaling pathway. Most likely this is due to an impaired activity of the viral nuclear export protein NEP (Pleschka et al., 2001). This indicates that active RNP export is an induced rather than a constitutive event, a hypothesis supported by a late activation of ERK in the viral life cycle. So far the detailed mechanism of how ERK regulates export of the RNPs is unsolved. It is most likely that this does not directly occur via phosphorylation of a viral protein involved in transport (Pleschka et al., 2001) but rather by control of a certain cellular export factor. It is striking that MEK inhibitors are not toxic for the cell while more general blockers of the active transport machinery, such as leptomycin B exert a high toxicity even in quite low concentrations. This may indicate that MEK inhibitors are no general export blockers but only block a distinct nuclear export pathway. Indeed there are first evidences that the classical mitogenic cascade specifically regulates nuclear export of certain cellular

RNA-protein complexes. In LPS treated mouse macrophages MEK-inhibition results in a specific retention of the TNF mRNA in the nucleus (Dumitru et al., 2000). This is also observed in cells deficient for Tpl-2, an activator of MEK and ERK. In these cells the failure to activate MEK and ERK by LPS again correlated with TNF mRNA retention while other cytokines are normally expressed (Dumitru et al., 2000). Thus the ERK-pathway may regulate a specific cellular export process but leaves other export mechanisms unaffected. It is likely that such a specific export pathway is employed by influenza A and B viruses.

The finding of an antiviral action of MEK inhibitors prompted further research showing that replication of other viruses, such as Borna disease virus (Planz et al., 2001), Visna virus (Barber et al., 2002) or Coxsackie B3 virus (Luo et al., 2002) is also impaired upon MEK inhibition.

Requirement of Raf/MEK/ERK activation for efficient virus replication may suggests that this pathway may be a cellular target for antiviral approaches. Besides the antiviral action against both, A and B type viruses, MEK inhibitors meet two further criteria which are a prerequisite for a potential clinical use. Although targeting an important signaling pathway in the cell the inhibitors showed a surprisingly little toxicity (a) in cell culture (Planz et al., 2001; Pleschka et al., 2001), (b) in an *in vivo* mouse model (Sebolt-Leopold et al., 1999), and (c) in clinical trials for the use as anti-cancer agent (Cohen, 2002). In the light of these findings it was hypothesized that the mitogenic pathway may only be of major importance during early development of an organism and may be dispensable in adult tissues (Cohen, 2002). Another very important feature of MEK inhibitors is that they showed no tendency to induce formation of resistant virus variants (Ludwig et al., 2004). Although targeting of a cellular factor may still raise the concern about side effects of a drug, it appears likely that local administration of an agent such as a MEK inhibitor to the primary site of influenza virus infection, the lung, is well tolerated. Here the drug primarily affects differentiated lung epithelial cells for which a proliferative signaling cascade like the Raf/MEK/ERK cascade may be dispensable. Following this approach it was recently demonstrated that the MEK inhibitor U0126 is effective in reducing virus titers in the lung of infected mice after local administration (Planz et al., 2003, personal communication).

PROTEIN KINASE C: A VIRAL ENTRY REGULATOR

Activation of the classical mitogenic Raf/MEK/ERK cascade is initiated by yet other phosphorylation events. The kinase Raf is known to be regulated by phosphorylation of different upstream kinases including members of the protein kinase C family (Cai et al., 1997; Kolch et al., 1993).

The protein kinase C superfamily consists of at least 12 different PKC isoforms that carry out diverse regulatory roles in cellular processes by linking into several downstream signaling pathways (Toker, 1998). Beside a regulation of the Raf/MEK/ERK cascade and other downstream pathways, PKCs may have additional functions during viral replication. A role of PKCs in the process of entry of several enveloped viruses has been proposed based on the action of protein kinase inhibitors H7 and staurosporine (Constantinescu et al., 1991). Influenza virus infection or treatment of cells with purified viral hemagglutinin results in rapid activation of PKCs upon binding to host-cell surface receptors (Arora and Gasse, 1998; Kunzelmann et al., 2000; Rott et al., 1995). In a recent study it was shown that the pan-PKC inhibitor bisindolylmaleimide I prevented influenza virus entry and subsequent infection in a dose dependent and reversible manner (Root et al., 2000). Using a dominant-negative mutant approach this function was assigned to the PKCßII isoform:

Overexpression of a phosphorylation-deficient mutant of PKCßII revealed that the kinase is a regulator of late endosomal sorting. Accordingly, expression of the PKCßII mutant resulted in a block of virus entry at the level of late endosomes (Sieczkarski et al., 2003a; Sieczkarski and Whittaker, 2003b).

INFLUENZA VIRUS AND THE IKK/NF-κB PATHWAY

Activation of the transcription factor NF-κB is a hallmark of most infections by viral pathogens (Hiscott et al., 2001) including influenza viruses (reviewed in Julkunen et al., 2000; Ludwig et al., 2003; Ludwig et al., 1999). Influenza viral NF-κB induction involves activation of IκB kinase (IKK) (Wurzer et al., 2004) and is also achieved with isolated influenza virus components. This includes dsRNA (Chu et al., 1999) or over-expression of the viral HA, NP or M1 proteins (Flory et al., 2000). Since gene expression of many proinflammatory or antiviral cytokines, such as IFNß or TNF-α, is controlled by NF-κB the concept emerged that IKK and NF-κB are essential components in the innate immune response to virus infections (Chu et al., 1999). Accordingly, influenza virus-induced IFNß promoter activity is strongly impaired in cells expressing transdominant negative mutants of IKK2 or IκBα (Wang et al., 2000; Wurzer et al., 2004). Recently it has been demonstrated using dsRNA or Sendai virus infection, that homologs of IKK, namely IKK-ε and TANK-binding kinase (TBK-1) are activators of IRF-3, another important regulator of the IFNß gene and other genes involved in the type I interferon response (Fitzgerald et al., 2003; Sharma et al., 2003). There is now evidence that influenza virus also activates these enzymes (Ehrhardt et al., 2004) which concomitantly results in a significant IRF-3 activation (Kim et al., 2002).

Nevertheless, IKK and NF-κB might not only have antiviral functions. If influenza virus titers from different host cells were compared to titers of viruses grown in cells with impaired NF-κB signaling, a dramatic reduction could be observed (Wurzer et al., 2004). This suggests that at least in cell culture NF-κB activity may also be required to some extent for efficient influenza virus replication.

INFLUENZA VIRUS-INDUCED PROGRAMMED CELL DEATH

Another cellular signaling response commonly observed upon virus infections, including influenza virus is the induction of the apoptotic cascade. Apoptosis is a morphologically and biochemically defined form of cell death (Kerr et al., 1972) and has been demonstrated to play a role in a variety of diseases including virus infections (Razvi and Welsh, 1995). Apoptosis is mainly regarded to be a host cell defense against virus infections since many viruses express anti-apoptotic proteins to prevent this cellular response. The central component of the apoptotic machinery is a proteolytic system consisting of a family of cysteinyl proteases, termed caspases (for review see (Cohen, 1997; Thornberry and Lazebnik, 1998). Two groups of caspases can be distinguished: upstream initiator caspases such as caspase 8 or caspase 9 which cleave and activate other caspases and downstream effector caspases, including caspase 3, 6 and 7 which cleave a variety of cellular substrates thereby disassembling cellular structures or inactivating enzymes (Thornberry and Lazebnik, 1998). Caspase 3 is the most intensively studied effector caspase. Work on MCF-7 breast carcinoma cells which are deficient in caspase 3 due to a deletion in the *Casp3* gene has revealed the existence of a crucial caspase 3 driven feedback loop which mediates the apoptotic process (Janicke et al., 1998; Slee et al., 1999). Thus, caspase 3 is a central player in apoptosis regulation and the level of procaspase 3 in the cell determines the impact of a given apoptotic stimulus.

It is long known that influenza virus infection with A and B type viruses results in the induction of apoptosis both in permissive and unpermissive cultured cells as well as *in vivo* (Fesq et al., 1994; Hinshaw et al., 1994; Ito et al., 2002; Lowy, 2003; Mori et al., 1995; Takizawa et al., 1993). Interestingly, viral activation of MAPKs or upstream kinases has been linked to the onset of apoptosis. In a mouse model for a neurovirulent influenza infection, JNK but not p38 activity correlated with apoptosis induction in the infected brain (Mori et al., 2003). In embryonic fibroblasts deficient for the MAPK kinase kinase ASK-1 virus-induced p38 and JNK activation was blunted concomitant with an inhibition of caspase 3 activation and virus-induced apoptosis (Maruoka et al., 2003). As an extrinsic mechanism of viral apoptosis induction it has been noted quite early on that the Fas receptor/ FasL apoptosis inducing system (Fujimoto et al., 1998; Takizawa et al., 1995; Takizawa et al., 1993; Wada et al., 1995) is expressed in a PKR dependent manner in infected cells (Takizawa et al., 1996). This most likely contributes to virus-induced cell death via a receptor mediated FADD/caspase 8-dependent pathway (Balachandran et al., 2000). Another mode of viral apoptosis induction might occur via activation of TGF-ß that is converted from its latent form by the viral neuraminidase (Schultz-Cherry and Hinshaw, 1996). Within the infected cell the apoptotic program is mediated by activation of caspases (Lin et al., 2002; Takizawa et al., 1999; Zhirnov et al., 1999) with a most crucial role of caspase 3 (Wurzer et al., 2003).

Although it is now well established that influenza virus infection induces caspses and subsequent apoptosis, the consequence of this activation for virus replication or host cell defense is still under a heavy debate (reviewed in (Lowy, 2003; Ludwig et al., 1999; Schultz-Cherry et al., 1998). Early studies demonstrated that overexpression of the anti-apoptotic protein Bcl-2 results in impaired virus production correlating with a misglycosylation of the viral surface protein hemagglutinin (Hinshaw et al., 1994; Olsen et al., 1996). Further, it has been shown by the same group that the viral non-structural protein NS1 has pro-apoptotic features and induces apoptosis when ectopically expressed (Schultz-Cherry et al., 2001). These data have been challenged recently by the finding that a recombinant influenza virus lacking the same protein, the delta NS1 virus, is a stronger apoptosis inducer than the wildtype suggesting an anti-apoptotic function of NS1 (Zhirnov et al., 2002a). These findings link viral apoptosis induction to the antiviral type I interferon (IFNα/β) response, since the NS1 protein was shown to be an efficient IFNα/β antagonist (Garcia-Sastre, 2001) and type I interferons are believed, besides the Fas/FasL system, to be main inducers of influenza virus induced apoptosis (Balachandran et al., 2000). Another finding in favour of an antiviral role of apoptosis is caspase-mediated cleavage of the influenza viral nucleoprotein (NP) (Zhirnov et al., 1999). The truncated form of the NP is not packaged into viral particles, suggesting that caspases act to limit amounts of virus protein for proper assembly. However, only the NPs of human virus strains are susceptible to this cleavage process (Zhirnov et al., 1999). Furthermore it has been demonstrated in an *in vitro* binding assay the the viral M1 protein specifically binds to caspase 8 and weakly to caspase 7, suggesting interference of M1 with a caspase 8 mediated apoptosis pathway (Zhirnov et al., 2002b).

With the identification of PB1-F2, a new influenza virus protein expressed from a +1 reading frame of the PB1 polymerase gene segment, another pro-apoptotic influenza virus protein has been discovered (Chen et al., 2001). PB1-F2 induces apoptosis via the mitochondrial pathway if added to cells and infection with recombinant viruses lacking the protein results in reduced apoptotic rates of lymphocytes (Chen et al., 2001). However, most of the avian virus strains are lacking the reading frame for this protein and PB1-F2-

deficient viruses do not affect apoptosis in a variety of other host cells (Chen et al., 2001). These results have let to the assumption that apoptosis induction by PB1-F2 may be required for the specific depletion of lymphocytes during an influenza virus infection, a process which is observed in infected animals (Tumpey et al., 2000; Van Campen et al., 1989a; Van Campen et al., 1989b).

Others suggest an antiviral function of apoptosis by providing apoptotic cells or material to be efficiently phagocytosed by macrophages (Watanabe et al., 2002) or to be taken up by dendritic cells, inducing a cytotoxic T-cell response (Albert et al., 1998). Furthermore, it was reported that viral induction of the apoptotic process limits the release of proinflammatory cytokines and thereby may reduce the severity of the inflammatory response to infection (Brydon et al., 2003). However, no direct proof for each of the suggested functions is given so far.

A recent study adds a new aspect to the open discussion by the surprising observation that influenza virus propagation was strongly impaired in the presence of caspase inhibitors (Wurzer et al., 2003). This dependence on caspase activity was most obvious in cells where caspase 3 was partially knocked-down by siRNA (Wurzer et al., 2003). Consistent with these findings, poor replication efficiencies of influenza A viruses in cells deficient for caspase 3 could be boosted 30-fold by ectopic expression of the protein. Mechanistically, the block in virus propagation appeared to be due to the retention of viral RNP complexes in the nucleus preventing formation of progeny virus particles (Wurzer et al., 2003). Interestingly the findings are consistent with a much earlier report showing that upon infection of cells overexpressing the anti-apoptotic protein Bcl-2 the viral RNP complexes were retained in the nucleus (Hinshaw et al., 1994) resulting in repressed virus titers (Olsen et al., 1996). Furthermore the recently identified pro-apoptotic PB1-F2 (Chen et al., 2001) is only expressed in later phases of replication consistent with a later step in the virus life cycle that requires caspase activity (Wurzer et al., 2003). The observation of a caspase requirement for RNP nuclear export was quite puzzling since this export process was shown before to be mediated by the active cellular export machinery involving the viral nuclear export protein (NS2/NEP) (Neumann et al., 2000; O'Neill et al., 1998) and the anti-apoptotic Raf/MEK/ERK cascade (Pleschka et al., 2001). Caspase activation does not support but rather inhibit the active nuclear export machinery by cleavage of transport proteins (Faleiro and Lazebnik, 2000). This suggests an alternate strategy by which caspases may regulate RNP export, e. g. by directly or indirectly increase the diffusion limit of nuclear pores (Faleiro and Lazebnik, 2000) to allow passive diffusion of larger proteins. Such a scenario is supported by the finding that isolated NPs or RNP complexes, which are nuclear if ectopically expressed, can partially translocate to the cytoplasm upon stimulation with an apoptosis inducer in a caspase 3 dependent manner (Wurzer et al., 2003). These findings can be merged into a model in which the RNPs are transported via an active export mechanism in intermediate steps of the virus life cycle. Once caspase activity increases in the cells, proteins of the transport machinery get destroyed, however, widening of nuclear pores may allow the viral RNPs to use a second mode of exit from the nucleus. That would be a likely mechanism to further enhance RNP migration to the cytoplasm in late phase of the viral life cycle and thereby support virus replication. Such a complementary use of both active Raf/MEK/ERK-dependent and inactive caspase-dependent transport mechanisms is supported by the observation that concentrations of MEK and caspase inhibitors which only poorly block influenza virus replication alone, efficiently impaired virus propagation if used in combination (Wurzer et al., 2003). Thus, while both pathways do not interfere which each other (Wurzer et al., 2003) they appear to synergize to mediate RNP export via different routes.

Thus one may conclude that influenza virus has acquired the capability to take advantage of supposedly antiviral host cell responses to support viral propagation. This includes early induction of caspase activity but not necessarily execution of the full apoptotic process that most likely is an antiviral response.

INFLUENZA VIRAL INDUCERS OR INHIBITORS OF SIGNALING AND APOPTOSIS

Unfortunately, there is only very limited information on the molecular interactions of influenza viral inducers with its cellular sensors. Some of the viral proteins have been reported to exhibit signal-inducing capacity. In early studies it was noted that different influenza strains differ in their ability to induce IFNα/β, which correlated with the NA activity of these strains (Chomik, 1981). Subsequently, it was shown that treatment of cells with recombinant NA, but not HA-induced the production of IL-1 and TNF-α in murine macrophages (Houde and Arora, 1990). In another study NA was shown to convert TGF-ß from its latent to its active form to an extent sufficient to induce TGF-ß dependent apoptosis (Schultz-Cherry and Hinshaw, 1996). The crucial role of NA for viral apoptosis induction has later been confirmed expressing the protein in a herpesvirus background (Morris et al., 2002). The exact molecular mechanism and the targeted pathways of NA-induced signaling still are not fully clear but most likely involve direct and indirect events.

HA can act as a signaling stimulus both inside the cell and at the cell surface. Binding of recombinant HA to cell surface receptors leads to a rapid induction of PKC signaling (Arora and Gasse, 1998). It should also be noted that treatment of primary cells prepared from human amnion tissue with purified HA was reported to result in the induction of apoptosis (Ohyama et al., 2003). When over-expressed inside the cell HA induces the activation of NF-κB (Pahl and Baeuerle, 1995), presumably by an ER-overload mechanism. NF-κB activation could also be achieved by expression of M1 and NP, which are not processed in the ER (Flory et al., 2000). This suggests different, yet unidentified cellular sensors for all three proteins, which are yet unknown. IKK and NF-κB induction by HA, NP and M1 seem to be relatively specific since transcription factors of the AP-1- and Ets-families dependent transcription or MAPK signaling pathways are not activated (Flory et al., 2000).

The most potent viral signaling component appears to be double-strand RNA (dsRNA). It is generally believed that most RNA viruses produce dsRNA-like intermediates representing a shared molecular pattern which that is sensed by the cell as an alert signal (Majde, 2000). Indeed, it is not surprising that dsRNA induces activation of pathways regulating the type I interferon response and expression of other antiviral cytokines. This includes the IKK/NF-κB pathway and both, the JNK and p38 MAPK cascades (Chu et al., 1999) (Ehrhardt et al., 2004, personal communication).

Among the viral signaling inducers, dsRNA is best characterized in terms of its cellular receptors. At least two cellular gene products have been identified to bind to dsRNA and signal for activation of NF-κB, the IFN-inducible RNA-dependent protein kinase PKR (Stark et al., 1998) and the Toll-like receptor 3 (Alexopoulou et al., 2001). While PKR acts as an intracellular sensor, TLR-3 is believed to detect dsRNA species released from dying cells (Alexopoulou et al., 2001). Such extracellular RNAs were indeed detected as relatively stable structures in the supernatants of influenza virus infected cells (Majde et al., 1998).

Since many of the signaling functions in a virus infected cell, especially those induced by dsRNA are most-likely part of an antiviral program it is not surprising that influenza viruses have evolved strategies to counteract this response. Although a surprising multitude of activities had been originally assigned to the RNA-binding influenza NS1 protein recent analyses have shown that a major *in vivo* function is to antagonize the antiviral type I IFN system (Garcia-Sastre, 2001). This included (*i*) inhibition of the polyadenylation and splicing of pre-mRNA which precludes its export to the cytoplasm ((Fortes et al., 1994; Lu et al., 1994; Wolff et al. 1998; Chen and Krug, 2000), (*ii*) enhancing the translational efficiency of viral mRNAs (de la Luna et al., 1994; Enami et al., 1994) and (*iii*) blocking the activation of the dsRNA-dependent protein kinase R (PKR) that via sustained phosphorylation of the eukaryotic translation initiation factor (eIF) 2α can decrease translation of cellular and viral mRNAs (Lu et al., 1995). However, the specific contributions of NS1 functions to viral replication and virulence could only be assessed until procedures for the genetic manipulation of the viral NS segment had been developed. A recombinant virus termed delNS1 with a deletion of the complete NS1 coding region had lost any virulence in a mouse influenza model system (Garcia-Sastre et al., 1998). Multicyclic replication of this mutant was strongly attenuated in IFN-competent but not in IFN-deficient tissue culture cells. The delNS1 virus regained virulence in mice with genetic knock-outs of either the STAT1 protein that transduces IFN-mediated signals to the transcriptional machinery (Garcia-Sastre et al., 1998) or of PKR whose expression is upregulated by IFN (Bergmann et al., 2000). Significantly, comparisons Comparisons of wild-type and delNS1 viruses revealed that NS1 inhibits the activation of the IFNβ gene that is controlled by the transcriptional activators NF-κB, IRF-3 and ATF-2/c-Jun (Ludwig et al., 2002; Talon et al., 2000; Wang et al., 2000). Significantly, the induction of the IFNβ gene in the absence of NS1 correlated with a strong activation of the three dsRNA-responsive transcription factors. These findings demonstrated that NS1 interferes with the intracellular signaling events that trigger activation of NF-κB, IRF-3 and ATF-2/c-Jun. This inhibitory activity was shown to depend on the NS1 RNA binding domain. Accordingly a recombinant influenza A virus expressing a RNA-binding-defective NS1 protein induces high levels of interferon beta and is attenuated in mice (Donelan et al., 2003). It is therefore believed that NS1 sequesters intracellular dsRNA produced during viral replication thereby keeping these molecules away from the antiviral enzyme PKR and other cellular dsRNA-sensor proteins.

The NS1 protein is also strongly interfering with the viral induction of apoptosis. Interestingly the protein has been both described to act pro- and anti-apoptotic. NS1 induces apoptosis if ectopically expressed (Schultz-Cherry et al., 2001) while infection of cells with the deltaNS1 virus mutant lacking NS1 also results in enhanced apoptotic levels (Zhirnov et al., 2002a). To date this discrepancy has not been properly addressed, however, it should be noted that both experimental settings may have its limitations. While the responses induced by ectopic over-expression of NS1 might not necessarily reflect NS1 function during a genuine infection, conclusions drawn upon deletion of the protein may also be misleading. The apoptotic response may simply be due to the strong interferon inducing activity of the deltaNS1 virus. However, wildtype influenza virus infection induces apoptosis also in interferon deficient cells to a similar degree as in interferon competent cell types (Wurzer et al., 2003). Thus, the function of NS1 in the apoptotic response has still to be fully elucidated.

Besides the RNA binding activity of NS1 the inhibition of the processing of residual IFN- or IFN-induced mRNAs by the protein further contribute to its antagonistic activity (Kim et al., 2002).

It should be noted that influenza patients actually secrete some IFNα (Hayden et al., 1998) and that some IFN-dependent genes were activated even by NS1-expressing wild-type virus (Geiss et al., 2002). Therefore, the IFN-antagonistic activity of NS1 might not be considered an absolute but rather a relative, exhaustible function that boosts viral replication in the early phase of infection.

In conclusion, the NS1 protein is a well characterised example of an IFN-antagonistic protein that strongly modulates viral signaling and apoptosis induction and which has a crucial role for the virulence of the pathogen. This was recently further supported by the finding, that the high virulence of avian influenza H5N1 viruses transmitted to humans in Hongkong 1997 was associated with a single amino acid in their NS1 proteins. These viruses were extremely resistant against the antiviral action of interferons or TNF-α and this resistance could be transmitted to other virus strains by transfer of the H5N1 virus NS gene segment (Heui Seo et al., 2002). although it is dispensable under certain laboratory conditions.

CONCLUSIONS AND PERSPECTIVES

A variety of signaling pathways induced by influenza viruses have been described in the last few years and evidences and suggestions for the activating components and functions in the cell have been provided. One common observation in these studies was, that the impact of inhibition of cellular factors on influenza viral replication was not too impressive. If viral factors are deleted or inhibited much stronger effects on the outcome of virus growth can be observed. In part this may be due to some of the experimental settings: Most inhibitors or dominant-negative mutants do not completely blunt the activity of the cellular enzymes, thus, in most cases there is some remaining residual activity left. But there are still other points to be considered. In contrast to the limited viral gene repertoire the cell has evolved genetic information for a variety of parallel pathways that may allow partial replacement of blocked functions. Furthermore, there are even different signaling pathways supporting the same step in the viral life cycle, e. g. the classical mitogenic cascade and caspases both support RNP export. Blockade of only one of these pathways only inefficiently impairs replication while prominent effects are only seen if both pathways are inhibited at the same time. Finally, one has to consider that cellular responses to virus infections are most likely supposed to be antiviral and are only misused by the virus to some extent. The findings merge into a scenario where a cell infected by a pathogenic invader responds with the activation of a variety of signaling events as an alarm program. Small RNA viruses such as influenza viruses have evolved strategies to misuse these antiviral cellular signaling programs to support their own replication. This may include even complementary responses such as anti-apoptotic activation of the Raf pathway or pro-apoptotic activation of caspases. It is easier for a viral invader to take advantage of existing cellular activities than to develop strategies to actively induce these activities in the infected host-cell. Thus, there appears to be no black or white situation. Cellular antiviral responses may also be supportive for the virus at some point. To fully evaluate the role of a given cellular component by interference with its function it will always be necessary to assess either the net outcome on virus-titers from cell culture or to monitor the disease progression in an infected animal.

References

Albert, M. L., Sauter, B., and Bhardwaj, N. (1998). Dendritic cells acquire antigen from apoptotic cells and induce class I-restricted CTLs, Nature *392*, 86–89.

Alexopoulou, L., Holt, A. C., Medzhitov, R., and Flavell, R. A. (2001). Recognition of double-stranded RNA and activation of NF-kappaB by Toll-like receptor 3, Nature *413*, 732–738.

Arora, D. J., and Gasse, N. (1998). Influenza virus hemagglutinin stimulates the protein kinase C activity of human polymorphonuclear leucocytes, Arch. Virol. *143*, 2029–2037.

Balachandran, S., Roberts, P. C., Kipperman, T., Bhalla, K. N., Compans, R. W., Archer, D. R., and Barber, G. N. (2000). Alpha/beta interferons potentiate virus-induced apoptosis through activation of the FADD/Caspase-8 death signaling pathway, J. Virol. *74*, 1513–1523.

Barber, S. A., Bruett, L., Douglass, B. R., Herbst, D. S., Zink, M. C., and Clements, J. E. (2002). Visna virus-induced activation of MAPK is required for virus replication and correlates with virus-induced neuropathology, J. Virol. *76*, 817–828.

Bergmann, M., Garcia-Sastre, A., Carnero, E., Pehamberger, H., Wolff, K., Palese, P., and Muster, T. (2000). Influenza virus NS1 protein counteracts PKR-mediated inhibition of replication, J. Virol. *74*, 6203–6206.

Brydon, E. W., Smith, H., and Sweet, C. (2003). Influenza A virus-induced apoptosis in bronchiolar epithelial (NCI-H292) cells limits pro-inflammatory cytokine release, J. Gen. Virol. *84*, 2389–2400.

Cai, H., Smola, U., Wixler, V., Eisenmann-Tappe, I., Diaz-Meco, M. T., Moscat, J., Rapp, U., and Cooper, G. M. (1997). Role of diacylglycerol-regulated protein kinase C isotypes in growth factor activation of the Raf-1 protein kinase, Mol. Cell Biol. *17*, 732–741.

Chen, W., Calvo, P. A., Malide, D., Gibbs, J., Schubert, U., Bacik, I., Basta, S., O'Neill, R., Schickli, J., Palese, P., Henklein, P., Bennink, J. R., and Yewdell, J. W. (2001). A novel influenza A virus mitochondrial protein that induces cell death, Nat. Med. *7*, 1306–1312.

Chen, Z. and Krug, R. M. (2000). Selective nuclear export of viral mRNAs in influenza-virus-infected cells. Trends Microbiol, 8, 376–383.

Chomik, M. (1981). Interferon induction by influenza virus: significance of neuraminidase, Arch. Immunol. Ther Exp. (Warsz) *29*, 109–104.

Chu, W. M., Ostertag, D., Li, Z. W., Chang, L., Chen, Y., Hu, Y., Williams, B., Perrault, J., and Karin, M. (1999). JNK2 and IKKbeta are required for activating the innate response to viral infection, Immunity *11*, 721–731.

Cohen, G. M. (1997). Caspases: the executioners of apoptosis, Biochem. J. *326*, 1–16.

Cohen, P. (2002). Protein kinases – the major drug targets of the twenty-first century?, Nat. Rev. Drug Discov *1*, 309–315.

Constantinescu, S. N., Cernescu, C. D., and Popescu, L. M. (1991). Effects of protein kinase C inhibitors on viral entry and infectivity, FEBS Lett. *292*, 31–33.

Courtois, G., Smahi, A., and Israel, A. (2001). NEMO/IKK gamma: linking NF-kappa B to human disease, Trends Mol. Med. *7*, 427–430.

de la Luna, S., Fortes, P., Beloso, A. and Ortin, J. (1995). Influenza virus NS1 protein enhances the rate of translation initiation of viral mRNAs. J Virol, 69, 2427–2433.

Delhase, M., and Karin, M. (1999). The I kappa B kinase: a master regulator of NF-kappa B, innate immunity, and epidermal differentiation, Cold Spring Harb Symp Quant Biol. *64*, 491–503.

Donelan, N. R., Basler, C. F., and Garcia-Sastre, A. (2003). A recombinant influenza A virus expressing an RNA-binding-defective NS1 protein induces high levels of beta interferon and is attenuated in mice, J. Virol. *77*, 13257–13266.

Dong, C., Davis, R. J., and Flavell, R. A. (2002). MAP kinases in the immune response, Annu. Rev. Immunol. *20*, 55–72.

Dumitru, C. D., Ceci, J. D., Tsatsanis, C., Kontoyiannis, D., Stamatakis, K., Lin, J. H., Patriotis, C., Jenkins, N. A., Copeland, N. G., Kollias, G., and Tsichlis, P. N. (2000). TNF-alpha induction by LPS is regulated posttranscriptionally via a Tpl2/ERK-dependent pathway, Cell *103*, 1071–1083.

Ehrhardt, C., Kardinal, C., Wolff, T., Wurzer, W. J., von Eichel-Streiber, C., Pleschka, S., Planz, O. and Ludwig, S. (2004). Rac1 and PAK1 are upstream of IKK-epsilon and TBK-1 in the viral activation of interferon regulatory factor-3 (IRF-3). FEBS Lett. 567, 230–238.

Enami, K., Sato, T. A., Nakada, S. and Enami, M. (1994). Influenza virus NS1 protein stimulates translation of the M1 protein. J Virol, 68, 1432–1437.

English, J., Pearson, G., Wilsbacher, J., Swantek, J., Karandikar, M., Xu, S., and Cobb, M. H. (1999). New insights into the control of MAP kinase pathways, Exp. Cell Res. *253*, 255–270.

Faleiro, L., and Lazebnik, Y. (2000). Caspases disrupt the nuclear-cytoplasmic barrier, J. Cell Biol. *151*, 951–959.

Fesq, H., Bacher, M., Nain, M., and Gemsa, D. (1994). Programmed cell death (apoptosis) in human monocytes infected by influenza A virus, Immunobiology *190*, 175–182.

Fitzgerald, K. A., McWhirter, S. M., Faia, K. L., Rowe, D. C., Latz, E., Golenbock, D. T., Coyle, A. J., Liao, S. M., and Maniatis, T. (2003). IKKepsilon and TBK1 are essential components of the IRF3 signaling pathway, Nat. Immunol. *4*, 491–496.

Flory, E., Kunz, M., Scheller, C., Jassoy, C., Stauber, R., Rapp, U. R., and Ludwig, S. (2000). Influenza virus-induced NF-kappaB-dependent gene expression is mediated by overexpression of viral proteins and involves oxidative radicals and activation of IkappaB kinase, J. Biol. Chem. *275*, 8307–8314.

Fortes, P., Beloso, A. and Ortin, J. (1994). Influenza virus NS1 protein inhibits pre-mRNA splicing and blocks mRNA nucleocytoplasmic transport. Embo J, 13, 704–712.

Fujimoto, I., Takizawa, T., Ohba, Y., and Nakanishi, Y. (1998). Co-expression of Fas and Fas-ligand on the surface of influenza virus-infected cells, Cell Death Differ *5*, 426–431.

Garcia-Sastre, A. (2001). Inhibition of interferon-mediated antiviral responses by influenza A viruses and other negative-strand RNA viruses, Virology *279*, 375–384.

Garcia-Sastre, A., Egorov, A., Matassov, D., Brandt, S., Levy, D. E., Durbin, J. E., Palese, P., and Muster, T. (1998). Influenza A virus lacking the NS1 gene replicates in interferon-deficient systems, Virology *252*, 324–330.

Garrington, T. P., and Johnson, G. L. (1999). Organization and regulation of mitogen-activated protein kinase signaling pathways, Curr. Opin. Cell Biol. *11*, 211–218.

Geiss, G. K., An, M. C., Bumgarner, R. E., Hammersmark, E., Cunningham, D., and Katze, M. G. (2001). Global impact of influenza virus on cellular pathways is mediated by both replication-dependent and -independent events, J. Virol. *75*, 4321–4331.

Geiss, G. K., Salvatore, M., Tumpey, T. M., Carter, V. S., Wang, X., Basler, C. F., Taubenberger, J. K., Bumgarner, R. E., Palese, P., Katze, M. G., and Garcia-Sastre, A. (2002). Cellular transcriptional profiling in influenza A virus-infected lung epithelial cells: The role of the nonstructural NS1 protein in the evasion of the host innate defense and its potential contribution to pandemic influenza, Proc. Natl. Acad. Sci. USA. *99*, 10736–10741.

Ghosh, S. (1999). Regulation of inducible gene expression by the transcription factor NF-kappaB, Immunol. Res. *19*, 183–189.

Gil, J., Alcami, J., and Esteban, M. (2000). Activation of NF-kappa B by the dsRNA-dependent protein kinase, PKR involves the I kappa B kinase complex, Oncogene *19*, 1369–1378.

Hayden, F. G., Fritz, R., Lobo, M. C., Alvord, W., Strober, W., and Straus, S. E. (1998). Local and systemic cytokine responses during experimental human influenza A virus infection. Relation to symptom formation and host defense, J. Clin. Invest *101*, 643–649.

Hazzalin, C. A., and Mahadevan, L. C. (2002). MAPK-regulated transcription: a continuously variable gene switch?, Nat. Rev. Mol. Cell Biol. *3*, 30–40.

Heui Seo, S., Hoffmann, E., and Webster, R. G. (2002). Lethal H5N1 influenza viruses escape host anti-viral cytokine responses, Nat. Med. *8*, 950–954.

Hinshaw, V. S., Olsen, C. W., Dybdahl-Sissoko, N., and Evans, D. (1994). Apoptosis: a mechanism of cell killing by influenza A and B viruses, J. Virol. *68*, 3667–3673.

Hiscott, J., Kwon, H., and Genin, P. (2001). Hostile takeovers: viral appropriation of the NF-kappaB pathway, J. Clin. Invest *107*, 143–151.

Hiscott, J., Pitha, P., Genin, P., Nguyen, H., Heylbroeck, C., Mamane, Y., Algarte, M., and Lin, R. (1999). Triggering the interferon response: the role of IRF-3 transcription factor, J. Interferon Cytokine Res. *19*, 1–13.

Houde, M., and Arora, D. J. (1990). Stimulation of tumor necrosis factor secretion by purified influenza virus neuraminidase, Cell Immunol. *129*, 104–111.

Ito, T., Kobayashi, Y., Morita, T., Horimoto, T., and Kawaoka, Y. (2002). Virulent influenza A viruses induce apoptosis in chickens, Virus Res. *84*, 27–35.

Iwamura, T., Yoneyama, M., Yamaguchi, K., Suhara, W., Mori, W., Shiota, K., Okabe, Y., Namiki, H., and Fujita, T. (2001). Induction of IRF-3/-7 kinase and NF-kappaB in response to double-stranded RNA and virus infection: common and unique pathways, Genes Cells *6*, 375–388.

Janicke, R. U., Sprengart, M. L., Wati, M. R., and Porter, A. G. (1998). Caspase-3 is required for DNA fragmentation and morphological changes associated with apoptosis, J. Biol. Chem. *273*, 9357–9360.

Julkunen, I., Melen, K., Nyqvist, M., Pirhonen, J., Sareneva, T., and Matikainen, S. (2000). Inflammatory responses in influenza A virus infection, Vaccine *19*, S32-37.

Karin, M. (1999a). The beginning of the end: IkappaB kinase (IKK) and NF-kappaB activation, J. Biol. Chem. *274*, 27339–27342.

Karin, M. (1999b). How NF-kappaB is activated: the role of the IkappaB kinase (IKK) complex, Oncogene *18*, 6867–6874.

Karin, M., and Ben-Neriah, Y. (2000). Phosphorylation meets ubiquitination: the control of NF-[kappa]B activity, Annu. Rev. Immunol. *18*, 621–663.

Karin, M., Liu, Z., and Zandi, E. (1997). AP-1 function and regulation, Curr. Opin. Cell Biol. *9*, 240–246.

Kerr, J. F., Wyllie, A. H., and Currie, A. R. (1972). Apoptosis: a basic biological phenomenon with wide-ranging implications in tissue kinetics, Br. J. Cancer *26*, 239–257.

Kim, M. J., Latham, A. G., and Krug, R. M. (2002). Human influenza viruses activate an interferon-independent transcription of cellular antiviral genes: Outcome with influenza A virus is unique, Proc. Natl. Acad. Sci. USA. *99*, 10096–10101.

Kolch, W., Heidecker, G., Kochs, G., Hummel, R., Vahidi, H., Mischak, H., Finkenzeller, G., Marme, D., and Rapp, U. R. (1993). Protein kinase C alpha activates RAF-1 by direct phosphorylation, Nature *364*, 249–252.

Kujime, K., Hashimoto, S., Gon, Y., Shimizu, K., and Horie, T. (2000). p38 mitogen-activated protein kinase and c-jun-NH2-terminal kinase regulate RANTES production by influenza virus-infected human bronchial epithelial cells, J. Immunol. *164*, 3222–3228.

Kunzelmann, K., Beesley, A. H., King, N. J., Karupiah, G., Young, J. A., and Cook, D. I. (2000). Influenza virus inhibits amiloride-sensitive Na+ channels in respiratory epithelia, Proc. Natl. Acad. Sci. USA. *97*, 10282–10287.

Lee, F. S., Peters, R. T., Dang, L. C., and Maniatis, T. (1998). MEKK1 activates both IkappaB kinase alpha and IkappaB kinase beta, Proc. Natl. Acad. Sci. USA. *95*, 9319–9324.

Lin, C., Holland, R. E., Jr., Donofrio, J. C., McCoy, M. H., Tudor, L. R., and Chambers, T. M. (2002). Caspase activation in equine influenza virus induced apoptotic cell death, Vet. Microbiol. *84*, 357–365.

Lin, R., Heylbroeck, C., Pitha, P. M., and Hiscott, J. (1998). Virus-dependent phosphorylation of the IRF-3 transcription factor regulates nuclear translocation, transactivation potential, and proteasome-mediated degradation, Mol. Cell Biol. *18*, 2986–2996.

Lowy, R. J. (2003). Influenza virus induction of apoptosis by intrinsic and extrinsic mechanisms, Int. Rev. Immunol. *22*, 425–449.

Lu, Y., Qian, X. Y. and Krug, R. M. (1994). The influenza virus NS1 protein: a novel inhibitor of pre-mRNA splicing. Genes Dev, 8, 1817–1828.

Lu, Y., Wambach, M., Katze, M. G. and Krug, R. M. (1995). Binding of the influenza virus NS1 protein to double-stranded RNA inhibits the activation of the protein kinase that phosphorylates the eIF-2 translation initiation factor. Virology, 214, 222–228.

Ludwig, S., Ehrhardt, C., Neumeier, E. R., Kracht, M., Rapp, U. R., and Pleschka, S. (2001). Influenza virus-induced AP-1-dependent gene expression requires activation of the JNK signaling pathway, J. Biol. Chem. 276, 10990–10998.

Ludwig, S., Planz, O., Pleschka, S., and Wolff, T. (2003). Influenza virus induced signaling pathways – Targets for antiviral therapy?, Trends Mol. Med. 9, 46–51.

Ludwig, S., Pleschka, S., and Wolff, T. (1999). A fatal relationship – influenza virus interactions with the host cell, Viral Immunol. 12, 175–196.

Ludwig, S., Wang, X., Ehrhardt, C., Zheng, H., Donelan, N., Planz, O., Pleschka, S., García-Sastre, A., Heins, G., and Wolff, T. (2002). The influenza A virus NS1 protein inhibits activation of Jun N-terminal kinase (JNK) and AP-1 transcription factors, J. Virol. 76, 11166–11171.

Ludwig, S., Wolff, T., Ehrhardt, C., Wurzer, W. J., Reinhardt, J., Planz, O. and Pleschka, S. (2004) MEK inhibition impairs influenza B virus propagation without emergence of resistant variants. FEBS Lett, 561, 37–43.

Luo, H., Yanagawa, B., Zhang, J., Luo, Z., Zhang, M., Esfandiarei, M., Carthy, C., Wilson, J. E., Yang, D., and McManus, B. M. (2002). Coxsackievirus B3 replication is reduced by inhibition of the extracellular signal-regulated kinase (ERK) signaling pathway, J. Virol. 76, 3365–3373.

Majde, J. A. (2000). Viral double-stranded RNA, cytokines, and the flu, J. Interferon Cytokine Res. 20, 259–272.

Majde, J. A., Guha-Thakurta, N., Chen, Z., Bredow, S., and Krueger, J. M. (1998). Spontaneous release of stable viral double-stranded RNA into the extracellular medium by influenza virus-infected MDCK epithelial cells: implications for the viral acute phase response, Arch. Virol. 143, 2371–2380.

Maniatis, T., Falvo, J. V., Kim, T. H., Kim, T. K., Lin, C. H., Parekh, B. S., and Wathelet, M. G. (1998). Structure and function of the interferon-beta enhanceosome, Cold Spring Harb Symp Quant Biol. 63, 609–620.

Maruoka, S., Hashimoto, S., Gon, Y., Nishitoh, H., Takeshita, I., Asai, Y., Mizumura, K., Shimizu, K., Ichijo, H., and Horie, T. (2003). ASK1 regulates influenza virus infection-induced apoptotic cell death, Biochem. Biophys Res. Commun. 307, 870–876.

Mizumura, K., Hashimoto, S., Maruoka, S., Gon, Y., Kitamura, N., Matsumoto, K., Hayashi, S., Shimizu, K., and Horie, T. (2003). Role of mitogen-activated protein kinases in influenza virus induction of prostaglandin E2 from arachidonic acid in bronchial epithelial cells, Clin. Exp. Allergy 33, 1244–1251.

Mogensen, T. H., and Paludan, S. R. (2001). Molecular pathways in virus-induced cytokine production, Microbiol. Mol. Biol. Rev. 65, 131–150.

Mori, I., Goshima, F., Koshizuka, T., Koide, N., Sugiyama, T., Yoshida, T., Yokochi, T., Nishiyama, Y., and Kimura, Y. (2003). Differential activation of the c-Jun N-terminal kinase/stress-activated protein kinase and p38 mitogen-activated protein kinase signal transduction pathways in the mouse brain upon infection with neurovirulent influenza A virus, J. Gen. Virol. 84, 2401–2408.

Mori, I., Komatsu, T., Takeuchi, K., Nakakuki, K., Sudo, M., and Kimura, Y. (1995). In vivo induction of apoptosis by influenza virus, J. Gen. Virol. 76, 2869–2873.

Morris, S. J., Smith, H., and Sweet, C. (2002). Exploitation of the Herpes simplex virus translocating protein VP22 to carry influenza virus proteins into cells for studies of apoptosis: direct confirmation that neuraminidase induces apoptosis and indications that other proteins may have a role, Arch. Virol. 147, 961–979.

Nemoto, S., DiDonato, J. A., and Lin, A. (1998). Coordinate regulation of IkappaB kinases by mitogen-activated protein kinase kinase 1 and NF-kappaB-inducing kinase, Mol. Cell Biol. 18, 7336–7343.

Neumann, G., Hughes, M. T., and Kawaoka, Y. (2000). Influenza A virus NS2 protein mediates vRNP nuclear export through NES-independent interaction with hCRM1, Embo J. *19*, 6751–6758.

O'Neill, R. E., Talon, J., and Palese, P. (1998). The influenza virus NEP (NS2 protein) mediates the nuclear export of viral ribonucleoproteins, Embo J, *17*, 288–296.

Ohyama, K., Nishina, M., Yuan, B., Bessho, T., and Yamakawa, T. (2003). Apoptosis induced by influenza virus-hemagglutinin stimulation may be related to fluctuation of cellular oxidative condition, Biol. Pharm Bull *26*, 141–147.

Olsen, C. W., Kehren, J. C., Dybdahl-Sissoko, N. R., and Hinshaw, V. S. (1996). bcl-2 alters influenza virus yield, spread, and hemagglutinin glycosylation, J. Virol. *70*, 663–666.

Pahl, H. L. (1999). Activators and target genes of Rel/NF-kappaB transcription factors, Oncogene *18*, 6853–6866.

Pahl, H. L., and Baeuerle, P. A. (1995). Expression of influenza virus hemagglutinin activates transcription factor NF-kappa B, J. Virol. *69*, 1480–1484.

Planz, O., Pleschka, S., and Ludwig, S. (2001). MEK-specific inhibitor U0126 blocks spread of Borna disease virus in cultured cells, J. Virol. *75*, 4871–4877.

Pleschka, S., Wolff, T., Ehrhardt, C., Hobom, G., Planz, O., Rapp, U. R., and Ludwig, S. (2001). Influenza virus propagation is impaired by inhibition of the Raf/MEK/ERK signalling cascade, Nat. Cell Biol. *3*, 301–305.

Razvi, E. S., and Welsh, R. M. (1995). Apoptosis in viral infections, Adv. Virus Res. *45*, 1–60.

Root, C. N., Wills, E. G., McNair, L. L., and Whittaker, G. R. (2000). Entry of influenza viruses into cells is inhibited by a highly specific protein kinase C inhibitor, J. Gen. Virol. *81*, 2697–2705.

Rott, O., Charreire, J., Semichon, M., Bismuth, G., and Cash, E. (1995). B cell superstimulatory influenza virus (H2-subtype) induces B cell proliferation by a PKC-activating, Ca(2+)-independent mechanism, J. Immunol. *154*, 2092–2103.

Schultz-Cherry, S., Dybdahl-Sissoko, N., Neumann, G., Kawaoka, Y., and Hinshaw, V. S. (2001). Influenza virus ns1 protein induces apoptosis in cultured cells, J. Virol. *75*, 7875–7881.

Schultz-Cherry, S., and Hinshaw, V. S. (1996). Influenza virus neuraminidase activates latent transforming growth factor beta, J. Virol. *70*, 8624–8629.

Schultz-Cherry, S., Krug, R. M., and Hinshaw, V. S. (1998). Induction of apoptosis by influenza virus, Semin Virol. *8*, 491–495.

Sebolt-Leopold, J. S., Dudley, D. T., Herrera, R., Van Becelaere, K., Wiland, A., Gowan, R. C., Tecle, H., Barrett, S. D., Bridges, A., Przybranowski, S., Leopold, W. R., and Saltiel, A. R. (1999). Blockade of the MAP kinase pathway suppresses growth of colon tumors in vivo, Nat. Med. *5*, 810–816.

Senftleben, U., Cao, Y., Xiao, G., Greten, F. R., Krahn, G., Bonizzi, G., Chen, Y., Hu, Y., Fong, A., Sun, S. C., and Karin, M. (2001). Activation by IKKalpha of a second, evolutionary conserved, NF-kappa B signaling pathway, Science *293*, 1495–1499.

Servant, M. J., Grandvaux, N., and Hiscott, J. (2002). Multiple signaling pathways leading to the activation of interferon regulatory factor 3, Biochem. Pharmacol *64*, 985–992.

Servant, M. J., Grandvaux, N., tenOever, B. R., Duguay, D., Lin, R., and Hiscott, J. (2003). Identification of the minimal phosphoacceptor site required for in vivo activation of interferon regulatory factor 3 in response to virus and double-stranded RNA, J. Biol. Chem. *278*, 9441–9447.

Servant, M. J., ten Oever, B., LePage, C., Conti, L., Gessani, S., Julkunen, I., Lin, R., and Hiscott, J. (2001). Identification of distinct signaling pathways leading to the phosphorylation of interferon regulatory factor 3, J. Biol. Chem. *276*, 355–363.

Sharma, S., tenOever, B. R., Grandvaux, N., Zhou, G. P., Lin, R., and Hiscott, J. (2003). Triggering the interferon antiviral response through an IKK-related pathway, Science *300*, 1148–1151.

Sieczkarski, S. B., Brown, H. A., and Whittaker, G. R. (2003a). Role of protein kinase C betaII in influenza virus entry via late endosomes, J. Virol. *77*, 460–469.

Sieczkarski, S. B., and Whittaker, G. R. (2003b). Differential requirements of Rab5 and Rab7 for endocytosis of influenza and other enveloped viruses, Traffic *4*, 333–343.

Slee, E. A., Adrain, C., and Martin, S. J. (1999). Serial killers: ordering caspase activation events in apoptosis, Cell Death Differ *6*, 1067–1074.

Smith, E. J., Marie, I., Prakash, A., Garcia-Sastre, A., and Levy, D. E. (2001). IRF3 and IRF7 phosphorylation in virus-infected cells does not require double-stranded RNA-dependent protein kinase R or Ikappa B kinase but is blocked by Vaccinia virus E3L protein, J. Biol. Chem. *276*, 8951–8957.

Stark, G. R., Kerr, I. M., Williams, B. R., Silverman, R. H., and Schreiber, R. D. (1998). How cells respond to interferons, Annu. Rev. Biochem. *67*, 227–264.

Takizawa, T., Fukuda, R., Miyawaki, T., Ohashi, K., and Nakanishi, Y. (1995). Activation of the apoptotic Fas antigen-encoding gene upon influenza virus infection involving spontaneously produced beta-interferon, Virology *209*, 288–296.

Takizawa, T., Matsukawa, S., Higuchi, Y., Nakamura, S., Nakanishi, Y., and Fukuda, R. (1993). Induction of programmed cell death (apoptosis) by influenza virus infection in tissue culture cells, J. Gen. Virol. *74*, 2347–2355.

Takizawa, T., Ohashi, K., and Nakanishi, Y. (1996). Possible involvement of double-stranded RNA-activated protein kinase in cell death by influenza virus infection, J. Virol. *70*, 8128–8132.

Takizawa, T., Tatematsu, C., Ohashi, K., and Nakanishi, Y. (1999). Recruitment of apoptotic cysteine proteases (caspases) in influenza virus-induced cell death, Microbiol. Immunol. *43*, 245–252.

Talon, J., Horvath, C. M., Polley, R., Basler, C. F., Muster, T., Palese, P., and Garcia-Sastre, A. (2000). Activation of interferon regulatory factor 3 is inhibited by the influenza A virus NS1 protein, J. Virol. *74*, 7989–7996.

Taniguchi, T., and Takaoka, A. (2002). The interferon-alpha/beta system in antiviral responses: a multimodal machinery of gene regulation by the IRF family of transcription factors, Curr. Opin. Immunol. *14*, 111–116.

Thanos, D., and Maniatis, T. (1995). Virus induction of human IFN beta gene expression requires the assembly of an enhanceosome, Cell *83*, 1091–1100.

Thornberry, N. A., and Lazebnik, Y. (1998). Caspases: enemies within, Science *281*, 1312–1316.

Toker, A. (1998). Signaling through protein kinase C, Front Biosci *3*, D1134–1147.

Tumpey, T. M., Lu, X., Morken, T., Zaki, S. R., and Katz, J. M. (2000). Depletion of lymphocytes and diminished cytokine production in mice infected with a highly virulent influenza A (H5N1) virus isolated from humans, J. Virol. *74*, 6105–6116.

Van Campen, H., Easterday, B. C., and Hinshaw, V. S. (1989a). Destruction of lymphocytes by a virulent avian influenza A virus, J. Gen. Virol. *70*, 467–472.

Van Campen, H., Easterday, B. C., and Hinshaw, V. S. (1989b). Virulent avian influenza A viruses: their effect on avian lymphocytes and macrophages in vivo and in vitro, J. Gen. Virol. *70*, 2887–2895.

Wada, N., Matsumura, M., Ohba, Y., Kobayashi, N., Takizawa, T., and Nakanishi, Y. (1995). Transcription stimulation of the Fas-encoding gene by nuclear factor for interleukin-6 expression upon influenza virus infection, J. Biol. Chem. *270*, 18007–18012.

Wang, X., Li, M., Zheng, H., Muster, T., Palese, P., Beg, A. A., and Garcia-Sastre, A. (2000). Influenza A virus NS1 protein prevents activation of NF-kappaB and induction of alpha/beta interferon, J. Virol. *74*, 11566–11573.

Watanabe, Y., Shiratsuchi, A., Shimizu, K., Takizawa, T., and Nakanishi, Y. (2002). Role of phosphatidylserine exposure and sugar chain desialylation at the surface of influenza virus-infected cells in efficient phagocytosis by macrophages, J. Biol. Chem. *277*, 18222–18228.

Widmann, C., Gibson, S., Jarpe, M. B., and Johnson, G. L. (1999). Mitogen-activated protein kinase: conservation of a three-kinase module from yeast to human, Physiol Rev. *79*, 143–180.

Wolff, T., O'Neill, R. E. and Palese, P. (1998). NS1-Binding protein (NS1-BP): a novel human protein that interacts with the influenza A virus nonstructural NS1 protein is relocalized in the nuclei of infected cells. J Virol, 72, 7170–7180.

Woronicz, J. D., Gao, X., Cao, Z., Rothe, M., and Goeddel, D. V. (1997). IkappaB kinase-beta: NF-kappaB activation and complex formation with IkappaB kinase-alpha and NIK, Science 278, 866–869.

Wurzer, W. J., Ehrhardt, C., Pleschka, S., Berberich-Siebelt, F., Wolff, T., Walczak, H., Planz, O. and Ludwig, S. (2004). NF-kappaB-dependent induction of tumor necrosis factor-related apoptosis-inducing ligand (TRAIL) and Fas/FasL is crucial for efficient influenza virus propagation. J Biol Chem, 279, 30931–30937.

Wurzer, W. J., Planz, O., Ehrhardt, C., Giner, M., Silberzahn, T., Pleschka, S., and Ludwig, S. (2003). Caspase 3 activation is essential for efficient influenza virus propagation, Embo J. 22, 2717–2728.

Xiao, G., Cvijic, M. E., Fong, A., Harhaj, E. W., Uhlik, M. T., Waterfield, M., and Sun, S. C. (2001). Retroviral oncoprotein Tax induces processing of NF-kappaB2/p100 in T cells: evidence for the involvement of IKKalpha, Embo J. 20, 6805–6815.

Yoneyama, M., Suhara, W., Fukuhara, Y., Fukuda, M., Nishida, E., and Fujita, T. (1998). Direct triggering of the type I interferon system by virus infection: activation of a transcription factor complex containing IRF-3 and CBP/p300, Embo J. 17, 1087–1095.

Zamanian-Daryoush, M., Mogensen, T. H., DiDonato, J. A., and Williams, B. R. (2000). NF-kappaB activation by double-stranded-RNA-activated protein kinase (PKR) is mediated through NF-kappaB-inducing kinase and IkappaB kinase, Mol. Cell Biol. 20, 1278–1290.

Zhirnov, O. P., Konakova, T. E., Garten, W., and Klenk, H. (1999). Caspase-dependent N-terminal cleavage of influenza virus nucleocapsid protein in infected cells, J. Virol. 73, 10158–10163.

Zhirnov, O. P., Konakova, T. E., Wolff, T., and Klenk, H. D. (2002a). NS1 protein of influenza A virus down-regulates apoptosis, J. Virol. 76, 1617–1625.

Zhirnov, O. P., Ksenofontov, A. L., Kuzmina, S. G., and Klenk, H. D. (2002b). Interaction of

Insights into influenza virus-host interactions through global gene expression profiling: cell culture systems to animal models

Marcus J. Korth, John C. Kash, Carole R. Baskin and Michael G. Katze

Abstract

Researchers attempting to study a biological process as complex as the host response to a pathogen have long been faced with difficult choices. Simple model systems, such as cultured cell lines, provide considerable control over experimental variables, and methods that focus on tightly defined parameters can yield results that are often readily interpretable. But simple systems may not be representative of a natural infection, and data generated by focusing on a single gene, protein, or pathway can be difficult to integrate into a global picture. Today, genomic technologies such as DNA microarrays make it possible to perform experiments that provide a near comprehensive view of even such intricate processes as pathogen-host interactions. Still, choosing an experimental infection system remains difficult, and data interpretation has never been more complicated. In this chapter, we describe how gene expression profiling is being used to examine the host response to influenza virus. We discuss how the application of this technology has progressed from simple experimental systems, such as the in vitro infection of HeLa cells, to highly complex systems such as mouse and nonhuman primate models of infection. We also discuss the challenges associated with interpreting huge volumes of data generated from microarray analyses, and describe how comparing the host's transcriptional response to infection by diverse wild type or engineered viruses is providing new insights into the characteristics of influenza virus that contribute to its variable virulence.

INTRODUCTION

Imagine having an inside view as a cell mounts its response to an invading virus. Intracellular signaling cascades kick into action, proteins interact and send their signals along specialized pathways, and the transcriptional machinery is mobilized to alter gene expression as the cell responds to infection. But the virus has come prepared, and as viral proteins enter the mix, signals are scrambled and cellular defenses are compromised. With the advent of DNA microarrays, remarkable pictures of this complex interplay are now available. By following cellular gene expression changes during the course of an infection, we have what amounts to an insider's view of this critical component of the cellular response to an invading virus.

In this chapter, we will focus our attention on the use of gene expression profiling to investigate diverse aspects of influenza virus infection. Much of this discussion will center on studies carried out in our laboratory, where we are applying high-throughput genomic technologies to study a wide variety of viral pathogens. Influenza virus has figured prominently in our studies, and the virus has many characteristics that make it especially interesting and challenging to study. Like many viruses, influenza has evolved mechanisms to evade the cellular interferon (IFN) response, but is unusual in that it appears to employ both viral and cellular proteins in the process. Influenza virus is also well known for its inhibitory effects on cellular mRNA translation, an event commonly referred to as the host cell shutoff, in which there is a dramatic decrease in the translation of cellular mRNAs while viral transcripts remain efficiently and selectively translated. Perhaps most importantly, influenza has the ability to occasionally generate particularly virulent strains that are capable of causing worldwide pandemics. As discussed extensively in this volume, there is considerable interest in determining the viral and cellular factors responsible for the increased virulence of pandemic strains, particularly for history's most deadly influenza virus, the infamous 1918 Spanish flu.

Each of these aspects of influenza virus infection can be examined by gene expression profiling, and the technology has been applied to diverse experimental systems that range from the in vitro infection of cell lines to the experimental infection of nonhuman primates. Comparing the host transcriptional response to infection with diverse wild type or engineered viruses has helped to determine the contribution of specific viral genes to innate immune evasion and viral virulence. When combined with polysome analysis, DNA microarrays have even been used to provide new insights into translational events such as the host-cell shutoff. We will begin by reviewing studies in which DNA microarrays have been used together with in vitro infections and progress to a discussion of their use in more complex systems, such as mouse and nonhuman primate models of infection. Although the focus of this chapter will be on the use of microarrays to study the cellular response to infection, we will also touch upon other applications of the technology, such as the use of DNA arrays to identify and type influenza viruses.

CELL CULTURE SYSTEMS

Cultured cell lines provide the simplest experimental system in which to profile cellular gene expression changes in response to virus infection. Their homogeneity, and the ease with which they can be cultured, infected, and otherwise manipulated continues to make them the mainstay for gene expression analyses. Fortunately, influenza virus infects a variety of cell lines and it is frequently possible to infect a very high percentage of cells, which greatly facilitates the ability to detect and measure gene expression changes. Of course, cell lines have their disadvantages too. Their transformed phenotype, chromosomal abnormalities, and their separation from their usual anatomical and chemical environment severely limit how well they represent a natural infection. With this in mind, we will begin by discussing our earliest gene expression profiling studies that used HeLa cells to look at the cellular response to an active infection or exposure to an inactivated virus. We will then move on to more recent studies in which recombinant viruses are being used to examine the contribution of specific viral genes to the cellular response to infection. We will also describe how cell culture and microarrays can be combined with polysome analysis to study changes in cellular mRNA translation during infection. Finally, we will discuss what may represent the cell culture system of the future: using primary differentiated human tracheobronchial epithelial cells to study influenza virus-host interactions.

Virus replication-dependent and –independent effects on cellular gene expression

In our first use of gene expression profiling to study the cellular response to influenza virus, we conducted a series of experiments in which cDNA microarrays were used to monitor changes in the cellular gene expression patterns of influenza virus-infected HeLa cells (Geiss et al., 2001). For these studies, we were also interested in determining the contribution of viral replication-dependent and –independent effects on the cellular response. Cells were therefore mock infected, infected with influenza virus (A/WSN/33), or exposed to a UV-inactivated virus that was unable to replicate, but which retained the ability to bind to the cell surface. Total RNA was then isolated at 4 and 8 hours after infection and used to prepare labeled probes for hybridization to microarrays containing approximately 4,500 human cDNAs. From these analyses, we obtained our first pictures of the ways in which influenza virus infection alters cellular gene expression. The majority of differentially expressed genes were involved in transcriptional regulation, interleukin and growth factor signaling, mRNA processing, protein synthesis, and protein degradation. Somewhat surprisingly, we detected only 39 genes that exhibited an increase in expression in response to active viral replication, while 351 genes exhibited a decrease in expression. As one might predict of gene expression changes that were elicited by viral replication, the majority of genes that were differentially expressed at the 4-hour time point were also differentially expressed at 8 hours post-infection.

Although to a much lesser extent, altered cellular gene expression was also observed in cells exposed to UV-inactivated virus. In contrast to the replication-dependent changes, the number of differentially expressed genes in cells exposed to UV-inactivated virus decreased from 84 genes at the 4-hour time point to only 13 genes at the 8-hour time point. It therefore appears that early events, such as attachment or fusion of the virus to the cell surface, are sufficient to transmit signals that alter cellular gene expression, but these changes are short lived in the absence of viral replication. Of particular note, we observed that exposure to UV-inactivated virus results in the up-regulation of several members of the metallothionein gene family. Although the physiological role of metallothioneins is not fully understood, they are induced by a variety of stimuli, including IL-6, zinc, or oxidative stress. Metallothionein gene induction may consequently represent a protective cellular response to the oxidative stress reported to be induced by virus attachment (Choi et al., 1996). It may therefore be interesting to examine whether the outcome of influenza virus infection is altered by inducing metallothionein gene expression before or after infection.

Recombinant viruses and the role of specific viral genes in pathogenesis

An important way in which gene expression profiling can be used to study virus-host interactions is to compare the cellular transcriptional response to various wild type or recombinant viruses that differ in their infectivity or virulence. These studies provide a means to better understand how specific viral genes contribute to influenza virus pathogenesis, and again, cell culture systems have been instrumental to these studies. We have been particularly interested in the influenza virus NS1 protein, because of the considerable evidence that NS1 acts as an IFN antagonist (García-Sastre, 2004). For example, an engineered influenza virus lacking the NS1 gene is able to replicate in cells deficient in IFN expression, such as Vero cells, but replicates poorly in MDCK cells or embryonated chicken eggs, most likely due to IFN-mediated effects (García-Sastre et al., 1998). We therefore carried out a series of experiments to examine whether NS1 plays a role in altering IFN-regulated gene expression (Geiss et al., 2002). For these studies, we infected A549

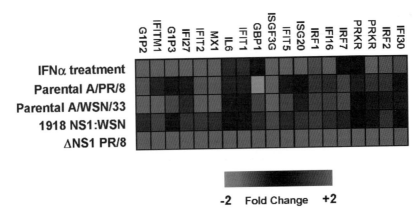

Figure 13.1 Gene expression profiling of the effects of NS1 on IFN-regulated gene expression. Shown is the expression of a group of IFN-regulated genes in cells treated with IFN (400 IU/ml for 16 hours) or infected with influenza virus containing wild-type NS1 (A/PR/8 and A/WSN/33), 1918 NS1 (in A/WSN/33), or a complete deletion of the NS1 gene (in A/PR/8). This analysis demonstrates the role of NS1 in antagonizing the IFN response and the enhanced function of the NS1 gene from the 1918 pandemic strain. Genes shown in red were up-regulated and genes shown in green were down-regulated in infected relative to uninfected A549 cells. Black indicates no change in gene expression. Primary analysis was performed using Resolver (Rosetta Biosoftware) and supplemental analysis using DecisionSite for Functional Genomics (Spotfire, Inc.). This figure is also reproduced in colour in the colour section at the end of the book

cells, a human alveolar type II epithelial cell line, with wild-type influenza virus (A/PR/8/34) or with engineered forms of the virus that lacked either all or part of the NS1 gene. Cells were harvested 8 hours after infection, and RNA was isolated for the production of cDNA probes that were hybridized to microarrays containing over 13,000 human cDNAs.

These studies revealed that cells infected with wild-type A/PR/8 virus exhibited the differential expression of 84 genes, including the induction of numerous genes associated with the IFN response (Figure 1). IFN-stimulated genes that increased in expression in response to virus infection included the influenza virus inhibitor gene MX1, as well as IFITM1, IFI27, IRF1, IRF7, IRF2, ISGF3G and others. The wild-type virus also induced the expression of other antiviral genes, including TNF-related apoptosis-inducing ligand (TRAIL), complement C1S component, and MYD88, an adapter molecule that participates in Toll-like receptor signaling. When cells were infected with a virus that lacked NS1 (delNS1), or with a virus that contained a C-terminal deletion in the NS1 gene (NS1 1-126), we observed an increase in the number of differentially expressed genes as well as an increase in the magnitude of expression changes. In particular, these viruses had a significant impact on genes in the signal transducer and activator of transcription (STAT)-signaling pathway, such as STAT1 and STAT3. In addition, genes encoding members of the suppressor of cytokine signaling (SOCS) family of proteins, SOCS2 and SOCS3, also exhibited an increase in expression in response to infection with delNS1, as did the gene encoding the antiviral kinase, PKR. These findings indicate that A549 cells respond to infection with A/PR/8 with an increase in the expression of genes associated with the IFN response, but that this response is at least partially attenuated by the NS1 gene.

We also examined the cellular response to an engineered virus that contained the NS1 gene from the 1918 pandemic strain. Because WSN was used as the background strain to construct the engineered virus, we also measured the effects of WSN on cellular gene expression in A549 cells. Interestingly, WSN elicited a lower level of expression of most IFN-regulated genes than did A/PR/8, although certain genes, such as MX1, IRF9, and ISG15, were expressed at high levels. In contrast, the engineered virus containing the NS1 gene from the 1918 pandemic strain was very efficient at blocking the induction of the IFN response, and no IFN-regulated genes were differentially expressed in response to infection with this virus (Figure 1). The engineered virus did, however, specifically elicit the differential expression of a number of other cellular genes, including the up-regulation of protein kinase inhibitor-β (PKIB) and the down-regulation of Toll-like receptor 5 (TLR5). Together, these findings suggest that the extreme virulence of the 1918 virus may have been due in part to the ability of its NS1 protein to act as a particularly efficient IFN antagonist. As described later, we are continuing with this line of studies by using similarly engineered viruses to examine the role of other genes from the 1918 strain, including the hemagglutinin (HA) and neuraminidase (NA) genes. We also note that the differences in the cellular response to A/PR/8 and WSN are an indicator of how the response to infection is markedly dependent upon both cell type and the strain of infecting virus.

Microarrays, polysomes, and selective mRNA translation

Cell lines also provided some of our first insights into the mechanisms by which influenza virus impacts cellular mRNA translation. Like all viruses, influenza hijacks the cellular protein synthesis machinery for the production of viral proteins. But one of the more remarkable aspects of influenza virus is its ability to selectively shut off the translation of most cellular mRNAs while viral transcripts remain efficiently translated. Although microarrays are used to profile transcriptional changes, we reasoned that by combining them with polysome analysis we could use these tools to gain new insights into the mechanisms underlying the phenomenon of selective mRNA translation. Before describing these studies, a bit of background is in order.

The translation of influenza virus mRNAs is a cap-dependent process that is stimulated by highly conserved sequences located within the 5' untranslated region (UTR) of the viral transcripts (reviewed in Gale, Jr. et al., 2000; Tan et al., 2000). The 5' UTRs are typically between 20 and 50 nucleotides in length and appear relatively unremarkable, apart from the 5' cap structure and 10 to 14 nucleotides that are obtained from cellular mRNAs via a 'cap-stealing' mechanism (Katze and Krug, 1990). We have demonstrated that a cellular mRNA-binding protein, GRSF-1, specifically interacts with a discrete and conserved region of the viral 5' UTR and is capable of stimulating protein synthesis in *in vitro* translation extracts prepared from infected HeLa cells (Park and Katze, 1995; Kash et al., 2002). Moreover, we know that enhanced translation of viral mRNAs *in vitro* is dependent upon the presence of the conserved 5' UTR, since reporter mRNAs containing mutated forms of the 5' UTR are translated at a lower efficiency and cannot be stimulated by the addition of recombinant GRSF-1. Related studies have shown that the 5' UTR-mediated stimulation of protein synthesis can be abolished by immunodepleting GRSF-1 from the *in vitro* translation lysate and can subsequently be rescued by the addition of exogenous GRSF-1 (Park et al., 1999). These findings strongly suggest that the interaction between GRSF-1 and the viral 5' UTR is an important determinant in the selective translation of viral mRNAs.

In order to better understand the relationship between GRSF-1 and influenza virus-mediated effects on cellular mRNA translation, we used polysome fractionation together with cDNA microarrays to identify cellular mRNAs that are selectively recruited to polyribosomes following infection (Kash et al., 2002). Our goal for these experiments was to identify cellular mRNAs that might be targets of GRSF-1-mediated translational regulation. Our prediction was that such genes would be transcriptionally induced during infection and that the corresponding mRNAs would then be recruited to polyribosomes. Although polyribosome fractionation is not a direct measurement of the translational state of a particular mRNA, the presence of high levels of the influenza viral mRNAs in polyribosomes, which are extensively translated during infection, validated the use of this procedure to study the small subset of transcriptionally and translationally activated cellular genes.

Our microarray analyses demonstrated that following influenza infection more than 85% of the cellular mRNAs (i.e., those that could be measured because of their presence on the microarray) were displaced from polyribosomes. Analysis of the 5' UTRs of the mRNAs that were recruited to polyribosomes following infection revealed that approximately 25% of the transcriptionally and translationally induced genes have 5' UTR sequences that contain variations of AGGGU, the consensus GRSF-1-binding sequence. Many of the cellular mRNAs that are recruited to polyribosomes during infection, and which contain a putative GRSF-1-binding site, are involved in the control of cell proliferation and survival, mitogenesis, or the inflammatory response. In contrast, GRSF-1-binding sites were not detected on genes that were transcriptionally induced, but whose mRNAs were displaced from the polyribosomes during infection. Together, these experiments reveal that the selective translation of influenza virus mRNAs is associated with dramatic changes in the polysomal distribution of mRNAs and suggest that recruitment of cellular and viral mRNAs to polyribosomes may in part arise from the GRSF-1-mediated discrimination of 5' UTR sequences. These studies also show how functional genomic analysis of polyribosome-fractionated mRNAs provides an important new tool for studying global changes in mRNA translation and the potential of this technique to reveal functional relationships between 5' UTR sequences and the regulation of polyribosome distribution during virus infection.

Primary cell culture systems

Despite their limitations, cell lines have clearly provided an abundance of data regarding virus-host interactions and the cellular response to infection, and their widespread use is likely to continue in the future. However, primary cell culture systems provide a measure of improvement in terms of mimicking a natural influenza infection, particularly the more complex systems that provide an approximation of normal human airways. In this regard, we are now working with Dr. Mikhail Matrosovich (Philipps-Universität Marburg), who is studying differences in human and avian influenza virus infection using primary human tracheobronchial epithelial cells (Matrosovich et al., 2004). In this culture system (described in more detail elsewhere in this volume), epithelial cells from bronchial tissues are seeded onto membrane supports, which are at first completely submerged in medium. Later, the apical surfaces of the superficial cell layer are exposed to air, which triggers cellular differentiation into ciliated (30%) and non-ciliated epithelial cells. Deeper layers include goblet and other types of secretory cells, which secrete mucin. These cultures provide a novel morphological and functional approximation of human tracheobronchial airways and therefore a unique experimental model in which to study influenza virus infection. Using this system, Dr. Matrosovich has demonstrated that influenza viruses enter the airway epithelium through specific target cells, which are different for human and avian viruses. Whereas human viruses preferentially infect nonciliated cells, which express cell-surface glycoproteins or glycolipids that contain 2–6-linked sialic acids [Neu5Ac(α2–6)Gal], avian viruses mainly target ciliated cells that express receptors containing terminal 2–3-linked sialyl-galactosyl moieties [Neu5Ac(α2–3)Gal].

By combining this system with gene expression profiling, we hope to gain a better understanding of the interplay between diverse influenza viruses and their target cells, as well as similarities and differences in the replication, transmission, and pathogenesis of avian and human viruses. We have recently performed preliminary microarray analyses using RNA isolated from cells after infection with either avian or human influenza viruses and have observed distinct differences in the propensity of these viruses to alter cellular gene expression. In particular, the avian virus induced a much stronger antiviral response, especially at early time points after infection. There also appeared to be a more intense shut-down of protein synthesis and general cell metabolism during infection with the avian virus, which would also be consistent with a strong IFN response. Although we are very enthusiastic about the benefits of this system, it also has its drawbacks, including the fact that not all cells become infected and that it is not currently possible to separate the different cell types for RNA extraction. Nevertheless, we are very hopeful that this cell culture system, used together with gene expression analysis, will provide important new clues to the differences between avian and human influenza viruses that account for rarer, yet often more pathogenic infections of avian viruses in humans.

ANIMAL MODELS

As discussed in the preceding sections, cultured cells provide a uniform and easy to manipulate system in which to study the cellular response to virus infection. However, even primary cells in culture are abnormal, in the sense that they do not function in their usual anatomical or physiological environment. In particular, cells in culture cannot interact with the immune system, which is a key player in viral clearance, tissue pathology, and the response and eventual fate of infected cells. The shortcomings of cell culture systems therefore make the use of appropriate animal models essential for a more complete understanding of virus-host interactions. Importantly, the use of animal models provides the ability to correlate gene expression changes with clinical data, and by obtaining samples at various times after infection, it is possible to gain insights into disease progression. Although influenza virus can infect a number of avian and mammalian species, the availability of comprehensive genomic tools currently limits gene expression profiling to two animal models: mice and nonhuman primates.

Mice

The mouse is perhaps the most commonly used animal for experimental virus infection and is frequently used to investigate various aspects of influenza virus virulence and pathogenesis. Of particular relevance to the topic of this chapter, the mouse is being used as model in which to evaluate the virulence of engineered viruses that contain one or more genes from the 1918 pandemic strain (Basler et al., 2001; Tumpey et al., 2002). Most recently, Tumpey, et al. studied the response of mice to infection with recombinant WSN expressing the HA and NA genes of the 1918 pandemic strain (1918 HA/NA:WSN) or with recombinant virus possessing the 1918 HA, NA, NS, matrix (M), and nucleoprotein (NP) genes (Tumpey et al., 2004). These studies showed that WSN and the recombinant 1918 HA/NA:WSN viruses are highly pathogenic in mice ($LD_{50} < 10^3$ pfu); whereas a recombinant WSN virus expressing the HA and NA genes of the human A/New Caledonia/99 H1N1 virus is significantly attenuated ($LD_{50} > 10^6$ pfu). These findings, together with previous studies that showed that a 1918 NS:WSN recombinant virus is actually less virulent than WSN in mice (Basler et al., 2001), suggest that the 1918 HA/NA genes may be primarily responsible for the increased pathogenicity of the recombinant virus in the mouse model. These results also suggest that the mouse may provide a useful (though not ideal) model for obtaining new insights into the high virulence of the 1918 virus.

To better understand differences in virulence at the level of the host transcriptional response, we are combining and correlating histopathology data with gene expression profiling (Kash et al., 2004). Histology performed on the lungs of mice infected with WSN or 1918 HA/NA:WSN virus revealed significantly more severe pathology than was observed in mice infected with the New Caledonia HA/NA:WSN recombinant virus. Whereas mice infected with New Caledonia HA/NA:WSN have minimal-to-moderate necrotizing bronchitis, mice infected with either WSN or 1918 HA/NA:WSN have pulmonary lesions consisting of moderate-to-severe necrotizing bronchitis and moderate-to-severe histiocytic alveolitis with associated pulmonary edema. Interestingly, the lungs of several mice infected with the parental WSN virus displayed some evidence of resolving bronchopneumonia at 72 h post-infection, although the acute tissue damage by this time would have been fatal had these animals not been sacrificed. In contrast, mice infected with 1918 HA/NA:WSN showed active diffuse pneumonitis, in aggregate suggesting that this virus elicits an active inflammatory process for a longer duration than does the wild-type WSN virus. Immunohistochemistry performed on the lungs of mice infected with either WSN or 1918 HA/NA:WSN revealed influenza viral antigen in necrotic cellular debris within bronchial lumina and alveoli. This staining was often associated within alveolar macrophages and was most intensely seen in mice infected with the 1918 HA/NA:WSN virus.

We then looked for correlations between the dramatic differences in lung pathology elicited by the high- and low-virulence influenza viruses and gene expression data obtained using RNA isolated from the lungs of infected mice. For these analyses, we used RNA that was pooled from the lungs of five mice from each infection group. Gene expression profiling revealed an increase in the expression of many common markers of an inflammatory response in all three infection groups. These genes included components of the IFN response, such as STAT1 and IRF1, as well as markers of activated immune cells, including the chemokine GRO1, the C-type lectin Clecsf9, and IL-2 receptor (IL2R). These studies also revealed that infection with WSN or 1918 HA/NA:WSN results in a more significant up-regulation of many pro-inflammatory, activated lymphocyte, reactive oxygen species (ROS) metabolism, and other stress-induced genes than is observed in the New Caledonia HA/NA:WSN infections (Kash et al., 2004).

Using these data, we have developed a model for the relationship between changes in gene expression and the severity of lung pathology during experimental influenza virus infection (Figure. 2). This model proposes that both virulent and attenuated influenza viruses elicit a mild necrotizing bronchitis and alveolitis that is associated with increases in the expression of common inflammation markers, including STAT1, IL2R, GRO1, and IRF1. The highly virulent influenza viruses trigger a markedly more robust inflammatory response, as indicated by high levels of expression of CSF1, IL2R, P-type selectin (Selp), and MHC class I expression (B2m). This is compounded by significant necrosis associated with the elevated expression of TNFα, inhibitor of NF-κB (Nfkbia), and GADD45g; coupled with increased oxidative stress, as indicated by the increased expression of HMOX1 and the down-regulation of genes encoding the antioxidants GPX3 and PRDX5. The significant activation of these pathology-associated pathways during a highly virulent influenza virus infection results in the development of more severe lung

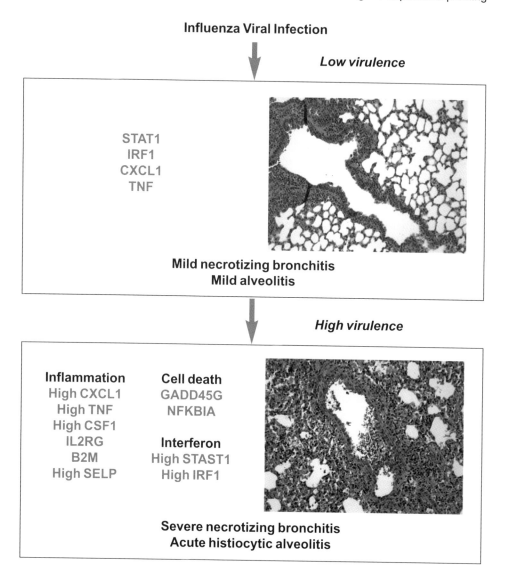

Figure 13.2 Functional genomic model of influenza virulence and pathogenesis. This model depicts important increases in gene expression that occur in the mouse lung during infection with low- or high-virulence influenza virus and their correlation with the severity of lung pathology. The model was developed by examining lung histology and gene expression patterns of mice infected with an attenuated, non-lethal influenza virus (New Cal. HA/NA:WSN) or a highly virulent, lethal virus (WSN or 1918 HA/NA:WSN).

pathology, as indicated by necrotizing bronchitis and acute alveolitis.

One of the difficulties of this analysis is that because these studies were performed using RNA isolated from whole lung, we are unable to define the contributions of individual cell types to the total gene expression changes we identified. It has been reported that human lungs contain approximately five major cell types in the parenchyma, including type I and II alveolar epithelial cells (15%), capillary epithelial cells (30%), interstitial cells (37%), and alveolar macrophages (5 to 20%) (Crapo et al., 1982). Since the primary sites of viral replication are the alveolar epithelial cells, which if one assumes a similar cellular composition in the mouse lung comprise only about 15% of the cell population, our ability to effectively measure gene expression changes in cells actually infected with virus is limited. Thus, the contribution of infected bronchial and alveolar epithelial cells to the overall transcriptional profile of the whole lung is difficult to determine. Rather, it is likely that many of the detected changes in gene expression correlate with the severity of the immune response and are due to the influx of inflammatory cells, such as T cells, macrophages, and neutrophils. Studies using bronchoalveolar lavages may allow for discrimination of the contribution that macrophages and neutrophils make to the gene expression profiles observed in different infection groups. Still, these studies demonstrate clear and significant differences in gene expression profiles in the lungs of mice infected with highly virulent or attenuated influenza viruses. Together, these data show that the severity of influenza virus infection correlates with dramatic differences in lung pathology and host gene expression, including a marked increase in the expression of genes involved in inflammation, oxidative stress, and necrosis.

Nonhuman primates

Although mice provide a convenient model in which to study influenza virus pathogenesis, and the experiments just described have yielded promising results, the model also suffers from an important disadvantage. Most notably, mice are not naturally infected with influenza virus and mouse-adapted strains are usually used for these studies. Indeed, it has been suggested that one reason an engineered WSN virus containing the NS1 gene from the 1918 pandemic strain is attenuated in mice may be because the 1918 gene is not from a mouse-adapted virus (Basler et al., 2001). Consequently, it is not clear how representative this model is of a human infection. The desire to use an animal model that more closely mimics a human infection therefore led us to explore the use of a nonhuman primate. Nonhuman primates are the closest living relatives of humans, both in evolutionary and genetic terms, and several species of Old World monkeys have successfully been used as models for influenza virus infection. In particular, influenza pathogenesis studies using various species of macaques date back to the 1940s (Berendt, 1974; Saslaw and Carlisle, 1965; Saslaw et al., 1946), and more recently, Drs Rimmelzwaan and Osterhaus have used cynomolgus macaques to study the 1997 Hong Kong virus [A/HK/156/97 (H5N1)] (Rimmelzwaan et al., 2001; Rimmelzwaan et al., 2003).

In light of this, we were interested in determining whether pigtailed macaques would also be an appropriate model for influenza virus infection. This species has proven useful for studying a variety of viruses and in some cases is more susceptible to experimental infection than its close relative, the rhesus macaque (Agy et al., 1992; Agy et al., 1997; Dewhurst et al., 1990). We therefore set out to determine how well the clinical course and pathology resulting from the experimental infection of pigtailed macaques would approximate a human influenza virus infection. We started these experiments using a mildly pathogenic strain of virus (A/Texas/36/91) with the goal of constructing a blueprint of an uncomplicated influenza infection, so that later experiments involving infections with potentially more pathogenic strains – such as recombinant viruses containing one more genes from the 1918 pandemic strain – will allow us to make new inferences into pathogenesis and to identify factors determining pathology and prognosis. Of course, we also wanted to generate gene expression data and determine how well this data would reflect the clinical picture during the course of infection.

In our initial study, two pigtailed macaques were used as mock-infected controls and two others were infected (intra-tracheally) with A/Texas/36/91 (Baskin et al., 2004). The two infected animals developed clinical signs that were highly consistent with human influenza in terms of type, intensity, and timing. Live virus was isolated from the lungs of the experimentally infected animal sacrificed at day 4, but not from the animal sacrificed at day 7. This confirmed the relatively low virulence of the Texas strain in this model and suggested that, at least in this instance, there was considerable clearance of the virus by day 7. Pathology was also a good match for what one would expect to see in humans infected with a mildly pathogenic strain of influenza. The experimental animal sacrificed at day 4 after inoculation had mild tracheitis and tracheobronchial lymphadenopathy, but the lungs and airways showed little gross pathology. In contrast, the lungs of the experimental animal sacrificed on day 7 showed multifocal-to-coalescing vascular congestion, edema, and mild-to-moderate consolidation of the lung parenchyma. Histopathology of lung tissue was also consistent with progressive primary viral pneumonia as seen in humans.

Gene expression profiling was performed using RNA isolated from lung tissue and tracheobronchial lymph nodes, and good hybridization was observed between the human cDNAs spotted on the microarrays and the labeled macaque cDNA probes. Gene expression data were then extensively mined in order to evaluate the response of infected respiratory epithelial cells as well as the presence of different types of immune cells within the tissues. These analyses revealed that numerous genes associated with the IFN response were differentially expressed in lung and tracheobronchial lymph nodes, particularly at day 4 after inoculation. We also observed an increase in the expression of genes encoding various mediators of chemotaxis, adhesion, and the transmigration of immune cells, suggesting the trafficking of these cells into the lungs and tracheobronchial lymph nodes, which was confirmed by histopathology. Genes encoding chemoattractants and other functions associated with neutrophils (such as IL8, CCL3, AOAH, C3AR1, CSF2RB, EBI2, and MNDA) were activated in the lungs at day 4. Interestingly, a different set of genes also associated with neutrophils and monocytes/macrophages was up-regulated at day 7, particularly in the tracheobronchial lymph nodes.

This study also revealed gene expression changes relevant to the processing of antigens on MHC Class I complexes, particularly in lung tissue. For example, many genes encoding proteins relevant to T-cell function were induced at one or more time points. Differential expression of genes related to cytotoxicity may also have been due to NK cells, which are usually considered more active than T cells early in infection. Other genes almost exclusively expressed by NK cells were up-regulated in lung tissue at day 4, including inhibitory lectin-like receptors (KLRC1, KLRC3, KLRC4, and KLRD1) and killer immunoglobulin-like receptors (KIR3DL2 and KIR2DL3). These receptors interact with different components of MHC Class I complexes that are normally expressed in larger numbers on the surface of uninfected cells (Achdout et al., 2003), and most were also induced at day 7 in lung tissue and day 4 in lymph nodes. It is therefore likely that NK cells and T cells were present and active as early as day 4 in lung tissue.

Although we used human cDNA microarrays to measure gene expression changes in macaque tissues, the high nucleotide sequence homology between humans and macaques makes this cross-species hybridization feasible. Still, it is not clear to what extent nucleotide sequence differences between these species may affect gene expression measurements. Therefore, we have recently developed a macaque oligonucleotide microarray, which should facilitate the use of the macaque model for future gene expression profiling experiments. Oligonucleotide sequences present on the array were designed based on sequence data obtained from the large-scale sequencing of cDNA libraries constructed from various macaque tissues. The array also includes a large number of human homologues and viral genes, including those of influenza and other respiratory viruses.

To further refine the macaque model, we are working to determine the variability of gene expression in the context of normal physiology by performing a series of microarray experiments using RNA isolated from tissues of healthy animals. The methodology for these experiments was developed in collaboration with biostatisticians and will allow us to determine biologically relevant gene expression changes in diseased tissue with greater accuracy. In addition, we intend to eventually expand our analyses to include tissues other than the lungs and lymph nodes, including circulating leukocytes and cells sampled during bronchoalveolar lavages. Gene expression profiling of immune cells present in the blood or other bodily fluids has greater diagnostic and prognostic potential because sampling methods are relatively noninvasive. The possibility of diagnosing infection early after exposure may be of particular consequence for the detection of highly virulent or pandemic strains and for biodefense or emerging infectious disease surveillance. We are confident the macaque model, together with gene expression profiling, will provide important new insights into influenza virus-host interactions, and in particular, the mechanisms underlying the high lethality associated with the 1918 virus and contemporary avian strains.

COMPARING GENE EXPRESSION DATA FROM DIVERSE EXPERIMENTAL SYSTEMS

One direction we see as critical to a better understanding of influenza virus virulence and pathogenesis is the integration of gene expression data from various model systems. As described, we use a variety of systems to study both low- and high-virulence influenza viruses, including established respiratory epithelial (A549) cells, primary human tracheobronchial cells, and mouse and nonhuman primate models of infection. As an example of this type of approach, we compared the gene expression profiles obtained from the lungs of mice and macaques that were experimentally infected with influenza virus (Figure 3A). In this analysis, we focused on the identification of gene orthologs that were ≥1.5-fold up-regulated at day 4 in macaque lung (corresponding to time with highest viral titer), and which also showed ≥1.5-fold up-regulation in mouse lung in at least two experiments. Using this tactic, we identified a number of genes that were induced in both animal models, including several lipid and cholesterol metabolism genes (LDLR, HMGCR, SOAT1, and LPL) and a large number of immune-related genes, such as STAT1, LY6E, MCL1, CTSB/C, CLIC4, LCP2, SCYA4 9 (also known as CCL4), SP100, and TNF-α (Figure 3B). This analysis further identified a group of immune-related genes (WARS, CLIC4, TNF, SCYA4, LCP2, and SP100) that were significantly more up-regulated in the lungs of animals infected with a high-virulence influenza virus (1918 HA/NA:WSN or A/WSN/33). The overall gene expression pattern obtained from macaques infected with A/Texas/91 was most similar to that obtained from the lungs of mice infected with New Cal. HA/NA:WSN. This is not unexpected, since A/Texas/91 and New Cal. HA/NA:WSN are both low-virulence viruses.

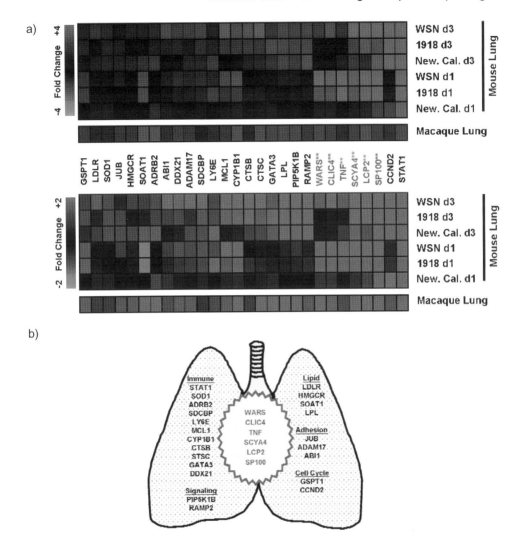

Figure 13.3 Comparison of macaque and murine gene expression patterns in response to influenza virus infection. a) The genes displayed exhibited a ≥1. 5-fold increase in expression ($P \leq 0.01$) in macaque lung infected with A/Texas/36/91 (day 4) and in at least two experiments in which mice were infected with New Caledonia HA/NA:WSN, 1918 HA/NA:WSN, or parental recombinant WSN virus. The bottom and top panels display the same intensity data, but with different color saturations (i. e., saturation is at 2 fold in the bottom panel and at 4 fold in the top panel). Gene names shown in red (**) were significantly up-regulated (≥4 fold) only in lungs from animals infected with a lethal dose of virus (1918 HA/NA:WSN and WSN). Genes depicted by red boxes were up-regulated and genes depicted by green boxes were down-regulated in infected relative to uninfected lung. Black indicates no change in gene expression. Primary analysis was performed using Resolver and supplemental analysis using Decision Site for Functional Genomics. b) Functional categories of genes listed in panel A. Gene names shown in the central section were significantly up-regulated (≥4 fold) only in lungs from animals infected with a lethal dose of virus (1918 HA/NA:WSN and WSN). Figure 3A is also reproduced in colour in the colour section at the end of the book

We are very interested in expanding these analyses by characterizing the transcriptional profile of macaque lungs in response to more pathogenic strains.

Another approach that we are using is the comparative study of cellular gene expression changes that occur in response to a variety of DNA or RNA viruses. As an example, we have compared the cellular gene expression response to infection with vaccinia virus, severe acute respiratory syndrome-associated coronavirus (SARS-CoV), Ebola Zaire, Marburg, and mouse-adapted (A/WSN/33), human (A/Texas/91 and A/Memphis/96) and avian (A/Mallard/Alberta/98) influenza viruses. Figure 4A depicts genes whose expression changed ≥1.5 fold in at least seven of the twelve experiments analyzed. This analysis shows gene expression patterns that are common and restricted to DNA and RNA viruses, as well as virus-specific gene expression signatures. A similar approach is depicted in Figure 4B, which shows IFN-regulated genes whose expression changed ≥1.5 fold in at least one of the ten experiments analyzed. This analysis shows a separation of DNA and RNA viruses at the root clade. Low-virulence influenza virus (A/Texas/91 and A/Memphis/96) infections are grouped together and elicit the up-regulation of a small number of IFN-regulated genes. In contrast, Ebola Zaire and Marburg virus infections showed very little activation of these genes, demonstrating the ability of these highly virulent viruses to antagonize activation of the antiviral response. Interestingly, profiles generated from human A549 cells infected with mouse-adapted (A/WSN/33) or avian influenza (A/Mallard/Alberta/98) viruses were grouped together, and these profiles contained the largest number of up-regulated IFN-responsive genes, which may be a reflection of cross-species infection.

One obvious and significant variable in these types of analyses is the influence of the chosen infection system on the measured gene expression changes. The use of a two-color array approach helps to minimize the influence of cell-line specific gene expression patterns that are independent of infection, since the measurements are analyzed as ratios between mock and infected cells of the same model system. Still, different cells or tissues might be expected to respond differently to infection with the same virus. In the end, we are of the opinion that in order to maximize the relevance of broad comparisons of the host response to infection, it is better to use different but appropriate model systems (e.g., influenza virus in respiratory cells and hepatitis C in liver cells) than to force all of the infectious agents into a common system. Comparative analyses will be useful in elucidating many aspects of the host response and should also lead to the identification of common and novel viral strategies for controlling the host response and for maximizing viral replication.

Additional insights into how the cellular response to influenza virus compares with that elicited by other pathogens come from a study that used microarrays to measure gene expression profiles of dendritic cells in response to *Escherichia coli*, *Candida albicans*, or influenza virus (Huang et al., 2001). This study revealed that the three pathogens elicit pathogen-specific responses as well as a common response that results in the differential expression of a common set of 166 genes. Immediately after infection all three pathogens elicit a transient increase in the expression of genes associated with immune cytokines, chemokines, and receptors that contribute to the recruitment of other immune cells to the site of infection. This is followed by the expression of genes involved in the generation of reactive oxygen species and finally by genes encoding chemokine receptors. Pathogen-specific programs of gene expression were also observed, and in the case of influenza virus this was marked by a strong induction of IFN-α and -β genes and IFN-inducible chemokine genes. In addition, influenza virus induced the expression of genes linked with the inhibition of the immune response and proapoptotic genes that may lead to the death of infected cells.

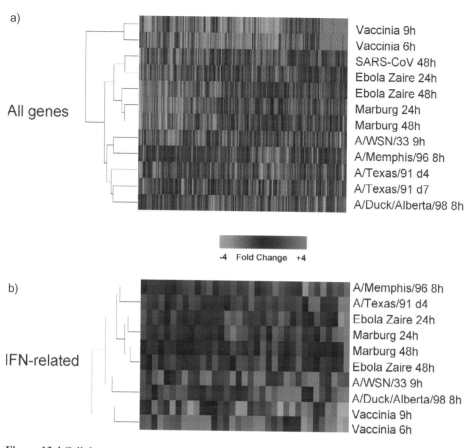

Figure 13.4 Cellular gene expression profiles elicited by influenza, vaccinia, SARS-Cov, Ebola Zaire, or Marburg virus infection. a) Matrix showing the two-dimensional clustering of 189 genes that showed a ≥1. 5-fold change in expression ($n = 4$; $P ≤ 0. 05$) in a least seven of the twelve infections. b) Matrix showing the two-dimensional clustering of 35 IFN-regulated genes that showed a ≥1. 5-fold change in expression ($n = 4$; $P ≤ 0. 05$) in a least one of the ten infections. Each of these hierarchical agglomerative clusters was created using complete link heuristics and Manhattan and Euclidean distance metrics for experiment and sequence axis, respectively. Vaccinia virus were performed using HeLa cells, SARS-CoV using Vero cells, Ebola Zaire and Marburg virus using Huh7 cells, and various Type A influenza viruses using macaques, A549 cells, or human tracheobronchial epithelial cells. Genes shown in red were up-regulated and genes shown in green were down-regulated in infected relative to uninfected cells. Black indicates no change in gene expression. Primary analysis was performed using Resolver and supplemental analysis using Decision Site for Functional Genomics. This figure is also reproduced in colour in the colour section at the end of the book

Therefore, at least in the case of dendritic cells, there appears to be a core response to a pathogen as well as pathogen-specific responses. Whether the response to influenza virus is a virus-specific response, an RNA virus-specific response, or unique to influenza virus awaits additional experimentation.

USING ARRAYS TO IDENTIFY AND TYPE INFLUENZA VIRUSES

Although the focus of this chapter is on how DNA microarrays are being used to profile cellular gene expression changes in response to influenza virus, it is worth noting that this technology can also being used to focus on the virus (Wang et al., 2002). In particular, there is interest in using array technology as a diagnostic tool for identifying and typing influenza viruses based on their hemagglutinin and neuraminidase genes. An early example of this approach was reported by Li et al., who spotted multiple influenza virus gene products – encoding approximately 500-bp fragments of 4 hemagglutinin, 3 neuraminidase, and 2 matrix genes – onto glass slides (Li et al., 2001). The arrays were then probed with labeled viral cDNAs generated by multiplex RT-PCR. Using this strategy, it was possible to accurately type and subtype several previously unsequenced influenza A and B virus strains.

A variation on this approach has been reported by Sengupta et al., who spotted onto glass slides a collection of 476 influenza virus-specific oligonucleotides averaging 21 nucleotides in length (Sengupta et al., 2003). The rationale behind this approach is that an array with more probes of shorter length should improve the detection of and discrimination among closely related virus strains. This study revealed considerable variation in the ability of different oligonucleotides to make distinctions among influenza viruses, and the intensities of spots complimentary to labeled probes varied from background to saturation. However, oligonucleotides that were monospecific for influenza virus type or hemagglutinin subtype were highly successful in making correct identifications. This was particularly the case for H1 and H3, which are well studied and for which the greatest number of signature oligonucleotides are available. Of the 9 NA subtypes, only N1 and N2 had signature oligonucleotides, which considerably limited the ability of the array to distinguish between NA subtypes. Still, improved oligonucleotide design is likely to enhance the ability of future arrays to make distinctions among influenza viruses.

More recently, Kessler et al. reported on the use of a variation of the traditional microarray called the DNA Flow-Thru Chip (Kessler et al., 2004). In this platform, molecular interactions occur within the three-dimensional volumes of ordered microchannels, rather than the two-dimensional format of microarrays. The technology has greater binding capacity, reduced hybridization times, and requires less sample amounts than glass-based arrays. Oligonucleotide probes were selected to recognize multiple fragments of the influenza A virus matrix protein gene, the influenza B virus NS gene, the H1, H3, and H5 genes, and the N1 and N2 genes. These oligonucleotides were immobilized in the microchannels of a silicon wafer and hybridized to cDNAs generated by multiplex or random RT-PCR. This method was capable of unambiguously identifying influenza A viruses H1N1, H3N2, H1N2, and H5N1. Of course, the actual clinical application of all these methods remains to be proven and may still be limited by expense and time constraints, as well as by the more complex nature of actual clinical samples. Nevertheless, it is likely that microarray-based methods for the identification and typing of influenza virus will continue to improve, as more and better oligonucleotide sequences become available and with continued technological advancements.

FUTURE PROSPECTS

One of the approaches we described in this chapter is the use of engineered viruses to examine the role that specific viral proteins play in modulating the cellular response to infection. In addition to examining the role of specific viral proteins, we are interested in exploring the role of certain cellular proteins as well. In particular, we are examining the role of the molecular chaperone P58[IPK], which we first identified as a cellular inhibitor of the IFN-induced protein kinase, PKR (Lee et al., 1990; Lee et al., 1992). P58^{IPK} is activated upon influenza virus infection (Melville et al., 1999), and we have suggested that the activation of P58^{IPK} results in the inhibition of PKR-mediated translational arrest, thereby ensuring that the cellular protein synthesis machinery remains available to synthesize viral proteins. It is unlikely, however, that P58^{IPK} evolved as a cellular gene to aid in viral replication. Indeed, we have since discovered that P58^{IPK} is transcriptionally induced during the unfolded protein response (UPR) and that it interacts with PERK to attenuate PERK-mediated eIF2α phosphorylation during endoplasmic reticulum (ER) stress (Yan et al., 2002). Thus, the P58^{IPK} gene is among a group of genes encoding molecular chaperones, protein-folding enzymes, and transcription factors that are induced following ER stress and which function to restore homeostasis to stressed cells.

Does this activity of P58^{IPK} play a role during influenza virus infection as well? It is likely that the synthesis of large amounts of viral glycoproteins during viral replication places a stress upon the ER, and the UPR-mediated transcriptional activation of P58^{IPK} may act synergistically with its post-transcriptional activation to offer the virus a measure of relief from both PERK and PKR-mediated translational repression. Although our previous microarray analyses of cells infected with influenza virus did not revealed a major activation of the UPR, we are further exploring this possibility using recently developed P58^{IPK} knockout mice. We are examining how these animals respond to influenza virus infection, including the changes in cellular gene expression that occur in response to infection.

Finally, while DNA microarrays provide a view of the transcriptional changes that occur in response to virus infection, a more complete picture of the response will require also looking at how the protein content of the cell is affected. This is especially significant when microarray analysis is performed on cells infected with a virus such as influenza, which is known to reduce the overall level of cellular mRNA translation. We are already using quantitative proteomic technologies to evaluate changes in the level of individual proteins in response to hepatitis C virus infection (Yan et al., 2004b; Yan et al., 2004a). In addition, we are using a targeted proteomics approach to identify cellular or viral proteins that interact with specific viral gene products, including the influenza virus NS1 protein. Because proteins typically interact with one another to perform a particular task, identifying the interaction partners of these proteins can be an important means to discovering their biological function or the molecular mechanism by which a function is carried out. Ultimately, we are working to integrate data obtained from gene expression profiling and proteomics to gain a comprehensive picture of the cellular response to infection and the various strategies used by viruses to counteract or evade cellular defense mechanisms. We believe that knowledge gained through the use of genomic and proteomic technologies will continue to contribute greatly to our understanding of virus-host interactions and will eventually lead to improved and rational-designed antiviral therapies as well as diagnostic and prognostic applications.

Acknowledgements

We thank our many collaborators and colleagues who contributed to the studies described in this chapter. Funding for gene expression studies in our laboratory is provided by Public Health Service grants R01AI22646, R01AI47304, R21AI53765, P01AI52106, U19AI48214, P30DA15625, R24RR16354, and P51RR00166 from the National Institutes of Health. Additional funding is provided by Illumigen Biosciences, Inc., Seattle, Washington.

References

Achdout, H., Arnon, T. I., Markel, G., Gonen-Gross, T., Katz, G., Lieberman, N., Gazit, R., Joseph, A., Kedar, E., and Mandelboim, O. (2003). Enhanced recognition of human NK receptors after influenza virus infection. J. Immunol. *171*, 915–923.

Agy, M. B., Frumkin, L. R., Corey, L., Coombs, R. W., Wolinsky, S. M., Koehler, J., Morton, W. R., and Katze, M. G. (1992). Infection of Macaca nemestrina by human immunodeficiency virus type-1. Science *257*, 103–106.

Agy, M. B., Schmidt, A., Florey, M. J., Kennedy, B. J., Schaefer, G., Katze, M. G., Corey, L., Morton, W. R., and Bosch, M. L. (1997). Serial in vivo passage of HIV-1 infection in Macaca nemestrina. Virology *238*, 336–343.

Baskin, C. R., García-Sastre, A., Tumpey, T. M., Bielefeldt-Ohmann, H., Carter, V. S., Nistal-Villán, E., and Katze, M. G. (2004). Integration of clinical data, pathology, and cDNA microarrays in influenza virus-infected pigtailed macaques (*Macaca nemestrina*). J. Virol. *78*, 10420–10432.

Basler, C. F., Reid, A. H., Dybing, J. K., Janczewski, T. A., Fanning, T. G., Zheng, H., Salvatore, M., Perdue, M. L., Swayne, D. E., García-Sastre, A., Palese, P., and Taubenberger, J. K. (2001). Sequence of the 1918 pandemic influenza virus nonstructural gene (NS) segment and characterization of recombinant viruses bearing the 1918 NS genes. Proc. Natl. Acad. Sci. USA *98*, 2746–2751.

Berendt, R. F. 1974. Simian model for the evaluation of immunity to influenza. Infect. Immun. *9*, 101–105.

Choi, A. M., Knobil, K., Otterbein, S. L., Eastman, D. A., and Jacoby, D. B. (1996). Oxidant stress responses in influenza virus pneumonia, gene expression and transcription factor activation. Am. J. Physiol. 271 (3 Pt 1), 383–391.

Crapo, J. D., Barry, B. E., Gehr, P., Bachofen, M., and Weibel, E. R. (1982). Cell number and cell characteristics of the normal human lung. Am. Rev. Respir. Dis. *126*, 332–337.

Dewhurst, S., Embretson, J. E., Anderson, D. C., Mullins, J. I., and Fultz, P. N. (1990). Sequence analysis and acute pathogenicity of molecularly cloned SIVSMM-PBj14. Nature *345*, 636–640.

Gale, M., Jr., Tan, S.-L., and Katze, M. G. (2000). Translational control of viral gene expression in eukaryotes. Microbiol. Mol. Biol. Rev. *64*, 239–280.

García-Sastre, A. (2004). Identification and characterization of viral antagonists of type I interferon in negative-strand RNA viruses. Curr. Top. Microbiol. Immunol. *283*, 249–280.

García-Sastre, A., Egorov, A., Matassov, D., Brandt, S., Levy, D. E., Durbin, J. E., Palese, P., and Muster, T. (1998). Influenza A virus lacking the NS1 gene replicates in interferon-deficient systems. Virology *252*, 324–330.

Geiss, G. K., An, M. C., Bumgarner, R. E., Hammersmark, E., Cunningham, D., and Katze, M. G. (2001). Global impact of influenza virus on cellular pathways is mediated by both replication-dependent and -independent events. J. Virol. *75*, 4321–4331.

Geiss, G. K., Salvatore, M., Tumpey, T. M., Carter, V. S., Wang, X., Basler, C. F., Taubenberger, J. K., Bumgarner, R. E., Palese, P., Katze, M. G., and García-Sastre, A. (2002). Cellular transcriptional profiling in influenza A virus-infected lung epithelial cells, the role of the nonstructural NS1 protein in the evasion of the host innate defense and its potential contribution to pandemic influenza. Proc. Natl. Acad. Sci. USA *99*, 10736–10741.

Huang, Q., Liu, D., Majewski, P., Schulte, L. C., Korn, J. M., Young, R. A., Lander, E. S., and Hacohen, N. (2001). The plasticity of dendritic cell responses to pathogens and their components. Science *294*, 870–875.

Kash, J. C., Basler, C. F., García-Sastre, A., Carter, V., Billharz, R., Swayne, D. E., Przygodzki, R. M., Taubenberger, J. K., Katze, M. G., and Tumpey, T. M. (2004). Global host immune response, pathogenesis and transcriptional profiling of type A influenza viruses expressing the hemagglutinin and neuraminidase genes from the 1918 pandemic virus. J. Virol. *78*, 9499–9511.

Kash, J. C., Cunningham, D. M., Smit, M. W., Park, Y.-W., Fritz, D., Wilusz, J., and Katze, M. G. (2002). Selective translation of eukaryotic mRNAs, functional molecular analysis of GRSF-1, a positive regulator of influenza virus protein synthesis. J. Virol. 76, 10417–10426.

Katze, M. G. and Krug, R. M. (1990). Translational control in influenza virus-infected cells. Enzyme 44, 265–277.

Kessler, N., Ferraris, O., Palmer, K., Marsh, W., and Steel, A. (2004). Use of the DNA flow-thru chip, a three-dimensional biochip, for typing and subtyping of influenza viruses. J. Clin. Microbiol. 42, 2173–2185.

Lee, T. G., Tomita, J., Hovanessian, A. G., and Katze, M. G. (1990). Purification and partial characterization of a cellular inhibitor of the interferon-induced protein kinase of M_r 68,000 from influenza virus-infected cells. Proc. Natl. Acad. Sci. USA 87, 6208–6212.

Lee, T. G., Tomita, J., Hovanessian, A. G., and Katze, M. G. (1992). Characterization and regulation of the 58,000-dalton cellular inhibitor of the interferon-induced, dsRNA-activated protein kinase. J. Biol. Chem. 267, 14238–14243.

Li, J., Chen, S., and Evans, D. H. (2001). Typing and subtyping influenza virus using DNA microarrays and multiplex reverse transcriptase PCR. J. Clin. Microbiol. 39, 696–704.

Matrosovich, M. N., Matrosovich, T. Y., Gray, T., Roberts, N. A., and Klenk, H. D. (2004). Human and avian influenza viruses target different cell types in cultures of human airway epithelium. Proc. Natl. Acad. Sci. USA 101, 4620–4624.

Melville, M. W., Tan, S.-L., Wambach, M., Song, J., Morimoto, R. I., and Katze, M. G. (1999). The cellular inhibitor of the PKR protein kinase, P58[IPK], is an influenza virus-activated co-chaperone that modulates heat shock protein 70 activity. J. Biol. Chem. 274, 3797–3803.

Park, Y.-W. and Katze, M. G. (1995). Translational control by influenza virus, identification of cis-acting sequences and trans-acting factors which may regulate selective viral mRNA translation. J. Biol. Chem. 270, 28433–28439.

Park, Y.-W., Wilusz, J., and Katze, M. G. (1999). Regulation of eukaryotic protein synthesis, selective influenza viral mRNA translation is mediated by the cellular RNA-binding protein GRSF-1. Proc. Natl. Acad. Sci. USA 96, 6694–6699.

Rimmelzwaan, G. F., Kuiken, T., van Amerongen, G., Bestebroer, T. M., Fouchier, R. A., and Osterhaus, A. D. (2001). Pathogenesis of influenza A (H5N1) virus infection in a primate model. J. Virol. 75, 6687–6691.

Rimmelzwaan, G. F., Kuiken, T., van Amerongen, G., Bestebroer, T. M., Fouchier, R. A., and Osterhaus, A. D. (2003). A primate model to study the pathogenesis of influenza A (H5N1) virus infection. Avian Dis. 47, 931–933.

Saslaw, S., Wilson, H. E., Doan, C. A., Woolpert, O. C., and Schwab, J. L. 1946. Reactions of monkeys to experimentally induced influenza virus A infection. Exp. Med. 84, 113–125.

Saslaw, S. and Carlisle, H. N. 1965. Aerosol exposure of monkeys to influenza virus. Proc. Soc. Exp. Biol. Med. 119, 838–843.

Sengupta, S., Onodera, K., Lai, A., and Melcher, U. (2003). Molecular detection and identification of influenza viruses by oligonucleotide microarray hybridization. J. Clin. Microbiol. 41, 4542–4550.

Tan, S.-L., Katze, M. G., and Gale, M. J. (2000). Translational reprogramming during influenza virus infection. In Translational control of gene expression, N. Sonenberg, J. W. B. Hershey, and M. B. Mathews, eds. (Plainview, N. Y., Cold Spring Harbor Laboratory Press), pp. 933–950.

Tumpey, T. M., García-Sastre, A., Mikulasova, A., Taubenberger, J. K., Swayne, D. E., Palese, P., and Basler, C. F. (2002). Existing antivirals are effective against influenza viruses with genes from the 1918 pandemic virus. Proc. Natl. Acad. Sci. USA 99, 13849–13854.

Tumpey, T. M., García-Sastre, A., Taubenberger, J. K., Palese, P., Swayne, D. E., and Basler, C. F. (2004). Pathogenicity and immunogenicity of influenza viruses with genes from the 1918 pandemic virus. Proc. Natl. Acad. Sci. USA 101, 3166–3171.

Wang, D., Coscoy, L., Zylberberg, M., Avila, P. C., Boushey, H. A., Ganem, D., and DeRisi, J. L. (2002). Microarray-based detection and genotyping of viral pathogens. Proc. Natl. Acad. Sci. USA *99*, 15687–15692.

Yan, W., Frank, C. L., Korth, M. J., Sopher, B. L., Novoa, I., Ron, D., and Katze, M. G. (2002). Control of PERK eIF2α kinase activity by the endoplasmic reticulum stress-induced molecular chaperone P58[IPK]. Proc. Natl. Acad. Sci. USA *99*, 15920–15925.

Yan, W., Lee, H., Deutsch, E. W., Lazaro, C. A., Tang, W., Chen, E., Fausto, N., Katze, M. G., and Aebersold, R. 2004a. A dataset of human liver proteins identified by protein profiling via isotope coded affinity tag (ICAT) and tandem mass spectrometry. Mol. Cell Proteomics. 3: 1039–1041

Yan, W., Lee, H., Yi, E. C., Reiss, D., Shannon, P., Kwieciszewski, B. K., Coito, C., Li, X. J., Keller, A., Eng, J., Galitski, T., Goodlett, D. R., Aebersold, R., and Katze, M. G. (2004b). System-based proteomic analysis of the interferon response in human liver cells. Genome Biol. *5*,R54.

Index

1918 'Spanish' virus   238–239
  epidemiology   304–306
  HA   306–311
  historical background   301–304
  mortality rate   302
  M segment   312–312
  NA   306–311
  NP   313–314
  NS gene   311–312, 345
  resistance to oseltamivir   180
  serology   304–306

Actin   7, 54
Adaptation   108
Adjuvant   206
Amantadine   66–67, 78–79,   170–172
  prophylaxis   175–177
  resistance   172–175, 258–259
  treatment   177–178
Animal model
  mice   347–350
  nonhuman primates   350–352
Antigen-presenting cells   120
Antiviral drugs   258–259. *See also*
      amantadine, rimantadine, oseltamivir
      phosphate, zanamivir
APC *See* antigen-presenting cells
Apical membrane   38, 53
  M2   70
Apoptosis   43, 328–329
  NA   331
  NS1   329
  PB1–F2   329
Assembly   53–55
Avian influenza   230–231, 239–245
  avian-to-human transmission   251–252
  avian-to-swine transmission   254
  ecology   241
  epidemiology   241–245
  H5N1 *See* H5N1 viruses
  HA cleavage   240–241
  history   239–240

Avian influenza (*cont.*)
  poultry, control in   259–262
  receptor specificity   109–113, 231
  vaccine   216–218
  virulence   240

Bacterial adhesion   122
B-cell activation   120
Bcl-2   329–330
BM2   42
  ion channel   82–86
B viruses   237–238

Ca See cold-adapted
Cap-binding   4–5, 17
Carbohydrates   104
CD8+ cytotoxic T lymphocytes *See* CTL
CD8+ T cell
  circulation   156
  cytokine production   159–161
  cytotoxicity   161–162
  epitope-specific CD8+ T cell repertoires
      158
  memory   162
  quantitation   157–158
  virus clearance   155–156
Cell culture-grown vaccine   207–208
Chemokine   325
Chicken viruses (H5N1), receptor specificity
      112–113
Chitosan   213
Clathrin-mediated endocytosis *See* endocytosis
CM2   42, 86
Cold-adapted phenotype   205–206
Corkscrew structure   10–13
CRM1   7, 48–52
cRNA   2, 19–22
  cRNA synthesis   19–22
  temporal regulation   24–25
CTL response   139–141
Cytokine   147, 159–161, 325
Cytoskeleton   43, 54

DC *See* dendritic cells
Dendritic cells
  CTL response   139–141
  cytokine production   147
  double-stranded RNA   145–146
  inflammatory mediators   146
  lymph nodes   141–144
  migration   141–144
  MxA   145–146
  NA   147
  NS1   146
  PKR   146
  plasmacytoid DC   148–149
  respiratory dendritic cells   140
  recruitment   145
DIVA (differentiate vaccinated from infected
    animals)   261
DNA vaccine *See* vaccine
Double-stranded RNA   145–146
Ds RNA *See* double-stranded RNA
Duck viruses   109
  host-range restriction   110
  receptor specificity   109–111

Ecology   230–231, 241
Egg-adaptation   117–119
Endocytosis   39
Endonuclease   1, 2–4, 17
Endosome   39
  uncoating   42–43, 67–68
Epidemic
  B viruses   237–238
  epidemic strains   233–238, 257
  excess deaths   233
  H1N1 viruses   235–236
  H1N2 viruses   236
  H3N2 viruses   236–237
  hospitalization rates   233
  surveillance   233
  vaccine   255–256
Epidemiology
  1918 'Spanish' virus   304–306
  avian viruses   241–245
  H5N1 viruses   244–245
Equine viruses, receptor specificity   113
ERK (extracellular signal regulated kinase)
    326–327
Evolutionary stasis   230, 281
Excess deaths   233

Filamentous virions   54
Fusion   39–42
Fusion peptide   42

GRSF-1   345–346
Gull viruses, receptor specificity   112

H1N1 viruses (human)   235–236
H1N1 viruses (swine)   245–246, 248–249
H1N2 viruses (human)   236
H1N2 viruses (swine)   248
H3N2 viruses (human)   236–237
H3N2 viruses (swine)   246–248
H5N1 viruses   112
  1997 human viruses   287–288
  2004 human and avian viruses   290–293
  A/goose/Guangdong/1/96 precursor
      viruses   286
  cleavage   240–241
  egg-adaptation   118–119
  epidemiology   244–245
  evolution   288–290
  interspecies transmission   251–252
  NS1   333
  vaccine   216–218
  Z genotype   289, 290
H5N2 viruses   285–286
H7 viruses   112
HA 97
  1918 'Spanish' virus   306–311
  antigen-presenting cells   120
  assembly   53
  avian influenza viruses   240–241
  bacterial adhesion   122
  B-cell activation   120
  carbohydrates   104
  fusion   39–42
  fusion peptide   42
  globular head   105
  H5N1 viruses   112
  H7 viruses   112
  helper T-cells   120
  host range   231–232
  H variant   120
  L variant   120
  natural killer (NK) cells   121
  neutrophils   121
  pathogenesis   282–283
  phylogenetic tree   108
  polyvalency   98
  raft association   53

HA 97 (*cont.*)
  receptor   37
  receptor-binding activity, assays   105–107
    – binding to gangliosides   106
    – binding to sialosides,
      sialylglycopolymers, and
      sialylglycoproteins   106
  receptor-binding site (RBS)   98
  receptor specificity   108, 231
    – of H5N1 chicken viruses   112–113
    – of duck viruses   109–111
    – of equine viruses   113
    – of gull viruses   112
    – of human viruses   115–119
    – of seal viruses   113
    – of swine viruses   114–115
    – of turkey viruses   112
    – tropism   119
  resistance to oseltamivir phosphate
    180–182
  respiratory burst   121
  sialic acid-binding pocket   98
  subtypes   108
  X-ray crystallography   98–101
hCLE   22
HEF   37
Helper T-cell   120
Hemagglutinin *See* HA
Hemagglutinin-esterase-fusion *See* HEF
Highly pathogenic avian influenza viruses
    112, 239–240. *See also* avian viruses
  ecology   241
  epidemiology   244–245
  HA cleavage   240–241
Hospitalization rates   233
Host-range restriction   231–232
  of duck viruses   110
Hsp90   22
Human viruses   115, 232–239
  1997 H5N1 viruses   287–288
  2004 H5N1 viruses   290–293
  B viruses   237–238
  epidemic strains   233–238
  excess deaths   233
  H1N1 viruses   235–236
  H1N2 viruses   236
  H3N2 viruses   236–237
  hospitalization rates   233
  human-to-swine transmission   254
  pandemic viruses   238–239
  reassortment   236, 239

Human viruses (*cont.*)
  receptor-binding sites   117–118
  receptor specificity   115–116
  receptor specificity and egg adaptation
    118–119
  receptor specificity and pandemics
    116–117
  H variant   120

IKK   324–325, 328
IFN *See* interferon
Importin alpha   7
Inactivated vaccine   205
Inflammatory mediators   146
Interferon   332–333
  microarray   343–345
Interspecies transmission
  H5N1 viruses   251–252
  avian-to-human   251–252
  avian-to-swine   254
  human-to-swine   254
  swine-to-human   253–254
Ion channel   42
  amantadine   170–178
  activation   77–78
  BM2   82–86
  CM2   86
  inhibition   79, 170–172
  ion channel activity   71–72
  ion channel structure   73–74
  ion selectivity   76–77
  M2   66–81
  rimantadine   170–178
ISAV (infectious salmon anaemia virus)
  hemagglutinin-esterase glycoprotein   38
ISCOMs   212

JNK (Jun-N-terminal kinase)   325–326

LAIV *See* live attenuated vaccine, vaccine
Lipid rafts *See* rafts
Liposomes   213
Live attenuated vaccine   205–206
Low-pathogenic avian influenza viruses   112,
    239. *See also* avian viruses
  ecology   241
  epidemiology   241–245
  HA cleavage   240–241
L variant   120
Lymph nodes   141–144

M1
  assembly   54
  nuclear localization signal   46
  RNP nuclear import   45
  RNP nuclear export   48–52
  Ts mutant   51
M2   42, 66
  1918 'Spanish' virus   312–313
  amantadine   66–69, 78–79, 170–178
  apical membrane   70
  assembly   53
  cytoplasmic tail   80–81
  deletion mutants   69
  ectodomain   81
  ion channel   66–81
    – activation   77–78
    – inhibition   79, 170–172
    – ion channel activity   71–72
    – ion channel structure   73–74
    – ion selectivity   76–77
  palmitoylation   69, 73
  phosphorylation   69, 73
  rafts   70 -71
  rimantadine   170–178
  TGN   80
  transmembrane domain   75–76
  vaccine   211
MAPK (mitogen-activated protein kinase)
  ERK (extracellular signal regulated
      kinase)   326–327
  JNK (Jun-N-terminal kinase)   325–326
  nuclear export   326
  overview   323–325
  p38   325–326
Membrane fusion *See* fusion
Messenger RNA   1
Mice, animal model   347–350
Microarray   325, 343–347, 356
Microtubules   43
'Mixing vessel'   114, 249
Mortality rate   233
  1819 'Spanish' virus   302
mRNA   1
  synthesis *See* transcription
Multivesicular body (MVB)   39
Mx   7, 8
  microarray   344
  RNP nuclear import   45
  DC response   146

N-acetylneuraminic acid (Neu5Ac)   96.
    See also sialic acid

N-glycolylneuraminic acid (Neu5Gc)   96.
    See also sialic acid
NA
  1918 'Spanish' virus   306–311
  apoptosis   331
  assembly   53
  DC Function   147
  inhibitors   178–188
  oseltamivir   178–188
  pathogenesis   283
  raft association   53
  resistance to oseltamivir phosphate
      182–185
  zanamivir   178–188
Natural killer cells   121
Natural reservoir   281
NB   82–84
NCR (noncoding region)   10, 345. *See also*
    UTR (untranslated region)
  GRSF-1   345
  Packaging of vRNA segments   10
NEP (nuclear export protein)   48–52
NES *See* nuclear export signal
Neutrophils   121
NFκB   324–325, 328
NK cells *See* natural killer cells
NLS *See* nuclear localization signal
Non-clathrin-dependent endocytosis *See*
    endocytosis
Noncoding region *See* NCR   10
Nonhuman primates, animal model   350–352
NP   7–8
  1918 'Spanish' virus   313–314
  assembly   54
  interaction with cellular proteins   7, 8
  interaction with M1   7
  interaction with PB1   2, 7
  interaction with PB2   4, 7
  intracellular localization   46–47
  nuclear localization signal   44
  phosphorylation   44
  replication   19–24
  RNA-binding   7
  RNP   8–9
  Ts mutants   19, 21, 23
NS1
  1918 'Spanish' virus   311–312, 345
  apoptosis   329, 332
  DC response   146
  H5N1 viruses   333
  interferon response   332–333

NS1 (*cont.*)
microarray   343
nuclear localization signal   46
pathogenesis   284–284
STAT1   332
vaccine   211
NS2 *See* NEP
NS gene, 1918 'Spanish' virus   311–312
Nuclear export of RNP   48–52
ERK   326
Nuclear export signal   48–52
Nuclear localization signal
M1   46
NP   44
NS1   46
PA   44
PB1   44
PB2   44
Nucleoprotein, see NP

Oseltamivir phosphate   178–188
mechanisms of action   178–179
prophylaxis   185–186
resistance   180–185
– mutations in HA   180–182
– mutations in NA   182–185
spectrum and potency   179–180
treatment   187–188

PA   5–7, 17–19, 21–22
interaction with PB1   2, 5
interaction with PB2   4
nuclear localization signal   44
RNP   9–10
phosphorylation   5
protease   5, 6–7, 21
replication   21–22
transcription   17–19
Ts mutants   5–6, 21, 23
Packaging of vRNA segments   10
Palmitoylation   69, 73
Pandemic
pandemic viruses   238–239
receptor-binding specificity and
pandemics   116–117
vaccine   256
Panhandle structure   1, 10–13
Pathogenesis
HA cleavage   282–283
NA   283
NS1   284–285

Pathogenesis (*cont.*)
PB2   284
receptor specificity   284
recombination   283
tropism   284
PB1   2–4, 17–19
interaction with NP   2, 7
interaction with PA   2
interaction with PB2   2, 7
nuclear localization signal   44
endonuclease activity   3
RNP   9–10
promoter binding   3
transcription   17–19
UV cross-linking   2–3
PB1–F2   43, 329–330
PB2   4–5, 17–19
interaction with NP   4
interaction with PA   4
cap-binding   4–5
nuclear localization signal   44
pathogenesis   284
RNP   9–10
transcription   5, 17–19
ts mutants   5, 23
UV cross-linking   4
Phosphorylation
M2   69, 73
NP   44
PA   5
PKC *See* protein kinase C
PKR *See* protein kinase R
Plasmacytoid DC *See* DC
Polyadenylation   2, 10, 18–19
Poultry   259–262
biosecurity   259–260
DIVA   261
surveillance and diagnostics   260
elimination of infected poultry   261
vaccination   261–262
Promoter   3, 10–16
corkscrew   10–13
core promoter   10
panhandle   10–13
polyA signal   10
role of the 3'-end   15–16
role of the 5'-end   13–15
RNA-fork   11, 13
Protease   5, 6–7, 21
Protein kinase C   327–328

Protein kinase R   146, 332
Proteosome   211–212

Rafts
  HA   53
  NA   53
  M2   70–71
RANTES   325
RBS *See* receptor-binding sites
Reassortment   236, 239
  reassortant H1N1 swine viruses   248
  reassortant H1N2 swine viruses   248
  reassortant H3N2 swine viruses   246–248
  'triple reassortant' viruses   248
Receptor   37–39
  influenza C virus   38
  receptor-mediated endocytosis *See*
    endocytosis
Receptor-binding activity
  assays   105–107
    – binding to gangliosides   106
    – binding to sialosides,
      sialylglycopolymers, and
      sialylglycoproteins   106
Receptor-binding sites   98
  avian-type   109
  carbohydrates   104
  human-type   115–116, 117–118
  H variant   120
  L variant   120
Receptor specificity   107
  receptor specificity of H5N1 chicken
    viruses   112–113
  receptor specificity of duck viruses
    109–111
  receptor specificity of equine viruses   113
  receptor specificity of gull viruses   112
  receptor specificity of human viruses
    115–119
  receptor specificity of seal viruses   113
  receptor specificity of swine viruses
    114–115
  receptor specificity of turkey viruses   112
  pathogenesis   284
  tropism   119
Recombinant DNA
  vaccine   208–209
Recombination   283
Replication   19–25
  cRNA synthesis   19–22
  PA-induced proteolysis   21–22

Replication (*cont.*)
  role of host cell factors   22
  role of NP   19–21
  role of PA   21–22
  temporal regulation   24–25
  vRNA synthesis   22–24
Reservoir   281
Respiratory burst   121
Reverse genetics
  vaccine   209–211
Ribonucleoprotein *See* RNP
Rimantadine   170–172
  prophylaxis   175–177
  resistance   172–175
  treatment   177–178
RNA-binding   7
RNA-dependent RNA polymerase   1, 2–4
RNPs   1, 8–10
  nuclear export   48–52
  nuclear Import   44–45
  transport   43

Seal viruses   113
Serology, 1918 'Spanish' virus   304–306
Sialic acid   95
  apical cell surface   37
  HA   37–39
  HA sialic acid-binding pocket   98
  host range   231–232
  NA   38
  N-acetylneuraminic acid (Neu5Ac)   96
  N-glycolylneuraminic acid (Neu5Gc)   96
  receptor   37–39
  receptor-binding site   98
  Sia(α2–3)Gal   101, 231
  Sia(α2–6)Gal   101, 231
  sialic acid linkage   37–39
  sialic acid species   96–97
Sialyloligosaccaride   95. *See also* sialic acid
'Spanish' virus *See* 1918 'Spanish' virus
Spherical particles   54
STAT1   332, 344
Subunit vaccine   206
Surveillance   233
Swine influenza   245–249
  classical H1N1 swine influenza viruses
    245–246
  H1N1 viruses   245–246, 248–249
  H1N2 viruses   248
  H3N2 viruses   246–248
  'Mixing vessel'   114, 249

Swine influenza (*cont*.)
  reassortant H1N1 viruses   248–249
  reassortant H1N2 viruses   248
  reassortant H3N2 viruses   246–248
receptor specificity   114–115
  swine-to-human-transmission   253–254
  'triple reassortant' viruses   248
  vaccine   256–258

Temperature-sensitive phenotype   205–206
TGN   80
Transcription   5, 10, 16, 17–19, 24–25
  initiation of transcription   17–18
  polyadenylation   18–19
  temporal regulation   24–25
Trans-Golgi network *See* TGN
Transmission
  avian-to-human   251–252
  avian-to-swine   254
  H5N1 viruses   251–252
  human-to-swine   254
  swine-to-human   253–254
Tropism   119, 284
Ts See temperature-sensitive phenotype
Ts mutants
  M1 *ts* mutant   51
  NP *ts* mutants   19, 21, 23
  PA *ts* mutants   5–6, 21, 23
  PB2 *ts* mutants   5, 23
Turkey viruses, receptor specificity   112

UAP56   7, 8, 22
Uncoating   42–43, 67–68
UTR (untranslated region)   10, 345. *See also*
      NCR (noncoding region)
  GRSF-1   345
  Packaging of vRNA segments   10

Vaccination, recommendations   204, 255, 257
Vaccine
  adjuvanted subunit vaccine   206
  antigenic variability   204–205
  avian influenza   216–218
  cell culture-grown   207–208
  chitosan
  cold-adapted (*ca*) phenotype   205–206
  delivery   207, 215–216
  DHEA   214
  DIVA   261
  DNA vaccine   214
  dose   207
  epidemic   255–256
  H5N1 viruses   216–218
  inactivated vaccine   205
  ISCOMs   212
  liposomes   213
  live attenuated vaccine   205–206
  M2   211
  NS1   211
  pandemic   256
  PCPP   214
  proteosome   211–212
  recombinant DNA   208–209
  reverse genetics   209–211
  strain selection   204–205, 257
  swine vaccine   256–258
  temperature-sensitive (*ts*) phenotype
      205–206
  TLR9 ligands   214
Virulence   240
Virus clearance   155–156
vRNA   22–24
  temporal regulation   24–25
  vRNA synthesis   22–24

Zanamivir   178–188
  mechanisms of action   178–179
  prophylaxis   185–186
resistance   180–185
  – mutations in HA   180–182
  – mutations in NA   182–185
  spectrum and potency   179–180
  treatment   187–188
Z genotype *See* H5N1 viruses

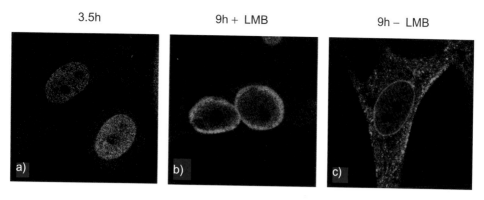

Figure 2.4 Timecourse of NP intracellular localisation. Baby hamster kidney fibroblasts were infected with influenza virus A/PR/8/34 virus and fixed and stained for NP (green) and nucleoporin 62 (red) at the indicated times post infection. Cells in (b) were treated with 11 nM leptomycin B (LMB) from 1 h post infection.

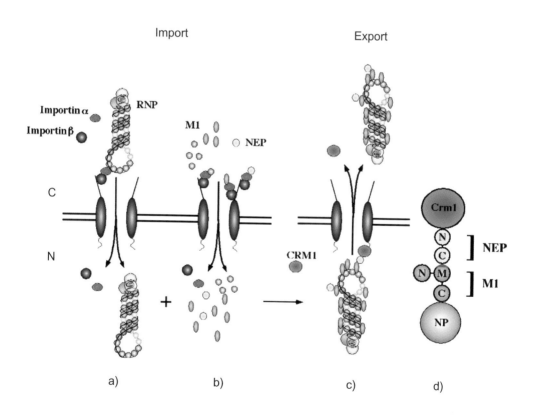

Figure 2.5 Influenza virus nucleocytoplasmic trafficking and the NEP hypothesis for RNP export. (a) RNPs and (b) monomeric NP, M1 and NEP are imported from the cytoplasm (C) to the nucleus (N) through NLS-mediated interactions with host cell importin α followed by importin ß docking with the nuclear pore complex (grey ovals). (c) An NEP-M1-RNP complex forms in the nucleus and interacts with host cell CRM1 vias the NES in NEP. (d) The daisy chain model for the NP-M1-NEP-Crm1 complex. The C-terminal domain of M1 interacts with NP, while the M domain interacts with the C-terminal domain of NEP through the basic NLS/RNA-binding sequence (indicated in red). The N-terminal domain of NEP contains the NES sequence and interacts with Crm1.

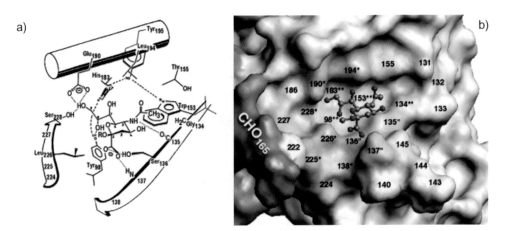

Figure 4.2 Scheme (**a**) and molecular model (**b**) of the HA sialic acid-binding pocket based on the crystal structure of X-31 virus HA complexes with sialosides (Weis et al., 1988; Sauter et al., 1992). Figure **a** is modified from Kelm et al., 1992. Dotted lines indicate possible hydrogen bonds between sialic acid and the RBS, dashed lines show potential hydrogen bond within the protein. Figure **b** shows position of sialic acid moiety (ball and stick model) in the binding pocket (solvent accessible surface). Different colors of amino acids are solely for the identification purposes. Two stars next to the amino acid number indicate that amino acid is conserved among all influenza A viruses. One star indicates that amino acid is conserved among avian viruses and changed in human influenza A viruses. '' indicates that atoms interacting with sialic acid moiety are conserved. CHO_{165} – a portion of N-linked glycan attached at Asn_{165} of the HA that was resolved by X-ray analysis. Figures 4.2b, 4.3, and 4.4 were generated using DS ViewerPro 5.0 (Accelrys Inc.)

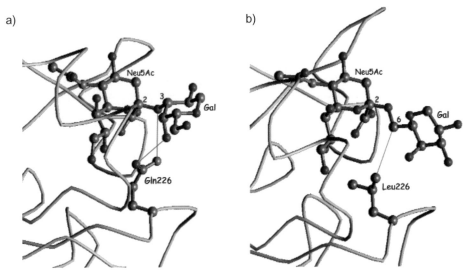

Figure 4.3 HA interactions with (α2-3)-linked and (α2-6)-linked galactose residues.
a. Neu5Ac(α2-3)Gal moiety of pentasaccharide LSTa [Neu5Ac(α2-3)Gal(β1-3)GlcNAc(β1-3)Gal(β1–4)Glc] (ball and stick model) in the receptor-binding site of A/Duck/Ukraine/63 (H3N8) (Ha et al., 2003). The galactose residue is bound in the minimum-energy *syn* conformation of the glycosidic linkage that allows hydrogen bonding (*red* lines) of the glycosidic oxygen and the 4-OH group of Gal to the side chain atoms of Gln_{226}. **b**. Neu5Ac(α2-6)Gal moiety of pentasaccharide LSTc [Neu5Ac(α2-6)Gal(β1-4)GlcNAc(β1-3)Gal(β1-4)Glc] in the RBS of X31 HA (Eisen et al., 1997). The Gal residue binds in the minimum-energy *anti* conformation, in which the C6-methylene group of Gal participates in van-der-Waals and hydrophobic interactions (*grey* line) with the nonpolar side chain of Leu_{226}. The HA protein backbone is depicted by cyan tubes on both figures.

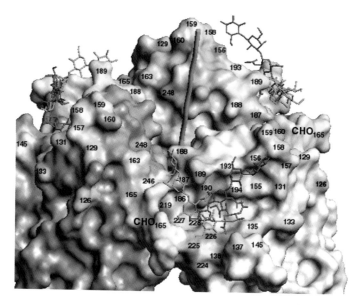

Figure 4.4 X31 HA complexes with pentasaccharides LSTc (*red*) and LSTa (*green*) (stick models) superimposed on the same HA model (solvent accessible surface). The sialic acid moiety is shown in magenta. Individual HA monomers are tinted in shades of gray. Positions of some amino acid residues that were shown to affect the receptor-binding activity and are discussed in this review are colored and numbered. The figure is based on crystallographic data of Eisen et al. (1997).

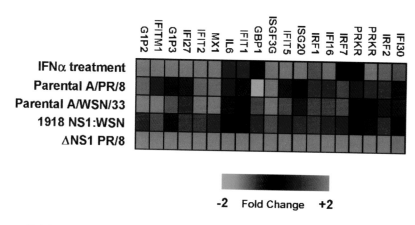

Figure 13.1 Gene expression profiling of the effects of NS1 on IFN-regulated gene expression. Shown is the expression of a group of IFN-regulated genes in cells treated with IFN (400 IU/ml for 16 hours) or infected with influenza virus containing wild-type NS1 (A/PR/8 and A/WSN/33), 1918 NS1 (in A/WSN/33), or a complete deletion of the NS1 gene (in A/PR/8). This analysis demonstrates the role of NS1 in antagonizing the IFN response and the enhanced function of the NS1 gene from the 1918 pandemic strain. Genes shown in red were up-regulated and genes shown in green were down-regulated in infected relative to uninfected A549 cells. Black indicates no change in gene expression. Primary analysis was performed using Resolver (Rosetta Biosoftware) and supplemental analysis using DecisionSite for Functional Genomics (Spotfire, Inc.).

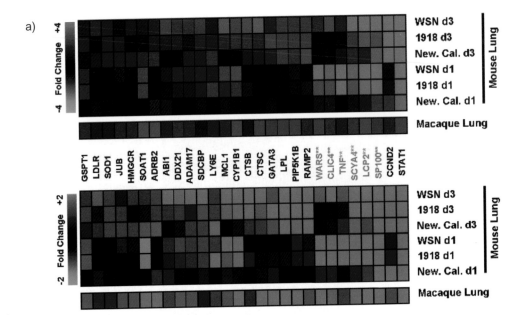

Figure 13.3 Comparison of macaque and murine gene expression patterns in response to influenza virus infection. a) The genes displayed exhibited a ≥1.5-fold increase in expression ($P ≤ 0.01$) in macaque lung infected with A/Texas/36/91 (day 4) and in at least two experiments in which mice were infected with New Caledonia HA/NA:WSN, 1918 HA/NA:WSN, or parental recombinant WSN virus. The bottom and top panels display the same intensity data, but with different color saturations (i. e., saturation is at 2 fold in the bottom panel and at 4 fold in the top panel). Gene names shown in red (**) were significantly up-regulated (≥4 fold) only in lungs from animals infected with a lethal dose of virus (1918 HA/NA:WSN and WSN). Genes depicted by red boxes were up-regulated and genes depicted by green boxes were down-regulated in infected relative to uninfected lung. Black indicates no change in gene expression. Primary analysis was performed using Resolver and supplemental analysis using Decision Site for Functional Genomics.

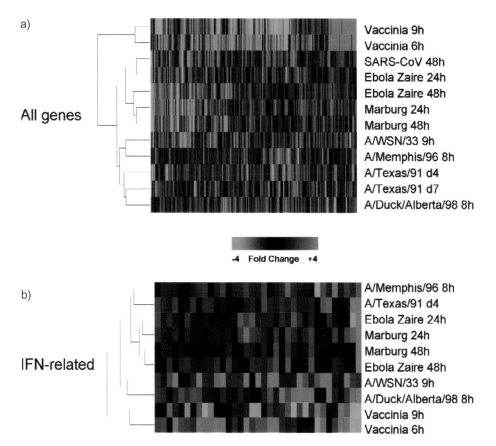

Figure 13.4 Cellular gene expression profiles elicited by influenza, vaccinia, SARS-Cov, Ebola Zaire, or Marburg virus infection. a) Matrix showing the two-dimensional clustering of 189 genes that showed a ≥1.5-fold change in expression ($n = 4$; $P \le 0.05$) in a least seven of the twelve infections. b) Matrix showing the two-dimensional clustering of 35 IFN-regulated genes that showed a ≥1.5-fold change in expression ($n = 4$; $P \le 0.05$) in a least one of the ten infections. Each of these hierarchical agglomerative clusters was created using complete link heuristics and Manhattan and Euclidean distance metrics for experiment and sequence axis, respectively. Vaccinia virus were performed using HeLa cells, SARS-CoV using Vero cells, Ebola Zaire and Marburg virus using Huh7 cells, and various Type A influenza viruses using macaques, A549 cells, or human tracheobronchial epithelial cells. Genes shown in red were up-regulated and genes shown in green were down-regulated in infected relative to uninfected cells. Black indicates no change in gene expression. Primary analysis was performed using Resolver and supplemental analysis using Decision Site for Functional Genomics.